AF323740

Recent Progress in Conformal Geometry

Imperial College Press Advanced Texts in Mathematics

Series Editor: Dennis Barden *(Univ. of Cambridge, UK)*

Vol. 1   Recent Progress in Conformal Geometry
by Abbas Bahri (Rutgers Univ. USA) & YongZhong Xu
(Courant Inst. for the Mathematical Sciences, USA)

ICP Advanced Texts in Mathematics – Vol. 1

Recent Progress in Conformal Geometry

Abbas Bahri
Rutgers University, USA

Yongzhong Xu
Courant Institute for the Mathematical Sciences, USA

Imperial College Press

Published by

Imperial College Press
57 Shelton Street
Covent Garden
London WC2H 9HE

Distributed by

World Scientific Publishing Co. Pte. Ltd.
5 Toh Tuck Link, Singapore 596224
USA office: 27 Warren Street, Suite 401-402, Hackensack, NJ 07601
UK office: 57 Shelton Street, Covent Garden, London WC2H 9HE

British Library Cataloguing-in-Publication Data
A catalogue record for this book is available from the British Library.

ISBN-13 978-1-86094-772-8
ISBN-10 1-86094-772-7

Printed in Singapore by World Scientific Printers (S) Pte Ltd

Preface

This book is divided into two parts. The first part is about Sign-Changing Yamabe-type problems. A Morse Lemma at infinity, under reasonable basic conjectures, is proved for such problems. This work is an attempt to define a new area of research for nonlinear analysts. We have tried in it to provide a family of estimates and techniques with the help of which the problem of finding infinitely many solutions to these equations on domains of $\mathbb{R}^3$ can be studied.

Our estimates and our work is a "cas d'ecole" in that we work on $\mathbb{R}^3$ or S^3, a framework where solutions are known to exist, in fact in infinite number; and we have chosen to study the asymptots generated by these solutions and their combinations under the action of the Conformal Group. This work **could** also be useful for other variational problems such as Einstein or Yang-Mills equations.

The second part of this book is about Contact Form Geometry via Legendrian curves. Given a three-dimensional compact orientable manifold M^3 and a contact form α on it, we have assumed in earlier works the existence of a "dual" contact form $\beta, \beta = d\alpha(v, \cdot)$, with the same orientation than α and we have introduced the variational problem $\int_0^1 \alpha_x(\dot{x})dt$ on $C_\beta = \{x \in H^1(S^1, M) | \beta_x(\dot{x}) \equiv 0\}$. We have defined a homology related to the periodic orbits of the Reeb vector-field ξ of α on C_β.

We prove in this framework two main results. First, we establish that the hypothesis that β is a contact form with the same orientation than α is not essential. The techniques involved in order to prove such a result (on a typical example) have the definite advantage that they are quantitative: as we allow regions where β is no longer a contact form with the same orientation than α, we track down the modifications of the variational problem and we provide bounds on a key quantity (denoted τ) as we introduce a large amount of rotation for $\ker \alpha$ along the orbits of v near these areas.

We then move to prove a compactness result about the flow-lines of this variational problem which originate at a periodic orbit of ξ. This — still slightly imperfect — compactness result indicates that all flow-lines originating at periodic orbits go to periodic orbits (at least if the difference of Morse indexes is 1), unless the number of zeros of the v-component of $\dot{x}$, the tangent vector to the curves under deformation, drops.

No critical point at infinity (asymptot)interferes with this homology. We strongly suspect that this homology is, in the case of the standard contact structure of S^3, the homology of $P\mathbb{C}^\infty$. We expect that we will be able to compute this homology in some easy cases at least.

We had been searching for a long time for such a result. This work entitled "Compactness" will be published independently by the first author and dedicated to his long time friend and collaborator Haim Brezis for his sixtieth birthday. Both directions of research i.e. Conformal Geometry, Einstein equations, Yang-Mills equations on one hand, Contact Form Geometry on the other hand, are also studied by other techniques due to "hard-chore" Geometry and Symplectic Geometry.

In fact, Geometers have always been our "co-area researchers". We view these areas which we have contributed to define — for Contact Form Geometry — in a different way and with different techniques.

This book is a book of collaboration and research. It also defines new goals and presents a new understanding. It is not (yet) a textbook for graduate students. It rather informs our collaborators about a definite progress in the two above mentioned areas.

This research has been long; and at times hard and difficult. It has been a strain on our friends and companions. Thanks are due: Abbas Bahri wishes to thank Haim Brezis, his long-time collaborator and friend, not only for his obvious support but more so for his friendship. Having a friend — and of such a quality — is a rare blessing in life.

Abbas Bahri wishes also to thank Diana Nunziante, his wife, for her patience, her understanding and her love as this book was being written.

Lines and equations are written, but only with the overwhelming intelligence and love of those closest to us.

Both of us extend our warmest thanks to Barbara Mastrian for her wonderful work as well as her wit and life. It has been a pleasure to work with her all these years.

Finally, we would like to thank H. Brezis, S. Chanillo, R. Nussbaum and Z. Han for giving up so much of their time and patiently listen to our arguments as they developed.

We also thank them, as well as our friends and colleagues of the Mathematics Department at Rutgers, for their thoughtful remarks and observations.

Contents

Preface   v

   A. Bahri and Y. Xu

1.   Sign-Changing Yamabe-Type Problems   1

  1.1   General Introduction   1

  1.2   Results and Conditions   2

  1.3   Conjecture 2 and Sketch of the Proof of Theorem 1; Outline   7

  1.4   The Difference of Topology   11

  1.5   Open Problems 14

      1.5.1   Understand the difference of topology   14

      1.5.2   Non critical asymptots 15

      1.5.3   The exit set from infinity 15

      1.5.4   Establishing Conjecture 2 and continuous forms of the

               discrete inequality   16

      1.5.5   The Morse Lemma at infinity, Part I, II, III 16

      1.5.6   Notations $\bar{v}, \bar{v}_i, \bar{h}_i$ 16

  1.6   Preliminary Estimates and Expansions, the Principal Terms   17

  1.7   Preliminary Estimates 18

      1.7.1   The equation satisfied by $\bar{v}$ 19

      1.7.2   First estimates on $\bar{v}_i$ and $\bar{h}_i$ 23

      1.7.3   The matrix A 25

      1.7.4   Towards an H_0^1-estimate on $\bar{v}_i$ and an L^∞-estimate

               on h_i 26

      1.7.5   The formal estimate on $\bar{h}_i$ 31

      1.7.6   Remarks about the basic estimates 35

1.7.7 Estimating the right hand side of Lemma 12 35
1.7.8 **R$_i$** and the estimate on $|\bar{v}_i|_{H_0^1}$ 45
1.8 Proof of the Morse Lemma at Infinity When the
Concentrations are Comparable 54
1.9 Redirecting the Estimates, Estimates on $|\bar{v}_i|_{H_0^1} + |\bar{h}_i^0|_\infty +$
$\sum_{\lambda_s > \lambda_i} \varepsilon_i^s |\bar{h}_i^s|_\infty$ 66
1.9.1 Content of Part II 66
1.9.2 Redirecting the estimates 68
1.10 Proof of the Morse Lemma at Infinity 108
1.10.1 Decomposition in groups, gradient and
L^∞-estimates on $\bar{v}$, proof of the Morse Lemma
at infinity 108
1.10.2 Content of Part III 108
1.10.3 Basic conformally invariant estimates 109
1.10.4 Estimates on $v - (v_I + v_{II})$ 121
1.10.5 The expansion 129
1.10.6 The coefficient in front of $\varepsilon_{k\ell}\delta_\ell$ 154
1.10.7 The σ_i-equation, the estimate on $\sum |\gamma_i^j|$ 159
1.10.8 The system of equations corresponding to the
variations of the points a_i 174
1.10.9 Rule about the variation of the points of
concentrations of the various groups 182
1.10.10 The basic parameters and the end of the expansion . 185
1.10.11 Remarks on the basic parameters 185
1.10.12 The end of the expansion and the concluding
remarks 189

Bibliography   199

2. Contact Form Geometry   201

2.1 General Introduction 201
2.2 On the Dynamics of a Contact Structure along a Vector Field
of its Kernel 205
2.2.1 Introduction 205
2.2.2 Introducing a large rotation 210
2.2.3 How γ is built 214
2.2.4 Modification of α into α_N 226
2.2.5 Computation of ξ_N 227

2.2.6 Conformal deformation 235
2.2.7 Choice of λ 240
2.2.8 First step in the construction of λ 241
2.3 Appendix 1 275
2.3.1 The normal form for (α, v) when α does not turn well 275
2.4 The Normal Form of (α, v) Near an Attractive Periodic Orbit
of v 276
2.5 Compactness 279
2.5.1 Some basic facts 280
2.5.2 A model for $W_u(x_m)$, the unstable manifold in C_β of
a periodic orbit of index m 282
2.5.3 Hypothesis (A), Hypothesis (B), Statement of the
result 288
2.5.4 The hole flow 291
2.5.4.1 Combinatorics 291
2.5.4.2 Normals 294
2.5.4.3 Hole flow and Normal (II)-flow on curves of
Γ_{4k} near x^∞ 296
2.5.4.4 Forced repetition 299
2.5.4.5 The Global picture, the degree is zero 301
2.5.5 Companions 304
2.5.5.1 Their definition, births and deaths 304
2.5.5.2 Families and nodes 305
2.5.6 Flow-lines for x_{2k+1} to x_{2k}^∞ 323
2.5.7 The S^1-classifying map 332
2.5.8 Small and high oscillation, consecutive characteristic
pieces 334
2.5.9 Iterates of critical points at infinity 354
2.5.10 The Fredholm aspect 359
2.5.11 Transversality and the compactness argument 364
2.6 Transmutations 384
2.6.1 Study of the Poincaré-return maps 402
2.6.2 Definition of a basis of $T_{\bar{x}_\infty}\Gamma_{2s}$ for the reduction of
$d^2 J(\bar{x}_\infty)$ 413
2.6.3 Compatibility 417
2.7 On the Morse Index of a Functional Arising in Contact Form
Geometry 420
2.7.1 Introduction 420
2.7.2 The Case of Γ_2 424

2.7.3　Darboux Coordinates 425

2.7.4　The v-transport maps 430

2.7.5　The equations of the characteristic manifold near x^∞;
the equations of a critical point 434

2.7.5.1　The characteristic manifold for the unperturbed problem 434

2.7.6　Critical points, vanishing of the determinant 436

2.7.7　Introducing the perturbation 437

2.7.8　The characteristic manifold for the perturbed
problem; the determinant equations 441

2.7.9　Reduction to the Case $k = 1$ 447

2.7.10 Modification of $d^2 J_\infty^\tau(\overline{x}_\infty)\,|_{span\{u_2,\cdots,u_{k-1}\}}$ 454

2.7.11 Calculation of $\partial^2 J_\infty(\overline{x}_\infty).u_2.u_3$ 459

2.8 Calculation of $\partial^2 J_\infty(\overline{x}_\infty).u_2.u_2$ 465

2.9 Calculation of $\partial^2 J_\infty(\overline{x}_\infty).u_2.u_4$ 471

2.10 Other Second Order Derivatives 474

2.11 Appendix 476

2.11.1 The Proof of Lemma 42 476

2.11.2 The proof of Lemma 47 480

2.11.3 Proof of the Lemmas in 2.7.11 483

2.11.4 Proof of Lemma 48 483

2.11.5 The Proof of Lemma 49 484

2.11.6 The proof of Lemma 50 485

2.11.7 Proof of Claim 1 487

2.11.8 Proof of Claim 3 489

2.11.9 The Final Details of the Calculation of
$\partial^2 J(\overline{x}_\infty).u_2.u_3$ 492

2.11.10 Details involved in 2.8 494

2.11.11 Proof of Lemma 52 494

2.11.12 Proof of Lemma 53 495

2.11.13 Proof of Claim 3 496

2.11.14 Proof of Claim 4 497

2.11.15 Details of the Calculation of $\partial^2 J.u_{2.}u_2$ 503

Bibliography　　　　　　　　　　　　　　　　　　　509

Chapter 1

Sign-Changing Yamabe-Type Problems

1.1 General Introduction

Let us consider a very simple (and classical) model in Nonlinear Analysis and Riemannian Geometry, namely the Yamabe problem on S^3, with the standard metric:

$$-3\Delta_{S^3} u + 4u = u^5 \quad u > 0. \tag{1.1}$$

Equivalently, we might consider

$$\Delta_{\mathbb{R}^3} u + u^5 = 0 \quad u > 0. \tag{1.2}$$

This problem has received a lot of attention in the 1980's and the 1990's because of the Yamabe conjecture and the scalar-curvature problem. Two or three main techniques have been devised to solve such problems: minimizing techniques combined with geometric results (positive mass conjecture *etc.*), variational techniques combined with topological techniques and study of critical points at infinity, maximum principle techniques to derive *a priori* estimates. We are interested here in the variational techniques and the study of the critical points at infinity.

In order to describe this technique, we recall that (1.1) or (1.2) are variational problems with defects. The defect relies in the "non-compactness" or failure of the Palais-Smale condition. Indeed, the set of solutions of (1.1) or (1.2) is non-compact because the conformal group of S^3 (which leaves (1.1) or (1.2) invariant) is non-compact. All the (for (1.2)) functions

$$\delta(a, \lambda) = \frac{c\sqrt{\lambda}}{(1 + \lambda^2 |x - a|^2)^{1/2}}$$

are solutions of (1.2). Combinations $\sum_{i=1}^{p} \delta(a_i, \lambda_i)$ of such functions ($\lambda_i \to +\infty$) are almost solutions. These problems have asymptots.

1

Our aim is to complete for them a Morse Lemma at infinity (see [Bahri 1989], [Bahri 2001], [Bahri and Coron (1988)]) from which formulae for the difference of topology in the level sets of the associated functional and (quite involved sometimes, see [Aubin (1976)]) mechanisms of existence could be derived. All the techniques — including the geometric techniques used by Aubin [Aubin (1976)], Schoēn [Schoēn (1988)], Schoēn and Yau [Schoēn and Yau (1988)] *etc.*— use heavily the assumption that the function u which is sought for should be positive. This further requirement, which should make the problem more difficult, makes it easier in some regard, since the variations (this is a variational problem) can be restricted (because of the special structure of (1.1)) to the set of positive functions. The non-compactness is then less stringent (limited to the $\delta(a, \lambda)$'s and their combiniations) and more controlled. The maximum principle can be used as well as Alexandrov reflection techniques. The positivity helps also considerably in the proof of the Morse Lemmas at infinity.

However, (1.1) and (1.2) are well-posed problems without the assumption $u > 0$ and it makes perfect sense to study these questions, as well as equations of this type on domains or with other metrics or manifolds without the positivity assumption.

There are several motivations to study (1.1) or (1.2) without the positivity assumption, besides the main ones, namely that it displays new phenomena, and a new and interesting direction of research.

One of the main mathematical motivations is that (1.1) or (1.2) has infinite many changing-sign solutions which are not explicitly known. Thus, understanding the non compactness in such problems is closer to the (two dimensional) harmonic map problem or the Yang-Mills equations (away from minima).

The Yamabe problem, without the positivity assumption, is a simpler model of less explicit non-compactness phenomena. (1.2) is "un cas d'ecole".

1.2   Results and Conditions

We state now the result which we wish to prove, then the result which we have established and the conditions under which we have completed this work. We then discuss these conditions.

The functional is

$$J(u) = \frac{1}{\int_{\mathbb{R}^3} u^6 dx}$$

for $u \in \sum = \{w; \int |\nabla w|^2 + \int w^6 < +\infty; \int |\nabla w|^2 = 1\}$.

Let

$$\bar{w}_1, \ldots, \bar{w}_p$$

be p (possibly changing) solutions of the Yamabe problem (on $\mathbb{R}^3$ or S^3; these are non singular solutions).

Let

$$a_1, \ldots, a_p \in \mathbb{R}^3 \text{ or } S^3$$

be p points away from infinity (we view them on S^3).

Let

$$\lambda_1, \ldots, \lambda_p > 0$$

be very large so that, $i \neq j$,

$$\varepsilon_{ij} = \frac{1}{\left(\frac{\lambda_i}{\lambda_j} + \frac{\lambda_j}{\lambda_i} + \lambda_i \lambda_j |a_i - a_j|^2 \right)^{1/2}}$$

is small $(< \varepsilon_0(p))$.

We wish to establish a formula for the difference of topology due to the asymptot of J which corresponds to

$$\sum_{i=1}^{p} \sqrt{\lambda_i} \bar{\omega}_i (\lambda_i (x - a_i)).$$

We thus need to find a normal form for the expansion of

$$J \left(\sum_{i=1}^{p} \alpha_i \sqrt{\lambda_i} \bar{\omega}_i (\lambda_i (x - a_i)) + v \right)$$

where v satisfies ($\omega_i = \sqrt{\lambda_i} \bar{\omega}_i (\lambda_i (x - a_i))$, σ_i are rotation parameters).

$$(\text{Vo}) \begin{cases} \int |\nabla v|^2 \text{ small} \\ \int \nabla v \nabla \omega_i = 0 \\ \int \nabla v \nabla \frac{\partial \omega_i}{\partial \lambda_i} = 0 \\ \int \nabla v \nabla \frac{\partial \omega_i}{\partial a_i} = 0 \\ \int \nabla v \nabla \frac{\partial \omega_i}{\partial \sigma_i} = 0 \end{cases} \quad .$$

Such functions

$$u = \sum_{i=1}^{p} \alpha_i \omega_i + v$$

where v satisfies (Vo) can be seen to span neighborhoods of the asymptot (neighborhoods of the critical points at infinity).

The conjecture is then:

Conjecture 1   *There is a change of variables in the $(\alpha_i, \lambda_i, a_i, v)$-space such that*

$$J\left(\sum_{i=1}^{p} \alpha_i \omega_i + v\right) = \frac{\left(\sum_{i=1}^{p} \tilde{\alpha}_i^2 \int |\nabla \bar{\omega}_i|^2\right)^3}{\sum_{i=1}^{p} \tilde{\alpha}_i^6 \int \bar{\omega}_i^6}$$

$$\times \left\{ 1 - \bar{c} \sum_{i \neq j} \bar{\omega}_i(\tilde{\tilde{a}}_j) \bar{\omega}_j^\infty \tilde{\varepsilon}_{ij} - \sum_{i \neq j} c_{ij}(\bar{\omega}_i, \bar{\omega}_j) \tilde{\varepsilon}_{ij}^3 + Q(V, V) \right\}$$

$(\tilde{\alpha}_i, \tilde{a}_i, \tilde{\lambda}_i, V)$ *are the new variables. Q is a nondegenerate quadratic form in V. V is in a small neighborhood of zero in a Hilbert space.*

The expression $\bar{\omega}_i(\tilde{\tilde{a}}_j) \bar{\omega}_j^\infty$ requires some explanations. $\bar{\omega}_j^\infty$ is the value of $\bar{\omega}_j(x)$ at the north pole when ω_j is concentrated at the south pole. When ω_i is totally deconcentrated, $\bar{\omega}_i(\tilde{\tilde{a}}_j)$ is the value of $\bar{\omega}$ (ω_i after deconcentration) at $\tilde{\tilde{a}}_j$ the concentration point of ω_j (see [Bahri 2001]). Index Q is the sum of the strict indexes of each $\bar{\omega}_i$. We assume that each of them is nondegenerate transversally to the conformal group.

The result which we prove here is close to Conjecture 1. Its proof relies on two additional assumptions which we conjecture to hold and which read:

Conjecture 2   *There exists a constant $c(p) > 0$ such that, for any $(a_1, \ldots a_p) \in \mathbb{R}^{3p}$ and $(u_1, \ldots, u_p) \in \mathbb{R}^p$,*

$$|Au|^2 + \sup_{i,j} |{}^t u \frac{\partial A}{\partial a_i^j} u| \geq c(p) \sum_{i=j} \frac{u_i^2}{|a_i - a_j|^2}$$

A is the matrix

$$\begin{pmatrix} 0 & & \frac{1}{|a_i - a_j|} \\ & \ddots & \\ \frac{1}{|a_i - a_j|} & & 0 \end{pmatrix}.$$

Conjecture 3   *Let σ_i be the compact (rotation-related) parameters of the conformal group acting on $\bar{\omega}_i$. Assume that, for each i and for a small constant c:*

If

$$
\begin{cases}
-\bar{c}\dfrac{\partial \bar{\omega}_i^\infty}{\partial \sigma_i}\left(\sum_{j\neq i}\bar{\omega}_j^\infty \varepsilon_{ij}\right) = \sum_{i\neq j}\dfrac{\partial c_{ij}}{\partial \sigma_i}\varepsilon_{ij}^3 \\[2mm]
\sum_i\left|\sum_{j\neq i}\left(\bar{\omega}_i^\infty\Big(\bar{\omega}_j^\infty + \dfrac{D\bar{\omega}_j^\infty(\frac{a_i-a_j}{|a_i-a_j|})}{\lambda_j|a_i-a_j|} + 3\dfrac{c_{ij}}{\bar{c}}\varepsilon_{ij}^2\Big)\right)\varepsilon_{ij}\right| \\[2mm]
\qquad \leq c\left(\sum_{i\neq j}\varepsilon_{ij}^3 + \sum \bar{\omega}_i^{\infty 2}\varepsilon_{ij}\right).
\end{cases}
$$

then,

$$
\sum_i\left|\sum_{j\neq i}\bar{\omega}_i^\infty\bar{\omega}_j^\infty\dfrac{\partial \varepsilon_{ij}}{\partial a_i} + \sum_{j\neq i}\varepsilon_{ij}\bar{\omega}_j^\infty\dfrac{\partial}{\partial a_i}\left(\dfrac{D\bar{\omega}_j^\infty|(\frac{a_i-a_j}{|a_i-a_j|})}{\lambda_j|a_i-a_j|}\right)\right.
$$

$$
\left.+\dfrac{3}{\bar{c}}\sum_{j\neq i}c_{ij}\varepsilon_{ij}^2\dfrac{\partial \varepsilon_{ij}}{\partial a_i}\right| \geq c\left(\sum_{j\neq i}\dfrac{\omega_i^{\infty 2}}{\lambda_i|a_i-a_j|} + \sum\dfrac{\varepsilon_{ij}^3}{|a_i-a_j|}\right)
$$

Observation.

Replacing ε_{ij} by $\dfrac{1}{\sqrt{\lambda_i\lambda_j|a_i-a_j|}}$, we can see that, if Conjecture 3 does not hold, we have more conditions than variables.

There are variants of Conjecture 2 and Conjecture 3 which we need also to introduce:

Conjecture 2′ Let φ and φ' be two different charts of $S^3-\{pt\}$ where the South and North poles are "far" from the points a_i. Let $\bar{u} = \begin{pmatrix}\vdots \\ \bar{\omega}_i^\infty/\sqrt{\lambda_i} \\ \vdots\end{pmatrix}$, in the first chart and $\bar{u}' = \begin{pmatrix}\vdots \\ \bar{\omega}_i^\infty/\sqrt{\lambda_i'} \\ \vdots\end{pmatrix}$, in the second chart be the corresponding vector associated to $\sum \alpha_i\omega_i$. Let A' be the matrix A at the points a_i' in the second chart.

There exists $c(p) > 0$ such that, for any $(a_1,\dots a_p) \in \mathbb{R}^{3p}$ and $(\omega_1,\dots,\omega_p)$ solutions of the Yamabe problem on S^3:

$$
\sup_{(i,j)}|{}^t\bar{u}\dfrac{\partial A}{\partial a_i^j}\bar{u}| + \sup_{(i,j)}|{}^t u'\dfrac{\partial A'}{\partial a_i'^j}\bar{u}'| \geq c(p)\sum_{i=j}\dfrac{\omega_i^{\infty 2}}{\lambda_i|a_i-a_j|^2}.
$$

Conjecture 3′   *Let σ_i be the compact rotation parameters corresponding to rotations preserving the polar axis (hence $\frac{\partial \bar{\omega}_i^\infty}{\partial \sigma_i} = 0$).*
   Assume that, for each i,

$$\left\{ \sum_{i \neq j} \frac{\partial c_{ij}}{\partial \sigma_i} \varepsilon_{ij}^3 = 0. \right.$$

The a_i-equations

$$\left\{ \sum_i \left| \sum_{j \neq i} \left(\bar{\omega}_i^\infty \bar{\omega}_j^\infty + 3\frac{c_{ij}}{\bar{c}} \varepsilon_{ij}^2 \right) \frac{\partial \varepsilon_{ij}}{\partial a_i} + \sum_{j \neq i} \varepsilon_{ij} \bar{\omega}_j^\infty \frac{\partial}{\partial a_i} \left(\frac{D\bar{\omega}_j^\infty(\frac{a_i - a_j}{|a_i - a_j|})}{\lambda_i |a_i - a_j|} \right) \right| = 0 \right.$$

hold in both charts

Then,

$$\sum_i \left| \sum_{j \neq i} \left(\bar{\omega}_i^\infty \left(\bar{\omega}_j^\infty + \frac{D\bar{\omega}_j^\infty(\frac{a_i - a_j}{|a_i - a_j|})}{\lambda_j |a_i - a_j|} \right) + 3\frac{c_{ij}}{\bar{c}} \varepsilon_{ij}^2 \right) \varepsilon_{ij} \right|$$

$$\geq c \left(\sum_{i \neq j} \varepsilon_{ij}^3 + \sum \bar{\omega}_i^{\infty 2} \varepsilon_{ij} \right).$$

We also define a **well-distributed packing** in groups of a configuration $\sum \alpha_i \omega_i + v$ to be a packing of the concentration points a_i of the ω_i's into groups $G_1, \ldots, G_\ell$ such that

i) $d(a_i, a_j) = o(d(a_i, a_k))$ if $i, j \in G_m$ and $k \in G_s, s \neq m$.

ii) If, for (i, j, m) pairwise distinct, $d(G_i, G_j) \leq d(G_i, G_m)$, then

$$\sum_{\substack{s \neq t \\ s \in G_i \\ t \in G_j}} \varepsilon_{st} \geq c \sum_{\substack{s \neq t \\ s \in G_i \\ t \in G_m}} \varepsilon_{st}.$$

We prove in this work the two following results:

Theorem 1   *Assume that Conjectures 2 and 3 hold and that*

$$\varepsilon_{ij} \sim \frac{1}{\sqrt{\lambda_i \lambda_j} |a_i - a_j|} \quad \text{for each } i \neq j. \tag{1.3}$$

Then Conjecture 1 holds for a **well-distributed** *packing of a configuration such that, for each i, $\sum\limits_{\substack{s\in G_i \\ t\in G_j \\ i\neq j}} \varepsilon_{st} \geq c \sum\limits_{s\neq t} \varepsilon_{st}$.*

Theorem 1$'$ *Assume that Conjectures 2, 2$'$, 3, 3$'$ hold and that (1.3) hold. Then Conjecture 1 holds for* **every** *configuration $\sum \alpha_i \omega_i + v$.*

These results open the gate to the finding of an existence mechanism for the solutions of (1.1) or (1.2). Such a mechanism has been found for positive solutions [Bahri 1989], [Bahri 2001], via variational techniques. The existence of solutions has been tied directly to the existence and behavior of the asymptots.

This project is however more difficult for changing-sign Yamabe-type problem since now once a solution u is found, an infinite number of genuine asymptots arise out of combinations $u + \sum \omega_i$, while for positive solutions $u + \sum \delta_i$ never forms an asymptot. Nevertheless, despite this complication, interesting "structures" emerge from our analysis, in particular we find a nice extension of the use of the symmetric matrix $A = \begin{pmatrix} 0 & & \frac{1}{|x_i - x_j|} \\ & \ddots & \\ & & 0 \end{pmatrix}$ already appearing for the study of positive solutions. It turns out that there is a way (see below) to combine the asymptots so that the $p \times p$ matrix A appears, acting on $\mathbb{R}^p - \{0\}$.

1.3 Conjecture 2 and Sketch of the Proof of Theorem 1; Outline

Conjecture 2 is a very interesting direction for research, which is easy to establish in the case of two masses i.e. $p = 2$. Much trickier is the following results due to Y. Xu [Xu].

Theorem A **(Y. Xu [Xu])** *Conjecture 2 holds for $p = 3$ i.e. for the case of three masses.*

Sketch of the proof of Theorem 1

We now sketch the proof of Theorem 1. The proof is divided in three steps which correspond to the subdivision of this work in three parts. Part I and Part III are basic and are required for the proof. Part II provides an intermediate result, a partial progress with respect to Part I. Part III goes much beyond Part II. However the results of Part II provide a natural insight in the problem. It is only after the work of Part II that we find out what should be done to overcome the problem of clusters of ω_i's having little interaction among themselves.

Part I

The main new ingredient in the proof of Theorem 1 with respect to the study of the same equation with $u > 0$ is, under (1.3), to think of equation (1.2) in a matrix way as u reads $\sum_{i=1}^{p} \alpha_i \omega_i + \bar{v}$. $\bar{v}$ is the optimal v under (Vo).

We define, for each i a domain of influence of ω_i, after setting:

$$\Omega_i = \{x \text{ such that } \lambda_i |x - a_i| \leq \operatorname*{Min}_{j \neq i} \frac{1}{8\varepsilon_{ij}} \text{ and } \lambda_j |x - a_j| \geq \frac{1}{\varepsilon_{ij}} \text{ for } \lambda_j \geq \lambda_i\}.$$

We then split $\bar{v}$ in Ω_i into two parts:

$$\bar{v}\big|_{\Omega_i} = \bar{v}_i + \bar{h}_i$$

where $\bar{v} \in H_0^1(\Omega_i)$, $\bar{h}_i$ is harmonic and we prove the estimate:

Theorem B $\quad |\bar{v}_i|_{H_0^1} + \frac{|h_i|_\infty}{\sqrt{\lambda_i}} \leq C \left(\sum_j |\bar{\omega}_i(\tilde{a}_i)| \varepsilon_{ij} + \varepsilon_{ij}^3 \right).$

This theorem does not assume (1.3). All the estimates, for every i, are tied to each other by (1.2) and the related equation satisfied by $\bar{v}$. After carefully splitting $\bar{v}$ in $\sum \bar{v}_i + \bar{h}_i$ and a remainder portion in $(U\Omega_i)^c$, we estimate each part separately. The estimate involves for i the other indexes j. We derive a matrix and bootstrap our arguments.

Theorem B is quite essential for the following reason: when we were working with positive functions, the ω_i's were δ_i's and the $\bar{\omega}_j^\infty$'s were positive quantities. Then, we used to combine ([Bahri 1989], [Bahri 2001], [Bahri and Coron (1988)]) estimates using λ_i-dervatives (along $\lambda_i \frac{\partial}{\partial \lambda_i}$) and estimates along $\frac{1}{\lambda_i} \frac{\partial}{\partial a_i}$ of the normal form

$$P = \sum \left(\bar{\omega}_i^\infty \bar{\omega}_j^\infty \varepsilon_{ij} + \frac{c_{ij}}{\bar{c}} \varepsilon_{ij}^3 \right).$$

The positivity of $\bar{\omega}_i^\infty, \bar{\omega}_j^\infty$ played a very strong role and allowed us to derive very good lowerbounds on $J'(u) = \operatorname{grad} J(u)$.

Once the positivity is removed, these lowerbounds disappear. We cannot use $\frac{1}{\lambda_i} \frac{\partial}{\partial a_i}$ anymore because the derivatives of the remainder terms can be large when compared to the derivatives of P which do not work together anymore.

We have to work with $\frac{\partial}{\partial a_i}$ instead of $\frac{1}{\lambda_i} \frac{\partial}{\partial a_i}$ and we have to track down the contribution of the remainder in a much better manner. Thus, the estimates of Part I set up the general framework and lead to the proof of

Theorem 1 under the additional condition:

$$\text{There exists } C > 0 \text{ such that } \quad \frac{1}{C} \leq \frac{\lambda_i}{\lambda_j} \leq C. \tag{1.4}$$

Part II

We derive in this part various estimates which improve, without getting rid of (1.4), the estimates of Part I.

The main improvements concern three quantities:

We first improve the estimate on

$$\left(\int_{\Omega_j} |\omega_j|^{24/5} |\omega_i|^{6/5} + \int_{\Omega_j} |\omega_i|^{24/5} |\omega_j|^{6/5} \right)^{5/6} \quad \text{for } \lambda_j \geq \lambda_i.$$

This expression is bounded by $C(|\omega_i(\bar{a}_j)|\varepsilon_{ij} + \varepsilon_{ij}^3)$.

This minor improvement considerably changes our estimates though. We then provide a better estimate on the optimal v, after using a boot-strapping procedure. Namely, we prove:

Proposition A *For every $h \in \mathbb{N}$, there exists a constant $C_h > 0$ and a function $\theta_h \in H^1$ such that*

(i) $|v - \theta_h|(x) \leq C_h \sum \delta_\ell(x) \quad \forall x \in S^3$
(ii) $|\theta_h|_{H^1} \leq C_h \sum \varepsilon_{k\ell}^h$
(iii) $|v - \theta_h|(x) \leq C_h \sum \left(\varepsilon_{k\ell} + |v_\ell|_{L^6} + \frac{|h_\ell|_{L^\infty}}{\sqrt{\lambda_\ell}} \right) \delta_\ell(x) \quad \forall x \in \Omega^c = (U\Omega_i)^c.$

Lastly, we start the proof of precise estimates, to the improved in Part III, on $\int \frac{w_\ell^4 |w_k|}{|x-y|}$. We establish:

Proposition B $\int \frac{w_\ell^4 |w_k|}{|x-y|} \leq C \sum\limits_{\substack{\lambda_i \geq \lambda_\ell \\ \lambda_i \geq \lambda_k}} \varepsilon_{ij} \delta_i$ *for $y \in (\Omega_\ell \cup \Omega_k)^c$.*

We then revisit all the estimates of Part I. They all improve as well as the derivatives of the remainder term. (1.3) − (1.4) can be weakened but the result is not decisive. We still have a problem with clusters of masses having little interaction between them.

Part III: Taking care of the clusters

We assume here that the ω_j's can be subdivided in several clusters, basically two clusters. Each cluster have a least concentrated ω_j.

The masses inside a given cluster have concentration points very close one to the other when compared to the distance between the average concentration points of the two clusters. Yet the interaction between the masses inside a cluster is small with respect to the interaction between the two basic masses of each cluster. This is the difficulty which we are facing in our expansions. We then rewrite $\sum \alpha_i \omega_i + \bar{v}$ under the form

$$\left(\sum_{i \in I} \alpha_i \omega_i + \bar{v}_I\right) + \left(\sum_{i \in II} \alpha_i \omega_i + \bar{v}_{II}\right) + (\bar{v} - (\bar{v}_I + \bar{v}_{II}))$$

where v_I and v_{II} are the optimal v's for each cluster. v is optimal for $\sum \alpha_i \omega_i$. In this part, we think of each cluster as a single mass and we thus move them together and we think of them as combined. Accordingly $\mathbb{R}^3$ is now derived in three regions Ω_I, Ω_{II} and $\mathbb{R}^3 - (\Omega_I \cup \Omega_{II})$. $\tilde{\Omega}_I$ and $\tilde{\Omega}_{II}$ are smaller versions of Ω_I and Ω_{II}. The distance between Ω_I and Ω_{II} is $|a|$. We then prove the three following basic estimates.

Proposition C

(i) $|\bar{v}(x)| \le C \sum (|\omega_k^\infty| + \sum |\omega_k(a_\ell)| \sqrt{\lambda_k} |a_k - a_\ell| + \varepsilon_{k\ell}) \varepsilon_{k\ell} \delta_\ell(x)$

(ii) $|\bar{v}_I(x)| \le C \sum_{k,\ell \in I} (|\omega_k^\infty| + \sum |\omega_k(a_\ell)| \sqrt{\lambda_k} |a_k - a_\ell| + \varepsilon_{k\ell}) \varepsilon_{k\ell} \delta_\ell$

(iii) $|\bar{v}_{II}(x)| \le C \sum_{k,\ell \in II} (|\omega_k^\infty| + \sum |\omega_k(a_\ell)| \sqrt{\lambda_k} |a_k - a_\ell| + \varepsilon_{k\ell}) \varepsilon_{k\ell} \delta_\ell$

(iv) $|\bar{v} - (\bar{v}_I + \bar{v}_{II})| \le \sum_{(k,\ell) \in (I,II)} \left[\left[(|\sum \omega_k^\infty| + \varepsilon_{k\ell}) \varepsilon_{k\ell} + \frac{|\sum \omega_k(a_\ell)|}{\sqrt{\lambda_\ell}} \right] \delta_\ell + \right.$

$\left. (|\sum \omega_\ell^\infty| + \varepsilon_{k\ell}) \varepsilon_{k\ell} + \frac{|\sum \omega_k(a_k)|}{\sqrt{\lambda_k}} \delta_k \right] + \sum_{\substack{(i,j) \in (I,II) \\ \text{or } (II,I)}} O(\sum \varepsilon_{\ell m}) \varepsilon_{ij} \delta_i.$

Proposition D   **(Gradient estimates)**

(i) *Assume that* y *is in* $\tilde{\Omega}_{II}$ *a smaller version of* Ω_{II} *such that* $d(\Omega_I, \tilde{\Omega}_{II}) \ge c|a|$.

Then, for $y \in \tilde{\Omega}_{II}$,

$$|\nabla \bar{v}_I(y)| \le \frac{C}{|a|} \left(\sum_{\substack{(k,j) \in I \\ k \ne j}} (|\omega_j^\infty| + |\omega_j(a_k)| \sqrt{\lambda_j} |a_j - a_k| + \varepsilon_{kj}) \varepsilon_{kj} \delta_k \right)$$

(ii) $\forall y \in S^3, |\nabla \bar{v}_I| \le C \sum_{i \in II} \sqrt{\lambda_i} \delta_i^2.$

Similar estimates hold for $\bar{v}_{II}$.

Lemma A

$$\int \frac{\omega_k^4 |\omega_\ell| + \omega_\ell^4 |\omega_k|}{|x - y|}$$

$$\leq C\big[(|\omega_\ell^\infty| + \varepsilon_{\ell k} + |\omega_\ell(a_k)|\sqrt{\lambda_\ell}|a_k - a_\ell|)\delta_k$$

$$+ (|\omega_k^\infty| + \varepsilon_{\ell k} + |\omega_k(a_\ell)|\sqrt{\lambda_k}|a_k - a_\ell|)\delta_\ell\big]\varepsilon_{\ell k}.$$

Ultimately, all these propositions allow us to prove the following key estimates (as usual, $\bar{v} - (\bar{v}_I + \bar{v}_{II})$ is split in each $\Omega_i - \Omega_i$ adapted to $\bar{\omega}_i$ - into an H_0^1 and a harmonic part $(\bar{v} - (\bar{v}_I + \bar{v}_{II}))_i + h_i^0$; h_i^s are the harmonic parts in domains Ω_s with $\lambda_s \geq \lambda_i$).

Theorem C

$$|(\bar{v} - \bar{v}_I + \bar{v}_{II})_i|_{H_0^1} + \frac{|h_i^0|_\infty}{\sqrt{\lambda_i}} + \sum_{\substack{\lambda_s \geq \lambda_i \\ s \neq i}} \varepsilon_{is}|h_i^s|_\infty$$

$$\leq C \left(\sum_{\substack{k \neq i \\ k \in I \ or \ k \in II}} \left| \sum_{\substack{j \in II \\ or \ j \in I}} \alpha_j \omega_j(a_k)\right| \frac{\varepsilon_{ik}}{\sqrt{\lambda_k}} + \frac{1}{\sqrt{\lambda_i}} \left| \sum_{\substack{j \in I \ if \ i \in II \\ j \in II \ if \ i \in I}} \alpha_j \omega_j(a_i)\right| \right)$$

$$+ O\left(\sum_{j \neq i} \frac{|\omega_j^\infty|\varepsilon_{ij}}{\lambda_i |a|} + \sum \varepsilon_{ij} \sum_{\substack{(j,m) \\ \in (I,II) \\ or \ (II,I)}} \varepsilon_{jm}^2 \right).$$

Observe that the estimate of Theorem C clearly shows that $\sum_{j \in I} \alpha_j \omega_j$ and $\sum_{j \in II} \alpha_j \omega_j$ behave as a single mass. Also the remainder term in this estimate is multiplied by $\frac{1}{\sqrt{\lambda_i |a|}}$ not by $\frac{1}{\sqrt{\lambda_j |a|}}$.

We then complete our expansion of J and of its gradient and we show that the remainder terms are now controlled by the corresponding in P. This establishes the Morse Lemma at infinity.

1.4   The Difference of Topology

We indicate here the main contributions in the formula for the difference of topology. We also indicate how this difference should be thought of if we need to compute it more precisely.

An explicit formula requires the removal of (1.3). This, in turn, would require a tedious checking that all estimates can be extended, unless a way of continuously rescaling all the configurations so that the concentration points would be pointwise distinct and (1.3) would be satisfied is found. This might work on S^3, not on domains of $\mathbb{R}^3$ though.

In addition, some more work would need to be completed to express the exit set (the contribution of $J_{c-\varepsilon}$ in $(J_{c+\varepsilon}, J_{c-\varepsilon})\infty$) when the ω_j's are so configured that $\bar{\omega}_i^\infty$ is zero for each of them. We have a preliminary good description of this set but we think that the general picture can be worked out far further.

The parameters set when the concentration point are distinct and the masses $\bar{\omega}_j$'s have no symmetry (which we assume for sake of simplicity) can be described as follows:

We choose a mass $\bar{\omega}_1$ as reference. $\bar{\omega}_2, \ldots, \bar{\omega}_p$ are then each defined up to the action of $O(4)$ on each of them. $\mathcal{P}$ thus designates the parameters' space component $O(4)^{p-1}$ which indicate in what positions are $\bar{\omega}_1, \bar{\omega}_{2,\sigma_2}, \ldots, \bar{\omega}_{p,\sigma_p}$.

We also choose p points (distinct at this stage of our work) on S^3 and p coefficients $\alpha_1, \ldots, \alpha_p, 0 \le \alpha_1 \le 1, \sum \alpha_i = 1$. This component of the parameter space is denoted $(S^3)_*^p \times \Delta_{p-1}$. Δ_{p-1} is the $(p-1)$-dimensional simplex. $(S^3)_*^p = \{(a_1, \ldots, a_p) \in (S^3)^p; a_i \neq a_j \text{ for } i \neq j\}$.

We also need p small disks $D_u', \ldots, D_u^p$ in the unstable manifold of each $\bar{\omega}_j$ (the strict ones) and p parameters $\lambda_1, \ldots, \lambda_p$ in $[A, +\infty)$ so that $\varepsilon_{ij} \sim$
$$\frac{1}{\sqrt{\lambda_i \lambda_j |a_i - a_j|}}.$$

The parameters' space reads them as a fiber bundle, with fiber $\mathcal{P}$, over a base which is $(S^3)_*^p \times \Delta_{p-1} \times \Pi_{j=1}^P D_u^j \times (A, +\infty)^p . \Pi_{j=1}^P D_u^j$ is the space for v.

$$\mathcal{P}$$
$$|$$
$$(S^3)_*^p \times \Delta_{p-1} \times \Pi D_u^j \times [A, +\infty)^p.$$

It is **extremely useful and fruitful** to think of all asymptots

$$\sum \varepsilon_i \bar{\omega}_i \qquad \varepsilon_i = \pm 1$$

together. The parameters' space glue up in a natural way via the Δ_{p-1}'s which rebuild then the sphere S^{p-1}, the unit sphere (up to rescaling) in $\mathbb{R}^p - \{o\}$. The glueing occurs when at least one α_i is zero i.e. below the critical level c_∞ (which is the same) of all the $\sum \varepsilon_i \bar{\omega}_i$. In terms of parameters' space, a configuration is below c_∞ if either it is in $\partial(\Delta_{p-1} \times \Pi D_u^j)$ or if the principal part P is negative (non positive) at this configuration.

P is made of two pieces, one usually of higher order than the other one. The first one is

$$-\bar{c}\sum \frac{\varepsilon_i\varepsilon_j\omega_i^\infty\omega_j^\infty}{\sqrt{\lambda_i\lambda_j}|a_i - a_j|}.$$

We are thus naturally led to the set where $\sum \frac{\varepsilon_i\varepsilon_j\omega_i^\infty\omega_j^\infty}{\sqrt{\lambda_i\lambda_j}|a_i-a_j|} \geq 0$. Since $\sum \varepsilon_i\alpha_i\omega_i$ is below c_∞ if α_{i/α_j} is not $1 + o(1)$, we can expand this set into

$$\sum \frac{\varepsilon_i\alpha_i\varepsilon_j\alpha_j\omega_i^\infty\omega_j^\infty}{\sqrt{\lambda_i\lambda_j}|a_i - a_j|} \geq 0.$$

The $\frac{\varepsilon_i\alpha_i}{\sqrt{\lambda_i}}\omega_i^\infty$'s for those ω_i^∞'s which are non zero form then a system of coordinates and this set becomes, with $\ell \leq p$,

$$N_\ell^- = \{u \in \mathbb{R}^\ell - \{0\} \text{ such that} {}^t uAu \geq 0\}$$

A is the $\ell \times \ell$ matrix $\begin{pmatrix} 0 & & \frac{1}{|a_i-a_j|} \\ & \ddots & \\ \frac{1}{|a_i-a_j|} & & 0 \end{pmatrix}$ corresponding to those indexes i such that ω_i^∞ is non zero.

In the description of the parameters' space, we have left a part largely not explicit; it is the part $\mathcal{P}$ corresponding to the relative positions of $\bar{\omega}_{i,\sigma_i}$, with respect to $\bar{\omega}_1$. We can think of $\mathcal{P}$ as follows: we choose a point $\hat{a}_i$ on S^3. We rotate $\bar{\omega}_i$ around $\hat{a}_i$ and we identify $\hat{a}_i$ and a_i. Then $\bar{\omega}_i$ is changed into $\bar{\omega}_{i,\sigma_i}$ through the rotation σ_i around $\hat{a}_i$ and $\bar{\omega}_i^\infty$ becomes $\bar{\omega}_i(\hat{a}_i)$. We then rescale around a_i the function $\bar{\omega}_{i,\sigma_i}$. In this way, the concentration point a_i is thought of as made of two points: a_i itself on S^3 and another point on S^3, with the standard $\bar{\omega}_i$ attached to it, which will be the point $\hat{a}_i$. We identify $\hat{a}_i$ and a_i through the action of an element of $O(4)$. We also rotate $\bar{\omega}_i$ around $\hat{a}_i$ and we rescale the function which we obtain (on the original S^3) in this way.

We thus derive a stratified S_p which can be described using the $\hat{a}_i$'s.

There is a natural map:

$$\Pi : S_p \to (S^3)^p \times (S^3)^p_*$$
$$s \to (\hat{a}_1, \ldots, \hat{a}_p) \times (a_1, \ldots, a_p)$$

$F = \Pi^{-1}((\hat{a}_1, \ldots, \hat{a}_p), (a_1, \ldots, a_p))$ is a set very similar to N_ℓ^- i.e. there are ℓ indexes $i_1, \ldots, i_\ell$ such that $\bar{\omega}_j^\infty = \bar{\omega}_j(\hat{a}_j)$ is now zero. F can be identified as $N_\ell^-(i_1, \ldots, i_\ell)$, where $N_\ell^-(i_1, \ldots, i_\ell)$ is defined as N_ℓ^- above, with the indexes $i_1, \ldots, i_\ell$.

Accordingly, the topology of S_p can be computed using the distribution of the signatures of A over $(S^3)^p_*$ and the zero sets of the $\bar{\omega}_i$'s.

We will discuss this more elsewhere.

The last piece of information about the difference of topology which we need to study comes in when all the $\bar{\omega}_i^\infty$'s are zero.

Indeed, under the hypothesis:

(H) For every p, the function $f(u, a_1, \ldots, a_p) = {}^t u A u$ defined on $S^{p-1} \times (S^3)^p_*$ does not have 0 as a critical value,

the principal part P of the expansion should reduce to

$$-\bar{c} \sum \frac{\bar{\omega}_i^\infty \bar{\omega}_j^\infty}{\sqrt{\lambda_i \lambda_j}|a_j - a_j|}$$

if the $\bar{\omega}_i^\infty$'s are not all $o(1)$.

If all $\bar{\omega}_i^\infty$'s are zero then the second piece of $P, -\sum c_{ij}\varepsilon_{ij}^3$ enters into play. We know [Bahri and Coron (1988)] that $c_{ij} = c \nabla \omega_i^\infty \cdot \nabla \omega_i^\infty$.

Thus, we derive a set of the type

$$-c \sum \nabla \bar{\omega}_i^\infty \cdot \nabla \bar{\omega}_i^\infty \varepsilon_{ij}^3 \leq 0$$

which should provide a condition on the relative positions of the $\bar{\omega}_i$'s involved in the configuration.

The $\hat{a}_i$'s are subject to the requirement $\bar{\omega}_i(\hat{a}_i) = 0$.

This gives a qualitative account of what we expect for the difference of topology. We think that this program is within reach.

1.5 Open Problems

1.5.1 *Understand the difference of topology*

As we explore the normal form in our expansion:

$$P = -\bar{c} \sum \bar{\omega}_i(\tilde{a}_j)\bar{\omega}_j^\infty \varepsilon_{ij} - \sum c_{ij}\varepsilon_{ij}^3$$

a quantity very close to P, which we denote P_∞, appears:

$$P_\infty = -\bar{c} \sum \bar{\omega}_i^\infty \bar{\omega}_j^\infty \varepsilon_{ij} - \sum c_{ij}\varepsilon_{ij}^3.$$

Thinking of ε_{ij} as $\dfrac{1}{\sqrt{\lambda_i \lambda_j}|a_i - a_j|}$, we find

$$P_\infty = -\bar{c} \left(\cdots \frac{\bar{\omega}_j^\infty}{\sqrt{\lambda_j}} \cdots \right) A \left(\begin{matrix} \vdots \\ \bar{\omega}_i^\infty/\sqrt{\lambda_i} \\ \vdots \end{matrix} \right) - \sum c_{ij}\varepsilon_{ij}^3.$$

Thus an important piece of information comes from the behavior of

$$S = \{u \in \mathbb{R}^p - \{0\} \, s.t. \, (Au, u) \geq 0\}.$$

This does not seem to be quite true at first glance since the u which appears in the definition of S is equal in our context to $\begin{pmatrix} \vdots \\ \bar{\omega}_i^\infty/\sqrt{\lambda_i} \\ \vdots \end{pmatrix}$, so that the sign of $u_i = \frac{\bar{\omega}_i^\infty}{\sqrt{\lambda_i}}$ is prescribed. However if all asymptots $\sum \pm \omega_i$ are combined, the various pieces combine and S is naturally found.

Another piece of information is provided by the zero sets of the $\bar{\omega}_i$'s and a last piece is less explicit, more related to the relative position of $\nabla \bar{\omega}_i^\infty$ with respect to $\nabla \bar{\omega}_j^\infty$.

1.5.2 *Non critical asymptots*

Assume that we have only "masses" $\bar{\omega}_i$ such that $\bar{\omega}_i^\infty$ is positive.

It is then fairly obvious that such combinations of $\bar{\omega}_i$ do not build a genuine asymptot since ${}^t u A u$ is negative on them. The same observation holds if the $\bar{\omega}_i^\infty$'s are all negative. The result holds as well when there are many more negative (or positive) contributions than contributions of the opposite sign. One would like to understand the behavior of the critical configurations as p tends to $\pm$ and relate them to discrete as well as continuous geometric problems on S^3.

1.5.3 *The exit set from infinity*

The expansion shows the use of the quadratic form ${}^t u A u$, with $u = \begin{pmatrix} \vdots \\ \bar{\omega}_i^\infty/\sqrt{\lambda_i} \\ \vdots \end{pmatrix}$. It is a quite striking fact that the expression of the normal form depends so little on the actual functions themselves. The main dependence is via the vector u i.e. via the signs of the $\bar{\omega}_i^\infty$'s. If we except the case when one $\bar{\omega}_i^\infty$ is zero, we could think of a model where the asymptot $\sum \omega_i$ would be replaced by $\sum \bar{\omega}_i^\infty \delta_i$. The critical levels, the indexes at infinity *etc.* would not match; but the value of ${}^t u A u$ which indicates whether with the preassigned concentrations $\lambda_1, \ldots, \lambda_p$ and with the preassigned values $\bar{\omega}_1^\infty, \ldots, \bar{\omega}_p^\infty$, the asymptot is genuine or not genuine would match. So would a decreasing flow defined at infinity.

We thus see that the exit set from infinity is independent of the actual asymptot. The simplest model i.e. the model involving $\sum \pm \delta_i$ intervenes as a basic factor in the understanding of the functional J at infinity.

1.5.4 *Establishing Conjecture 2 and continuous forms of the discrete inequality*

Conjecture 2 is a key hypothesis in the proof of Theorem 1. The proof provided in the case $p = 3$ [Xu] is quite involved. Y. Xu has also provided in [Xu] heuristic reasons why this result should hold in general.

A continuous form of this inequality should also be derived as the number p of points tends to $\pm\infty$.

1.5.5 *The Morse Lemma at infinity, Part I, II, III*

The proof of the Morse Lemma at infinity is divided in three distinct parts. Part I and Part III provide the proof. Part II is an intermediate step which on one hand provides insight in what should be completed to remove the restrictions involved in Part I (where the Morse Lemma at infinity is established when the concentration are comparable) and on the other hand provides a key estimate (Theorem 1).

1.5.6 *Notations $\bar{v}, \bar{v}_i, \bar{h}_i$*

The ω_i's are solutions of the Yamabe problem on $\mathbb{R}^3$. They are rescaled version of $\bar{\omega}_i$'s which all have concentration equal to $O(1)$.

For each ω_i, a domain of influence

$$\Omega_i = \{x \in \mathbb{R}^3 s.t. \lambda_i |x - x_i| \leq \text{Min} \frac{1}{8\varepsilon_{ij}} \forall j,$$

$$\lambda_j |x - x_j| \geq \frac{1}{\varepsilon_{ij}} \text{ if } \lambda_j \geq \lambda_i\}.$$

Any u is defined close to $\sum_{i=1}^{p} \omega_i$ $\left(\int |\nabla(u - \sum \omega_i)|^2 \text{ is small}\right)$ can be uniquely written as $u = \sum_{i=1}^{p} \alpha_i \omega_i + v$ where v satisfies a family of orthogonality conditions which read

$$(\text{Vo}) \quad \begin{cases} \int \nabla \omega_i \nabla v = 0 \\ \int \nabla \frac{\partial \omega_i}{\partial a_i} \nabla v = 0 \\ \int \nabla \frac{\partial \omega_i}{\partial \lambda_i} \nabla v = 0 \\ \int \nabla \frac{\partial \omega_i}{\partial \sigma_i} \nabla v = 0 \quad (\sigma_i \text{ is in } O(4)) \end{cases} \quad . \quad (\text{Vo})$$

The functional $J(\sum_{i=1} \alpha_i \omega_i + v)$ can be max-minimized with respect to the variation of v (under (Vo)) if we assume that each ω_i **is a non degenerate critical point of J transversally to the parameters of the conformal group**. We will work under this assumption. Then, following the results of [Bahri and Coron (1988)], the expansion of J in v reads as $(f, v) + Q(v, v) + o((|\nabla v|^2)^{3/2})$, where Q is non degenerate of index equal to $p - 1 + \sum_{i=1}^{p} \text{index}$ $J''(\omega_i)$. Such an expression can be max-minimized. We derive an optimal $\bar{v}$.

This $\bar{v}$ can be decomposed in each Ω_i into $\bar{v}_i + \bar{h}_i$, where $\bar{h}_i$ is harmonic and $\bar{v}_i$ is in $H_0^1(\Omega_i)$.

1.6 Preliminary Estimates and Expansions, the Principal Terms

The Content of Part I.

Part I provides the framework for the completion of the Morse Lemma at infinity. After two preliminary estimates (to be improved later), the equation satisfied by $\bar{v}$ is extracted and it is decomposed accordingly to the Ω_i's. $\bar{v}$ is split in $\sum(\bar{v}_i + \bar{h}_i)$. Due to the constraints $(V0)$ on $\bar{v}$, this equation involves projection terms which are **global by nature** (they express orthogonality conditions). Thus, the various $\bar{v}_i$'s, $\bar{h}_i$'s *etc.* are tied although their supports are disjoint. The main quantity tying them is the matrix A of the H_0^1-scalar product in

$$\text{Span}_{i=1}^{p} \left\{ \omega_i, \frac{1}{\lambda_i} \frac{\partial \omega_i}{\partial a_i}, \lambda_i \frac{\partial \omega_i}{\partial \lambda_i}, \frac{\partial \omega_i}{\partial \sigma_i} \right\}.$$

We thus need to estimate A and also $\bar{v}_i, \bar{h}_i$.

In a first step (Lemmas 3-11), the estimate on $|\bar{v}_i|_{H_0^1}$ is shown to depend on an estimate on $|\bar{h}_i|_\infty$.

Next, using the Green's function of an annulus-type domain, a pointwise estimate is derived on $\bar{h}_i$ (Lemma 12).

The estimate is complex because it involves (via the right hand side of the equation satisfied by $\bar{v}$) an enormous amount of terms and expressions related to Ω_i but also Ω_j, for $j \neq i$, $(U\Omega_i)^c$. It ties $\bar{v}_i, \bar{h}_i$ with $\bar{v}_j, \bar{h}_j$ *etc.*

We need to estimate carefully all of these expressions. This is what we complete in Lemmas 13—30.

We then are in position to derive an estimate (not yet optimal) on $|\bar{v}_i|_{H_0^1}$ and $|\bar{h}_i|_\infty$ (Lemmas 31—34).

Using this estimate, we derive in the last section of Part I, after establishing two further estimates (Lemmas 35—36) on the contribution of $\bar{v}$ in $J'(\sum \alpha_j \omega_j + \bar{v})$. $\partial \omega_i$, our Morse Lemma at infinity under (1.3)—(1.4).

1.7 Preliminary Estimates

We will denote in what follows

$$\bar{\partial}\omega_i$$

any of $\omega_i, \frac{1}{\lambda_i}\frac{\partial \omega_i}{\partial a_i}, \lambda_i \frac{\partial \omega_i}{\partial \lambda_i}$ and by ψ_i the Laplacian of any of those functions. We will work on $\mathbb{R}^3$ or S^3. $\Delta_{\mathbb{R}^3}$ transforms into the Yamabe operator L on S^3. We differenciate (V_0) to get

$$\int \nabla \bar{\partial}\omega_i \nabla \frac{\partial \bar{v}}{\partial a_i} = -\int \nabla \frac{\partial^2 \omega_i}{\partial a_i \bar{\partial}} \nabla \bar{v}.$$

We decompose $\bar{v}$ into $\sum \bar{v}_j + \bar{h}$, where each $\bar{v}_j$ has support into Ω_j; the $\Omega'_j s$ are disjoint and $\bar{h}$ is harmonic in $\cup \Omega_j$.

We thus need to estimate

$$\alpha_{ij} = \int \nabla \frac{\partial^2 \omega_i}{\partial a_i \bar{\partial}} \nabla \bar{v}_j \quad i \neq j$$

$$\alpha_{ii} = \int \nabla \frac{\partial^2 \omega_i}{\partial a_i \partial} \nabla \bar{v}_i$$

$$\beta_i = \int \nabla \frac{\partial^2 \omega_i}{\partial a_i \bar{\partial}} \nabla \bar{h}.$$

Observe that

$$\frac{\partial^2 \omega_i}{\partial a_i \bar{\partial}} = 0(\lambda_i \delta_i), \Delta \frac{\partial^2 \omega_i}{\partial a_i \bar{\partial}} = 0(\lambda_i \delta_i^5).$$

It is difficult to estimate α_{ij} if $\lambda_j \geq \lambda_i$, in particular α_{ii}. We need more insight into the equations satisfied by the $\bar{v}_k$'s and $\bar{h}$.

We start with the following preliminary estimates:

Lemma A $\qquad \left(\int_{\Omega_i^c} |\bar{\partial}\omega_i|^6\right)^{5/6} + \left(\int |\bar{\partial}\omega_i|^{6/5} \sum_{\ell \neq i} |\bar{\partial}\omega_\ell|^{24/5}\right)^{5/6} \leq$

$C(\sum(|\omega_i^\infty| + |\omega_j^\infty|)\varepsilon_{ij} + \varepsilon_{ij}^{5/2})$.

Proof. The estimate on $(\int_{\Omega_i^c} |\bar{\partial}\omega_i|^6)^{5/6}$ will follow from 1. of Lemma 13 below and from the definition of Ω_i. In order to upperbound

$(\int |\bar{\partial}\omega_i|^{6/5}|\bar{\partial}\omega_\ell|^{24/5})^{5/6}$, we observe that we may assume that ω_i is the most concentrated mass ($\lambda_i \geq \lambda_\ell$). We define $\Omega_{i,\ell} = \{x \; s.t \lambda_i |x - x_i| \leq \frac{1}{\varepsilon_{i\ell}}\}$. Then, on $\Omega_{i,\ell}^c$, $\bar{\partial}\omega_i$ reads as (see [Bahri 2001]):

$$\bar{\partial}\omega_i = c\omega_i^\infty 0(\delta_i) + \frac{0(\delta_i)}{1 + \lambda_i^2 |x - a_i|^2}.$$

We thus need to estimate

$$\left(\int_{\Omega_{i,\ell}^c} \frac{\delta_i^{6/5}}{(1 + \lambda_i^2 |x - a_i|^2)^{6/5}} |\omega_\ell|^{24/5}\right)^{5/6} \leq C \left(\int_{r \geq \frac{1}{\text{Max } \varepsilon_{ij}}} \frac{r^2 dr}{(1 + r^2)^9}\right)^{1/6}$$

$$\leq C \text{ Max } \varepsilon_{ij}^{5/2}.$$

On $\Omega_{i,\ell}$, we know, from [Bahri 2001, proof of Lemma 3.2], that the contribution of $\partial\omega_i - c\omega_i^\infty 0(\delta_i)$ yields $0(\varepsilon_{i\ell}^3)$. The claim follows. $\qquad\square$

Lemma B $\displaystyle\sum_{\ell \neq i} \int \omega_\ell^4 \omega_i^2 + \omega_i^4 \omega_\ell^2 \leq C \left((\omega_\ell^{\infty 2} + \omega_i^{\infty 2})\varepsilon_{\ell i}^2 + \varepsilon_{\ell i}^6\right).$

Proof. We come back to [Bahri 2001, p. 461] and we observe that, for $r \leq 1$ we used the upperbound

$$\omega_1^2 \leq \omega_1(0)^2 + Cr$$

while we could have used

$$\omega_1^2 \leq C(\omega_1(0)^2 + r^2).$$

With this upperbound, the estimate improves into the one provided by Lemma B. We will use this estimate in this work. $\qquad\square$

1.7.1 *The equation satisfied by $\bar{v}$*

$\bar{v}$ satisfies

$$Q_{F^\perp}\left(J'\left(\sum \alpha_j \omega_j + \bar{v}\right)\right) = 0$$

where $Q_{F^\perp}$ is the H_0^1-orthogonal projection onto $\text{Span}_{i=1}^P \{\partial\omega_i\}^\perp$. This reads ($L$ is the Yamabe operator on the standard S^3)

$$Q_{F^\perp}\left(J'\left(\sum \alpha_j \omega_j\right)\right) + Q_{F^\perp}(J''(\sum \alpha_j \omega_j) \cdot \bar{v})$$

$$+ Q_{F^\perp} L^{-1}\left(0\left(\sum |\omega_j|^3 \bar{v}_k^2 + \sum |\omega_j|^3 \bar{h}^2 + |\bar{h}|^5 + \sum |\bar{v}_k|^5\right)\right) = 0.$$

On the other hand

$$J'' \left(\sum \alpha_j \omega_j \right) \cdot \overline{v} = \sum \overline{v}_i + \overline{h} + 5L^{-1} \left(\left(\sum \alpha_j \omega_j \right)^4 \left(\sum \overline{v}_i + \overline{h} \right) \right).$$

Applying L and observing that $LQ_{F\perp} = Q^*L$, where Q^* is the dual to $Q_{F\perp}$ in the H^{-1}/H^1 duality, we derive:

$$Q^*L \left(J' \left(\sum \alpha_j \omega_j \right) \right)$$
$$+ Q^* \left(\sum \Delta \overline{v}_i + \Delta \overline{h} \right) + 5Q^* \left(\left(\sum \alpha_j \omega_j \right)^4 \left(\sum \overline{v}_i + \overline{h} \right) \right)$$
$$+ Q^*0 \left(\sum |\omega_j|^3 \left(\overline{v}_k^2 + \overline{h}^2 \right) + |\overline{h}|^5 + \sum |\overline{v}_k|^5 \right) = 0.$$

Observe that, since $\overline{v}$ satisfies (Vo):

$$Q^* \left(\Delta \left(\sum \overline{v}_i + \overline{h} \right) \right) = \Delta \left(\sum \overline{v}_i + h \right)$$

so that, since the Ω_i's are disjoint, the above equation can be seen as a family of equations on the various $\overline{v}_i$'s, with very little interaction between them, except for $\overline{h}$. We need to write the equation satisfied by $\overline{h}$. Multiplying the equation by ψ, we have:

$$\int \Delta \left(\sum \overline{v}_i + \overline{h} \right) \psi = \int f\psi = \sum \int_{\Omega_i} f\psi + \int_{(\cup \Omega_i)^c} f\psi.$$

Thus,

$$\sum_i \int_{\partial \Omega_i} \frac{\partial \overline{v}_i}{\partial \nu_i} \psi + \sum_i \left(\int \overline{v}_i \Delta \psi - \int_{\Omega_i} f\psi \right) + \int_{(\cup \Omega_i)^c} (\Delta \overline{h} - f)\psi = 0.$$

Thus,

$$\sum_i \int_{\partial \Omega_i} \left(\frac{\partial \overline{v}_i}{\partial \nu_i} - \frac{\partial}{\partial \nu_i} \Delta_{\Omega_i}^{-1} f \right) \psi + \sum_i \int_{\Omega_i} (\overline{v}_i - \Delta_{\Omega_i}^{-1} f) \Delta \psi + \int_{(\cup \Omega_i)^c} (\Delta \overline{h} - f)\psi = 0.$$

Hence,

Lemma 1 $(\overline{v}_i, \overline{h})$ *satisfy*

$$\overline{v}_i = \Delta_{\Omega_i}^{-1} f, \Delta \overline{h} = f \ in \ (\cup \Omega_i)^c, \frac{\partial \overline{v}_i}{\partial \nu_i} = \frac{\partial}{\partial \nu_i} \Delta_{\Omega_i}^{-1} f.$$

Proof.   Straightforward. $\square$

On the other hand, $\sum \bar{v}_i + \bar{h}$ is smooth, so that

$$\frac{\partial \bar{v}_i}{\partial \nu_i} + \frac{\partial \bar{h}}{\partial \nu_i} = -\frac{\partial \bar{h}}{\partial \nu_i^-}$$

where $\frac{\partial}{\partial \nu_i}$ is the outwards normal derivative of Ω_i and $\frac{\partial}{\partial \nu_i^-}$ is the inwards one.

f reads as

$$f = -Q^*(\Delta J' \left(\sum \alpha_j \omega_j\right) + 5 \left(\sum \alpha_j \omega_j\right)^4 \left(\sum \bar{v}_i + \bar{h}\right)$$
$$+ O\left(\sum |\omega_j|^3 \left(\bar{v}_k^2 + \bar{h}^2\right) + |\bar{h}|^5 + \sum |\bar{v}_k|^5\right) \tag{1.5}$$

so that the equation on $\bar{v}_i$ reads:

$$\begin{cases} \Delta \bar{v}_i + Q^*(5(\sum \alpha_j \omega_j)^4 \bar{v}_i + 0(\sum_j |\omega_j|^3 \bar{v}_i^2 + |\bar{v}_i|^5)) = -Q^*(\Delta J'(\sum \alpha_j \omega_j) \\ +5(\sum \alpha_j \omega_j)^4(\sum_{j \neq i} \bar{v}_j + \bar{h}) + O(\sum_{k \neq i} |\omega_j|^3(\bar{v}_k^2 + \bar{h}^2) + |\bar{h}|^5 + \sum_{k \neq i} |\bar{v}_k|^5)) \\ \bar{v}_i = 0|_{\partial \Omega_i} \end{cases} \cdot$$
$$\tag{1.6}$$

Let

$$e_i = L \partial \omega_i \, ; \quad A = \begin{pmatrix} \ddots & & -\int e_j L^{-1} e_i \\ & \ddots & \\ & & \ddots \end{pmatrix} . \tag{1.7}$$

Then

$$Q^*(g) = g - \sum \left(A^{-1} \left(\begin{matrix} \vdots \\ -\int g L^{-1} e_i \\ \vdots \end{matrix} \right) \right)_j e_j. \tag{1.8}$$

We write $\bar{h}$ as

$$\bar{h} = \sum \bar{h}_i + k^* \tag{1.9}$$

when $\bar{h}_i = \bar{h} \chi_{\Omega_i}$ then rereads:

$$\begin{cases} \Delta \overline{v}_i + Q^*(5(\sum \alpha_j \omega_j)^4 \overline{v}_i + O(\sum_j |\omega_j|^3 \overline{v}_i^2 + |\overline{v}_i|^5)) = -Q^*(\Delta J'(\sum \alpha_j \omega_j) \\[2mm] -Q^*(5(\sum \alpha_j \omega_j)^4 \overline{h}_i + O(\sum |\omega_j|^3 (\overline{h}_i^2 + |\overline{h}_i|^5)) \\[2mm] + \sum \left(A^{-1} \left(\begin{pmatrix} \vdots \\ -\int (\sum\limits_{j \neq i} 5(\sum \alpha_s \omega_s)^4 (\overline{v}_j + \overline{h}_j + k^*) + O(\sum\limits_{k \neq i} |\omega_j|^3 \overline{v}_k^2 + \\ +\overline{h}_k^2 + k^{*2}) + \sum\limits_{k \neq i} (|\overline{v}_k|^5 + |\overline{h}_k|^5 + |k^*|^5)) L^{-1} e_m \\ \vdots \end{pmatrix} \right) \right)_\ell e_\ell \\[2mm] \overline{v}_i = 0|_{\partial \Omega_i}. \end{cases}$$

$$\tag{1.10}$$

In addition, we have

$$\frac{\partial \overline{h}}{\partial \nu_i} + \frac{\partial \overline{h}}{\partial \nu_i^-} = -\frac{\partial}{\partial \nu_i} \left(\Delta_{\overline{\Omega}_i}^{-1} f_i \right) \big|_{\partial \Omega_i}. \tag{1.11}$$

with

$$f_i = -Q^* \left(\Delta J'(\sum \alpha_j \omega_j) - 5(\sum \alpha_j \omega_j)^4 (\overline{v}_i + \overline{h}_i) \right.$$
$$\left. + O(\sum |\omega_j|^3 (\overline{v}_i^2 + \overline{h}_i^2) + |\overline{h}_i|^5 + |\overline{v}_i|^5) \right)$$
$$+ \sum \left(A^{-1} \left(\begin{pmatrix} \vdots \\ -\int [5(\sum \alpha_\ell \omega_\ell)^4 (\sum\limits_{j \neq i} \overline{v}_j + \overline{h}_j + k^*) \\ +O(\sum\limits_{k \neq i} |\omega_j|^3 (\overline{v}_k^2 + \overline{h}_k^2 + k^{*2}) + \sum\limits_{k \neq i} (|\overline{h}_k|^5 \\ +|\overline{v}_k|^5 + |k^*|^5))] L^{-1} e_m \\ \vdots \end{pmatrix} \right)_\ell e_\ell \right).$$

$$\tag{1.12}$$

Our aim is to find a good estimate on each $\overline{v}_i$ in H_0^1 and an estimate on $|\overline{h}_i|_\infty$. The two estimates turn out to be tied. The estimates for two different indexes $i \neq j$ are also tied. We will derive these estimates after a careful analysis of the contribution of each term of f_i in (1.12). These terms are either projection terms due to Q^* or other terms involving the interaction of the ω_j's between each other or the $\overline{v}_k$'s, $\overline{h}_k$'s.

We need to estimate each of them carefully. This is what we complete below:

1.7.2 *First estimates on $\overline{v}_i$ and $\overline{h}_i$*

A key step in the estimate about $\overline{v}_i$ is to obtain a good estimate on the mean value of $\overline{h}_i$. This will lead us to a good estimate on $\overline{h}_i$ in a sizable ball around x_i. We start with

Lemma 2   *Let $\psi_i = \Delta \partial \omega_i$.*

$$\int \overline{v}_i \psi_i = - \int \overline{h}_i \psi_i + Max \; \varepsilon_{im}^{5/2} 0 \left(\sum \varepsilon_{k\ell} \right).$$

Proof.

$$\int \overline{v}_i \psi_i = \int \overline{v} \psi_i - \int \overline{h}_i \psi_i - \int (\overline{v} - \overline{v}_i - \overline{h}_i) \psi_i$$

$\int \overline{v} \psi_i$ is zero and since $\overline{v} - \overline{v}_i - \sum \overline{h}_j$ is zero on Ω_i,

$$\left| \int (\overline{v} - \overline{v}_i - \overline{h}_i) \psi_i \right| \leq C \left(\int |\nabla \overline{v}|^2 \right)^{1/2} \left(\int_{\Omega_i^c} |\psi_i|^{6/5} \right)^{5/6}$$

$$\leq C \left(\int |\nabla \overline{v}|^2 \right)^{1/2} Max \; \varepsilon_{im}^{5/2}$$

from which Lemma 2 follows. $\qquad\qquad\qquad\qquad\qquad\qquad\qquad\qquad\qquad$ □

The above estimate on $(\int_{\Omega_i^c} |\psi_i|^{6/5})^{5/6}$ will be established later. (1. of Lemma 13)

A weaker estimate can be seen to hold as follows: Ω_i^c is made of an exterior part
$\{x \; s.t \; \lambda_i |x - x_i| \geq \; Min \; \frac{1}{\varepsilon_{ij}} \}$ where the estimate is straightforward and interior parts
$\{x \; s.t \; \lambda_j |x - x_j| \leq \frac{c}{\varepsilon_{ij}}, \text{ for } j \neq i, \lambda_j \geq \lambda_i \}.$

On such parts, $|\psi_i| \leq C \delta_i^5$ and because $\lambda_j \geq \lambda_i, \lambda_j |x - x_j| \leq \frac{c}{\varepsilon_{ij}}$, by [Bahri 1989, (3.64)],

$$\delta_i \leq C \delta_j$$

for such values of x.

Thus

$$\left(\int_{\lambda_j |x - x_j| \leq \frac{c}{\varepsilon_{ij}}} |\psi_i|^{6/5} \right)^{5/6} \leq C \left(\int \delta_i^3 \delta_j^3 \right)^{5/6} \leq C \varepsilon_{ij}^{5/2} \log^{5/6} \varepsilon_{ij}^{-1}$$

using [Bahri 1989].

Let B_i be the natural ball containing Ω_i

$$B_i = \left\{ x / \lambda_i |x - x_i| < \frac{1}{8} \, \text{Min} \, \frac{1}{\varepsilon_{im}} \right\} \tag{1.13}$$

and let $B_i/2$ be the ball of half radius.

Since $\int \delta_i^5 = \frac{c}{\sqrt{\lambda_i}}$, we have:

Lemma 3

$$\left| \int_{\Omega_i} \bar{h}_i \psi_i \right| \leq \frac{C}{\sqrt{\lambda_i}} \, \underset{B_i/2}{\text{Max}} |\bar{h}_i| + \int_{\Omega_{i-B_i/2}} |\bar{h}_i| |\psi_i| \leq \frac{C}{\sqrt{\lambda_i}} \, \underset{B_i/2}{\text{Max}} |\bar{h}_i|$$

$$+ 0 \left(\left(\int |\nabla \bar{v}|^2 \right)^{1/2} \times \text{Max} \varepsilon_{im}^{5/2} \right).$$

This implies:

Lemma 4

$$\left| \int \bar{v}_i \psi_i \right| \leq C \left(\sum \varepsilon_{k\ell} \right) \left(\text{Max} \, \varepsilon_{im}^{5/2} + \sum \varepsilon_{ij}^{5/2} \log^{5/6} \varepsilon_{ij}^{-1} \right) + \frac{C}{\sqrt{\lambda_i}} \, \underset{B_i/2}{\text{Max}} |\bar{h}_i|.$$

Next, we estimate:

Lemma 5

$$\text{If } \ell \neq i, \left| \int_{\Omega_i} |e_\ell| w \right| \leq C \varepsilon_{\ell i}^{5/2} |w|_{H_0^1} \quad \forall w \in H_0^1(\Omega_i) \text{ if } \lambda_\ell \geq \lambda_i$$

$$\leq C \varepsilon_{\ell i}^{5/2} \log^{5/6} \varepsilon_{\ell i}^{-1} |w|_{H_0^1} \qquad \text{if } \lambda_\ell \geq \lambda_i.$$

Proof. If $\lambda_\ell \geq \lambda_i, \ell \neq i,$

$$\left| \int_{\Omega_i} e_\ell w \right| \leq \left(\int_{\Omega_i} |e_\ell^{6/5}| \right)^{5/6} |w|_{H_0^1} \leq C \varepsilon_{\ell i}^{5/2} |w|_{H_0^1}$$

since $\Omega_i \subset \{ \lambda_\ell |x - x_\ell| \geq \frac{c}{\varepsilon_{\ell i}} \}$.

If $\lambda_\ell \leq \lambda_i, \ell \neq i$, then, on $\Omega_i, \delta_\ell \leq C \delta_i$ since $\Omega_i \subset \{ \lambda_i |x - x_i| < \frac{1}{\varepsilon_{i\ell}} \}$.

Thus,

$$|e_\ell| \leq C \delta_i^5$$

and

$$\left(\int e_\ell^{6/5} \right)^{5/6} \leq \left(\int \delta_i^3 \delta_\ell^3 \right)^{5/6} \leq C \left(\varepsilon_{\ell i}^3 \log \varepsilon_{\ell i}^{-1} \right)^{5/6}$$

and the estimate follows again. $\qquad\qquad \square$

We then have:

Lemma 6

$$\int_{\Omega_i} Q^* \left(\Delta J'(\textstyle\sum \alpha_j \omega_j) \right) w$$

$$\leq C \left(\sum \left(|\omega_j^\infty| \varepsilon_{ij} + |\omega_i^\infty| \varepsilon_{ij} \right) + \sum \varepsilon_{ij}^{5/2} \log^{5/6} \varepsilon_{ij}^{-1} \right) |w|_{H_0^1}.$$

Proof. The operator Q^* removes from $\Delta J'(\sum \alpha_j \omega_j)$ the contribution of ω_i^5. We are left with ω_ℓ^5 for $\ell \neq i$ or $\omega_\ell^4 \omega_k$ or the like. Q^* of such terms involve first these terms themselves. Their contribution is given for ℓ and $k \neq i$ by the previous lemma and for ℓ or $k = i$ by [Bahri 2001, Lemma 3.2]. In addition, there are the projection terms related to A^{-1}. Those corresponding to e_ℓ, with $\ell \neq i$, are again controlled by the previous lemma. The term corresponding to e_i is typically:

$$\left(A^{-1} \left(\begin{array}{c} \vdots \\ -\int \omega_\ell^4 \omega_k L^{-1} e_m \\ \vdots \end{array} \right) \right)_i e_i. \qquad \square$$

Here, we need to understand more the matrix A and its inverse.

1.7.3 *The matrix A*

A can be written as

$$B + C$$

where C is an almost diagonal matrix which separates the block i from the block j:

$$B = \begin{pmatrix} () & 0 & 0 \\ 0 & () & 0 \\ 0 & 0 & () \end{pmatrix}.$$

Each block is nearly the identity matrix and, in fact, after a change of basis which does not affect our estimates, can be considered to be the identity matrix; so that $B = Id$.

C corresponds to the interactions between these various blocks. Typically, it involves terms such as

$$-\int e_j L^{-1} e_i$$

which, by [Bahri 2001, Lemma 5 and Lemma 3.2], after splitting the contribution onto Ω_i^c and Ω_i, is

$$O\left(\left(|\omega_i^\infty| + |\omega_j^\infty|\right)\varepsilon_{ij} + \varepsilon_{ij}^{5/2}\log^{5/6}\varepsilon_{ij}^{-1}\right)$$

A^{-1} is $(Id + C)^{-1} = Id - C - C^2 \ldots$.

We then claim:

Lemma 7   *The coefficients of A^{-1} at the line i, b_{im}, for m in a different block than i, are*

$$O(\sum(|\omega_i^\infty| + |\omega_m^\infty|)\varepsilon_{im} + \sum \varepsilon_{im}^{5/2}\log^{5/6}\varepsilon_{im}^{-1}).$$

Proof of Lemma 6 completed. In view of Lemma 7, we need to understand only $\int \omega_\ell^4 \omega_k L^{-1} e_i$. Either ℓ or $k = i$ and [Bahri 2001, Lemma 3.2] provides the result. Or $\ell, k \neq i$. We then split the integral between its contribution on Ω_i and on Ω_i^c and use Lemma A and Lemma 5. $\qquad\square$

Proof of Lemma 7. First observe that the coefficients b_{im} of $Id - C - C^2 \ldots$ build convergent series as each multiplication adds p terms, but which are all multiplied by an additional $o(1)$. These series are, in absolute values, bounded above by a geometric series.

Furthermore, the coefficient of the line i of C^r are obtained after multiplication of the line i of C with the columns of C^{r-1}. The estimate on $-\int e_j L^{-1} e_i$ provided above yields then the result. $\qquad\square$

1.7.4   *Towards an H_0^1-estimate on $\overline{v}_i$ and an L^∞-estimate on h_i*

We would like to derive an H_0^1-estimate on v_i and an L^∞-estimate on h_i. We thus need to estimate all projection terms in (1.10)–(1.12) and we also need to estimate each term in $\int f_i w$, where $w \in H_0^1(\Omega_i)$. We start with:

Lemma 8

$$\left| \int O\left(\sum_{k \neq i} |\omega_\ell|^4 (|\overline{v}_k| + |\overline{h}_k| + |k^*|) + |\overline{v}_k|^5 + |\overline{h}_k|^5 + |k^*|^5 \right) \partial\omega_i \right| \leq$$

$$C\left(\sum \varepsilon_{k\ell}^5 \varepsilon_{ij}^{1/2} + \sum \varepsilon_{km}\varepsilon_{ij}^{5/2} + \sum \varepsilon_{km}\left(\sum (|\omega_i^\infty| + |\omega_j^\infty|)\varepsilon_{ij} + \varepsilon_{ij}^{5/2} \right) \right).$$

Proof. $v_k, \overline{h}_k, k^*$ have all supports in Ω_i^c. $\qquad\square$

Thus, denoting

$$s = O\left(\sum_{k \neq i} |\omega_\ell|^4(|\overline{v}_k| + |\overline{h}_k| + |k^*|) + |\overline{v}_k|^5 + |\overline{h}_k|^5 + |k^*|^5\right)$$

we have:

$$\left|\int s\partial\omega_i\right| \leq C \int_{\Omega_i^c} |\partial\omega_i| \left(\sum_{k \neq i} |\omega_\ell|^4(|\overline{v}_k| + |\overline{h}_k| + |k^*|) + |\overline{v}_k|^5 + |\overline{h}_k|^5 + |k^*|^5\right)$$

$$\leq C \left(\left(\int |\nabla\overline{v}|^2\right)^{5/2} \left(\int_{\Omega_i^c} |\varphi_i|^6\right)^{1/6}\right.$$

$$\left. + \left(\int_{\Omega_i^c} |\partial\omega_i|^{6/5} \sum |\omega_\ell|^{24/5}\right)^{5/6} \left(\int |\nabla\overline{v}|^2\right)^{1/2}\right).$$

It is then clear that

$$\left(\int_{\Omega_i^c} |\varphi_i|^6\right)^{1/6} \leq C \operatorname{Max} \varepsilon_{ij}^{1/2}$$

and, by Lemma A, that

$$\left(\int_{\Omega_i^c} |\partial\omega_i|^{6/5} \sum |\omega_\ell|^{24/5}\right)^{5/6} \leq C \left(\sum(|\omega_i^\infty| + |\omega_j^\infty|)\varepsilon_{ij} + \varepsilon_{ij}^{5/2}\right).$$

The result follows.

We move now to estimate the contribution of $\overline{h}_i$. We observe that

$$5\left(\sum \alpha_j\omega_j\right)^4 \overline{h}_i + O\left(\sum |\omega_j|^3\overline{h}_i^2 + |\overline{h}_i|^5\right) = O\left(\sum |\omega_j|^4|\overline{h}_i| + |\overline{h}_i|^5\right)$$

and we then have:

Lemma 9 $\forall w \in H_0^1(\Omega_i)$,

$$\sum \int_{\Omega_i} (|\omega_j|^4|\overline{h}_i| + |\overline{h}_i|^5)|w| \leq C \left(\int |\nabla w|^2\right)^{1/2}$$

$$\times \left(\sum \varepsilon_{k\ell}^5 + \frac{|\overline{h}_i|_\infty}{\sqrt{\lambda_i}} B_{i/2} + \left(\sum \varepsilon_{km}\right)\left(\sum \varepsilon_{\ell i}^2 \log^{2/3} \varepsilon_{\ell i}^{-1}\right)\right).$$

Observation. $|\overline{h}_i|_{\infty, B_{i/2}} = \operatorname{Sup}_{x \in B_{i/2}} |\overline{h}_i(x)|.$

Proof.   The contribution of $\int_{\Omega_i} |w||\overline{h}_i|^5$ is clear.

For $\lambda_j \geq \lambda_i, j \neq i$, we have:

$$\int \omega_j^4 |\overline{h}_i||w| \leq C \left(\int |\nabla w|^2 \right)^{1/2} \left(\int |\nabla \overline{h}_i|^2 \right)^{1/2} \left(\int_{\Omega_i} \omega_j^6 \right)^{2/3}$$

and

$$\int_{\Omega_i} \omega_j^6 \leq \int_{\lambda_j |x - x_j| \geq \frac{1}{\varepsilon_{ij}}} \omega_j^6 \leq C\varepsilon_{ij}^3.$$

For $\lambda_j \leq \lambda_i, \lambda_j \neq \lambda_i, |\omega_j| \leq C\delta_i$ on Ω_i, so that

$$\int_{\Omega_i} \omega_j^4 |\overline{h}_i||w| \leq C \left(\int |\nabla w|^2 \right)^{1/2} \left(\int \nabla \overline{h}_i|^2 \right)^{1/2} \left(\int \delta_i^3 \delta_j^3 \right)^{2/3}$$

and the estimate follows again.

For $j = i$,

$$\int \omega_i^4 |\overline{h}_i||w| \leq C \left(\int |\nabla w|^2 \right)^{1/2} \left(|\overline{h}_i|_{\infty B_{i/2}} \left(\int |\omega_i|^{24/5} \right)^{5/6} \right.$$

$$\left. + \left(\int |\nabla \overline{v}|^2 \right)^{1/2} \left(\int_{\lambda_i |x - x_i| \geq \text{ Min} \frac{1}{\varepsilon_{ki}}} \omega_i^6 \right)^{2/3} \right)$$

and the estimate follows again. $\qquad \square$

Lemma 4 and Lemma 9 can be improved as follows: for $j \neq i$ and $\lambda_j \geq \lambda_i$,

$$\Omega_j \subset B_j = \left\{ x \; s.t \; \lambda_j |x - x_j| < \frac{2}{\varepsilon_{ij}} \right\}.$$

We claim that:

Lemma 10   *In Lemma 4 and Lemma 9, $\text{Max}_{B_{i/2}} |\overline{h}_i|$ can be replaced by $\text{Max}_{B_{i/2} - B_j} |\overline{h}_i|$ if the remainder terms is replaced by $O(\sum \varepsilon_{ij}^{5/2} \log^{5/6} \varepsilon_{ij}^{-1})$.*

Proof.   The main fact is that we are now missing the contribution of $\overline{h}_i$ on B_j.

We observe that

$$\int_{B_j} \partial \omega_i^6 = O \left(\int_{B_j} \partial \omega_i^3 \delta_j^3 \right) = O(\varepsilon_{ij}^3 \log \varepsilon_{ij}^{-1})$$

and the results follows. $\qquad \square$

In order to complete our estimates, we need some additional work on $\overline{h}_i$. We have:

Lemma 11

$$\sum \int_{\Omega_i} |\omega_j|^4 (|\overline{v}_i| + |\overline{h}_i|) + |\overline{h}_i|^5 + |\overline{v}_i|^5$$

$$+ O\left(O\left(\int \omega_i^4 \overline{v}_i \partial \omega_i\right) + \int |\omega_i|^3 |\partial \omega_i| \overline{v}_i^2 + \int |\overline{v}_i|^5 |\partial \omega_i| + \sum_{j \neq i} \int |\omega_j|^4 |\overline{v}_i| |\partial \omega_i| \right)$$

$$\times \int |\partial \omega_i|^5 \leq C \left(\frac{|\overline{v}_i|_{H_0^1}}{\sqrt{\lambda_i}} + \frac{|\overline{v}_i|_{H_0^1}^5}{\sqrt{\lambda_i} \ \mathrm{Max} \ \varepsilon_{ij}} + \frac{|\overline{h}_i|_\infty}{\lambda_i} B_{i/2} - B_j \right.$$

$$\left. + \frac{\sum \varepsilon_{k\ell}^5}{\sqrt{\lambda_i} \ \mathrm{Max} \ \varepsilon_{ij}} + \frac{\sum \varepsilon_{k\ell}}{\sqrt{\lambda_i}} \ \mathrm{Max} \ \varepsilon_{ij}^{3/2} \right).$$

Proof. Using [Bahri 2001, Lemma 3.2], we have:

$$\left(\int |\omega_i|^3 |\partial \omega_i| \overline{v}_i^2 + \int \overline{v}_i^5 |\partial \omega_i| + \sum_{j \neq i} |\omega_j|^4 |\overline{v}_i| |\partial \omega_i| \right) \int |\partial \omega_i|^5$$

$$\leq C \left(\frac{|\overline{v}_i|_{H_0^1}^2}{\lambda_i} + \frac{\varepsilon_{ij} |\overline{v}_i|_{H_0^1}}{\sqrt{\lambda_i}} \right).$$

$\square$

Through Lemmas 2, 3, 4 we have estimated $\int \omega_i^4 \overline{v}_i \partial \omega_i$. We have

$$\left| \int \omega_i^4 \overline{v}_i \partial \omega_i \right| \int |\partial \omega_i|^5$$

$$\leq \frac{C}{\sqrt{\lambda_i}} \left(\frac{1}{\sqrt{\lambda_i}} \ \underset{B_{i/2} - B_j}{\mathrm{Max}} \ |\overline{h}_i| + \left(\sum \varepsilon_{k\ell} \right) \left(\mathrm{Max} \ \varepsilon_{im}^{5/2} + \varepsilon_{ij}^{5/2} \log^{5/6} \varepsilon_{ij}^{-1} \right) \right).$$

Observe now that either $\lambda_j \geq \lambda_i$ or $|\omega_j| \leq C\delta_i$ on Ω_i. Observe also that

$$\Omega_i \subset B_i \left(x_i, \frac{1}{\lambda_i \ \mathrm{Max} \ \varepsilon_{ij}} \right)$$

so that

$$\int_{\Omega_i} \left(\omega_j^4 |\overline{v}_i| + |\overline{v}_i|^5 \right) \leq C |\overline{v}_i|_{H_0^1} \left(\frac{1}{\sqrt{\lambda_i}} + |\overline{v}_i|_{H_0^1}^4 \times \frac{1}{\sqrt{\lambda_i} \ \mathrm{Max} \ \varepsilon_{ij}} \right).$$

On the other hand,

$$\int_{\Omega_i} \omega_j^4 |\overline{h}_i| + |\overline{h}_i|^5 \leq C \left(\frac{|\overline{h}_i|_\infty}{\lambda_i} B_{i/2} - B_\ell \right.$$

$$\left. + \sum \varepsilon_{k\ell} \sum_j \left(\int_{\Omega_i - (B_{i/2} - B_\ell)} \omega_j^{24/5} \right)^{5/6} + \frac{\sum \varepsilon_{k\ell}^5}{\sqrt{\lambda_i} \, \mathrm{Max} \, \varepsilon_{ij}} \right).$$

Either $\lambda_j \geq \lambda_i, j \neq i$. Then,

$$\left(\int_{\Omega_i - (B_{i/2} - B_\ell)} \omega_j^{24/5} \right)^{5/6} \leq \left(\int_{\lambda_j |x - x_j| \geq \frac{1}{\varepsilon_{ij}}} \omega_j^{24/5} \right)^{5/6} \leq \frac{1}{\sqrt{\lambda_i}} O(\varepsilon_{ij}^{3/2}).$$

Or $\lambda_j \leq \lambda_i, |\omega_j| \leq C\delta_i$ on Ω_i and

$$\left(\int_{\Omega_i - (B_{i/2} - B_\ell)} \omega_j^{24/5} \right)^{5/6} \leq C \left(\int_{\Omega_i - (B_{i/2} - B_\ell)} \delta_j^{24/5} \right)^{5/6}$$

$$\leq C \left(\frac{1}{\sqrt{\lambda_i}} \, \mathrm{Max} \, \varepsilon_{im}^{3/2} + \left(\int_{B_\ell \cap \Omega_i} \delta_i^{24/5} \right)^{5/6} \right).$$

On $B_\ell (\lambda_\ell \geq \lambda_i, \ell \neq i)$,

$$\delta_i \leq C\delta_\ell$$

and

$$\left(\int_{\Omega_i \cap B_\ell} \delta_i^{24/5} \right)^{5/6} \leq \left(\int_{\lambda_\ell |x - x_\ell| \geq \frac{1}{\varepsilon_{i\ell}}} \delta_\ell^{24/5} \right)^{5/6} \leq \frac{1}{\sqrt{\lambda_i}} O(\varepsilon_{i\ell}^{3/2}).$$

Next, we assume for sake of simplicity that Ω_i is:

$$B_i = B(x_i, \rho_i)$$

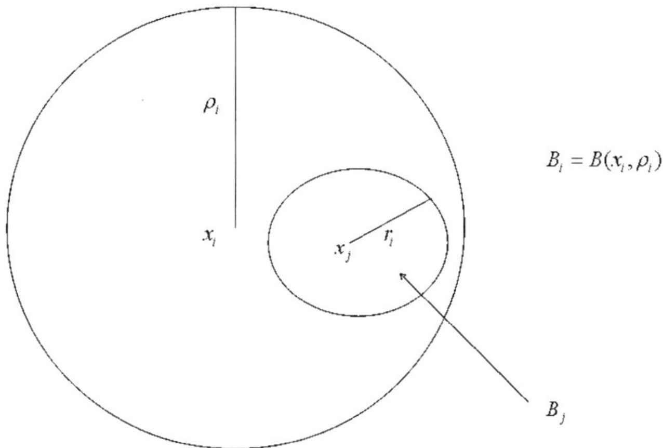

1.7.5 *The formal estimate on $\overline{h}_i$*

Using the Green function on annuli-type domains, we derive in what follows a preliminary estimate on $\overline{h}_i$ (Lemma 12).

This estimate involves the contribution of f i.e. of the f_j's on Ω_i but also on Ω_i^c; it is an intermediate result which displays the expressions which we need to estimate in order to upperbound $|\overline{h}_i|_\infty$ and $|v_i|_{H_0^1}$.

We give an estimate for $\overline{h}_i$ in $\tilde{\Omega}_i$.

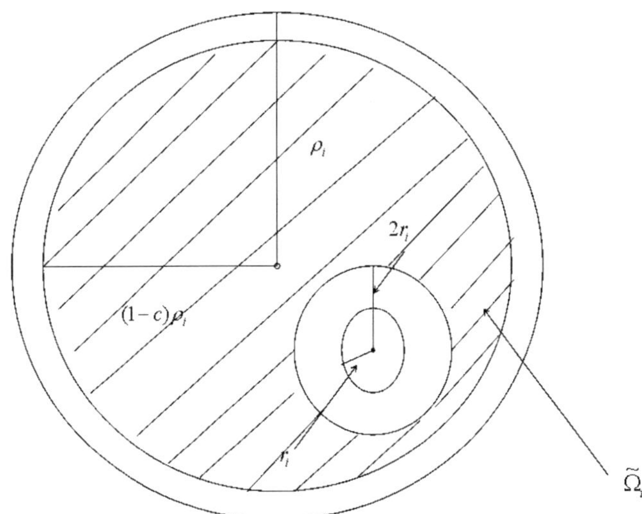

Lemma 12   *For $y \in \widetilde{\Omega}_i$,*

$$|\overline{h}_i(y)| \le C \left(\frac{1}{\rho_i} \int_{\Omega_i} |f| + \frac{r_i}{|y - x_j|} \int_{\Omega_i} \frac{|f|}{|x - x_j|} \right.$$
$$\left. + \sqrt{\lambda_i} \int_{B_i^c} |f| \delta_i + \sqrt{\lambda_j} \delta_j(y) \int_{B_j} |f| \right).$$

Proof.   Let A_t be an annulus-type domain such as Ω_i but with outside radius $\left(1 - \frac{c}{2}t\right) \overline{\rho}_i$ and inside radius (around x_j) $\left(1 + \frac{t}{2}\right) \overline{r}_i, t \in [0, 1]$, $\overline{\rho}_i = \left(1 - \frac{c}{10}\right) \rho_i, \overline{r}_i = \frac{11}{10} r_i.$ $\square$

Since $\overline{h}_i$ is harmonic in Ω_i, we have

$$\overline{h}_i(y) = - \int_{\partial A_t} \frac{\partial G_t}{\partial \nu}(x, y) \overline{h}_i(x) dx \tag{1.14}$$

where G_t is the Greens function of A_t.

Averaging and changing variables, we find

$$\overline{h}_i(y) = \frac{2}{c\rho_i} \int O\left(\frac{1}{\rho_i^2}\right) \chi_{(1-\frac{c}{2})\overline{\rho}_i \le |x - x_i| \le \overline{\rho}_i} \overline{h}_i(x) dx$$
$$+ \frac{2}{r_i} \int O\left(\frac{1}{r_i|y - x_j|}\right) \chi_{\overline{r}_i \le |y - x_j| \le \frac{3}{2}\overline{r}_i} \overline{h}_i(x) dx.$$

Indeed, for $y \in \widetilde{\Omega}_i$,

$$\frac{\partial G_t}{\partial \nu}(x, y)\Big|_{\partial A_{t,ext}} = O\left(\frac{1}{\rho_i^2}\right)$$

and if $|y - x_j| \ge 2r_i$

$$-\frac{\partial G_t}{\partial \nu}(x, y)\Big|_{\partial A_{t,int}} \le C\left(\frac{1}{|x - y|^2} + \frac{1}{r_i|x - y|}\right)\Big|_{\partial A_{t,int}} \le \frac{C_1}{r_i|y - x_j|}.$$

We solve

$$-\Delta\psi = \frac{2}{c\rho_i} O\left(\frac{1}{\rho_i^2}\right) \chi_{(1-\frac{c}{2})\overline{\rho}_i \le |x - x_i| \le \overline{\rho}_i} + \frac{2}{r_i} O\left(\frac{1}{r_i|y - x_j|}\right) \chi_{\overline{r}_i \le |x - x_j| \le \frac{3}{2}\overline{r}_i}$$

ψ is positive, splits naturally into two pieces:

$$\psi = \psi_1 + \psi_2$$

and it is easy to check that

$$\rho_i\psi_1 \text{ and } r_i\psi_2 \text{ are } L^\infty - \text{ bounded independently of } i \text{ for } y \in \widetilde{\Omega}_i. \quad (1.15)$$

Thus, using (1.14)

$$\overline{h}_i(y) = -\int_{\mathbb{R}^3} \Delta\psi\overline{h} = -\int_{(\cup\Omega_\ell)^c} \psi\Delta\overline{h} + \sum\int_{\partial\Omega_\ell} \left(\frac{\partial\overline{h}_\ell}{\partial\nu_\ell} + \frac{\partial\overline{h}_\ell}{\partial\nu_\ell^-}\right)\psi$$

$$= -\int_{(\cup\Omega_\ell)^c} \psi\Delta\overline{h} - \sum_\ell \int_{\partial\Omega_\ell} \frac{\partial}{\partial\nu}(\Delta_{\Omega_\ell}^{-1}f_\ell)\psi = -\int_{(\cup\Omega_\ell)^c} \psi\Delta\overline{h} - \sum_\ell \int_{\Omega_\ell} f_\ell\widetilde{\psi}_\ell$$

where $\widetilde{\psi}_\ell$ is the harmonic extension of $\psi|_{\partial\Omega_\ell}$.

Using (1.5), we find

$$\overline{h}_i(y) = -\int_{(\cup\Omega_\ell)^c} \psi f - \sum_\ell \int_{\Omega_\ell} f_\ell\widetilde{\psi}_\ell$$

$\psi|_{(\cup\Omega_\ell)^c}$ and the $\widetilde{\psi}_\ell$ are harmonic positive. They are all upperbounded by the original ψ. Thus,

$$|\overline{h}_i(y)| \le \int_{(\cup\Omega_\ell)^c} \psi|f| + \sum_\ell \int_{\Omega_\ell} \psi|f_\ell|$$

ψ reads as

$$\psi(z) = \int_{\mathbb{R}^3} \frac{c_1}{|x-z|}$$

$$\times \left\{ \frac{2}{c\rho_i}O\left(\frac{1}{\rho_i^2}\right)\chi_{(1-\frac{c}{2}\overline{\rho}_i\le|x-x_i|\le\overline{\rho}_i)} + \frac{2}{r_i}O\left(\frac{1}{r_i|y-x_j|}\right)\chi_{\overline{r}_i\le x-x_j|\le\frac{3}{2}\overline{r}_i} \right\}$$

$$= \quad I \qquad\qquad\qquad + \qquad II.$$

If $z \in \Omega_i$, then

$$I \le \frac{C}{\rho_i}.$$

If $z \in B_j$, $\quad I \le \frac{C}{\rho_i}\chi_{z\in B_j} \le C\delta_j(y)\sqrt{\lambda_j}\chi_{z\in B_j}$.

If $z \in B_i^c$, then by choice of $\overline{\rho}_i$,

$$\frac{1}{|x-z|}\chi_{(1-\frac{c}{2})\overline{\rho}_i\le|x-x_i|\le\overline{\rho}_i} \le \frac{C}{|z-x_i|}.$$

Since $\lambda_i |z - x_i| \geq \lambda_i \rho_i$ is large, this is upperbounded by $C\sqrt{\lambda_i}\delta_i(z)$ so that

$$I \leq C\sqrt{\lambda_i}\delta_i(z).$$

Thus, the contribution of I to the upperbound on $|\bar{h}_i(y)|$ is

$$\frac{C}{\rho_i}\int_{\Omega_i}|f| + C\sqrt{\lambda_i}\int_{B_i^c}|f|\delta_i + C\delta_j(y)\sqrt{\lambda_j}\int_{B_j}|f|.$$

If $z \in B_i^c$, the contribution of (II) is upperbounded as for I since $|y - x_j| \geq 2r_i$.

If $z \in B_j$

$$\frac{1}{|x - z|}\chi_{\bar{r}_i \leq |x - x_j| \leq \frac{3}{2}\bar{r}_i} \leq \frac{C}{r_i}\chi_{\bar{r}_i \leq |x - x_j| \leq \frac{3}{2}\bar{r}_i}$$

and

$$\frac{1}{|y - x_j|} \leq C\delta_j(y)\sqrt{\lambda_j} \quad \text{since } |y - x_j| \geq 2r_i.$$

Thus, if $z \in B_j$, the contribution of II is upperbounded by $C\sqrt{\lambda_j}\delta_j(y)\int_{B_j}|f|$.

Finally, if $z \in \Omega_i$, either $|z - x_j| \geq 2r_i$. Then,

$$\frac{1}{|x - z|}\chi_{\bar{r}_i \leq |x - x_j| \leq \frac{3}{2}\bar{r}_i} \leq \frac{C}{|z - x_j|}\chi_{\bar{r}_i \leq |x - x_j| \leq \frac{3}{2}\bar{r}_i}.$$

This contribution of II is bounded by

$$\frac{Cr_i}{|y - x_j|}\int_{\Omega_i}\frac{|f|}{|z - x_j|}\chi_{|z - x_j| \geq 2r_i}.$$

Or $|z - x_j| \leq 2r_i$ and $|x - z| \leq 4r_i$ if $\chi_{\bar{r}_i \leq |x - x_j| \leq \frac{3}{2}\bar{r}_i}$ is non zero.

So that

$$\int_{|z - x_j| \leq 2r_i}\frac{1}{|x - z|}\chi_{\bar{r}_i \leq |x - x_j| \leq \frac{3}{2}\bar{r}_i} \leq Cr_i^2$$

and this contribution of II is bounded by

$$\frac{C}{|y - x_j|}\int_{r_i \leq |z - x_j| \leq 2r_i}|f| \leq \frac{C_1 r_i}{|y - x_j|}\int_{\Omega_i}\frac{|f|}{|z - x_j|}.$$

Lemma 12 follows.

1.7.6 *Remarks about the basic estimates*

(1) We note that the values ω_i^∞ are intrinsic. They are related to the a_i's.

(2) We therefore note that

$$\int \omega_i L\omega_j = O\left((|\omega_i^\infty| + |\omega_j^\infty|)\varepsilon_{ij} + \varepsilon_{ij}^3\right).$$

(3) We note that, in $\left(\int |\omega_1^{24/5}||\omega_2^{6/5}|\right)^{5/6}$, we may always assume that ω_2 is the most concentrated. Hence, we derive the estimate $|\omega_2^\infty|\varepsilon_{ij} + O(\varepsilon_{ij}^{5/2})$. This is a direct estimate.

(4) $\int \omega_1^4 \partial\omega_1 \partial\omega_2$.

We may assume that ω_2 is the new concentrated one. We split between

$$\int_{\Omega_1} \omega_1^4 \partial\omega_1 \partial\omega_2 + \int_{\Omega_1^c} \omega_1^4 \partial\omega_1 \partial\omega_2 = \underbrace{\int_{\Omega_1} \omega_1^4 \partial\omega_1 \partial\omega_2}_{(I)} + O\left(\varepsilon_{12}^{5/2}\right)$$

ω_1 is completely deconcentrated. On Ω_1, $\lambda r = \lambda|x - a_2| \geq \frac{1}{\varepsilon_{12}}$ where $\lambda = \frac{1}{\varepsilon_{12}}$ (due to rescaling).
Then

$$|(I)| \leq \underbrace{\left|\int_{\Omega_1} \omega_1^4 \partial\omega_1 \omega_2^\infty 0(\delta_2)\right|}_{0(|\omega_2^\infty|\varepsilon_{12})} + C\int_{\Omega_1} \delta_1^5 \frac{\delta_2}{1 + \lambda^2|x - a_2|^2} \leq$$

$$O(|\omega_2^\infty|\varepsilon_{12}) + C\left(\int_{r \geq \frac{1}{\varepsilon_{12}}} \frac{r^2 dr}{(1 + r^2)^9}\right)^{1/6} = O(|\omega_2^\infty|\varepsilon_{12} + \varepsilon_{12}^{5/2}).$$

1.7.7 *Estimating the right hand side of Lemma 12*

The expression of f can be read on (1.10), (1.12). It is quite complicated. Lemma 12 provides an inequality, a gateway in order to estimate $|\overline{h}_i|_\infty$. But this inequality will become effective only if we estimate all the quantities involved in its right hand side.

We thus need a series of very basic estimates, on each of these quantities. We start with:

Lemma 13

(1) *For $\ell \neq i$, $\int_{\Omega_i} \omega_\ell^6 \leq C \varepsilon_{i\ell}^3$*

(2) *If $\ell \neq k$, $\int_{\Omega_i} \omega_\ell^4 |\omega_k| \leq \frac{1}{\sqrt{\lambda_i}\, Max\, \varepsilon_{is}} \left((|\omega_i^\infty| + |\omega_m^\infty|)\varepsilon_{im} + \sum \varepsilon_{is}^{5/2} \right).$
(m here stands for whichever of ℓ or k is not i if the other one is i).*

(3) *$\int \delta_\ell^4 \delta_i = O\left(\frac{\varepsilon_{i\ell}}{\sqrt{\lambda_\ell}} \right)$ for $\ell \neq i$*

(4) *$\int_{\Omega_i} (\sum \omega_\ell^4)\delta_j = O\left(\frac{\varepsilon_{ij}}{\sqrt{\lambda_i}} + \sum_{\lambda_\ell \geq \lambda_i} \frac{\varepsilon_{\ell i}^2 \sqrt{\varepsilon_{ij}}}{\sqrt{\lambda_i}\, Max\, \varepsilon_{is}} \right)$*

(5) *$\sum_{k \neq \ell} \int_{B_j^c} \omega_k^4 |\omega_\ell| \delta_j = \varepsilon_{kj} O\left((|\omega_k^\infty| + |\omega_\ell^\infty|)\varepsilon_{k\ell} + \varepsilon_{k\ell}^3 \right) + \varepsilon_{ij}^{5/2}$*

(6) *$\int_{B_j^c} |e_\ell| \delta_j = O(\varepsilon_{\ell j})$ for $\ell \neq j$*
              $= O(\sum \varepsilon_{mj}^{5/2})$ for $\ell = j$

(7) *$\sum_{\substack{k \neq \ell \\ k \neq i}} \int_{\Omega_i} \omega_k^4 |\omega_\ell| \delta_j = O(\varepsilon_{ik}^2 \varepsilon_{ij}^{1/2})$*

(8) *$\int_{\Omega_i} |e_\ell| \delta_j = O(\varepsilon_{ij})$ for $\ell = i$*
              $= O(\varepsilon_{ij}^{1/2} \varepsilon_{\ell i}^{5/2})$ for $\ell \neq i$

(9) *$\sum_{\ell \neq i} \int_{\Omega_i} \omega_i^4 |\omega_\ell| \delta_j \leq C \varepsilon_{ij} (\sum (|\omega_i^\infty| + |\omega_\ell^\infty|)\varepsilon_{i\ell} + \varepsilon_{i\ell}^3).$*

Proof of 1. If $\lambda_\ell \geq \lambda_i$, the estimate follows from the definition of $\Omega_i (\Omega_i \subset \{x \; s.t \; \lambda_\ell |x - x_\ell| \geq \frac{1}{\varepsilon_{i\ell}}\})$. We thus assume that $\lambda_\ell \leq \lambda_i$. On $\Omega_i, |x - x_i| < \frac{1}{8\lambda_i \varepsilon_{i\ell}}$, so that

$$\big| |x - x_\ell| - |x_\ell - x_i| \big| < \frac{1}{8\lambda_i \varepsilon_{i\ell}}.$$

$\square$

Thus,

$$\big| \lambda_\ell |x - x_\ell| - \lambda_\ell |x_\ell - x_i| \big| < c\lambda_\ell \sqrt{\frac{\lambda_\ell}{\lambda_i}} |x_\ell - x_i|.$$

If c is small enough, $\lambda_\ell |x - x_\ell|$ and $\lambda_\ell |x_\ell - x_i|$ are of the same order. Then,

$$\int_{\Omega_i} \delta_\ell^6 \leq \frac{C}{\lambda_i^3 |x_\ell - x_i|^6} |\Omega_i| \leq \frac{C}{\lambda_i^3 |x_\ell - x_i|^6} \frac{1}{(\lambda_i\, Max\, \varepsilon_{im})^3} \leq C \varepsilon_{i\ell}^3.$$

Proof of 2. Observe that $\int_{\Omega_i} \omega_\ell^4 |\omega_k| \leq |\Omega_i|^{1/6} \left(\int_{\Omega_i} \omega_\ell^{24/5} |\omega_k|^{6/5} \right)^{5/6}$

$\leq \frac{C}{\sqrt{\lambda_i} \operatorname{Max} \varepsilon_{is}} \left(\int_{\Omega_i} |\omega_\ell|^{24/5} |\omega_k|^{6/5} \right)^{5/6}.$ $\qquad \square$

Assume first that ℓ or $k = i$. The claim then follows from [Bahri 2001, Lemma 3.2]. Next, if ℓ and k are different from i, we use 1. and Hölder to conclude.

Proof of 3. Either $|x - x_i| \geq \frac{1}{10}|x_\ell - x_i|$. Then, $\delta_i(x) \leq c\sqrt{\lambda_\ell}\varepsilon_{i\ell}$ and $\int_D \delta_\ell^4 \delta_i = 0\left(\frac{\varepsilon_{i\ell}}{\sqrt{\lambda_\ell}} \right)$ where D is the domain where $|x - x_i| \geq \frac{|x_\ell - x_i|}{10}$.

Or $|x - x_\ell| \geq \frac{1}{2}|x_\ell - x_i|$ and $|x - x_i| \leq \frac{1}{10}|x_\ell - x_i|$. Then, $\delta_\ell(x) \leq C\sqrt{\lambda_i}\varepsilon_{i\ell}$ and

$$\int_{D^c} \delta_\ell^4 \delta_i \leq C\lambda_i^2 \varepsilon_{i\ell}^4 \int_{|x-x_i| \leq \frac{1}{10}|x_\ell - x_i|} \frac{\sqrt{\lambda_i}}{(1 + \lambda_i^2|x - x_i|^2)^{1/2}}$$

$$= C\frac{\lambda_i^2 \sqrt{\lambda_i}}{\lambda_i^3} \varepsilon_{i\ell}^4 \int_{r \leq \frac{\lambda_i|x_\ell - x_i|}{10}} \frac{r^2 dr}{(1 + r^2)^{1/2}}$$

$$\leq \frac{C_1 \varepsilon_{i\ell}^4}{\sqrt{\lambda_i}} \lambda_i^2 |x_\ell - x_i|^2 = \frac{C_1}{\sqrt{\lambda_i \lambda_\ell}|x_\ell - x_i|} \cdot \frac{1}{\lambda_\ell^{3/2}|x_\ell - x_i|} = o\left(\frac{\varepsilon_{i\ell}}{\sqrt{\lambda_\ell}} \right).$$

$\qquad \square$

Proof of 4. Either $\lambda_\ell \leq \lambda_i$. Then $|\omega_\ell| \leq C\delta_i$ on Ω_i and the estimate follows from 3. Or $\lambda_\ell > \lambda_i$ and the estimate follows, after the use of the Hölder inequality, from the definition of Ω_i. $\qquad \square$

Proof of 5. Either $k = j$ and the estimate follows from the fact that $B_j^c \subset \{\lambda_j|x - x_j| \geq \frac{1}{\varepsilon_{ij}}\}$. Or $k \neq j$. Then

$$\int \omega_k^4 |\omega_\ell| \delta_j \leq \left(\int \omega_k^4 \omega_\ell^2 \right)^{1/2} \left(\int \omega_k^4 \delta_j^2 \right)^{1/2}.$$

$\qquad \square$

The estimate then follows from Lemma B.

Proof of 6. Straightforward. $\qquad \square$

Proof of 7 and 8. Follows from 1, after the use of the Holder inequality (for $\ell \neq i$ in the case of 8.) The estimate on $\int_{\Omega_i} |e_i| \delta_j$ is straightforward. $\qquad \square$

Proof of 9. Use Hölder and [Bahri 2001, Lemma 3.2]. $\qquad \square$

2. can be modified as follows if $\lambda_\ell \leq \lambda_i, \ell \neq i$ or if $\lambda_\ell > \lambda_j$:

Lemma 14   • *If $\ell \neq k$ and $\lambda_\ell \leq \lambda_i, \ell \neq i$,*

$$\int_{B_j} \omega_\ell^4 |\omega_k| \leq \frac{C\varepsilon_{i\ell}^2}{\sqrt{\lambda_j}\varepsilon_{ij}}$$

• *If $\ell \neq k$ and $\lambda_\ell \geq \lambda_j$,*

$$\int_{B_j} \omega_\ell^4 |\omega_k| \leq \frac{C}{\sqrt{\lambda_\ell}} \left(\sum (|\omega_\ell^\infty| + |\omega_k^\infty|)\varepsilon_{k\ell} + \varepsilon_{k\ell}^3 \right).$$

Proof.   If $\ell \neq k$ and $\lambda_\ell \leq \lambda_i, \ell \neq i$, we use 1. of Lemma 13 and Hölder. If $\ell \neq k$ and $\lambda_\ell \geq \lambda_j$,

$$\int_{B_j} \omega_\ell^4 |\omega_k| \leq \left(\int_{B_j} \omega_\ell^4 \omega_k^2 \right)^{1/2} \left(\int \omega_\ell^4 \right)^{1/2}.$$

$\square$

Lemma 15

$$\sum \int_{\Omega_i^c} \omega_\ell^4 |\omega_k| \delta_i + \int_{\Omega_i} \omega_\ell^4 |\omega_k| \delta_j + \sqrt{\lambda_j} O(\varepsilon_{ij}) \int_{B_j} \omega_\ell^4 |\omega_k|$$

$$\leq \sum_{s \neq i} \varepsilon_{si} O \left(\sum_k (|\omega_k^\infty| + |\omega_s^\infty|)\varepsilon_{ks} + \varepsilon_{ks}^3 \right) + \varepsilon_{ij}^{5/2} + \sum_k O(\varepsilon_{ik}^2 \varepsilon_{ij})$$

$$+ \frac{O(\varepsilon_{ij})}{\sqrt{Max\ \varepsilon_{js}}} (\sum (|\omega_j^\infty| + |\omega_m^\infty|)\varepsilon_{jm} + \varepsilon_{jm}^{5/2}).$$

Proof.   We use 5 of Lemma 13 for the first term, 7 and 9 of Lemma 13 for the second term and 2 of Lemma 13 for the third term. $\square$

Lemma 16

$$\int_{\Omega_i^c} |\bar{v}|^5 \delta_i + \int_{\Omega_i} |\bar{v}|^5 \delta_j + O(\varepsilon_{ij}\sqrt{\lambda_j}) \int_{B_j} |\bar{v}|^5 = O \left(\sqrt{\varepsilon_{ij}} \left(\int |\nabla\bar{v}|^2 \right)^{5/2} \right).$$

Proof.   Straightforward, via Hölder. $\square$

Lemma 17   *For $\mathcal{O} = \delta_i$ or δ_j,*

$$\int_{(\cup\Omega_m)^c} \left(\sum \alpha_\ell \omega_\ell\right)^4 |k^*| \mathcal{O} + \sqrt{\lambda_j} O(\varepsilon_{ij}) \times \int_{(\cup\Omega_m)^c \cap B_j} \left(\sum \alpha_\ell \omega_\ell\right)^4 |k^*|$$

$$\leq C \left(\int |\nabla\bar{v}|^2\right)^{1/2} \sum \varepsilon_{s\ell}^2 \left(\sum_t \sqrt{\varepsilon_{ti}} + \sum_t \sqrt{\varepsilon_{tj}}\right).$$

Proof.   Observe that $\int_{\Omega_\ell^c} \omega_\ell^6 = O\left(\sum \varepsilon_{\ell s}^3\right)$ and that $|B_j|^{1/6} \leq \dfrac{C}{\sqrt{\lambda_j}\,\mathrm{Max}\,\varepsilon_{js}}$. $\qquad\square$

The proof follows in a straightforward way from these estimates.

Lemma 18   *Let s be bounded in L^6. If $m \neq \ell$,*

$$\left|\left(A^{-1}\left(\begin{array}{c}\vdots \\ \int sL^{-1}e_m \\ \vdots\end{array}\right)\right)_\ell\right| \left|\left(\int_{\Omega_i^c} |e_\ell|\delta_i + \int_{\Omega_i} |e_\ell|\delta_j + 0(\varepsilon_{ij})\sqrt{\lambda_j}\int_{B_j} |e_\ell|\right)\right|$$

$$\leq C \left(\sum_{\ell \neq i} \varepsilon_{i\ell}\left(\sum(|\omega_\ell^\infty| + |\omega_m^\infty|)\varepsilon_{\ell m} + \varepsilon_{\ell m}^{5/2} \log^{5/6} \varepsilon_{\ell m}^{-1}\right)\right)$$

$$+ \left(\sum \varepsilon_{\ell j}^{5/2}\sqrt{\varepsilon_{ij}} + \varepsilon_{is}^3\right)\left((\omega_\ell^\infty| + |\omega_m^\infty|)\varepsilon_{\ell m} + \varepsilon_{\ell m}^{5/2}\log^{5/6}\varepsilon_{\ell m}^{-1}\right)$$

$$+ \varepsilon_{ij}\left(\sum_{m \neq i}(|\omega_i^\infty| + |\omega_m^\infty|)\varepsilon_{im} + \varepsilon_{im}^{5/2}\log^{5/6}\varepsilon_{im}^{-1}\right).$$

Proof.   Observe that $\int_{\Omega_i^c} |e_\ell|\delta_i = O(\varepsilon_{\ell i})$ for $\ell \neq i, O\left(\sum \varepsilon_{is}^3\right)$ for $\ell = i$; that $\int_{\Omega_i} |e_\ell|\delta_j = O(\varepsilon_{i\ell})$ if $\ell \neq i$ since either $\lambda_\ell \leq \lambda_i$ and $\delta_\ell \leq c\delta_i$ on Ω_i, $\int_{\Omega_i} |e_\ell|\delta_j \leq c\int_{\Omega_i}\delta_\ell^4\delta_i\delta_j \leq C\left(\int_{\Omega_i}\delta_\ell^{24/5}\delta_i^{6/5}\right)^{5/6} \leq C\varepsilon_{i\ell}$; or $\lambda_\ell \geq \lambda_i, \ell \neq i$ and the estimate follows from the definition of Ω_i after the use of Hölder.

Observe that $\int_{\Omega_i} |e_i|\delta_j = O(\varepsilon_{ij})$. Observe finally that

$$O(\varepsilon_{ij})\sqrt{\lambda_j}\int_{B_j} |e_\ell| = O(\varepsilon_{ij}) \text{ if } \ell = j$$

$$= O(\varepsilon_{\ell j}^{5/2}\sqrt{\varepsilon_{ij}}) \text{ after the use of Hölder if } \ell \neq j.$$

We also know that, for $\ell \neq m$, the coefficient of the matrix A is

$$0(\sqrt{(\omega_m^{\infty 2} + \omega_\ell^{\infty 2})}\varepsilon_{\ell m} + \varepsilon_{\ell m}^{5/2} \log^{5/6} \varepsilon_{\ell m}^{-1}).$$

$\square$

Lemma 18 follows.

A corollary of the proof of the above lemma is the following estimate:

Lemma 19

$$\int_{\Omega_i^c} |e_\ell|\delta_i + \int_{\Omega_i} |e_\ell|\delta_j + \sqrt{\lambda_j}O(\varepsilon_{ij}) \int_{B_j} |e_\ell| = O(\varepsilon_{i\ell} + \sqrt{\varepsilon_{ij}} \sum \varepsilon_{\ell m}^{5/2}) \text{ if } \ell \neq i$$

$$= O(\varepsilon_{ij} + \sum \varepsilon_{im}^3) \text{ if } \ell = i.$$

We then have:

Lemma 20

$$\int \omega_k^4 |\omega_s||L^{-1}e_\ell|$$

$$= O((|\omega_k^\infty| + |\omega_\ell^\infty|)\varepsilon_{k\ell} + \varepsilon_{k\ell}^{5/2})((|\omega_k^\infty| + |\omega_s^\infty|)\varepsilon_{ks} + \varepsilon_{ks}^3) \text{ if } k \neq \ell$$

$$= O((|\omega_\ell^\infty| + |\omega_s^\infty|)\varepsilon_{\ell s} + \varepsilon_{\ell s}^3) \text{ if } k = \ell.$$

Proof. If $k = \ell \neq s$, the estimate follows from [Bahri 2001, Lemma 3.2]. If $k \neq \ell$, we have:

$$\int \omega_k^4 |\omega_s||L^{-1}e_\ell| \leq \left(\int \omega_k^4 (L^{-1}e_\ell)^2\right)^{1/2} \left(\int \omega_k^4 \omega_s^2\right)^{1/2}.$$

$L^{-1}e_\ell$ expands as $\partial \omega_\ell$ does. We may assume that ω_k is less concentrated than ω_ℓ after the use of the conformal group on S^3. Lemma 20 follows from the estimate on $\int \overline{\omega}_1^4 \omega_2^2$ in [Bahri 2001] verbatim. The precise expression of ω_2 is never used in this proof, only expansions, which $L^{-1}e_\ell$ satisfies, are used. $\square$

Lemma 21

$$\int |\overline{v}|^5|L^{-1}e_\ell| = O\left(\left(\int |\nabla \overline{v}|^2\right)^{5/2}\right).$$

Proof. Straightforward. $\square$

Lemma 22

$$\int \left(\sum \alpha_s \omega_s\right)^4 |k^*||L^{-1}e_\ell| \leq C \sum \varepsilon_{st}^{5/2} \left(\int |\nabla \overline{v}|^2\right)^{1/2}.$$

Proof. Straightforward. $\square$

Lemma 23

$$\left| \int \omega_\ell^4 L^{-1} e_\ell (\overline{v}_\ell + \overline{h}_\ell) \right| \le C \left(\int |\nabla \overline{v}|^2 \right)^{1/2} \left(\sum \varepsilon_{\ell m}^{5/2} + \sum \varepsilon_{\ell m}^{5/2} \log^{5/6} \varepsilon_{\ell m}^{-1} \right).$$

Proof. Follows from Lemmas 2, 3, 4 after gathering the contribution of $\overline{v}_\ell$ and $\overline{h}_\ell$. $\qquad\square$

We then have:

Lemma 24 *If $\ell \ne t$*

$$\int |\omega_\ell|^3 |\omega_t| |\overline{v}_m + \overline{h}_m| |L^{-1} e_\ell| \le C \left(\int |\nabla \overline{v}|^2 \right)^{1/2} \left(\int (\omega_\ell^3 L^{-1} e_\ell)^{6/5} |\omega_t|^{6/5} \right)^{5/6}$$

$$\le C \left(\int |\nabla \overline{v}|^2 \right)^{1/2} \left((|\omega_t^\infty| + |\omega_\ell^\infty|) \varepsilon_{t\ell} + \varepsilon_{t\ell}^{5/2} \log^{5/6} \varepsilon_{t\ell}^{-1} \right).$$

Proof. A small variant of the proof of Lemma A. $\qquad\square$

Lemma 25 *If $\ell \ne t$*

$$\int |\omega_\ell| |\omega_t|^3 |\overline{v}_m + \overline{h}_m| |L^{-1} e_\ell|$$

$$\le C \left(\int (|\omega_t|^{6/5} (|\omega_\ell|^3 |L^{-1} e_\ell|)^{6/5} + |\omega_t|^{24/5} |L^{-1} e_\ell|^{6/5} \right)^{5/6} \left(\int |\nabla \overline{v}|^2 \right)^{1/2}$$

$$\le C \left(\int |\nabla \overline{v}|^2 \right)^{1/2} \left((|\omega_t^\infty| + |\omega_\ell^\infty|) \varepsilon_{t\ell} + \varepsilon_{t\ell}^{5/2} \log^{5/6} \varepsilon_{t\ell}^{-1} \right).$$

Proof. We split between the case when $|\omega_\ell(x)| \le |\omega_t(x)|$ and the case when $|\omega_t(x)| \le |\omega_\ell(x)|$ at a given x and we use Hölder. $\qquad\square$

Lemma 26 *If $k \ne \ell$,*

$$\int \omega_k^4 |\overline{v}_m + \overline{h}_m| |L^{-1} e_\ell|$$

$$\le C \left(\int |\nabla \overline{v}|^2 \right)^{1/2} \left(\sum (|\omega_k^\infty| + |\omega_\ell^\infty|) \varepsilon_{\ell k} + \varepsilon_{\ell k}^{5/2} \log^{5/6} \varepsilon_{k\ell}^{-1} \right).$$

Proof. Straightforward. $\qquad\square$

Next, we have the following key Lemma which displays the quantities that we need to estimate:

Lemma 27

$$\int |\omega_\ell|^3 |\overline{v}_\ell + \overline{h}_\ell|^2 |L^{-1}e_\ell|$$

$$\leq C \left(|\overline{v}_\ell|^2_{H^1_0} + \frac{1}{\lambda_\ell \rho^2_\ell} \left(\int_{\Omega_\ell} |f| \right)^2 + \lambda_k \varepsilon^2_{\ell k} \left(\int_{B_k} |f| \right)^2 + \left(\int_{B^c_\ell} |f|\delta_\ell \right)^2 \right.$$

$$\left. + \left(\int_{B_\ell} |f|\delta_f \right)^2 \right) + C \int |\nabla \overline{v}|^2 \sum \varepsilon^2_{\ell m}.$$

Proof.   Using Lemma 12, we have:

$$\int |\omega_\ell|^3 |\overline{v}_\ell + \overline{h}_\ell|^2 |L^{-1}e_\ell| \leq C|\overline{v}_\ell|^2_{H^1_0} + \int |\omega_\ell|^3 |L^{-1}e_\ell| \left(\frac{1}{\rho_\ell} \int_{\Omega_\ell} |f| \right.$$

$$\left. + \sqrt{\lambda_k}\delta_k \int_{B_k} |f| + \frac{\delta_k}{\mathrm{Max}\ \varepsilon_{\ell s}} \int_{B_\ell} |f|\delta_k + \sqrt{\lambda_\ell} \int_{B^c_\ell} |f|\delta_\ell \right)^2$$

$$+ C \int |\nabla v|^2 \sum \varepsilon^2_{\ell m} \text{ (for the contribution in } \Omega_\ell - \widetilde{\Omega}_\ell; \text{ observe that } \int_D |\omega_\ell|^6$$

$\leq C\varepsilon^3_{\ell k}$ on the inner radius of $\Omega_\ell - \widetilde{\Omega}_\ell$. This follows from 1 of Lemma 13).

□

Since $\int |\omega_\ell|^3 |L^{-1}e_\ell| = 0 \left(\frac{1}{\lambda_\ell} \right)$ and $\int |\omega_\ell|^3 |L^{-1}e_\ell|\delta^2_k = O(\varepsilon^2_{\ell k})$, Lemma 23 follows.

Observe that the above proof implies the estimate:

$$\left(\int \omega^4_\ell (\overline{v}_\ell + \overline{h}_\ell)^2 \right)^{1/2} \leq C \left(\frac{1}{\sqrt{\lambda_\ell}\rho_\ell} \int_{\Omega_\ell} |f| + \sqrt{\lambda_k}\ \varepsilon_{\ell k} \int_{B_k} |f| \right.$$

$$+ \int_{B^c_\ell} |f|\delta_\ell + \int_{B_\ell} |f|\delta_k + |\overline{v}_\ell| + \left(\int |\nabla v|^2 \right)^{1/2}$$

$$\left. \times \left(\sum \varepsilon_{\ell m} \right) \right) = C\gamma_\ell.$$

(1.16)

We then have:

Lemma 28

$$\int_{\Omega_i} \omega^4_i |\overline{v}_i + \overline{h}_i|\delta_j + \int_{\Omega_\ell} \omega^4_\ell |\overline{v}_\ell + \overline{h}_\ell|\delta_i + \sqrt{\lambda_j}\, O(\varepsilon_{ij}) \int_{B_j} \omega^4_j |\overline{v}_j + \overline{h}_j|$$

$$\leq O(\varepsilon_{ij})(\gamma_i + \gamma_j) + O(\varepsilon_{i\ell}\gamma_\ell).$$

Proof. we observe that

$$\int \omega_j^4|\overline{v}_j+\overline{h}_j| \le \left(\int \omega_j^4\right)^{1/2}\left(\int \omega_j^4|\overline{v}_j+\overline{h}_j|^2\right)^{1/2} \le \frac{C}{\sqrt{\lambda_j}}\gamma_j$$

$$\int \omega_j^4|\overline{v}_\ell+\overline{h}_\ell|\delta_i \le \left(\int \omega_\ell^4\delta_i^2\right)^{1/2}\left(\int \omega_\ell^4|\overline{v}_\ell+\overline{h}_\ell|^2\right)^{1/2} \le C\gamma_\ell\varepsilon_{i\ell}$$

$$\int \omega_i^4|\overline{v}_i+\overline{h}_i|\delta_j \le \left(\int \omega_i^4\delta_j^2\right)^{1/2}\left(\int \omega_i^4|\overline{v}_i+\overline{h}_i|^2\right)^{1/2} \le C\varepsilon_{ij}\gamma_i.$$

$\square$

Lemma 28 follows.

Lemma 29

$$\int_{\Omega_j} \omega_\ell^4|\omega_m| \le C\frac{|\omega_m^\infty|\varepsilon_{m\ell}}{\sqrt{\lambda_\ell}} + \frac{O(\varepsilon_{m\ell}^2)}{\sqrt{\lambda_m}} \tag{a}$$

$$\int_{\Omega_j} \omega_\ell^4|\omega_m| \le \frac{c}{\sqrt{\lambda_j}\,\mathrm{Max}\,\varepsilon_{js}}\sum \varepsilon_{jt}^{5/2} \text{ if } \ell,m \ne j. \tag{b}$$

Proof. (b) follows in a straightforward way from 1. of Lemma 13 and Hōlder. To prove (a), we expand

$$\omega_m = c\omega_m^\infty\delta_m + \frac{O(\delta_m^2)}{\sqrt{\lambda_m}}$$

so that

$$\int_{\Omega_j} \omega_\ell^4|\omega_m| \le c|\omega_m^\infty|\int_{\Omega_j}\omega_\ell^4\delta_m+\frac{1}{\sqrt{\lambda_m}}\int_{\Omega_j}\omega_\ell^4\delta_m^2 \le C\frac{\varepsilon_{\ell m}^2}{\sqrt{\lambda_m}}+c|\omega_m^\infty|\int_{\Omega_j}\omega_\ell^4\delta_m.$$

We split $\int_{\Omega_j}\omega_\ell^4\delta_m$ into two pieces A and B. A is the portion on $\{x\ s.t\ |x-a_m| \le \frac{1}{4}|a_\ell-a_m|\}$. On B, $\frac{|x-a_m|}{|a_\ell-a_m|} \ge \frac{1}{4}$. We then have

$$c|\omega_m^\infty|\int_{\Omega_j}\omega_\ell^4\delta_m \le \frac{c|\omega_m^\infty|}{\lambda_\ell^2|a_\ell-a_m|^4}\cdot\frac{1}{\lambda_m^{5/2}}\int_{r\le\frac{\lambda_m}{4}|a_\ell-a_m|}\frac{r^2dr}{\sqrt{1+r^2}}$$

$$+\frac{c|\omega_m^\infty|}{\sqrt{\lambda_m}|a_\ell-a_m|}\int_{\Omega_j}\omega_\ell^4 \le \frac{c|\omega_m^\infty|\varepsilon_{\ell m}}{\sqrt{\lambda_\ell}\lambda_\ell|a_\ell-a_m|}$$

$$+\frac{c|\omega_m^\infty|}{\sqrt{\lambda_m}|a_\ell-a_m|}\inf\left(\frac{1}{\lambda_j},\frac{1}{\lambda_\ell}\right).$$

(Observe that if $\lambda_\ell \leq \lambda_j, \ell \neq j, |\omega_\ell| \leq C\delta_j$ on Ω_j). $\qquad\qquad\qquad\square$

The claim follows.

Lemma 15 is modified accordingly. Indeed, in this estimate of this lemma, we have used 2 of Lemma 13 to estimate $\sqrt{\lambda_j}O(\varepsilon_{ij})\int_{B_j}\omega_\ell^4|\omega_k|$.

For $\ell, k \neq j$, we use (b) of Lemma 29. For ℓ or $m = j$, we use (a) of Lemma 29 or 2 of Lemma 13. We observe that if $\ell = j$, by (a) of Lemma 29:

$$\sqrt{\lambda_j}\,O(\varepsilon_{ij})\int_{B_j}\omega_j^4|\omega_k| \leq C|\omega_k^\infty|\varepsilon_{jk}\varepsilon_{ij} + \sqrt{\lambda_j}\,\frac{0(\varepsilon_{kj}^2)}{\sqrt{\lambda_k}}\varepsilon_{ij}.$$

This works for $\lambda_k \geq \lambda_j$ since we then derive:

$$\sqrt{\lambda_j}\,0(\varepsilon_{ij})\int_{B_j}\omega_j^4|\omega_k| \leq C|\omega_k^\infty|\varepsilon_{jk}\varepsilon_{ij} + 0(\varepsilon_{kj}^2\varepsilon_{ij}).$$

If $\lambda_k \leq \lambda_j$, we come back to 2 of Lemma 13 and we derive:

$$\sqrt{\lambda_j}\,O(\varepsilon_{ij})\int_{B_j}\omega_j^4|\omega_k| \leq \frac{O(\varepsilon_{ij})}{\sqrt{\text{Max }\varepsilon_{js}}}\left((|\omega_j^\infty| + |\omega_k^\infty|)\varepsilon_{jk} + \sum\varepsilon_{js}^{5/2}\right).$$

This provides the estimate if $k = i$. We thus assume now that $k \neq i$.

Since $|a_i - a_k| \leq |a_i - a_j| + |a_j - a_k|$,

$$\frac{1}{|a_i - a_k|} \geq \frac{1}{2|a_i - a_j|} \quad\text{or}\quad \frac{1}{|a_i - a_k|} \geq \frac{1}{2|a_j - a_k|}$$

so that either $(\lambda_k \leq \lambda_j)$

$$\varepsilon_{ij}\varepsilon_{jk} \leq \frac{c}{\lambda_j\sqrt{\lambda_k}\sqrt{\lambda_i}|a_i - a_k||a_j - a_k|} \leq C\varepsilon_{ik}\varepsilon_{jk}$$

or

$$\varepsilon_{ij}\varepsilon_{jk} \leq \frac{c}{\lambda_j\sqrt{\lambda_k}\sqrt{\lambda_i}|a_i - a_j||a_i - a_k|} \leq C\varepsilon_{ik}\varepsilon_{ij} \text{ since } \lambda_j \geq \lambda_i$$

$(B_j \subset B_i, k \neq i)$.

At any rate,

$$\frac{\varepsilon_{ij}\varepsilon_{jk}}{\sqrt{\text{Max }\varepsilon_{js}}} \leq C\varepsilon_{ik}\sqrt{\varepsilon_{jk}}.$$

Thus,

$$\sqrt{\lambda_j}\,O(\varepsilon_{ij})\int_{B_j}\omega_j^4|\omega_k| \leq C\left((|\omega_j^\infty|\varepsilon_{ij} + |\omega_k^\infty|\varepsilon_{ik})\sqrt{\varepsilon_{jk}} + \varepsilon_{ij}\sum\varepsilon_{js}^2\right).$$

Lastly, if $k = j$, we have after using (a) of Lemma 29:

$$\sqrt{\lambda_j}\, O(\varepsilon_{ij}) \int_{B_j} \omega_\ell^4 |\omega_j| \leq C |\omega_j^\infty| \frac{\sqrt{\lambda_j}\, \varepsilon_{j\ell}\varepsilon_{ij}}{\sqrt{\lambda_\ell}} + O(\varepsilon_{ij})\varepsilon_{j\ell}^2$$

$$\leq \frac{C |\omega_j^\infty| \varepsilon_{ij}}{\lambda_\ell |a_j - a_\ell|} + O(\varepsilon_{ij})\varepsilon_{j\ell}^2.$$

Summarizing, we have:

Lemma 30

$$\sqrt{\lambda_j}\, O(\varepsilon_{ij}) \int_{B_j} \omega_\ell^4 |\omega_k| \leq o(|\omega_\ell^\infty|\varepsilon_{\ell i} + |\omega_k^\infty|\varepsilon_{ki}) + C\varepsilon_{ij}\sum \varepsilon_{js}^2.$$

1.7.8 R_i *and the estimate on* $|\overline{v}_i|_{H_0^1}$

We are now ready to derive an estimate on $|\overline{v}_i|_{H_0^1} + \frac{|h_i|_\infty}{\sqrt{\lambda_i}}$.

We denote

$$R_i = \int_{\Omega_i} |f||\delta_j + \int_{B_i^c} |f||\delta_i + \sqrt{\lambda_j}\, \varepsilon_{ij} \int_{B_j} |f|$$

$$\Gamma_s = \sum (|\omega_t^\infty| + |\omega_s^\infty|)\varepsilon_{ts} + \sum \varepsilon_{tm}^{5/2} \log^{5/6} \varepsilon_{tm}^{-1}.$$

We then have:

Lemma 31

$$R_i \leq o\left(\Gamma_i\right) + O\left(\frac{1}{\rho_i \sqrt{\lambda_i}} \int_{\Omega_i} |f| + \sum \varepsilon_{i\ell} \left(\frac{1}{\rho_\ell \sqrt{\lambda_\ell}} \int_{\Omega_\ell} |f| + |\overline{v}_\ell|_{H_0^1}\right)\right)$$

$$+ \left(\sum \varepsilon_{ij}\right) |\overline{v}_i|_{H_0^1}\right) + O\left(\varepsilon_{ij} \sum \varepsilon_{st}^{3/2} + \sqrt{\varepsilon_{it}} \sum \varepsilon_{st}^{5/2}\right).$$

Proof of Lemma 31. The expression of f is provided in (1.5).

By Lemma 15 improved by Lemma 29, the contribution of $\omega_\ell^4 \omega_k$ is bounded by

$$\sum_{s \neq i} \varepsilon_{si}\Gamma_s + \varepsilon_{ij}^{5/2} + \sum_k O(\varepsilon_{ik}\varepsilon_{ij}) + o\left(\Gamma_i\right) + C\varepsilon_{ij}\sum \varepsilon_{js}^2.$$

Since $\varepsilon_{si}\varepsilon_{st} = o(\varepsilon_{it})$, this can be replaced by

$$o\left(\Gamma_i\right) + O\left(\varepsilon_{ij} \sum \varepsilon_{st}^{3/2}\right).$$

The contribution of $|\overline{v}|^5$ by Lemma 16 is $0\left(\sqrt{\varepsilon_{ij}}\left(\int |\nabla \overline{v}|^2\right)^{5/2}\right)$. The contribution of $(\sum \alpha_\ell \omega_\ell)^4 |k^*|$ is, by Lemma 17,

$$O\left(\left(\int |\nabla \overline{v}|^2\right)^{1/2} \sum \varepsilon_{s\ell}^2 \left(\sum_t \sqrt{\varepsilon_{ti}} + \sum_t \sqrt{\varepsilon_{tj}}\right)\right).$$

The contribution of $\sum (\sum \alpha_\ell \omega_\ell)^4 (\overline{v}_m + \overline{h}_m)$ and the like (the direct contribution, not the one due to Q^*) splits between the terms where $\ell = m$ and the terms where one of the indexes ℓ is not m. We may upperbound this total direct contribution with

$$C\left(\omega_m^4 |\overline{v}_m + \overline{h}_m| + \sum_{k \neq m} \omega_k^4 |\overline{v}_m + \overline{h}_m|\right).$$

The contribution to R_i of $\omega_m^4 |\overline{v}_m + \overline{h}_m|$ is provided by Lemma 28: it is

$$0(\varepsilon_{ij})\left(R_i + R_j + \frac{1}{\rho_i \sqrt{\lambda_i}}\int_{\Omega_i} |f| + \frac{1}{\rho_j \sqrt{\lambda_j}}\int_{\Omega_j} |f| + |\overline{v}_j|_{H_0^1}\right.$$
$$\left. + \left(\int |\nabla \overline{v}|^2\right)^{1/2} \sum \varepsilon_{\ell s}\right)$$
$$+ 0(\varepsilon_{i\ell})\left(R_\ell + \frac{1}{\rho_\ell \sqrt{\lambda_\ell}}\int_{\Omega_\ell} |f| + |\overline{v}_\ell| + \left(\int |\nabla \overline{v}|^2\right)^{1/2} \sum \varepsilon_{ts}\right).$$

The contribution of $\sum\limits_{k \neq m} \omega_k^4 |\overline{v}_m + \overline{h}_m|$ is

$$\int_{\Omega_i} \omega_k^4 |\overline{v}_i + \overline{h}_i| \delta_j + \int_{\Omega_\ell} \omega_t^4 |\overline{v}_\ell + \overline{h}_\ell| \delta_i + \sqrt{\lambda_j}\, 0(\varepsilon_{ij})\int_{B_j} \omega_s^4 |\overline{v}_j + \overline{h}_j|$$

with $k \neq i, t \neq \ell, s \neq j, j \neq i, \ell \neq i$.

The first two terms are easily upperbounded using 1 of Lemma 13 by

$$C\varepsilon_{ki}^2 \sqrt{\varepsilon_{ij}}\left(\int |\nabla \overline{v}|^2\right)^{1/2} + C\sqrt{\varepsilon_{i\ell}}\, \varepsilon_{t\ell}^2 \left(\int |\nabla \overline{v}|^2\right)^{1/2}.$$

The third term is upperbounded by

$$C\sqrt{\lambda_j}\varepsilon_{ij}\varepsilon_{sj}^2 \left(\int |\nabla \overline{v}|^2\right)^{1/2} \frac{1}{\sqrt{\lambda_j}\, \text{Max } \varepsilon_{jn}} = O\left(\varepsilon_{ij}\varepsilon_{sj}^{3/2} \sum \varepsilon_{k\ell}\right).$$

The direct contribution of $(\sum \alpha_\ell \omega_\ell)^3 (\bar{v}_m + \bar{h}_m)^2$ and the like is upper-bounded by the contribution of

$$C\left(\left(\sum \alpha_\ell \omega_\ell\right)^4 |\bar{v}_m + \bar{h}_m| + |\bar{v}|^5\right).$$

We are left with the contribution of the projection terms. For $m \neq \ell$, the contribution of these terms to R_i is provided in Lemma 18. It is

$$O\left(\sum \varepsilon_{i\ell} \Gamma_\ell\right) + \left(\varepsilon_{is}^3 + \sqrt{\varepsilon_{ij}} \sum \varepsilon_{\ell j}^{5/2}\right) \Gamma_\ell.$$

For $m = \ell$, Lemmas 23, 26, 27, 28 provide the contribution to R_i. It is

$$\text{if } \ell = i \qquad O\left(\varepsilon_{ij} + \sum \varepsilon_{im}^3\right) \times C_0,$$

$$\text{if } \ell \neq i \qquad O\left(\varepsilon_{i\ell} + \sqrt{\varepsilon_{ij}} \sum \varepsilon_{\ell m}^{5/2}\right) \times C_0.$$

with

$$C_0 \leq C\left(\Gamma_\ell + \left(\int |\nabla \bar{v}|^2\right)^{5/2} + \sum \varepsilon_{st}^{5/2} \left(\int |\nabla \bar{v}|^2\right)^{1/2} \log^{5/6} \varepsilon_{st}^{-1} + |\bar{v}_\ell|_{H_0^1}^2 \right.$$

$$\left. + R_\ell^2 + \frac{1}{\lambda_\ell \rho_\ell^2} \left(\int_{\Omega_\ell} |f|\right)^2 + \int |\nabla v|^2 \sum \varepsilon_{\ell m}^2\right).$$

Observe that

$$R_i = o(1), \Gamma_i = o(1), \frac{1}{\sqrt{\lambda_i}\rho_i} \int_{\Omega_i} |f| = o(1), \varepsilon_{i\ell}\Gamma_\ell = o(\Gamma_i).$$

Thus, our estimates read:

$$\begin{pmatrix} 1 & & O(\varepsilon_{ij}) \\ & \ddots & \\ & & 1 \end{pmatrix} \begin{pmatrix} R_1 \\ \vdots \\ R_p \end{pmatrix} = \left(o(\Gamma_i) \times \frac{1}{\rho_i \sqrt{\lambda_i}} \int_{\Omega_i} |f| + \sum_{i \neq \ell} |\bar{v}_\ell|_{H_0^1} \varepsilon_{i\ell} \right.$$

$$\left. + \sum_{j \neq i} \varepsilon_{ij} |\bar{v}|_{H_0^1} + O(\varepsilon_{ij} \sum \varepsilon_{st}^{3/2} \sqrt{\varepsilon_{it}} \sum \varepsilon_{st}^{5/2}) \right).$$

An equation goes here

It is not difficult to see that

$$\left(Id + \begin{pmatrix} 0 & & O(\varepsilon_{ij}) \\ & \ddots & \\ & & 0 \end{pmatrix} \right)^{-1} = Id + \begin{pmatrix} 0 & & O(\varepsilon_{ij}) \\ & \ddots & \\ & & 0 \end{pmatrix}$$

because $\varepsilon_{it}\varepsilon_{tm} = o(\varepsilon_{im})$.

Since $\varepsilon_{i\ell}\Gamma_\ell = o(\Gamma_i)$, the claim follows. $\qquad\qquad\square$

Lemma 32

$$\frac{1}{\sqrt{\lambda_i}\rho_i} \int_{\Omega_i} |f| = o\left(\Gamma_i + |\bar{v}_i|_{H_0^1} \right) + O\left(\sum \varepsilon_{i\ell} |\bar{v}_\ell|_{H_0^1} \right).$$

Proof.   Using 2 of Lemma 13 as well as Lemma 14 and Lemma 29 (for $\ell \neq i$), using the fact that $\int_{\Omega_i} e_\ell \leq \dfrac{C}{\sqrt{\lambda_i}\ \mathrm{Max}\ \varepsilon_{is}}\, \varepsilon_{i\ell}^{5/2}$ (for $\ell \neq i$) $= C\sqrt{\rho_i}\, \varepsilon_{i\ell}^{5/2}$, we find that the contribution of $Q^*(\Delta J'(\sum \alpha_j \omega_j))$ is $o(\Gamma_i)$.

Using then Lemma 14 and Lemma 12, which we revisit after Lemma 31, we find that the remainder of the contribution of f is:

$$o\left(|\bar{v}_i|_{H_0^1} + \Gamma_i + R_i \right) + O\left(\frac{1}{\lambda_i^{3/2}\rho_i^2} \int_{\Omega_i} |f| \right)$$

$$= o\left(|\bar{v}_i|_{H_0^1} + \Gamma_i \right) + o\left(\frac{1}{\rho_i\sqrt{\lambda_i}} \int_{\Omega_i} |f| + \sum \varepsilon_{i\ell} |\bar{v}_\ell|_{H_0^1} \right)$$

$$+ o\left(\sum \varepsilon_{i\ell} \frac{1}{\rho_\ell\sqrt{\lambda_\ell}} \int_{\Omega_\ell} |f| \right).$$

Inverting our matrix as usual, we derive the estimate (use $\varepsilon_{i\ell}\Gamma_\ell = o(\Gamma_i)$). $\qquad\square$

Finally, we have:

Lemma 33

$$|\bar{v}_i|_{H_0^1} = O(\Gamma_i) + o(\Gamma_i).$$

Proof. $\qquad\qquad\qquad\qquad\qquad\qquad\qquad\qquad\qquad\qquad\qquad\quad\square$

We multiply (1.2) by w where $w \in H_0^1(\Omega_i)$ and w is H_0^1-orthogonal to the small eigenvalues-eigenspace of $-\Delta - 5\omega_i^4$ on $H_0^1(\Omega_i)$.

We derive

$$\int_{\Omega_i} (\Delta v_i + 5\omega_i^4 v_i) w = \int_{\Omega_i} gw$$

g is made of several terms. The first contribution comes from $Q^*(\Delta J'(\sum \alpha_j \omega_j))$. It is estimated in Lemma 6. We have:

$$\int Q^* \left(\Delta J'\left(\sum \alpha_j \omega_j\right)\right) w = O(\Gamma_i)|w|_{H_0^1}.$$

Next, we have the contribution of

$$O\left(\sum_{k \neq i} \omega_\ell^4 (|\overline{v}_k| + |\overline{h}_k| + |k^*|) + |\overline{v}_k|^5 + |\overline{h}_k|^5 + |k^*|^5\right)$$

which, by Lemma 8, is

$$o(\Gamma_i)|w|_{H_0^1}.$$

Next, we have the contribution of $\overline{h}_i$ which we trace back to Lemmas 9–10 and Lemma 11. It is $o(\Gamma_i)|w|_{H_0^1}$ except for $\int \omega_i^4 |\overline{h}_i||w|$ which yielded a contribution equal to

$$|w|_{H_0^1} o(\Gamma_i) + |w|_{H_0^1} 0 \left(\frac{1}{\sqrt{\lambda_i}} \underset{B_i/2 - B_j}{\mathrm{Max}} |\overline{h}_i|\right).$$

We revisit this estimate using Lemma 12. We obtain:

$$\left(o(\Gamma_i) + O\left(\frac{1}{\sqrt{\lambda_i \rho_i}} \int_{\Omega_i} |f| + \int_{B_i^c} |f| \delta_i\right)\right) |w|_{H_0^1} +$$

$$O\left(\left(\left(\int_{\Omega_i} |f| \delta_j\right) \frac{1}{\mathrm{Max}\, \varepsilon_{is}} + \sqrt{\lambda_j} \int_{B_j} |f|\right) \times \int_{\Omega_i} \omega_i^4 \delta_j |w| =\right.$$

$$\left(o(\Gamma_i) + O\left(\frac{1}{\sqrt{\lambda_i \rho_i}} \int_{\Omega_i} |f| + \int_{B_i^c} |f| \delta_i + \int_{\Omega_i} |f| \delta_j + \sqrt{\lambda_j}\, \varepsilon_{ij} \int_{B_j} |f|\right)\right) |w|_{H_0^1}.$$

Using Lemmas 31 and 32, we find that it is $\left(o(\Gamma_i + |v_i|_{H_0^1}) + \right.$ $o\left(\sum \varepsilon_{i\ell} |\overline{v}_\ell|_{H_0^1}\right)\right) |w|_{H_0^1}$.

We are left with the projection terms associated to $\overline{v}_i, \overline{h}_i$ and with the contribution of $\overline{v}_i$ to g.

The contribution of $\overline{v}_i$ to g comes from $\left(\left(\sum_{j \neq i} \alpha_j \omega_j\right)^4 - \alpha_i^4 \omega_i^4\right) \overline{v}_i +$

$0(|\overline{v}_i|^2)$ and thus yields

$$o(|\overline{v}_i|_{H_0^1})|w|_{H_0^1}.$$

For the projection terms, the contribution of $\overline{v}_i$ and $\overline{h}_i$ is gathered. For $\ell \neq i$,

$$\left| \int_{\Omega_i} e_\ell w \right| \leq C\varepsilon_{i\ell}^{5/2} |w|_{H_0^1}.$$

For $\ell = i$, we use Lemmas 20–24 and derive that these projection terms contribute (use also Lemmas 25–27)

$$\left(O\left(\sum \varepsilon_{i\ell} |\overline{v}_\ell|_{H_0^1} + o(|\overline{v}_i|_{H_0^1} + \Gamma_i) \right) |w|_{H_0^1} \right).$$

We thus have:

$$\int_{\Omega_i} (\Delta v_i + 5\omega_i^4 v_i) w = \left(O(\Gamma_i) + o(\Gamma_i + |\overline{v}_i|_{H_0^1}) \right) |w|_{H_0^1} \quad \forall w \in H_0^1(\Omega).$$

$$(1.17)$$

On the other hand, revisiting the proof of Lemma 12 after Lemmas 31–32, we find

$$\int h_i \psi_i = \left(o(\Gamma_i + |v_i|_{H_0^1}) + O\left(\sum \varepsilon_{i\ell} |\overline{v}_\ell|_{H_0^1} \right) \right).$$

Thus

$$\int v_i \psi_i = o\left(\Gamma_i + |v_i|_{H_0^1} \right) + O\left(\sum \varepsilon_{i\ell} |\overline{v}_\ell|_{H_0^1} \right). \qquad (1.18)$$

Let $\overline{\partial \omega_i}$ be the H_0^1-orthogonal projection of $\partial \omega_i$ onto $H_0^1(\Omega_i)$. We have:

$$\int \nabla v_i \nabla \overline{\partial \omega_i} = \left(o(\Gamma_i + |v_i|_{H_0^1}) + O\left(\sum \varepsilon_{i\ell} |\overline{v}_\ell|_{H_0^1} \right) \right) |w|_{H_0^1}. \qquad (1.19)$$

We assume as in [Bahri 2001] that the operator

$$-\Delta - 5\omega_i^4$$

is not degenerate on Span $\{\partial \omega_i\}^\perp = W$.

Let $w \in W$. Let $\overline{w}$ be the orthogonal projection of w on $H_0^1(\Omega_i)$. Using [Bahri 1989, Lemma 3.2]:

$$\int \nabla w_i \nabla w - 5 \int \omega_i^4 v_i w = \int \nabla v_i \nabla \overline{w} - 5 \int \omega_i^4 v_i \overline{w}$$

$$+ o\left(\left(\int |\nabla v_i|^2 \right)^{1/2} \left(\int |\nabla w|^2 \right)^{1/2} \right)$$

$$= \left(O(\Gamma_i) + o(\Gamma_i + |v_i|_{H_0^1}) + O\left(\sum \varepsilon_{i\ell} |\overline{v}_\ell|_{H_0^1} \right) \right) |w|_{H_0^1}.$$

Adding up the norm of the projection $\tilde{v}_i$ of v_i onto $\mathrm{Span}\{\partial\omega_i\}$, which we know by (1.18) to be of the same order we find, after inverting the matrix:

$$|v_i|_{H_0^1} = O(\Gamma_i) + o(\Gamma_i).$$

$\square$

Lemma 34 *Given $\varepsilon_o > 0$, there exists a constant C_{ε_o} such that*

$$|v_i|_{H_0^1}^2 \leq C_{\varepsilon_o} \left(\sum_{k \neq i} \omega_k^\infty \varepsilon_{ik} \right)^2$$

$$+ \varepsilon_o \left(\sum \left(\omega_k^{\infty^2} + \omega_i^{\infty^2} \right) \varepsilon_{ki}^2 + \sum \varepsilon_{st}^4 \varepsilon_{ki} \log^{5/3} \varepsilon_{st}^{-1} \right) + O_m \left(\sum \varepsilon_{\ell i}^{5/2} \right).$$

Proof. Let ψ be an eigenfunction on $\mathbb{R}^3$ for the linearized operator with eigenvalue $\theta_i \neq 5, \theta_i \neq 1$

$$-L_i = -\Delta - 5\omega_i^4.$$

$\square$

Thus ψ_i satisfies

$$-\theta_i \Delta \psi_i - 5\omega_i^4 \psi_i = o \quad , \theta_i \neq 5, \theta_i \neq 1.$$

We estimate

$$\sum_{k \neq i} \int \omega_i^4 \omega_k \psi_i.$$

We pick up a special value of k. We rescale ω_k so that

$$\gamma_{\varepsilon_o} \leq \frac{\lambda_i}{\lambda_k} \leq \gamma_{\varepsilon_o}^2$$

where γ_{ε_o} is a fixed and large constant. We also assume, changing stereographic projection if needed, that $|a_i - a_k|$ is bounded.

Let

$$A_k = \{x / \lambda_i |x - a_i| < \frac{1}{\varepsilon_{ik}}\}.$$

Since γ_{ε_o} is large,

$$|x - a_k| = |a_i - a_k|(1 + o_{\varepsilon_o}(1)) \quad \text{for } x \in A_k$$

ω_k expands on A_k into (L is a bounded linear form)

$$\omega_k(x) = \omega_k^\infty \frac{\sqrt{\lambda_k}}{(1 + \lambda_k^2|x - a_k|^2)^{1/2}} + \frac{\sqrt{\lambda_k}\, L(\lambda_k(x - a_k))}{(1 + \lambda_k^2|x - a_k|^2)^{3/2}}$$

$$+ \frac{O(\sqrt{\lambda_k})}{(1 + \lambda_k^2|x - a_k|^2)^{3/2}} = \frac{\omega_k^\infty}{\sqrt{\lambda_k}|a_k - a_i|}(1 + o(1)) + \frac{O(\gamma_{\varepsilon_o}^2)}{\lambda_i \lambda_k^{3/2}|a_i - a_k|^3}$$

$$+ \sqrt{\lambda_k}\, L(\lambda_k(x - a_k)) \times \left(\frac{1}{(1 + \lambda_k^2|x - x_k|^2)^{3/2}} - \frac{1}{(1 + \lambda_k^2|a_i - a_k|^2)^{3/2}} \right)$$

$$+ \frac{\sqrt{\lambda_k}\, L(\lambda_k(x - a_k))}{(1 + \lambda_k^2|a_i - a_k|^2)^{3/2}} = \frac{\omega_k^\infty(1 + o(1))}{\sqrt{\lambda_k}\,|a_k - a_i|} + \frac{O(\gamma_{\varepsilon_o}^2)}{\lambda_i \lambda_k^{3/2}|a_i - a_k|^3}$$

$$+ \lambda_k \sqrt{\lambda_k} \frac{O(|x - a_i|)}{(1 + \lambda_k^2|a_i - a_k|)^{3/2}} + \frac{\sqrt{\lambda_k}\, L(\lambda_k(x - a_k))}{(1 + \lambda_k^2|a_i - a_k|^2)^{3/2}}.$$

Calculating

$$\int_{A_k} \omega_i^4 \omega_k \psi_i = -\frac{\theta_i}{5} \int_{A_k} \Delta\psi_i \omega_k,$$

observing that ψ_i expands as an ω_i does at infinity,

$$-\theta_i \int_{A_k} \Delta\psi_i = \frac{c_i}{\sqrt{\lambda_i}} - \theta_i \int_{A_k^c} \Delta\psi_i = \frac{c_i}{\sqrt{\lambda_i}} + \frac{1}{\sqrt{\lambda_i}} O(\varepsilon_{ki}^2)$$

and that

$$-\theta_i \int_{A_k} \Delta\psi_i\, L(x - a_k) = -\theta_i \int_{\mathbb{R}^3} \Delta\psi_i L(x - a_k) + \theta_i \int_{A_k^c} \Delta\psi_i L(x - a_i)$$

$$+ \theta_i L(a_i - a_k) \int_{A_k^c} \Delta\psi_i = O\left(\frac{1}{\lambda_i^{3/2}} \right) \varepsilon_{ik} + \frac{1}{\sqrt{\lambda_i}} O(\varepsilon_{ik}^2),$$

we derive

$$\int_{A_k} \omega_i^4 \omega_k \psi_i = c_i \omega_k^\infty \varepsilon_{ik}(1 + o(1)) + O(\varepsilon_{ik}^3).$$

Observe now that

$$\int_{A_k^c} \omega_i^4 \omega_k \psi_i = O\left(\left(\int_{A_k^c} \delta_i^6 \right)^{5/6} \right) = O\left(\varepsilon_{ik}^{5/2} \right).$$

We thus derive that

$$\sum_{k \neq i} \int \omega_i^4 \omega_k \psi_i = c_i \sum_{k \neq i} \omega_k^\infty \varepsilon_{ik}(1 + o(1)) + O(\varepsilon_{ik}^{5/2}).$$

Observe also that for every e_i,

$$\int e_i \psi_i = o,$$

since $L^{-1} e_i$ is an eigenvalue of $-\Delta - 5\omega_i^4$ with eigenvalue 5 or 1. We thus see that

$$\sum_{k \neq 1} \int Q^*(\omega_i^4 \omega_k)\psi_i = c_i \sum_{k \neq i} \omega_k^\infty \varepsilon_{ik}(1 + o(1)) + O(\varepsilon_{ik}^{5/2}).$$

This holds for every single eigenfunction $\psi_i^1, \ldots, \psi_i^m$, with $\theta_j \neq 5, \theta_j \neq 1$. As m tends to $+\infty$, θ_m tends to zero. We split

$$H^1 = E_m \oplus E_m^\perp$$

where

$$E_m = \text{Span } \{\psi_i^1, \ldots, \psi_i^m, L^{-1} e_i\}$$

and $E_m^\perp$ is its orthogonal for the scalar - product $\int \nabla h \nabla k$ as well as $\int \omega_i^4 h k$. Let $p_m^\perp : H \to E_m^\perp$ be the orthogonal projection.

For $w \in E_m^\perp$, we have:

$$o\left(\sum \varepsilon_{\ell i}^{5/2}\right) + \int Q^*(\omega_i^4 \omega_k)w = \int \omega_i^4 \omega_k w \leq \left(\int \omega_i^4 \omega_k^2\right)^{1/2} \left(\int \omega_i^4 w^2\right)^{1/2}$$

$$\leq \theta_m \left(\int \omega_i^4 \omega_k^2\right)^{1/2} \left(\int |\nabla w|^2\right)^{1/2}.$$

This implies that $(|\varphi|_{H^1} = \left(-\int |L|\varphi\varphi\right)^{1/2}$ for φ satisfying (Vo)

$$|L_i^{-1} p_m^\perp(Q^*(\omega_i^4 \omega_k))|_{H^1} = o\left((|\omega_i^\infty| + |\omega_k^\infty|)\varepsilon_{ik} + \varepsilon_{ik}^3\right)$$

and thus,

$$|L_i^{-1}(Q^*(\omega_i^4 \omega_k))|_{H^1}$$

$$\leq C_m \left|\sum_{k \neq i} \omega_k^\infty \varepsilon_{ik}\right| + o\left((|\omega_i^\infty| + |\omega_k^\infty|)\varepsilon_{ik} + \varepsilon_{ik}^3\right) + O_m\left(\sum \varepsilon_{\ell i}^{5/2}\right).$$

If we consider other terms in $Q^*(\Delta J'(\sum \alpha_j \omega_j))$, we can drop Q^* in view of Lemma 18, the observations on the coefficients of A and Lemma 20.

We are left estimating

$$\int_{\Omega_i} (|\omega_i|^3 w_k^2 + \omega_k^4 |w_i|)|w|.$$

We have:

$$\int_{\Omega_i} |\omega_i^3| |w_k^2| |w| \leq \left(\int_{\Omega_i} \omega_i^4 w_k^2\right)^{1/2} \left(\int_{\Omega_i} \omega_i^2 w_k^2 w^2\right)^{1/2}$$

$$\leq C \left(\int_{\Omega_i} \omega_i^4 w_k^2\right)^{1/2} \left(\int_{\Omega_i} |\omega_i|^3 |w_k|^3\right)^{1/3} \left(\int |\nabla w|^2\right)^{1/2}$$

$$= o((|\omega_i^\infty| + |\omega_k^\infty|)\varepsilon_{ki} + \varepsilon_{ki}^2)$$

$$\int_{\Omega_i} \omega_k^4 |\omega_i| |w| \leq \left(\int_{\Omega_i} \omega_k^4 \omega_i^2\right)^{1/2} \left(\int_{\Omega_i} \omega_k^4 w^2\right)^{1/2}$$

$$\leq C \left(\int_{\Omega_i} \omega_k^4 \omega_i^2\right)^{1/2} \left(\int_{\Omega_i} \omega_k^6\right)^{1/3} \left(\int |\nabla w|^2\right)^{1/2}$$

$$= o\left((|\omega_i^\infty| + |\omega_k^\infty|)\varepsilon_{ki} + \varepsilon_{ki}^3\right).$$

We thus see ($\tilde{L}_i$ is L_i acting on $H_0^1(\Omega_i)$).

$$|\tilde{L}_i^{-1} Q^*(\Delta J'(\sum \alpha_j \omega_j))|_{H_0^1} \leq C_m |\sum \omega_k^\infty \varepsilon_{ik}|$$

$$+ o\left(\sum(|\omega_k^\infty| + |\omega_i^\infty|)\varepsilon_{ki} + \sum \varepsilon_{st}^2 \sqrt{\varepsilon_{ki}} \log^{5/6} \varepsilon_{st}^{-1}\right) + O_m\left(\sum \varepsilon_{\ell i}^{5/2}\right).$$

This combined with Lemma 31, Lemma 32 and Lemma 33 yields the result.
$\square$

1.8   Proof of the Morse Lemma at Infinity When the Concentrations are Comparable

After estimating the contribution of $\overline{v}$ in $J'(\sum \alpha_j \omega_j + \overline{v})$. $\partial \omega_i$, we establish the Morse Lemma at infinity under (1.3) (which we always assume) and (1.4) which we will eventually suppress. (1.4) states that the concentrations of all masses are comparable.

In what follows, we will denote

$$S_i^2 = C_{\varepsilon_0} \left(\sum_{k \neq i} \omega_k^\infty \varepsilon_{ik} \right)^2$$
$$+ \varepsilon_o \left(\sum \left(\omega_k^{\infty^2} + \omega_i^{\infty^2} \right) \varepsilon_{ki}^2 + \sum \varepsilon_{st}^4 \varepsilon_{ki} \log^{5/3} \varepsilon_{st}^{-1} \right) \tag{1.20}$$
$$+ O_m \left(\sum \varepsilon_{\ell i}^5 \right).$$

In the sequel, $\tilde{\partial}\omega_i$ is $\frac{\partial \omega_i}{\partial a_i}$.

Lemma 35

$$\sum_{\ell \neq i} \int_{\widetilde{\Omega}_i^c} \omega_\ell^4 |\tilde{\partial}\omega_i||\overline{v}| \leq o \left(\sum \lambda_m \Gamma_m^2 \right).$$

Proof. For a given ℓ, we decompose the above integral into two pieces, one on $\widetilde{\Omega}_\ell \cap \widetilde{\Omega}_i^c$ and the other one on $(\widetilde{\Omega}_\ell \cup \widetilde{\Omega}_i)^c$. $\widetilde{\Omega}_\ell$ and $\widetilde{\Omega}_i$ are the subdomains of Ω_ℓ and Ω_i where the estimate of Lemma 12 holds. $\square$

We first take care of

$$\int_{\widetilde{\Omega}_\ell \cap \widetilde{\Omega}_i^c} \omega_\ell^4 |\tilde{\partial}\omega_i||\overline{v}_\ell + \overline{h}_\ell|.$$

We start with $\int_{\widetilde{\Omega}_\ell \cap \widetilde{\Omega}_i^c} \omega_\ell^4 |\partial \omega_i||\overline{v}_\ell|$ which we split into two pieces $A + B$.

$$A = \int_{\widetilde{\Omega}_\ell \cap \widetilde{\Omega}_i^c \cap \{|x - a_i| \leq \frac{1}{4}|a_i - a_\ell|\}} \omega_\ell^4 |\tilde{\partial}\omega_i||\overline{v}_\ell|$$

$$B = \int_{\widetilde{\Omega}_\ell \cap \widetilde{\Omega}_i^c \cap \{|x - a_i| \geq \frac{1}{4}|a_i - a_\ell|\}} \omega_\ell^4 |\tilde{\partial}\omega_i||\overline{v}_\ell|.$$

Since $|x - a_i||\partial \omega_i| \leq C|\tilde{\omega}_i|$, where $\tilde{\omega}_i$ behaves just as ω_i does:

$$B \leq \left(\int_{\widetilde{\Omega}_\ell \cap \widetilde{\Omega}_i^c} \omega_\ell^4 |\tilde{\omega}_i||\overline{v}_\ell| \right) \frac{1}{|a_\ell - a_i|} \leq \frac{1}{|a_\ell - a_i|} S_\ell \times \left((|\omega_\ell^\infty| + |\omega_i^\infty|)\varepsilon_{\ell i} + \varepsilon_{\ell i}^3 \right)$$

$$\leq \left(\sum_{m \neq \ell} |\omega_m^\infty \varepsilon_{\ell m}| \right) \left((|\omega_\ell^\infty| + |\omega_i^\infty|)\varepsilon_{\ell i} + \varepsilon_{\ell i}^3 \right) +$$

$$o \left(\sum (|\omega_m^\infty| + |\omega_\ell^\infty|)\varepsilon_{m\ell} + \sum \varepsilon_{st}^4 \varepsilon_{\ell m} \log^{5/6} \varepsilon_{st}^{-1} + \sum 0(\varepsilon_{\ell s}^{5/2}) \right)$$

$$\times((|\omega_\ell^\infty| + |\omega_i^\infty|)\varepsilon_{\ell i} + \varepsilon_{\ell i}^3))\frac{1}{|a_\ell - a_i|} = \sum_m B_m.$$

Either

$$|a_\ell - a_i| \geq \frac{1}{10}|a_\ell - a_m|.$$

Then

$$B_m \leq C\left((|\omega_\ell^\infty|^2 + |\omega_i^\infty|^2)\varepsilon_{\ell i}^2 + \varepsilon_{\ell i}^6\right)\frac{1}{|a_\ell - a_i|}$$

$$+ \frac{C}{|a_\ell - a_m|}\left((|\omega_m^\infty|^2 + |\omega_\ell^\infty|^2)\varepsilon_{m\ell}^2 + O(\varepsilon_{m\ell}^2\varepsilon_{st}^6 + \varepsilon_{s\ell}^5)\right).$$

Or $|a_\ell - a_i| \leq \frac{1}{10}|a_\ell - a_m|$ so that

$$\frac{9}{10} \leq \frac{|a_\ell - a_m|}{|a_i - a_m|} \leq \frac{11}{10}.$$

Then,

$$\varepsilon_{\ell m} \cdot \varepsilon_{i\ell} \leq \frac{10}{9}\,\varepsilon_{im} \cdot \frac{1}{\lambda_\ell|a_i - a_\ell|}$$

and

$$B_m \leq C\lambda_i\left(\left(|\omega_\ell^\infty|^2 + |\omega_i^\infty|^2 + \sum|\omega_m^\infty|(|\omega_\ell^\infty| + |\omega_i^\infty|)\right)\varepsilon_{im}\varepsilon_{i\ell}^2\right)$$

$$+ \frac{1}{|a_\ell - a_i|}\sum\varepsilon_{st}^{5/2}(|\omega_\ell^\infty| + |\omega_i^\infty|)\varepsilon_{\ell i} + \frac{\varepsilon_{\ell i}^3\sum\varepsilon_{st}^{5/2}}{|a_\ell - a_i|}$$

$$+ \sqrt{\lambda_\ell\lambda_i}\left(\sum(|\omega_\ell^\infty| + |\omega_m^\infty|)\varepsilon_{\ell m}\right)\varepsilon_{\ell i}^5.$$

Thus

$$B_m \leq o\left((\lambda_i + \lambda_\ell)(|\omega_\ell^\infty|^2 + |\omega_i^\infty|^2)\varepsilon_{i\ell}^2\right) + o\left(\lambda_i\sum|\omega_m^\infty|^2\varepsilon_{im}^2\right) + C\sqrt{\lambda_\ell\lambda_i}\sum\varepsilon_{st}^5$$

This settles B for $\bar{v}_\ell$.

Now considering A, we have:

$$\frac{3}{4}|a_i - a_\ell| \leq |x - a_\ell| \leq \frac{5}{4}|a_i - a_\ell|.$$

Thus,

$$A \leq \frac{C}{\lambda_\ell^2|a_i - a_\ell|^4}|\bar{v}_\ell|_{H_0^1}\left(\int_{\tilde{\Omega}_i^c \cap \{|x-a_i| \leq \frac{1}{4}|a_i-a_\ell|\}}|\partial\omega_i|^{6/5}\right)^{5/6}.$$

$\tilde{\partial}\omega_i$ expands (see [Bahri 2001]) as follows:

$$\tilde{\partial}\omega_i = \lambda_i \left(\omega_i^\infty \tilde{\partial}\delta_i + \frac{O(\delta_i)}{1 + \lambda_i^2 |x - a_i|^2} \right) = \lambda_i \omega_i^\infty \tilde{\partial}\delta_i + O(\delta_i^3)$$

where $\tilde{\partial}\delta_i$ is the derivative of δ_i with respect to rescaled variables $\frac{1}{\lambda_i}\frac{\partial}{\partial a_i^j} \cdot \tilde{\Omega}_i^c$ is made of two parts $E_i \cup I_i$, where

$$E_i = \{x \notin \tilde{\Omega}_i, \lambda_i|x - x_i| \geq c \operatorname{Min}\left(\frac{1}{\varepsilon_{ij}}\right)\}$$

$$I_i = \{x; \lambda_m|x - x_m| \leq \frac{c}{\varepsilon_{im}} \text{ for some } \lambda_m \geq \lambda_i, \lambda_i, \neq \lambda_m \neq \lambda_i\}$$

and we have accordingly:

$$A \leq \frac{C|\overline{v}_\ell|_{H_o^1}}{\lambda_\ell^2 |a_i - a_\ell|^4} \times \frac{1}{\lambda_i} \left(|\omega_i^\infty| \left(\int_{r \leq \frac{1}{4}\lambda_i|a_i - a_\ell|} \frac{r^2 dr}{(1 + r^2)^{6/5}} \right)^{5/6} \right.$$

$$\left. + \left(\int_{r \geq \frac{c}{\operatorname{Max} \varepsilon_{ij}}} \frac{r^2 dr}{(1 + r^2)^{12/5}} \right)^{5/6} + \frac{C|\overline{v}_\ell|_{H_0^1}}{\lambda_\ell |a_i - a_\ell|^2} \left(\int_{I_i} \omega_\ell^{12/5} \delta_i^{18/5} \right)^{5/6} \right)$$

$$\leq C\lambda_i |\overline{v}_\ell|_{H_0^1} \left\{ \varepsilon_{i\ell}^4 \left(|\omega_i^\infty|\sqrt{\lambda_i|a_i - a_\ell|} + \operatorname{Max}\varepsilon_{ij}^{3/2} \right) \right.$$

$$\left. + \varepsilon_{i\ell}^2 \left(\sum_{\substack{\lambda_m \geq \lambda_i \\ \lambda_m \neq \lambda_i}} \varepsilon_{mi}^{5/2} \log^{5/6} \varepsilon_{mi}^{-1} + \sum_{\substack{\lambda_m \geq \lambda_i \\ \lambda_m \neq \lambda_i}} \varepsilon_{\ell m}\varepsilon_{im}^{3/2} \right) \right\}$$

$$\leq \frac{C\sqrt{\lambda_i}\varepsilon_{i\ell}|\overline{v}_\ell|_{H_0^1} \cdot \sqrt{\lambda_i}|\omega_i^\infty|\varepsilon_{i\ell}^2}{\lambda_\ell^{1/2}|a_i - a_\ell|^{1/2}} + o\left(\lambda_i \left(\varepsilon_{i\ell}^2|\overline{v}_\ell|_{H_0^1}^2 + \sum \varepsilon_{im}^5 \right) \right)$$

$$\leq o\left(\lambda_i\varepsilon_{i\ell}^2|\overline{v}_\ell|_{H_0^1}^2 + \lambda_i\omega_i^\infty\varepsilon_{i\ell}^4 \right) + C\lambda_i \sum \varepsilon_{im}^5.$$

We now take care of the contribution of $\overline{h}_\ell$. δ_k is the notation for the generic mass (with $\lambda_k \geq \lambda_\ell$) contained in $B_\ell = \{x \text{ s.t } \lambda_\ell|x - x_\ell| \leq c \operatorname{Min} \frac{1}{\varepsilon_{\ell s}}\}$.

We are studying

$$\int_{\tilde{\Omega}_\ell \cap \tilde{\Omega}_i^c} \omega_\ell^4 |\overline{h}_\ell||\tilde{\partial}\omega_i| \leq \theta_\ell \int_{\tilde{\Omega}_\ell \cap \tilde{\Omega}_i^c} \omega_\ell^4 \delta_k |\tilde{\partial}\varphi_i| + \beta_\ell \int_{\tilde{\Omega}_\ell \cap \tilde{\Omega}_i^c} \omega_\ell^4 |\tilde{\partial}\omega_i|$$

with

$$\theta_\ell\varepsilon_{k\ell} + \frac{\beta_\ell}{\sqrt{\lambda_\ell}} = o\left(\sum(|\omega_m^\infty| + |\omega_\ell^\infty|)\varepsilon_{\ell m} + \varepsilon_{\ell m}^{5/2}\log^{5/6}\varepsilon_{\ell m}^{-1} \right).$$

We start with

$$\theta_\ell \int_{\widetilde{\Omega}_\ell \cap \widetilde{\Omega}_i^c} \omega_\ell^4 \delta_k |\widetilde{\partial} \omega_i| \leq \lambda_i \theta_\ell \left(\int_{\widetilde{\Omega}_\ell \cap \widetilde{\Omega}_i^c} \omega_\ell^4 |\delta_k^2| \right)^{1/2} \times \left(\int_{\widetilde{\Omega}_\ell \cap \widetilde{\Omega}_i^c} \omega_\ell^4 |\widetilde{\partial}\omega_i|^2 \right)^{1/2}$$

$$\leq \lambda_i \theta_\ell \varepsilon_{\ell k} \left(\int_{\widetilde{\Omega}_\ell \cap \widetilde{\Omega}_i^c} \omega_\ell^4 |\widetilde{\partial}\omega_i|^2 \right)^{1/2}.$$

We split the integral into two pieces A and B as above; $\widetilde{\Omega}_i^c$ is made of two pieces E_i and I_i; $\widetilde{\Omega}_\ell$ is connected, hence contained in one of them.

Then,

$$A \leq \frac{C}{\lambda_i \lambda_\ell |a_i - a_\ell|^2} \left\{ \int_{E_i} \left(\frac{\omega_i^{\infty^2}}{(1 + \lambda_i^2 |x - a_i|^2)^2} + \frac{1}{(1 + \lambda_i^2 |x - a_i|^2)^3} \right) \lambda_i^3 r^2 dr \right.$$

$$\left. + \int_{\widetilde{\Omega}_\ell \cap I_i} \delta_i^6 \right\}^{1/2} \leq C \varepsilon_{i\ell}^2 \left\{ |\omega_i^\infty| \varepsilon_{is}^{1/2} + \varepsilon_{is}^{3/2} + \varepsilon_{i\ell}^{3/2} \right\}$$

and

$$\lambda_i \theta_\ell \varepsilon_{\ell k} A \leq C \sqrt{\lambda_i} \, (\theta_\ell \varepsilon_{\ell k}) \varepsilon_{i\ell} \left\{ \sqrt{\lambda_i} \, |\omega_i^\infty| \varepsilon_{i\ell} \varepsilon_{is}^{1/2} + \sqrt{\lambda_i} \, \varepsilon_{i\ell} \varepsilon_{is}^{3/2} \right\}$$

$$\leq o(\lambda_i (\theta_\ell \varepsilon_{\ell k})^2 \varepsilon_{i\ell}^2 + \lambda_i \omega_i^{\infty^2} \varepsilon_{i\ell}^2) + C \lambda_i \varepsilon_{i\ell}^2 \varepsilon_{is}^3.$$

We now have $\left(\text{for } B, \ \frac{|x - a_i|}{|a_i - a_\ell|} \geq c > o; \lambda_i |\widetilde{\partial}\omega_i||x - a_i| \leq C\left(|\omega_i^\infty| \delta_i + \frac{\delta_i}{(1 + \lambda_i^2 |a_i - a_\ell|^2)^{1/2}} \right) \right)$

$$B \leq \frac{C}{\sqrt{\lambda_i} \, |a_i - a_\ell|} \left\{ \frac{\omega_i^{\infty^2}}{1 + \lambda_i^2 |a_i - a_\ell|^2} + \frac{1}{(1 + \lambda_i^2 |a_i - a_\ell|^2)^2} \right\}^{1/2} \left(\int \omega_\ell^4 \right)^{1/2}$$

$$\leq \frac{C}{\lambda_i |a_i - a_\ell|} \left(|\omega_i^\infty| \varepsilon_{i\ell} + \frac{\varepsilon_{i\ell}}{\lambda_i |a_i - a_\ell|} \right)$$

and

$$\lambda_i \theta_\ell \varepsilon_{\ell k} B \leq C \left(\frac{\theta_\ell \varepsilon_{\ell k} |\omega_i^\infty| \varepsilon_{i\ell}}{|a_i - a_\ell|} + \frac{\theta_\ell \varepsilon_{\ell k} \varepsilon_{i\ell}}{\lambda_i |a_i - a_\ell|^2} \right)$$

$$\leq C \left(\sqrt{\lambda_\ell} \, \theta_\ell \varepsilon_{\ell k} \sqrt{\lambda_i} \, |\omega_i^\infty| \varepsilon_{i\ell}^2 + \sqrt{\lambda_\ell} \, \theta_\ell \varepsilon_{\ell k} \cdot \sqrt{\lambda_\ell} \, \varepsilon_{i\ell}^3 \right)$$

$$\leq O \left(\lambda_\ell (\theta_\ell \varepsilon_{\ell k})^2 + \lambda_i \omega_i^{\infty^2} \varepsilon_{i\ell}^2 + \lambda_\ell \varepsilon_{i\ell}^5 \right).$$

We study now

$$\beta_\ell \int_{\tilde{\Omega}_\ell \cap \tilde{\Omega}_i^c} \omega_\ell^4 |\tilde{\partial}\omega_i|.$$

We split again $\int_{\tilde{\Omega}_\ell \cap \tilde{\Omega}_i^c} \omega_\ell^4 |\tilde{\partial}\omega_i|$ into two pieces A and B.
We have:

$$A \le \frac{C\lambda_i}{\lambda_i^{5/2}\lambda_i^2 |a_i - a_\ell|^4}|\omega_i^\infty| \int_{\le c\lambda_i |a_i - a_\ell|} \frac{r^2 dr}{1 + r^2} + \frac{C}{\sqrt{\lambda_\ell}\,|a_i - a_\ell|}\int_{\tilde{\Omega}_\ell} \omega_\ell^3 \delta_i^3$$

$$\le \frac{C|\omega_i^\infty|\varepsilon_{i\ell}}{\lambda_\ell^{3/2}|a_i - a_\ell|^2} + C\sqrt{\lambda_i}\,\varepsilon_{i\ell}(\varepsilon_{i\ell}^3 \log \varepsilon_{i\ell}^{-1}).$$

Thus,

$$\beta_\ell A \le C\left(\sqrt{\lambda_\ell}\left(\frac{\beta_\ell}{\sqrt{\lambda_\ell}}\right)\frac{\sqrt{\lambda_\ell}\,|\omega_i^\infty|\varepsilon_{i\ell}}{\lambda_\ell^2 |a_i - a_\ell|^2} + \sqrt{\lambda_\ell}\left(\frac{\beta_\ell}{\sqrt{\lambda_\ell}}\right)\sqrt{\lambda_i}\varepsilon_{i\ell}^4 \log \varepsilon_{i\ell}^{-1}\right)$$

$$\le O\left(\lambda_\ell\left(\frac{\beta_\ell^2}{\lambda_\ell}\right) + \lambda_\ell \omega_i^{\infty^2}\varepsilon_{i\ell}^2 + \lambda_i \varepsilon_{i\ell}^7\right).$$

We study now B: $(|x - a_i| \ge c|a_i - a_\ell|)$

$$B \le \frac{C}{|a_i - a_\ell|}\left(\int_{\tilde{\Omega}_\ell} \omega_\ell^4\right)\left(\frac{|\omega_i^\infty|\sqrt{\lambda_i}}{(1 + \lambda_i^2 |a_i - a_\ell|^2)^{1/2}} + \frac{\sqrt{\lambda_i}}{1 + \lambda_i^2 |a_i - a_\ell|^2}\right)$$

$$\le \frac{C}{\sqrt{\lambda_\ell}}\left(\frac{|\omega_i^\infty|\varepsilon_{i\ell}}{|a_i - a_\ell|} + \frac{\varepsilon_{i\ell}}{\lambda_i |a_i - a_\ell|^2}\right)$$

$$\beta_\ell B \le C\left(\frac{\beta_\ell}{\sqrt{\lambda_\ell}}\frac{|\omega_i^\infty|\varepsilon_{i\ell}}{|a_i - a_\ell|} + C\sqrt{\lambda_\ell}\left(\frac{\beta_\ell}{\sqrt{\lambda_\ell}}\right)\cdot\sqrt{\lambda_\ell}\varepsilon_{i\ell}^3\right)$$

$$\le C\left(\frac{\beta_\ell^2}{\lambda_\ell}\frac{1}{|a_i - a_\ell|} + O\left(\frac{\omega_i^{\infty^2}\varepsilon_{i\ell}^2}{|a_i - a_\ell|}\right) + \lambda_\ell\left(\frac{\beta_\ell^2}{\lambda_\ell}\right) + \lambda_\ell\varepsilon_{i\ell}^6\right).$$

To complete the proof of Lemma 35, we need to estimate

$$\int_{(\tilde{\Omega}_\ell \cup \tilde{\Omega}_i)^c} \omega_\ell^4 |\tilde{\partial}\omega_i||\bar{v}|.$$

In $(\tilde{\Omega}_\ell \cup \tilde{\Omega}_i)^c$, there are three portions: the first one is exterior to the two balls $\tilde{B}_\ell$ and $\tilde{B}_i$. We denote it A. We then have B and C which are the portions in $\tilde{B}_\ell \cap \tilde{\Omega}_\ell^c$ and in $\tilde{B}_i \cap \tilde{\Omega}_i^c$.

We first estimate

$$\int_A \omega_\ell^4 |\tilde{\partial}\omega_i||\overline{v}|$$

which we upperbound as follows:

$$\int_A \omega_\ell^4 |\tilde{\partial}\omega_i||\overline{v}| \leq C\omega_\ell^{\infty^4} \int_A \delta_\ell^4 |\tilde{\partial}\omega_i||\overline{v}| + C\int_A \frac{\delta_\ell^4}{(1+\lambda_\ell^2|x-a_\ell|^2)^2}|\tilde{\partial}\omega_i||\overline{v}|$$

$$\leq C\omega_\ell^{\infty^4}\lambda_i\varepsilon_{\ell s}^2 \left(\int |\nabla\overline{v}|^2\right)^{1/2} \left(\int_A \frac{\delta_i^6}{(1+\lambda_i^2|x-a_i|^2)^3}\right)^{1/6} \cdot |\omega_i^\infty|$$

$$+ C\lambda_i \left(\int_A \frac{\delta_\ell^6}{(1+\lambda_\ell^2|x-a_\ell|^2)^2}\right)^{2/3} \left(\int |\nabla\overline{v}|^2\right)^{1/2} \left(\int_A \delta_i^6\right)^{1/6}.$$

Observe that

$$A \subset \left\{\lambda_i|x-a_i| \geq \frac{c}{\text{Max } \varepsilon_{is}}\right\}$$

$$A \subset \left\{\lambda_\ell|x-a_\ell| \geq \frac{c}{\text{Max } \varepsilon_{\ell s}}\right\}$$

so that

$$\int_A \omega_\ell^4 |\tilde{\partial}\omega_i||\overline{v}| \leq C|\omega_i^\infty||\omega_i^\infty|^4\lambda_i\varepsilon_{\ell s}^2 \left(\int |\nabla\overline{v}|^2\right)^{1/2} \cdot \varepsilon_{is}^{3/2}$$

$$+ C\lambda_i\varepsilon_{\ell s}^6 \left(\int |\nabla\overline{v}|^2\right)^{1/2} \varepsilon_{is}^{1/2} \leq o(\lambda_i|\omega_i^\infty|^2\varepsilon_{is}^2)$$

$$+ C\lambda_i\varepsilon_{is}^{1/2}\varepsilon_{\ell s}^4 \left(\int |\nabla\overline{v}|^2\right)^{1/2}.$$

This settles A.

B is typically made of balls $\cup B_k$ corresponding to "masses" ω_k whose center a_k lies in $\widetilde{B}_\ell \cap \widetilde{\Omega}_\ell^c$ with $\lambda_\ell \geq \lambda_\ell$.

On such balls,

$$\lambda_k|x-a_k| \leq \frac{1}{4\varepsilon_{\ell k}} \sim \sqrt{\lambda_k\lambda_\ell}\frac{|a_\ell-a_k|}{4}$$

so that on such balls,

$$\frac{3}{4}|a_\ell-a_k| \leq |x-a_\ell| \leq \frac{5}{4}|a_\ell-a_k|$$

and, therefore,

$$\int_{B_k} \omega_\ell^4 |\tilde{\partial}\omega_i||\bar{v}| \leq C \left(\frac{|\omega_\ell^\infty|^4}{\lambda_\ell^2 |a_\ell - a_k|^4} + \frac{1}{\lambda_\ell^6 |a_\ell - a_k|^8} \right) \int_{B_k} |\tilde{\partial}\omega_i||\bar{v}|$$

$$\leq C\lambda_i \left(\frac{|\omega_\ell^\infty|^4}{\lambda_\ell^2 |a_\ell - a_k|^4} + \frac{1}{\lambda_\ell^6 |a_\ell - a_k|^8} \right) \left(\int |\nabla\bar{v}|^2 \right)^{1/2} \left(\int_{B_k} |\tilde{\partial}\omega_i|^{6/5} \right)^{5/6}$$

$$\leq C\lambda_i \left(\frac{|\omega_\ell^\infty|^4}{\lambda_\ell^2 |a_\ell - a_k|^4} + \frac{1}{\lambda_\ell^5 |a_\ell - a_k|^8} \right) \left(\int |\nabla\bar{v}|^2 \right)^{1/2} \times \frac{1}{\sqrt{\lambda_i}}$$

$$\times \left(\int_{B_k} \frac{(\sqrt{\lambda_i})^{12/5}}{(1 + \lambda_i^2 |x - a_i|^2)^{6/5}} \right)^{5/6} \leq C\lambda_i \left(\frac{|\omega_\ell^\infty|^4}{\lambda_\ell^2 |a_\ell - a_k|^4} + \frac{1}{\lambda_\ell^6 |a_\ell - a_k|^8} \right)$$

$$\times \left(\int |\nabla\bar{v}|^2 \right)^{1/2} \frac{1}{\sqrt{\lambda_i}} \left(\int_{B_k} \delta_i^6 \right)^{1/3} |B_k|^{3/5}.$$

Thus, with $\varepsilon_{kt} = \text{Max } \varepsilon_{km}$

$$\int_{B_k} \omega_\ell^4 |\tilde{\partial}\omega_i||\bar{v}|$$

$$\leq C\sqrt{\lambda_i} \left(\frac{|\omega_\ell^\infty|^4}{\lambda_\ell^2 |a_\ell - a_k|^4} + \frac{1}{\lambda_\ell^6 |a_\ell - a_k|^8} \right) \left(\int |\nabla\bar{v}|^2 \right)^{1/2} \varepsilon_{ik} \frac{\lambda_k^{1/5} \varepsilon_{kt}^{1/5}}{\lambda_k^2 \varepsilon_{kt}^2}$$

$$\leq C\sqrt{\lambda_i}\lambda_k^{1/5} \frac{\varepsilon_{\ell k}^4}{\varepsilon_{kt}^2} \left(\int |\nabla\bar{v}|^2 \right)^{1/2} \varepsilon_{ik}\varepsilon_{kt}^{1/5} = o \left(\lambda_i\varepsilon_{ik}^2 \sum \varepsilon_{st}^3 + \lambda_k\varepsilon_{\ell k}^2 \sum \varepsilon_{st}^3 \right).$$

This settles B.

Arguing in the same way on C, we have:

$$\int_{C_k} \omega_\ell^4 |\tilde{\partial}\omega_i||\bar{v}| \leq C\lambda_i \left(\frac{|\omega_\ell^\infty|\sqrt{\lambda_i}}{(1 + \lambda_i^2 |a_i - a_k|^2)} + \frac{\sqrt{\lambda_i}}{(1 + \lambda_i^2 |a_i - a_k|^2)^{3/2}} \right)$$

$$\times \left(\int |\nabla\bar{v}|^2 \right)^{1/2} \left(\int_{C_k} \omega_\ell^6 \right)^{2/3} \cdot \frac{1}{\sqrt{\lambda_k}\text{Max } \varepsilon_{kt}}$$

$$\leq C \frac{|\omega_i^\infty|\varepsilon_{ik}}{|a_i - a_k|} \varepsilon_{\ell k}^2 \frac{(\int |\nabla\bar{v}|^2)^{1/2}}{\sqrt{\text{Max } \varepsilon_{kt}}} + \frac{(\int |\nabla\bar{v}|^2)^{1/2}\varepsilon_{\ell k}^2}{\sqrt{\text{Max } \varepsilon_{kt}}} \frac{\varepsilon_{ik}}{\lambda_i|a_i - a_k|^2}$$

$$\leq C\sqrt{\lambda_i}|\omega_i^\infty|\varepsilon_{ik} \cdot \sqrt{\lambda_k}\varepsilon_{\ell k}^{3/2}\varepsilon_{ik} \left(\int |\nabla\bar{v}|^2 \right)^{1/2} + \lambda_k \left(\int |\nabla\bar{v}|^2 \right)^{1/2} \varepsilon_{\ell k}^{3/2}\varepsilon_{ik}^3$$

$$\leq o \left(\lambda_i\omega_i^{\infty^2}\varepsilon_{ik}^2 + \lambda_k\varepsilon_{\ell k}^3\varepsilon_{ik}^2 \int |\nabla\bar{v}|^2 \right).$$

This settles C.

Lemma 36 $\sum_{\ell \neq i} \int_{\Omega_i} \omega_\ell^4 |\overline{v}| |\tilde{\partial}\omega_i| \leq C\lambda_i S_i^2.$

Proof.   In view of Lemma 35, we just need to estimate

$$\sum_{\ell \neq i} \int_{\tilde{\Omega}_i} \omega_\ell^4 |\overline{v}_i + \overline{h}_i| |\partial\omega_i| \leq C \left(\sum_{\ell \neq i} |\overline{v}_i|_{H_0^1} \left(\int_{\tilde{\Omega}_i} \omega_\ell^{24/5} |\tilde{\partial}\omega_i|^{6/5} \right)^{5/6} \right.$$

$$+ \sum_{\ell \neq i} \left(\frac{1}{\rho_i} \int_{\Omega_i} |f| + \sqrt{\lambda_i} \int_{B_i^c} |f|\delta_i \right) \int_{\tilde{\Omega}_i} \omega_\ell^4 |\tilde{\partial}\omega_i|$$

$$+ \left. \sum_{\ell \neq i} \left(\frac{1}{\text{Max } \varepsilon_{is}} \int_{\Omega_i} |f|\delta_j + \sqrt{\lambda_i} \int_{B_j} |f| \right) \int_{\tilde{\Omega}_i} \omega_\ell^4 |\tilde{\partial}\omega_i|\delta_j \right). \qquad \square$$

Observe that $|\tilde{\partial}\omega_i| \leq C \left(\sqrt{\lambda_i} |\omega_i^\infty| \delta_i^2 + \delta_i^3 \right)$ so that

$$\int_{\tilde{\Omega}_i} \omega_\ell^4 |\tilde{\partial}\omega_i| \leq C\sqrt{\lambda_i} |\omega_i^\infty| \varepsilon_{i\ell}^2 + C\sqrt{\lambda_\ell} \int_{\tilde{\Omega}_i} \omega_\ell^3 \delta_i^3$$

$$\leq C\sqrt{\lambda_i} |\omega_i^\infty| \varepsilon_{i\ell}^2 + C\sqrt{\lambda_\ell} \, \varepsilon_{i\ell}^3 \log \varepsilon_{i\ell}^{-1}$$

while

$$\int_{\tilde{\Omega}_i} \omega_\ell^4 |\tilde{\partial}\omega_i|\delta_j \leq \lambda_i \left(\int_{\tilde{\Omega}_i} \omega_\ell^4 \delta_j^2 \right)^{1/2} \left(\int \omega_\ell^4 |\tilde{\partial}\omega_i|^2 \right)^{1/2}$$

$$\leq \lambda_i \varepsilon_{\ell j} \left((|\omega_\ell^\infty| + |\omega_i^\infty|)\varepsilon_{\ell i} + \varepsilon_{\ell i}^4 \right).$$

Hence,

$$\sum_{\ell \neq i} \int_{\tilde{\Omega}_i} \omega_\ell^4 |\overline{v}_i + \overline{h}_i| |\tilde{\partial}\omega_i| \leq C \left(\sum_{\ell \neq i} \lambda_i |\overline{v}_i|_{H_0^1} ((|\omega_\ell^\infty| + |\omega_i^\infty|)\varepsilon_{\ell i} + \varepsilon_{\ell i}^3) \right)$$

$$+ \sum_{\ell \neq i} \sqrt{\lambda_i} \left(\frac{1}{\rho_i\sqrt{\lambda_i}} \int_{\Omega_i} |f| + \int_{B_i^c} |f|\delta_i \right) \left(\sqrt{\lambda_i} |\omega_i^\infty| \varepsilon_{i\ell}^2 + \sqrt{\lambda_\ell} \, \varepsilon_{i\ell}^3 \log \varepsilon_{i\ell}^{-1} \right)$$

$$+ \sum_{\ell \neq i} \sqrt{\lambda_i} \left(\sqrt{\lambda_j} \, \varepsilon_{\ell j} \int_{B_j} |f| \right) \cdot \sqrt{\lambda_i} \left((|\omega_\ell^\infty| + |\omega_i^\infty|)\varepsilon_{\ell i} + \varepsilon_{\ell i}^4 \right)$$

$$+ \sum_{\ell \neq i} \frac{\lambda_i}{\text{Max } \varepsilon_{is}} \int_{\Omega_i} |f|\delta_j \left(\int_{\tilde{\Omega}_i} \omega_\ell^4 \delta_j^2 \right)^{1/2} \times \left((|\omega_\ell^\infty| + |\omega_i^\infty|)\varepsilon_{\ell i} + \varepsilon_{\ell i}^4 \right).$$

Either $\lambda_\ell \le \lambda_i$. Then $|\omega_\ell| \le C\delta_i$ on $\widetilde{\Omega}_i$ and $\left(\int_{\widetilde{\Omega}_i} \omega_\ell^4 \delta_j^2\right)^{1/2} \le C\left(\int_{\widetilde{\Omega}_i} \delta_i^4 \delta_j^2\right)^{1/2} \le C\varepsilon_{ij}$.

Or $\lambda_\ell \ge \lambda_i (\ell \ne i)$. Then,

$$\left(\int_{\widetilde{\Omega}_i} \omega_\ell^4 \delta_j^2\right)^{1/2} = O(\varepsilon_{\ell i}\sqrt{\varepsilon_{ij}}).$$

Thus

$$\frac{\lambda_i}{\text{Max } \varepsilon_{is}} \int_{\widetilde{\Omega}_i} |f|\delta_j \left(\int_{\widetilde{\Omega}_i} \omega_\ell^4 \delta_j^2\right)^{1/2} ((|\omega_\ell^\infty| + |\omega_i^\infty|)\varepsilon_{\ell i} + \varepsilon_{\ell i}^4)$$

$$\le C\left(\sqrt{\lambda_i} \int_{\Omega_i} |f|\delta_j\right) \sqrt{\lambda_i} \left((|\omega_\ell^\infty| + |\omega_i^\infty|)\varepsilon_{\ell i} + \varepsilon_{\ell i}^4\right)$$

and the estimate follows.

Lemma 37 $\sum\limits_{\ell \ne i} \int_{\widetilde{\Omega}_i} |\omega_\ell|^3 \overline{v}^2 |\tilde{\partial}\omega_i| \le C\lambda_i S_i^2.$

Proof. If $\ell \ne i$, either $|\omega_\ell(x)| \le |\overline{v}(x)|$ or $|\overline{v}(x)| \le |\omega_\ell(x)|$ so that

$$\int_{\widetilde{\Omega}_i} |\omega_\ell|^3 \overline{v}^2 |\tilde{\partial}\omega_i| \le \int_{\widetilde{\Omega}_i} \omega_\ell^4 |\overline{v}||\tilde{\partial}\omega_i| + \lambda_i \left(\int |\nabla\overline{v}|^2\right)^{5/2}$$

and the estimate follows from the previous lemma. $\square$

Lastly, we have:

Lemma 38 $\int |\omega_i|^3 |\overline{v}|^2 |\tilde{\partial}\omega_i| \le C\lambda_i S_i^2.$

Proof. We split the integral in $\int_{\widetilde{\Omega}_i} |\omega_i|^3 \overline{v}^2 |\tilde{\partial}\omega_i|$ and $\int_{\widetilde{\Omega}_i^c} |\omega_i|^3 \overline{v}^2 |\tilde{\partial}\omega_i|$. $\square$

For the first part, we have

$$\int_{\widetilde{\Omega}_i} |\omega_i|^3 \overline{v}^2 |\tilde{\partial}\omega_i| \le C\int_{\widetilde{\Omega}_i} |\omega_i|^3 \overline{v}_i^2 |\tilde{\partial}\omega_i| + C\int_{\widetilde{\Omega}_i} |\omega_i|^3 \overline{h}_i^2 |\tilde{\partial}\omega_i|$$

$$\le C\lambda_i |\overline{v}_i|_{H_0^1}^2 + C\left(\frac{1}{\rho_i}\int_{\Omega_i} |f| + \sqrt{\lambda_i}\int_{B_i^c} |f|\delta_i\right)^2$$

$$+ C\left(\sqrt{\lambda_j}\int_{B_j} |f| + \frac{1}{\text{Max } \varepsilon_{it}}\int_{\Omega_i} |f|\delta_j\right)^2 \times \int_{\widetilde{\Omega}_i} |\omega_i|^3 \delta_j^2 |\partial\omega_i|$$

$$\leq C\lambda_i \left(|\bar{v}_i|_{H_0^1}^2 + \left(\frac{1}{\sqrt{\lambda_i}\rho_i} \int_{\Omega_i} |f| + \int_{B_i^c} |f|\delta_i \right)^2 \right.$$

$$\left. + C\left(\sqrt{\lambda_j}\varepsilon_{ij} \int_{B_j} |f| + \int_{\Omega_i} |f|\delta_j \right)^2 \right).$$

The first part is thus settled.

Next, we estimate

$$\int_{\widetilde{\Omega}_i^c} |\omega_i|^3 |\tilde{\partial}\omega_i| |\bar{v}|^2$$

$|\tilde{\partial}\omega_i|$ is bounded by $\lambda_i \left(|\omega_i^\infty|\delta_i + \frac{\delta_i}{(1+\lambda_i^2|x-a_i|^2)^{1/2}} \right)$.

Thus,

$$\int_{\widetilde{\Omega}_i^c} |\omega_i|^3 |\tilde{\partial}\omega_i| \bar{v}^2 \leq C\lambda_i |\omega_i^\infty| \int_{\widetilde{\Omega}_i^c} \delta_i^4 \bar{v}^2 + C\lambda_i \int_{\widetilde{\Omega}_i^c} \frac{\delta_i^4}{(1+\lambda_i^2|x-a_i|^2)^2} |\bar{v}|^2$$

$$\leq C\lambda_i |\omega_i^\infty| \left(\int |\nabla\bar{v}|^2 \right) \sum \varepsilon_{i\ell}^2 + C\lambda_i \int |\nabla\bar{v}|^2 \left(\int_{\widetilde{\Omega}_i^c} \frac{\delta_i^6}{(1+\lambda_i^2|x-a_i|^2)^3} \right)^{2/3}.$$

The first term is easily upperbounded by

$$C\lambda_i |\omega_i^\infty|^2 \sum \varepsilon_{i\ell}^2 + C\lambda_i \left(\int |\nabla\bar{v}|^2 \right)^2 \sum \varepsilon_{i\ell}^2.$$

For the second term, we split the integral over $\widetilde{\Omega}_i^c$ into two pieces.

The first one is

$$C\lambda_i \int |\nabla\bar{v}|^2 \left(\int_{\lambda_i|x-a_i| \geq \frac{1}{\mathrm{Max}\varepsilon_{it}}} \frac{\delta_i^6}{(1+\lambda_i^2|x-a_i|^2)^3} \right)^{2/3}$$

$$\leq C\lambda_i \int |\nabla\bar{v}|^2 \, \mathrm{Max}\, \varepsilon_{it}^6.$$

The second one involves integration over subdomains B_k of B_i which corresponds to masses "ω_k" having $\lambda_k \geq \lambda_i$.

On B_k

$$\lambda_i|x-a_i| \geq \frac{1}{2}\lambda_i|a_k-a_i|$$

and we have:

$$C\lambda_i \int |\nabla\overline{v}|^2 \left(\int_{B_k} \frac{\delta_i^6}{(1+\lambda_i^2|x-a_i|^2)^3} \right)^{2/3} \leq C\lambda_i \int |\nabla\overline{v}|^2 \cdot \frac{1}{\lambda_i^4|a_k-a_i|^4}\varepsilon_{ik}^2$$

$$\leq \frac{C\sqrt{\lambda_i\lambda_k}}{\lambda_i^3|a_k-a_i|^3} \int |\nabla\overline{v}|^2\varepsilon_{ik}^3 = o(\lambda_i \int |\nabla\overline{v}|^2\varepsilon_{ik}^3 + \lambda_k \int |\nabla\overline{v}|^2\varepsilon_{ik}^3).$$

Lemma 38 follows.

Next we observe that:

Lemma 39

$$\sum_{\ell\neq i} \int |\omega_i|^3|\omega_\ell||\tilde{\partial}\omega_i||\overline{v}| + |\omega_\ell|^3|\omega_i||\overline{v}||\tilde{\partial}\omega_i|$$

$$\leq C \left(\int |\omega_i|^3|\tilde{\partial}\omega_i||\overline{v}|^3 + \sum_{\ell\neq i} \int |\omega_i|^3|\tilde{\partial}\omega_i|\omega_\ell^2 + \sum_{\ell\neq i} \int \omega_\ell^4|\tilde{\partial}\omega_i||\overline{v}| \right).$$

Proof.   Observe that

$$\int_{|\omega_i|\geq|\omega_\ell|} |\omega_\ell|^3|\omega_i||\tilde{\partial}\omega_i||\overline{v}| \leq \int_{\substack{|\omega_i|\geq|\omega_\ell|\\|\overline{v}|\geq|\omega_\ell|}} |\omega_i|^3|\tilde{\partial}\omega_i||\overline{v}|^2 + \int_{|\omega_i|\geq|\omega_\ell|} \omega_\ell^4|\omega_i||\tilde{\partial}\omega_i|$$

$$\leq \int |\omega_i|^3|\tilde{\partial}\omega_i|\overline{v}^2 + \int |\omega_i|^3|\tilde{\partial}\omega_i|\omega_\ell^2.$$

Lemma 40 follows. The estimate of Lemma 8 on $\int|\omega_i|^4\omega_\ell^2$ extends of course to $\frac{1}{\lambda_i}\int|\omega_i|^3|\partial\omega_i|\omega_\ell^2$. $\square$

We expand

$$J'\left(\sum\alpha_j\omega_j+\overline{v}\right)\cdot\partial\omega_i = J'\left(\sum\alpha_j\omega_j\right)\cdot\partial\omega_i + J''\left(\sum\alpha_j\omega_j\right)\cdot\overline{v}\cdot\partial\omega_i$$

$$+ O\left(\sum\int\omega_j^4 v^2|\partial\varphi_i| + \int|\overline{v}|^5|\partial\varphi_i|\right).$$

The expansion of $J'(\sum\alpha_j\omega_j)\frac{\partial\omega_i}{\partial a_i}$ is straightforward and yields as principal terms

$$\sum_{i\neq j}\omega_j^\infty\frac{\partial}{\partial a_i}\left(\left(\omega_i^\infty + \frac{D\bar{\omega}_i(\frac{a_i-a_j}{|a_i-a_j|})}{\lambda_i|a_i-a_j|}\right)\varepsilon_{ij}\right) - \sum c_{ij}\frac{\partial\varepsilon_{ij}^3}{\partial a_i}$$

$\bar{\omega}_i$ is the rescaled version of ω_i. There are additional terms which are

$$o\left(\sum\frac{\omega_i^{\infty 2}}{\lambda_i|a_i-a_j|^2} + \frac{\varepsilon_{ij}^3}{|a_i-a_j|} + \cdots\right).$$

We split the various integrals involving v, after the use of (Vo) to get rid of all $\int \omega_i^4 \frac{\partial \omega_i}{\partial a_i} v$ *etc.* between the various contributions over each domains Ω_i and the contribution over $(U\Omega_i)^c$. We use **Lemmas 32 to 39** and we find that these contributions can be thrown into a remainder. However, the remainder does not behave as a small o of the principal terms for two main reasons: first we find, because our estimates over $|v_i|_{H_0^1}$ are not suitable, contributions of the type $o\left(\lambda_i \omega_i^{\infty 2} \varepsilon_{ij}\right)$, which are

$$o\left(\frac{\omega_i^{\infty 2}}{\lambda_j |a_i - a_j|^2}\right), \text{ not } o\left(\frac{\omega_i^{\infty 2}}{\lambda_i |a_i - a_j|^2}\right).$$

Second, because the remainders in these estimates (and in the estimates on $\frac{|\bar{h}_i^o|_\infty}{\sqrt{\lambda_i}} + \sum_{\lambda_s \geq \lambda_i} \varepsilon_{is} |\bar{h}_s^i|_\infty$) are only $O(\varepsilon_{ij} \sum \varepsilon_{st}^2)$ at best, we find terms such as $o\left(\frac{\varepsilon_{ij} \sum \varepsilon_{st}^2}{|a_i - a_j|}\right)$ which are not $o\left(\frac{\varepsilon_{ij}^3}{|a_i - a_j|}\right)$.

Nevertheless, if the λ_i's are comparable as well as the ε_{ij}'s, these remainder terms are all

$$o\left(\sum \frac{\omega_i^{\infty 2}}{\lambda_i |a_i - a_j|^2} + \sum \frac{\varepsilon_{ij}^3}{|a_i - a_j|}\right).$$

Using then Conjecture 2 and the σ_i-equation (see Proposition 8 in the third part of the book), we can establish the Morse Lemma at infinity.

The details about this proof and this expansion are provided at the end of the third part of the first half of the book. We refer the reader to "Basic Conformal Estimates" and more precisely to Lemmas 62–67 and to the sections "The system of equations corresponding to the variations of the points a_i's" and the following sections for all details on the expansions and proof of the Morse Lemma at infinity.

If the λ_i's are comparable as well as the ε_{ij}'s, the estimates provided above are sufficient. But these are strong hypotheses which we remove, step by step, over the remainder of this work.

1.9 Redirecting the Estimates, Estimates on $|\bar{v}_i|_{H_0^1} + |\bar{h}_i^0|_\infty + \sum_{\lambda_s > \lambda_i} \varepsilon_i^s |\bar{h}_i^s|_\infty$

1.9.1 *Content of Part II*

Part II contains three new estimates:

First, a simple observation allows to improve the estimate on

$$\left(\int_{\Omega_j} |\omega_j|^{24/5} |\omega_i|^{6/5} + |\omega_i|^{24/5} |\omega_j|^{6/5}\right)^{3/2} \text{ for } \lambda_j \geq \lambda_i.$$

Instead of $C\left(\left(|\overline{\omega}_i^\infty| + |\overline{\omega}_j^\infty|\right)\varepsilon_{ij} + \varepsilon_{ij}^3\right)$, we find

$$C\left(|\overline{\omega}_i(\overline{a}_j)|\varepsilon_{ij} + \varepsilon_{ij}^3\right) \ i.e. \ \text{basically} \ C(|\overline{\omega}_i^\infty|\varepsilon_{ij} + \varepsilon_{ij}^3).$$

Second, we derive an L^∞-estimate on $\overline{v}$ after bootstrapping in the equation that $\overline{v}$ satisfies (derived in Part I). This L^∞-estimate reads

$$|\overline{v} - \theta_h|(x) \leq C_h \sum \delta_\ell(x) \quad \forall x \in S^3$$

with

$$|\theta_h|_{H^1(S^3)} \leq C_h \sum \varepsilon_{k\ell}^h$$

h is an arbitrary exponent. This pointwise estimate can be improved in $(U\Omega_i)^c$.

Third, we derive a better estimate on

$$\int \frac{\omega_\ell^4 |\omega_k|}{|x - y|}.$$

We prove (Lemma 40) that:

$$\int \frac{\omega_\ell^4 |\omega_k|}{|x - y|} \leq C \sum_{\substack{\lambda_i \geq \lambda_\ell \\ \text{or} \\ \lambda \geq \lambda_k}} \varepsilon_{ij}\delta_i \ \text{for} \ y \in (\Omega_\ell \cup \Omega_k)^c.$$

After these three estimates, we revisit all the estimates of Part I, Lemmas 13 to 30 in particular, which we improve considerably. (Lemmas 41–56)

The remainder terms in our expression, on each Ω_i involve now a power of $|\overline{v}_i|_{H_0^1} + \frac{|\overline{h}_i|_\infty}{\sqrt{\lambda_i}}$; this power is however too weak to allow us to prove the Morse Lemma at infinity without (1.3). (See Lemma 57).

After Lemma 41–56, we derive the main improvement of Part II which is provided in Theorem 1: we establish that a much better estimate on $|\overline{v}_i|_{H_0^1} + \frac{|\overline{h}_i|_\infty}{\sqrt{\lambda_i}}$ when compared to the ones provided in Lemmas 33–34 holds, up to a remainder term equal to $C_h \sum \varepsilon_{st}^h$ (due to bootstrapping; we will get rid of it in Part III).

If we remove the remainder term, the estimate on $|\overline{v}_i|_{H_0^1}$ is optimal because the remainder is then (essentially) $O\left(\sum_\ell \varepsilon_{ik}^3\right)$.

Lemma 57 of Part II can be skipped if one wants to go directly to the proof of the Morse Lemma at infinity.

1.9.2 *Redirecting the estimates*

In the previous sections, we have devised a method and a number of estimates. We have thought of S^3 as subdivided in various regions $\Omega_i, \Omega_c = (U\Omega_i)^c$ adapted to our family of solutions ω_i and we have thought of v as decomposed in $\sum(v_i + h_i) + k^*$. We have estimated the various v_i, h_i, k^*, thinking of them at single entries related by "a matrix of interaction". However, these estimates are not enough to cover the general case of any possible configuration. So far, we can complete our Morse Lemma at infinity only if the concentrations λ_i are of the same order and the points are not "too close".

Here we redirect our estimates, keeping the same general framework and work out in this way improvements in the estimates. Eventually, we will work out the general case.

We start with an observation. Namely,

Proposition 1   *Assume that $\lambda_j \geq \lambda_i$. Then,*

$$\left(\int_{\Omega_j} |\omega_j|^{24/5}|\omega_i|^{6/5} + \int_{\Omega_j} |\omega_i|^{24/5}|\omega_j|^{6/5} \right)^{5/6}$$

$$\leq C \left(|\underline{\omega}_i(\bar{a}_j)|\varepsilon_{ij} + \varepsilon_{ij}^3 \right)$$

where $\underline{\omega}_i(\bar{a}_j)$ is the value of ω_i set in its standard form (with concentration $0(1)$) at $\bar{a}_j$, the (then new) concentration point (on S^3, after dilation) of ω_j.

Proof.   We revisit [Bahri 2001, proof of Lemma 3.2].

For $\left(\int_{\Omega_j} |\omega_j|^{24/5}|\omega_i|^{6/5} \right)^{5/6}$, there is nothing to change.

For $\left(\int_{\Omega_j} |\omega_i|^{24/5}|\omega_j|^{6/5} \right)^{5/6}$, we notice that

$$\Omega_j \subset \left\{ \lambda_j|x - a_j| \leq \frac{C}{\varepsilon_{ij}} \right\}.$$

All our domains in fact are defined on the sphere, using the geodesic distance. When needed, we complete a stereographic projection with respect to a North Pole which keeps all points of concentration in a given compact set and all concentrations roughly of the same size than they were on the sphere if needed.

Thinking of a_j as being the South Pole (we are here only estimating $\int_{\Omega_j} |\omega_i|^{24/5}|\omega_j|^{6/5}$, not considering all the mases together, our observations which we just stated do not apply) after rescaling so that ω_i has concentration 1, performing a stereographic projection, Ω_j becomes:

$$\widetilde{\Omega}_j = \left\{ \bar{\lambda}_j |x| \leq \frac{C}{\varepsilon_{ij}} \right\}.$$

Now, ε_{ij} is of the size of $\bar{\lambda}_j^{-1/2}$, hence $\widetilde{\Omega}_j$ is a very small ball of radius $0\left(\frac{1}{\bar{\lambda}_j^{1/2}}\right)$ around the origin.

$\underline{\omega}_i$ expands as $\underline{\omega}_i(0) + 0(|x|)$ and we derive that

$$\left(\int_{\widetilde{\Omega}_j} |\underline{\omega}_i|^{24/5} |\tilde{\omega}_j|^{6/5} \right)^{5/6} \leq C\underline{\omega}_i(0)^4 \left(\int_{r \leq \frac{1}{\sqrt{\bar{\lambda}_j}}} \left(\frac{\sqrt{\bar{\lambda}_j}}{(1 + \bar{\lambda}_j^2 r^2)^{1/2}} \right)^{6/5} \right)^{5/6}$$

$$+ C \left(\int_{r \leq \frac{1}{\sqrt{\bar{\lambda}_j}}} r^{24/5} \left(\frac{\sqrt{\bar{\lambda}_j}}{(1 + \bar{\lambda}_j^2 r^2)^{1/2}} \right)^{6/5} \right)^{5/6} \leq C \left(\frac{|\underline{\omega}_i(0)|}{\sqrt{\bar{\lambda}_j}} + \frac{1}{\bar{\lambda}_j^{3/2}} \right).$$

$\square$

We now move to prove a much finer estimate on v, namely:

Proposition 2 $\forall h \in \mathbb{N}$, *there exists a constant* $C_h > 0$ *and a function* $\theta_h \in H^1$ *such that*

(i) $|v - \theta_h|(x) \leq C_h \sum \delta_\ell(x) \quad \forall x \in S^3$

(ii) $|\theta_h|_{H^1} \leq C_h \sum \varepsilon_{k\ell}^h$

(iii) $|v - \theta_h|(x) \leq C_h \sum \left(\varepsilon_{k\ell} + |v_\ell|_{L^6} + |h_\ell|_\infty / \sqrt{\lambda_\ell} \right) \delta_\ell(x) \quad \forall x \in \Omega_c = (U\Omega_i)^c.$

Proof.
Proof of (i) and (ii). v satisfies

$$-\Delta v - \left(\left(\sum \alpha_j \omega_j + v \right)^5 - \left(\sum \alpha_j \omega_j \right)^5 - 5 \left(\sum \alpha_j \omega_j \right)^4 v \right)$$

$$= f + 5 \left(\sum \alpha_j \omega_j \right)^4 v$$

which we rewrite as

$$-\Delta v - b(x, v) = f + 5 \left(\sum \alpha_j \omega_j \right)^4 v.$$

$\square$

Let

$$\varphi = -\Delta^{-1}\left(f + 5\left(\sum \alpha_j \omega_j\right)^4 v\right).$$

It is easy to see that, with a suitable constant C:

$$|f(x)| \le C\left(\sum \delta_\ell^5\right).$$

We consider

$$\int \frac{\omega_i^4 |v|}{|x - y|}.$$

Either $|y - x_i| \ge \frac{C}{\lambda_i}$. Then

$$\int \frac{\omega_i^4 |v|}{|x - y|} \le \int_{|x-y|\le\frac{c}{2\lambda_i}} \frac{\omega_i^4 |v|}{|x - y|} + C_1\sqrt{\lambda_i} \int_{|x-y|\ge\frac{c}{2\lambda_i}} \omega_i^4 \delta_i(y, \lambda_i)(x)|v|.$$

Clearly, if $|x - y| \le \frac{c}{2\lambda_i}$,

$$\omega_i^4 \le C\delta_i^4(x_i, \lambda_i)(y)$$

and

$$\int_{|x-y|\le\frac{c}{2\lambda_i}} \frac{|v|}{|x - y|} \le \frac{C|v|_{L^6}}{\lambda_i^{3/2}}.$$

Thus,

$$\int_{|x-y|\le\frac{c}{2\lambda_i}} \frac{\omega_i^4 |v|}{|x - y|} \le C\frac{\delta_i^4(x_i, \lambda_i)(y)|v|_{L^6}}{\lambda_i^{3/2}} \le C\delta_i(x_i, \lambda_i)(y)|v|_{L^6}.$$

On the other hand,

$$\int \omega_i^4 \delta_i(y, \lambda_i)(x)|v| \le C|v|_{L^6}\left(\int \delta_i^{24/5}(x_i, \lambda_i)\delta_i^{6/5}(y, \lambda_i)(x)\right)^{5/6} \le \frac{C|v|_{L^6}}{\lambda_i|y - x_i|}.$$

Thus,

$$\int \frac{\omega_i^4 |v|}{|x - y|} \le C_2|v|_{L^6}\delta_i(x_i, \lambda_i)(y) \text{ if } \lambda_i|y - x_i| \ge C.$$

If $|y - x_i| \le \frac{C}{\lambda_i}$, then (observe that $|x - x_i| \ge \frac{2C}{\lambda_i} \Rightarrow |x - y| \ge \frac{|x-x_i|}{2}$)

$$\int \frac{\omega_i^4 |v|}{|x - y|} \le \int_{|x-x_i| \le \frac{2C}{\lambda_i}} \frac{\omega_i^4 |v|}{|x - y|} + C\sqrt{\lambda_i} \int \delta_i^5 |v|$$

$$\le C_1 \delta_i^4(x_i, \lambda_i)(y) \int_{|x-y| \le \frac{3C}{\lambda_i}} \frac{|v|}{|x - y|} + C_1 \delta_i(x_i, \lambda_i)(y)|v|_{L^6}$$

$$\le C_2 \frac{\delta_i^4 |v|_{L^6}}{\lambda_i^{3/2}} + C_2 \delta_i(x_i, \lambda_i)(y)|v|_{L^6} \le C_3 \delta_i |v|_{L^6}.$$

Combining these estimates and the one of f, we derive that

$$|\varphi(x)| = O\left(\sum \delta_\ell\right).$$

Then,

$$-\Delta(v - \varphi) - (b(x, v) - b(x, \varphi)) = b(x, \varphi).$$

Since b is of second order in the second variable,

$$|b(x, v) - b(x, \varphi)|_{L^{6/5}} = o(|v - \varphi|_{L^6})$$

and

$$\int |\nabla(v-\varphi)|^2 \le C \int b(x, \varphi)(v-\varphi) \le C|v-\varphi|_{L^6}|b(x, y)|_{L^{6/5}} \le C_1|v-\varphi|_{L^6}|\varphi|_{L^6}^2.$$

Thus,

$$\left(\int |\nabla(v - \varphi)|^2\right)^{1/2} \le C_2|\varphi|_{L^6}^2.$$

Set

$$\varphi_1 = -\Delta^{-1}\left(b(x, \varphi)\right).$$

Clearly,

$$|\varphi_1|_{L^6} \le C_3|\varphi|_{L^6}^2.$$

Then,

$$-\Delta\left(v - (\varphi + \varphi_1)\right) - (b(x, v) - b(x, \varphi + \varphi_1)) = b(x, \varphi + \varphi_1) - b(x, \varphi)$$

$$|b(x, v) - b(x, \varphi + \varphi_1)|_{L^{6/5}} = o\left(|v - (\varphi + \varphi_1)|_{L^6}\right)$$

and

$$|b(x, \varphi + \varphi_1) - b(x, \varphi)|_{L^{6/5}} \leq C|\varphi_1|_{L^6}(|\varphi|_{L^6} + |\varphi_1|_{L^6}) \leq C_4|\varphi|_{L^6}^3$$

so that

$$\left(\int |\nabla(v - (\varphi + \varphi_1))|^2\right)^{1/2} \leq C_5|\varphi|_{L^6}^3$$

and

$$|\varphi_1(x)| = |-\Delta^{-1}(b(x, \varphi))| \leq C\sum \delta_\ell(x).$$

Bootstrapping and observing that

$$f = \left(\sum \alpha_j \omega_j\right)^5 - \left(\sum \alpha_j \omega_j + \bar{v}\right)^5$$
$$- Q^*\left(\left(\sum \alpha_j \omega_j + \bar{v}\right)^5 - \left(\sum \alpha_j \omega_j\right)^5\right)$$
$$- Q^*\left(\left(\sum \alpha_j \omega_j\right)^5\right).$$

So that, since $|\bar{v}|_{H^1} \leq C\sum \varepsilon_{ij}$ and since $|Q * ((\sum \alpha_j \omega_j)^5)|_{L^{6/5}} \leq C\sum \varepsilon_{ij}$,

$$|f|_{L^{6/5}} \leq C\sum \varepsilon_{ij},$$

we derive that

$$|\varphi|_{L^6} \leq C\sum \varepsilon_{ij}.$$

(i) and (ii) of Proposition 2 follow. $\qquad\qquad\qquad\qquad\qquad\qquad\square$

Proof of (iii). v satisfies

$$-\Delta v - Q^*\left(\left(\sum \alpha_j \omega_j + v\right)^5\right) = 0$$

which we rewrite:

$$-\Delta v - Q^*\left(\left(\sum \alpha_j \omega_j\right)^5 + \gamma(x)v\right) = 0.$$

Using (i) and (ii), we can estimate γ as follows:

$$\begin{cases} |\gamma - \gamma_h|(x) \leq C_h \sum \delta_\ell^4 \\ |\gamma_h|_{L^{3/2}} \leq C_h \sum \varepsilon_{k\ell}^h. \end{cases}$$

Thus,

$$-\Delta v - Q^* \left(\left(\sum \alpha_j \omega_j \right)^5 + O_h \left(\sum \delta_\ell^4 \right) v + \gamma_h v \right) = 0$$

and

$$v = \int \frac{Q^* \left(\left(\sum \alpha_j \omega_j \right)^5 + O_h \left(\sum \delta_\ell^4 \right) v + \gamma_h v \right)}{|x - y|}.$$

$\square$

We now prove the following:

Lemma 40

$$\int \frac{\omega_\ell^4 |\omega_k|}{|x - y|} \leq C \sum_{\substack{\lambda_i \geq \lambda_\ell \\ or \\ \lambda_i \geq \lambda_k}} \varepsilon_{ij} \delta_i \ \textit{for } y \in (\Omega_\ell \cup \Omega_k)^c.$$

Observation. The dependence on δ_i for i also different from ℓ and k is due to the definition of Ω'_ℓ, Ω'_k which involves other δ'_i's.

Proof. We first establish the lemma with $\lambda_\ell \geq \lambda_k, \lambda_\ell |y - x_k|, \lambda_k |y - x_k| \geq 1$, with only the two masses δ_ℓ and δ_k involved in the definition of Ω_ℓ, Ω_k.

We reduce, after the use of some arguments which include the action of the conformal group, the general case to this one.

We first claim that $(\Omega'_\ell, \Omega'_k$ are two smaller versions of $\Omega_\ell, \Omega_k)$:

$$\int_{\Omega'_\ell} \frac{\omega_\ell^4 |\omega_k|}{|x - y|} \leq \frac{C}{|y - x_\ell|} \int_{\Omega_\ell} \omega_\ell^4 |\omega_k|.$$

Indeed, in Ω'_ℓ,

$$\lambda_\ell |x - x_\ell| \leq \frac{1}{18\varepsilon_{\ell k}}$$

while

$$\lambda_\ell |y - x_\ell| \geq \frac{1}{8\varepsilon_{\ell k}}$$

so that, if $x \in \Omega'_\ell$, $|y - x_\ell|$ and $|x - y|$ are of the same order while $|x - x_k|$ is of the order of $|x_k - x_\ell|$.

Thus,

$$\int_{\Omega_\ell'} \frac{\omega_\ell^4 |\omega_k|}{|x-y|} \le \frac{C}{|y-x_\ell|} \int_{\Omega_\ell'} \omega_\ell^4 |\omega_k| \le \frac{C_1}{|y-x_\ell| \sqrt{\lambda_k} |x_k - x_\ell|} \int_{\Omega_\ell'} \omega_\ell^4$$

$$\le \frac{C_2}{\sqrt{\lambda_\ell} |y - x_\ell|} \varepsilon_{k\ell} \sim C_2' \varepsilon_{\ell k} \delta_\ell.$$

In Ω_k', $\lambda_\ell |x - x_\ell| \ge \frac{1}{18\varepsilon_{\ell k}}$ and $\lambda_k |x - x_k| < \frac{c}{2\varepsilon_{ks}}$, while $\lambda_k |y - x_k| > \frac{1}{8\varepsilon_{ks}}$ or $\lambda_k |y - x_k| < \frac{1}{8\varepsilon_{ks}}$ and $\lambda_\ell |y - x_\ell| < \frac{c}{\varepsilon_{\ell k}}$.

Thus, either $|x - y|$ is of the order of $|y - x_k|$ or $|x - x_\ell|$ is large when compared to $|y - x_\ell|$ so that

$$|y - x| \ge c|y - x_\ell|.$$

In all,

$$\int_{\Omega_k'} \frac{\omega_\ell^4 |\omega_k|}{|x-y|} \le C \left(\frac{1}{|y - x_\ell|} + \frac{1}{|y - x_k|} \right) \int_{\Omega_k'} \omega_\ell^4 |\omega_k|$$

$$\le C \left(\varepsilon_{\ell k}^2 \delta_k + \frac{1}{|y - x_\ell|} \frac{\varepsilon_{k\ell}^2}{\sqrt{\lambda_k}} \right) \le C \varepsilon_{k\ell} (\delta_k + \delta_\ell).$$

For $x \in (\Omega_\ell' \cup \Omega_k')^c$,

$$\int_{(\Omega_k' \cup \Omega_\ell')^c} \frac{\omega_\ell^4 |\omega_k|}{|x-y|} \le \int_{\substack{|x-y| \le \frac{1}{4} \inf(|y-x_k|, |y-x_\ell|) \\ (\Omega_k' \cup \Omega_\ell')^c}} \frac{\omega_\ell^4 |\omega_k|}{|x-y|}$$

$$+ C \left(\frac{1}{|y - x_k|} + \frac{1}{|y - x_\ell|} \right) \int_{(\Omega_k' \cup \Omega_\ell')^c} \omega_\ell^4 |\omega_k|.$$

Since $|x - y|$ is small when compared to $|y - x_k|$ and $|y - x_\ell|$ for the first integral, we have:

$$\begin{cases} \omega_\ell^4(x) \le C \delta_\ell^4(y) \\ |\omega_k(x)| \le C \delta_k(y) \end{cases}$$

and

$$\int_{(\Omega_k' \cup \Omega_\ell')^c} \frac{\omega_\ell^4 |\omega_k|}{|x-y|} \le C \delta_\ell^4(y) \delta_k(y) \inf(|y - x_k|^2, |y - x_\ell|^2)$$

$$+ C \left(\frac{1}{|y - x_k|} + \frac{1}{|y - x_\ell|} \right) \frac{1}{\sqrt{\lambda_k}} \varepsilon_{k\ell}^2$$

$$\leq C\varepsilon_{k\ell}(\delta_k + \delta_\ell) + \frac{C}{\lambda_\ell^2 |y - x_\ell|^4 \sqrt{\lambda_k} |y - x_k|} \inf(|y - x_k|^2, |y - x_\ell|^2)$$

$$\leq C\varepsilon_{k\ell}(\delta_k + \delta_\ell) + \frac{C\delta_k}{\lambda_\ell^2 |y - x_\ell|^2} \cdot \inf\left(\frac{|y - x_k|}{|y - x_\ell|}, 1\right).$$

$|y - x_\ell|$ or $|y - x_k|$ is lowerbounded by $c|x_k - x_\ell|$. Thus,

$$\frac{\delta_k}{\lambda_\ell^2 |y - x_\ell|^2} \leq \frac{1}{C}\left(\frac{\delta_k}{\lambda_\ell^2 |y - x_\ell||x_\ell - x_k|} \times \frac{|y - x_k|}{|y - x_\ell|} + \frac{\varepsilon_{k\ell}}{\lambda_\ell^{3/2} |y - x_\ell|^2}\right)$$

$$\leq \frac{1}{c}\varepsilon_{k\ell}\delta_\ell.$$

This provides the estimate.

Suppose now that the definition of Ω_ℓ, Ω_k involves a third mass ω_i. Then, y could also be in the domain Ω_i^1 of this mass and we have to prove the estimate on $\int \frac{\omega_\ell^4 |\omega_k|}{|x - y|}$ for y in $\Omega_i^c \cap \Omega_i^1$.

We rename $\tilde{\Omega}_\ell, \tilde{\Omega}_k$ the domains related to ω_ℓ and ω_k when there is no additional mass. Our estimate is established for $y \in \left(\tilde{\Omega}_\ell \cup \tilde{\Omega}_k\right)^c$. We need to do some more work if Ω_i^1 has some intersection with $\tilde{\Omega}_\ell \cup \tilde{\Omega}_k$. Then,

$$\lambda_i \geq \lambda_k$$

and

$$|\omega_k|(x) \leq C\delta_i(x) \text{ on } \Omega_i^1 \supset \Omega_i \supset \Omega_i'.$$

Thus,

$$\int \frac{\omega_\ell^4 |\omega_k|}{|x - y|} \leq \int_{\Omega_i^1} \frac{\omega_\ell^4 |\omega_k|}{|x - y|} + \int_{\Omega_i^{1c}} \frac{\omega_\ell^4 |\omega_k|}{|x - y|} \leq C \int_{\Omega_i^1} \frac{\omega_\ell^4 \delta_i}{|x - y|} + \int_{\Omega_i^{1c}} \frac{\omega_\ell^4 |\omega_k|}{|x - y|}$$

$\int \frac{\omega_\ell^4 \delta_i}{|x-y|}$ can now be estimated as above, since $y \in (\Omega_\ell \cup \Omega_i)^c$.

We need to consider

$$\int_{\Omega_i^{1c}} \frac{\omega_\ell^4 |\omega_k|}{|x - y|}.$$

Assume for sake of simplicity that Ω_i reduces to a ball $\{x | \lambda_i |x - x_i| < \frac{1}{8} \text{Min} \frac{1}{\varepsilon_{ij}}\}$ around x_i which does not interest $\tilde{\Omega}_\ell$ and is entirely contained in $\tilde{\Omega}_k$. The general case is only a variation of this one. We might shrink some domains, enlarge others when there is some intersection, but the arguments remain the same.

Then, y is in an annular region $\Omega_i^c \cap \widetilde{\Omega}_k$ around x_i included in Ω_i^1, (in the remainder of $(\Omega_\ell \cup \Omega_k \cup \Omega_i)^c$, the estimate is already established) and x is outside an even larger ball

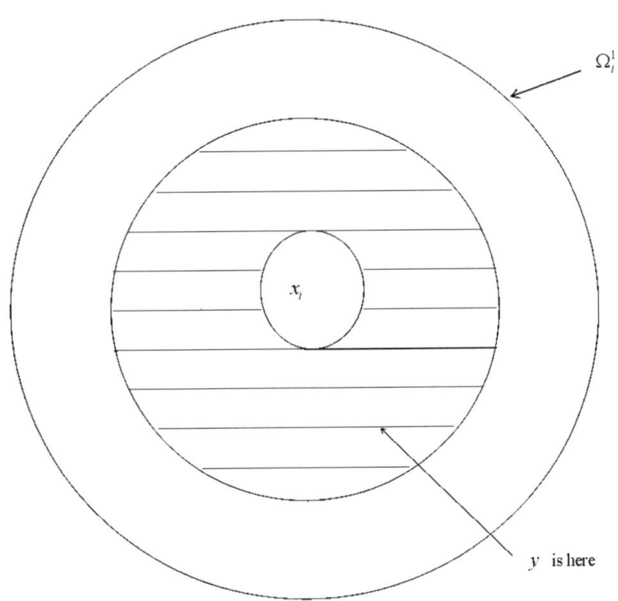

Thus,

$$\frac{1}{c}\lambda_i|x - x_i| \geq \lambda_i|x - y| \geq c\lambda_i|x - x_i|$$

and

$$\int_{\Omega_i^{1c}} \frac{\omega_\ell^4|\omega_k|}{|x - y|} \leq c^{-1}\sqrt{\lambda_i}\int_{\Omega_i^{1c}} \omega_\ell^4|\omega_k|\delta_i(x) \leq C\sqrt{\lambda_i}\varepsilon_{\ell i}\varepsilon_{k\ell}.$$

Observe that:

$$|x_\ell - x_i| \geq c|x_\ell - y|$$

since $x_\ell \notin \Omega_i^1$.

Thus,

$$\sqrt{\lambda_i}\varepsilon_{\ell i} \leq C\delta_\ell(y)$$

and

$$\int_{\Omega_i^{1c}} \frac{\omega_\ell^4 |\omega_k|}{|x-y|} \leq C\varepsilon_{\ell k}(\delta_\ell + \delta_k).$$

The estimate follows. There are other intermediate, overlapping domains, but these estimates extend to them in a natural way.

We discuss now the assumptions $\lambda_\ell \geq \lambda_k, \lambda_\ell|y - x_\ell|, \lambda_k|y - x_k| \geq 1$. We come back to the sphere where our domains have the following typical profile (there are others, intermediate ones, but they are of the same type, with small variations):

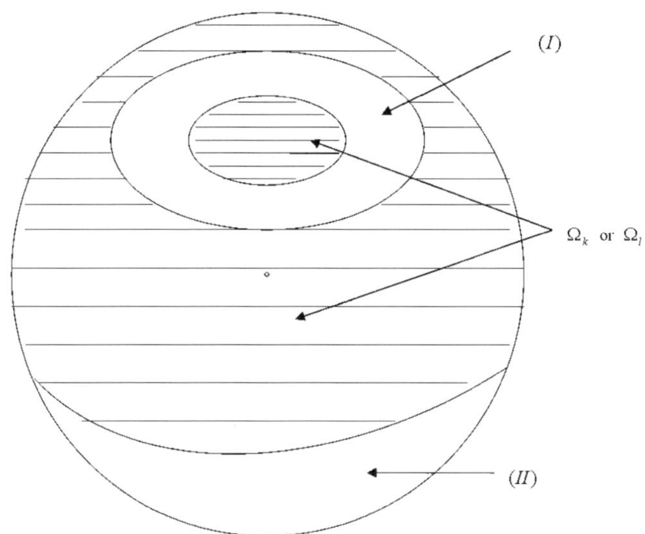

$$(I)\Omega_\ell \text{ or } \Omega_k$$

For $y \in (II)$, the estimate on $\int \frac{|\omega_k||\omega_\ell^4|}{|x-y|}$ does not require $\lambda_\ell \geq \lambda_k$ since then

$$\frac{1}{|x-y|} \leq \frac{C}{|y-x_\ell|} + \frac{C}{|y-x_k|}$$

for $x \in \Omega_\ell' \cup \Omega_k'$ and if $x \in (\Omega_\ell^1 \cup \Omega_k^1)^c$, the proof stated above does not use the fact that $\lambda_\ell \geq \lambda_k$.

Removing (II), we derive that we need to prove the estimate only for $y \in (I)$. But this rescales easily and we may assume, after rescaling, that $\lambda_k \geq \lambda_k$. In fact, we can even asume that $\lambda_\ell = 2\lambda_k$. Then, in (I), $\lambda_\ell|y - x_\ell|$ and $\lambda_k|y - x_k|$ are larger than 1. $\qquad\square$

Next, we have the estimate:

Lemma 41

$$\int \frac{\delta_i^4 |v_i + h_i|}{|x - y|} \leq C \left(|v_i|_{L^6} + \frac{|h_i|_\infty}{\sqrt{\lambda_i}} \right) \delta_i.$$

Proof. The estimate on $\int \frac{\delta_i^4}{|x-y|}$ has been completed in the proof of Proposition 2.

For h_i, we split between the case

    a. $|y - x_i| \geq \frac{C}{\lambda_i}$. Then,

$$\int \frac{\delta_i^4}{|x - y|} \leq \int_{|x-y| \leq c|y-x_i|} \frac{\delta_i^4}{|x - y|} + \int_{|x-y| \geq c|y-x_i|} \frac{\delta_i^4}{|x - y|} \leq C\delta_i^4(y)|y - x_i|^2$$

$$+ \frac{1}{c|y - x_i|} \int \delta_i^4 \leq \frac{C}{\lambda_i^2 |y - x_i|^2} + \frac{C\delta_i}{\sqrt{\lambda_i}} \leq \frac{C_1 \delta_i}{\sqrt{\lambda_i}}$$

    b. $|y - x_i| \leq \frac{C}{\lambda_i}$.
    Then,

$$\int \frac{\delta_i^4}{|x - y|} \leq \int_{|x-y| \geq c|y-x_i|} \frac{\delta_i^4}{|x - y|} + \int_{|x-y| \leq c|y-x_i| \leq \frac{Cc}{\lambda_i}} \frac{\delta_i^4}{|x - y|}$$

$$\leq \int_{|x-y| \geq c|y-x_i|} \frac{\delta_i^4}{|x - y|} + C\delta_i^4(y)|y - x_i|^2$$

$$\leq \int_{\substack{|x-y| \geq c|y-x_i| \\ |x-y| \geq \frac{C}{\lambda_i}}} \frac{\delta_i^4}{|x - y|} + \int_{\substack{|x-y| \geq c|y-x|_i| \\ |x-y| \leq \frac{C}{\lambda_i}}} \frac{\delta_i^4}{|x - y|} + C\delta_i^4(y)|y - x_i|^2$$

$$\leq \frac{C\lambda_i}{(1 + \lambda_i^2 |y - x_i|^2)^{1/2}} \times \frac{1}{\lambda_i} + \frac{C\delta_i^4(y)}{\lambda_i^2} + C\delta_i^4(y)|y - x_i|^2 \leq \frac{C\delta_i}{\sqrt{\lambda_i}}.$$

Lemma 41 follows. $\qquad\qquad\qquad\qquad\qquad\qquad\qquad\qquad\qquad\quad$ $\square$

Next, we have

Lemma 42

$$\left| \tilde{\theta}_h - O\left(\int_{\Omega_i'^c} \frac{\delta_i^4 |v|}{|x - y|} + \int_{\Omega_i'^c} \frac{\delta_i^5}{|x - y|} \right) \right| \leq C_h \sum \varepsilon_{j\ell} \delta_j$$

with $|\tilde{\theta}_h|_{H^1} \leq C_h(\sum \varepsilon_{k\ell}^h)$ *for* $Y \in \Omega_c = (U\Omega_\ell)^c$.

Proof. We use (i) of Proposition 2 and replace $|v|$ by $\sum \delta_\ell$. The price to pay is the introduction of the function $\tilde{\theta}_h$.

We thus need to take care of

$$\int_{\Omega_i'^c} \frac{\delta_i^4 \delta_\ell}{|x-y|}.$$

For $\ell \neq i$, since $y \in \Omega^c = (U\Omega_k)^c$, Lemma 40 provides the estimate.

For $\ell = i$, we notice that $\Omega_i'^c$ is made of two parts; one where $\lambda_i |x - x_i| \geq \frac{1}{8\varepsilon_{is}}$ and another one where $\lambda_j |x - x_j|$ with $\lambda_j \geq \lambda_i$. On the latter sets,

$$\delta_i \leq C\delta_j.$$

Thus,

$$\int_{\Omega_i'^c} \frac{\delta_i^5}{|x-y|} \leq C \int \frac{\delta_j^4 \delta_i}{|x-y|}$$

and the estimate follows from Lemma 40.

We thus assume that $x \in \Omega_i^{*c} = \{x \text{ such that } \lambda_i |x - x_i| \geq \frac{1}{8\varepsilon_{is}}\}$.

We then have

$$\int_{\Omega_i^*} \frac{\delta_i^5}{|x-y|} \leq \int_{\Omega_i^{*c} \cap \{|x-y| \geq \frac{1}{8}|y-x_i|\}} \frac{\delta_i^5}{|x-y|}$$

$$+ C\delta_i^5(y) \int_{\Omega_i^{*c} \cap \{|x-y| \leq \frac{1}{8}|y-x_i|\}} \frac{1}{|x-y|} = (1) + (2).$$

Let

$$A = \Omega_i^{*c} \cap \{|x - y| \geq \frac{1}{8}|y - x_i|\}.$$

Then,

$$(1) \leq \left(\int_A \delta_i^5\right) \sqrt{\lambda_i}\delta_i(y) = O(\varepsilon_{i\ell}\delta_i) \text{ if } |y - x_i| \geq \frac{C}{\lambda_i}$$

$$(1) \leq C\sqrt{\lambda_i} \int_{\Omega_i^{*c}} \delta_i^6 \leq C\varepsilon_{i\ell}^3 \sqrt{\lambda_i} = O(\varepsilon_{i\ell}\delta_i) \text{ if } |y - x_i| \leq \frac{C}{\lambda_i}$$

since, on Ω_i^{*c},

$$|x - y| \geq |x - x_i| - |y - x_i| \geq \frac{1}{\lambda_i}\left(\frac{1}{\varepsilon_{is}} - C\right) \geq \frac{1}{2\lambda_i\varepsilon_{is}}.$$

On the other hand,

$$(2) \leq C\delta_i^5(y)|y - x_i|^2 \leq C_1\delta_i \frac{\lambda_i^2 |y - x_i|^2}{(1 + \lambda_i^2 |y - x_i|^2)^2} \leq C_1'\delta_i\varepsilon_{is}$$

since, if

$$|x - y| \leq \frac{1}{8}|y - x_i|$$

then,

$$\frac{9}{8}|y - x_i| \geq |x - x_i|$$

and

$$\lambda_i |y - x_i| \geq \frac{8}{9}\lambda_i |x - x_i| \geq \frac{8}{9\varepsilon_{is}}.$$

$\square$

In the sequel, we introduce on Ω_i the two quantities:

$$|h_i^0|_\infty = C\left(\frac{1}{\rho_i}\int_{\Omega_i}|f| + \sqrt{\lambda_i}\int_{B_i^c}|f|\delta_i\right). \tag{1.21}$$

$$|h_{i,j}^1|_\infty = C\left(\sqrt{\lambda_j}\int_{B_j}|f| + \frac{1}{\varepsilon_{ij}}\int_{\Omega_i}|f|\delta_j\right) \tag{1.22}$$

so that, according to Lemma 12, we have

$$|\bar{h}_i(y)| \leq |h_i|_\infty + \sum_{\substack{\lambda_j \geq \lambda_i \\ \lambda_j \neq i}} |h_{ij}^1|_\infty \delta_j(y) \quad \text{for } y \in \widetilde{\Omega}_i. \tag{1.23}$$

We then improve (iii) of Proposition 2 into:

Proposition 2′

$$|v - \theta_h|(x) \leq C_h\left(\sum \varepsilon_{k\ell} + |v_\ell|_{L^6} + \frac{|h_\ell^0|_\infty}{\sqrt{\lambda_\ell}} + \sum_{\substack{\lambda_s \geq \lambda_\ell \\ s \neq \ell}} \varepsilon_{s\ell}|h_{\ell s}^1|_\infty\right)\delta_\ell(x)$$

$$\forall x \in \Omega_c = (U\Omega_i)^c \tag{iii′}$$

The proof of (iii) requires a slight modification for (iii').
We now have a new term to estimate:

$$|h_{is}^1|_\infty \int_{\Omega_i}\frac{\delta_\ell^4 \delta_s}{|x - y|}.$$

We will see later (Lemma 68, below) that $\delta_\ell \leq C\delta_i$ on Ω_i so that our estimate becomes an estimate on

$$|h^1_{is}|_\infty \int_{\Omega_i} \frac{\delta^4_i \delta_s}{|x-y|}$$

which we split as usual between the various cases $\lambda_i|y-x_i| \geq C$, $\lambda_i|y-x_i| \leq C$ etc. For $\lambda_i|y-x| \geq 1$, this becomes an estimate on $\sqrt{\lambda_i}|h^1_{is}|_\infty \int_{\Omega_i} \delta^4_i \delta_{y,\lambda_i}(x)\delta_s \leq \sqrt{\lambda_i}|h^1_{is}|_\infty \left(\int \delta^4_i \delta^2_{y,\lambda_i}\right)^{1/2} \left(\int \delta^4_i \delta^2_s\right)^{1/2} \leq C|h^1_{is}|_\infty \varepsilon_{is}\delta_i(y)$.

If $\lambda_i|y-x| \leq 1$, the estimate is straightforward.

We now establish the following sequence of Lemmas and directly improve of the estimates of Section 1.1. The Morse Lemma at infinity does not follow from these improvements, but this will bring us a step closer.

We will prove:

Lemma 43   *Let $\widetilde{\Omega}_c$ be the complement of $U_i\widetilde{\Omega}_i$. then,*

$$\int_{\widetilde{\Omega}_c} \delta^5_\ell \delta_i \leq C \sum_{\lambda_j \geq \lambda_\ell} \varepsilon_{ij}\varepsilon^2_{js}.$$

We will then prove:

Lemma 44   *For $\ell \neq i$,*

$$\int_{\Omega_\ell} \delta_i|v|^5 \leq C\varepsilon_{i\ell}\left(|v_\ell|_{H^1_0} + \frac{|h^0_\ell|_\infty}{\sqrt{\lambda_\ell}} + \sum_{\substack{\lambda_s \geq \lambda_\ell \\ s\neq \ell}} \varepsilon_{s\ell}|h^1_{\ell,s}|_\infty\right)$$

$$+ C \sum_{\substack{\lambda_k \geq \lambda_\ell \\ k\neq \ell}} \varepsilon_{ik}\varepsilon^2_{\ell k} + C\sum \varepsilon_{i\ell}\varepsilon^2_{k\ell} + C_h\left(\sum \varepsilon_{st}\right)^h.$$

Lemma 45   *Assume $\lambda_k \geq \lambda_\ell, \lambda_k \neq \lambda_\ell$. Then*

$$\int_{\Omega_\ell \cap \Omega^c_i} \delta_i \delta^5_k \leq C(\varepsilon_{i\ell}\varepsilon^2_{k\ell} + \varepsilon_{ik}\varepsilon^2_{k\ell}) \leq C_1\varepsilon_{i\ell}\varepsilon_{k\ell}.$$

Lemma 46   *Let $\lambda_k \geq \lambda_\ell$. Then,*

$$\int_{\Omega_\ell} \delta_i \delta^4_k \leq \frac{C\varepsilon_{i\ell}}{\sqrt{\lambda_\ell}}.$$

We turn them to our previous estimates of Section 1 and derive the following intermediate improvements:

Lemma 47

$$(i) \quad \sqrt{\lambda_j}\varepsilon_{ij} \int_{B_j} |v_j + h_j|^5 \leq C\varepsilon_{ij}\left(|v_j|_{H_0^1} + \frac{|h_j^0|_\infty}{\sqrt{\lambda_j}} + \sum_{\lambda_s \geq \lambda_j} \varepsilon_{sj}|h_{j,s}^1|_\infty\right)$$

$$+ C\varepsilon_{ij}\left(\sum_{\lambda_k \geq \lambda_j} \varepsilon_{kj}^{5/2} + \varepsilon_{jt}^{5/2}\right) + C_h\left(\sum \varepsilon_{st}\right)^h \sqrt{\varepsilon_{ij}}$$

$$(ii) \quad \frac{1}{\rho_i\sqrt{\lambda_i}} \int_{\Omega_i} |v_i + h_i|^5 \leq C\varepsilon_{iq}\left(|v_i|_{H_0^1} + \frac{|h_i^0|_\infty}{\sqrt{\lambda_i}} + \varepsilon_{ij}|h_{ij}^1|_\infty\right)$$

$$+ C\varepsilon_{ij}\sum_{\lambda_k \geq \lambda_j}\varepsilon_{kj}^{5/2} + C_h(\sum \varepsilon_{st})^h\sqrt{\varepsilon_{ij}}.$$

Lemma 48

$$\int |\bar{v}|^5|L^{-1}e_\ell| \leq C(|\bar{v}_\ell|_{H_0^1}^2 + o\left(\frac{|h_i^0|_\infty}{\sqrt{\lambda_\ell}}\right) + o\left(\sum \varepsilon_{\ell s}|h_{\ell,s}^1|_\infty\right)$$

$$+ C\sum \varepsilon_{\ell k}^3 + \sum \varepsilon_{\ell m}\left(|v_n|_{H_0^1} + \frac{|h_m^0|_\infty}{\sqrt{\lambda_m}} + \sum_{\lambda_m \geq \lambda_\ell} \varepsilon_{m\ell}|h_{\ell,m}^1|_\infty\right)$$

$$+ C\sum \varepsilon_{\ell k}\varepsilon_{ks}^2 + C_h\left(\sum \varepsilon_{st}\right)^h.$$

Lemma 49

$$\int \omega_s^4|k^*||L^{-1}e_\ell| \leq C\sum \varepsilon_{\ell k}\varepsilon_{ks} + \left(\sum \varepsilon_{st}\right)^h.$$

Lemma 50

$$\int |\omega_\ell|^3|\bar{v}_\ell + \bar{h}_\ell|^2|L^{-1}e_\ell| \leq C\left(|\bar{v}_\ell|_{H_0^1}^2 + \frac{1}{\lambda_\ell\rho_\ell^2}\left(\int_{\Omega_\ell} |f|\right)^2\right)$$

$$+ \sum\nolimits_{\substack{\lambda_k \geq \lambda_\ell \\ k \neq \ell}} \left[\lambda_k\varepsilon_{\ell k}^2\left(\int_{B_k} |f|\right)^2 + \left(\int_{B_\ell^c} |f|\delta_\ell\right)^2 + \left(\int_{B_\ell} |f|\delta_k\right)^2\right]$$

$$+ C\sum \varepsilon_{\ell m}^2\varepsilon_{k\ell} + C_h\left(\sum \varepsilon_{st}\right)^h \varepsilon_{\ell m}\right).$$

Lemma 51

$$\int_{\Omega_i^c} |\bar{v}|^5 \delta_i + \int_{\Omega_i} |\bar{v}|^5 \delta_j + 0\left(\varepsilon_{ij}\sqrt{\lambda_j}\right) \int_{B_j} |\bar{v}|^5$$

$$\leq C \sum_{\ell \neq i} \varepsilon_{i\ell} \left(|v_\ell|_{H_0^1} + \frac{|h_\ell^0|_\infty}{\sqrt{\lambda_\ell}} + \sum_{\substack{\lambda_s \geq \lambda_\ell \\ s \neq \ell}} \varepsilon_{s\ell} |h_{\ell,s}^1|_\infty \right)$$

$$+ C\varepsilon_{ij} \left(|v_i|_{H_0^1} + \frac{|h_i^0|_\infty}{\sqrt{\lambda_i}} + \sum_{\substack{\lambda_s \geq \lambda_i \\ s \neq i}} \varepsilon_{si} |h_{i,s}^1|_\infty \right)$$

$$+ C\left(\sum \left(\varepsilon_{jk}\varepsilon_{ks}^2 + \varepsilon_{i\ell}\varepsilon_{k\ell}^2\right)\right) + C_h \sum \varepsilon_{st}^h.$$

Lemma 52   *For $\theta = \delta_i$ or δ_j :*

$$\int_{(U\Omega_m)^c} \left(\sum \alpha_\ell \omega_\ell\right)^4 |k^*| \theta + \sqrt{\lambda_j} 0(\varepsilon_{ij}) \int_{(U\Omega_m)^c \cap B_j} \left(\sum \alpha_\ell \omega_\ell\right)^4 |k^*|$$

$$\leq C\left(\sum \varepsilon_{i\ell}\varepsilon_{\ell s}^2 + \sum \varepsilon_{j\ell}\varepsilon_{\ell s}^2\right) + C_h \left(\sum \varepsilon_{st}\right)^h.$$

We then improve Lemmas 8, 9, 11:

Lemma 53

$$\sum \int_{\Omega_i} \omega_j^4 |\bar{h}_i| + |\bar{h}_i|^5 \leq \frac{C}{\sqrt{\lambda_i}} \left(\frac{|\bar{h}_i^0|_\infty}{\sqrt{\lambda_i}} + \sum_{\substack{\lambda_s \geq \lambda_i \\ s \neq i}} \varepsilon_{is} |\bar{h}_{i,s}^1|_\infty \right)$$

$$+ C \sum \frac{\varepsilon_{si}^2}{\sqrt{\lambda_i}} + C_h \frac{\left(\sum \varepsilon_{st}\right)^h}{\sqrt{\lambda_i}\varepsilon_{im}}.$$

Lemma 54

$$\int (\omega_j^4 |\bar{h}_i| + |\bar{h}_i|^5)|w|$$

$$\leq C|w|_{H_0^1} \left(\frac{|\bar{h}_i^0|_\infty}{\sqrt{\lambda_i}} + \sum_{\substack{\lambda_s \geq \lambda_i \\ s \neq i}} \varepsilon_{is} |\bar{h}_{i,s}^1|_\infty + \varepsilon_{it}^{5/2} + C_h \left(\sum \varepsilon_{st}\right)^h \right).$$

Lemma 55

$$\int_{\Omega_i^c} \omega_\ell^4 (|\bar{v}_k| + |\bar{h}_k| + |k^*| + |\bar{v}_k|^5 + |\bar{h}_k| + |k^*|^5)|\partial\omega_i|$$

$$\leq C \left(|\omega_i^\infty| + \varepsilon_{it} + \sum_{\substack{\lambda_j \geq \lambda_i \\ j \neq i}} \frac{1}{\lambda_i |a_i - a_j|} \right) \sum \varepsilon_{i\ell}\varepsilon_{\ell s}^2 + C_h \left(\sum \varepsilon_{st} \right)^h.$$

In view of all these new estimates, we revisit the proof of Lemmas 31–34 of Section 1.1. Observe that now we can use Proposition 1 which yields (for $\lambda_i \geq \lambda_j$)

$$\left(\int_{\Omega_i} |\omega_j|^{24/5}|\omega_i|^{6/5} + |\omega_i|^{24/5}|\omega_j|^{6/5} \right)^{5/6} \leq C(|\underline{\omega}_j(\bar{a}_i)|\varepsilon_{ij} + \varepsilon_{ij}^3)$$

while for $\ell, k \neq i$,

$$\left(\int_{\Omega_i} |\omega_\ell|^{24/5}|\omega_k|^{6/5} + |\omega_k|^{24/5}|\omega_\ell|^{6/5} \right)^{5/6} = 0(\varepsilon_{ki}^2 + \varepsilon_{\ell i}^2).$$

Observe also that if $\lambda_k \geq \lambda_\ell$:

$$\bullet \sqrt{\lambda_i}\varepsilon_{i\ell}(|\underline{\omega}_k(\underline{a}_\ell)| + |\underline{\omega}_\ell(\underline{a}_k)|)\varepsilon_{k\ell} = o\left(\sqrt{\lambda_\ell}|\underline{\omega}_k(\underline{a}_\ell)|\varepsilon_{k\ell} + \sqrt{\lambda_k}|\underline{\omega}_\ell(\underline{a}_k)|\varepsilon_{k\ell} \right).$$

For $\lambda_k \leq \lambda_\ell$, Proposition 1 yields an improvement throughout our estimates since $(|\underline{\omega}_k(\underline{a}_\ell)| + |\underline{\omega}_\ell(\underline{a}_k)|)\varepsilon_{k\ell}$ is replaced by $|\underline{\omega}_k(\underline{a}_\ell)|\varepsilon_{k\ell}$ for the contribution on Ω_ℓ and this yields for

$$\bullet \sqrt{\lambda_i}\varepsilon_{i\ell}\underline{\omega}_k(\underline{a}_\ell)|\varepsilon_{k\ell} = o\left(\sqrt{\lambda_\ell}|\underline{\omega}_k(\underline{a}_\ell)\varepsilon_{k\ell} \right).$$

Next, we have:

Lemma 56   *Assume $k \neq s$, $\int \omega_k^4|\omega_s||\omega_\ell| = 0(\varepsilon_{s\ell}^2)$ if one of k or s is ℓ. If both are different from ℓ, then*

$$\int \omega_k^4|\omega_s||\omega_\ell| \leq C\varepsilon_{s\ell}(|\underline{\omega}_k(\underline{a}_s)|\varepsilon_{ks} + \varepsilon_{ks}^2) + C\varepsilon_{\ell k}(|\underline{\omega}_s(\underline{a}_k)|\varepsilon_{ks} + \varepsilon_{ks}^2)$$

$$+ C(|\underline{\omega}_k(\underline{a}_\ell)|\varepsilon_{k\ell} + \varepsilon_{k\ell}^3)(\ if\ \lambda_k \leq \lambda_\ell) + C((|\underline{\omega}_k(\underline{a}_\ell)| + |\underline{\omega}_\ell(\underline{a}_k)|)\varepsilon_{k\ell} + \varepsilon_{k\ell}^3)(\ if\ \lambda_\ell \leq \lambda_k)$$

$$+ C\varepsilon_{s\ell}\left((|\underline{\omega}_s(\underline{a}_k)| + |\underline{\omega}_k(\underline{a}_s)|)\varepsilon_{sk} + \varepsilon_{sk}^2\right)(\ if\ \lambda_k \geq \lambda_s)$$

$$+ C\sum_{\lambda_j \geq \lambda_k} \varepsilon_{\ell j}\varepsilon_{jm}^2 + C\sum_{\lambda_j \geq \lambda_s} \varepsilon_{j\ell}\varepsilon_{j\ell}^2 + C\varepsilon_{\ell k}((|\underline{\omega}_s(\underline{a}_k)| + |\underline{\omega}_k(\underline{a}_s)|)\varepsilon_{sk} + \varepsilon_{jk}^2)(\ if\ \lambda_k \leq \lambda_s).$$

This allows us to derive the better estimate:

Theorem 1

$$|\bar{v}_i|_{H_0^1} + \frac{1}{\rho_i\sqrt{\lambda_i}}\int_{\Omega_i}|f| + \int_{\Omega_i}|f|\delta_j + \int_{B_i^c}|f|\delta_1 + \sqrt{\lambda_j}\varepsilon_{ij}\int_{B_j}|f| \leq$$

$$C\left(\sum_\ell(|\underline{\omega}_\ell(\bar{a}_i)|\varepsilon_{i\ell} + \varepsilon_{i\ell}^3) + \sum_\ell \varepsilon_{i\ell}\varepsilon_{\ell s}^2\right) + C_h(\sum \varepsilon_{st})^h.$$

We move now to control better the remainder terms in the expansion of $\partial J(\sum \alpha_i\omega_1 + \bar{v})\cdot\partial\varphi_i\cdot\partial\varphi_i$ is $\tilde{\omega}_i = \frac{\partial\omega_i}{\partial a_i}$. We establish:

Lemma 57 *For $\ell \neq i$:*

(1) $\int_{\Omega_\ell}|\partial\varphi_i|\delta_\ell^4(|v_\ell| + |h_\ell|)$

$$\leq C\sqrt{\lambda_\ell}\left(|v_\ell|_{L^6} + \frac{|h_\ell^0|_\infty}{\sqrt{\lambda_\ell}} + \sum_{\subset\lambda_s\geq\lambda_\ell s\neq\ell}\varepsilon_{\ell s}|h_\ell^s|_\infty\right)\sqrt{\lambda_i}\varepsilon_{i\ell}^2 + \frac{C}{\lambda_\ell|a_\ell-a_s|}$$

$$\left(\frac{|\omega_i^\infty|^2}{\lambda_i|a_i-a_s|^2} + \sum_{\subset\lambda_s\geq\lambda_\ell s\neq\ell}\lambda_\ell\varepsilon_{\ell s}^2|h_{\ell,s}^1|_\infty^2 + \lambda_i\varepsilon_{i\ell}^4\right)$$

$$+C\sum\lambda_k\left(|v_k|_{H_0^1}^2 + \frac{|h_k^0|_\infty^2}{\lambda_k} + \sum_{\subset\lambda_s\geq\lambda_k s\neq k}|h_{k,s}^1|_\infty^2\varepsilon_{ks}^2\right)$$

$$+C\left(\sum_{k\neq i}\lambda_i|\omega_k(a_i)|^2\varepsilon_{ki}^2 + \varepsilon_{ki}^2\varepsilon_{k\ell}^4 + \sum_{k\neq\ell}\lambda_\ell|\omega_k(a_\ell)|^2\varepsilon_{k\ell}^2 + \varepsilon_{k\ell}^2\varepsilon_{kt}^4\right)$$

$$+C\lambda_\ell\left(\sum\varepsilon_{\ell m}(\sum\varepsilon_{kt})^2 + \sum\varepsilon_{k\ell}^2\varepsilon_{kt}^4\right)$$

$$+C\lambda_i\left(\sum\varepsilon_{ki}^2\varepsilon_{k\ell}^4 + \sum\varepsilon_{im}^4(\sum\varepsilon_{st})^2\right) + C\lambda_k(\sum\varepsilon_{mk}^2\varepsilon_{kt}^4 + \sum\varepsilon_{mk}^4\varepsilon_{st}^2).$$

(2) For $s \neq \ell$, $\int_{\Omega_\ell}|\partial\varphi_i|\delta_s^4(|v_\ell| + |h_\ell|) \leq C\sqrt{\lambda_s}\varepsilon_{s\ell}^2\sqrt{\lambda_i}\varepsilon_{i\ell}(|v_\ell|_{H_0^1} + \frac{|h_\ell^0|_{L^\infty}}{\sqrt{\lambda_\ell}}$

$$+ \sum_{\lambda_s\geq\lambda_\ell}\varepsilon_{\ell s}|h_\ell^s|_\infty) + \frac{C}{\lambda_\ell|a_\ell-a_s|}\left(\frac{|\omega_i^\infty|^2}{\lambda_i|a_i-a_s|^2} + \sum_{\substack{\lambda_s\geq\lambda_\ell\\s\neq\ell}}\lambda_\ell\varepsilon_{\ell s}^2|h_{\ell,s}^1|_\infty^2 + \lambda_i\varepsilon_{i\ell}^4\right)$$

$$+C\sum\lambda_k\left(|v_k|_{H_0^1}^2 + \frac{|h_k^0|_\infty^2}{\lambda_k} + \sum_{\substack{\lambda_s\geq\lambda_k\\s\neq k}}|h_{k,s}^1|_\infty^2\varepsilon_{ks}^2\right)$$

$$+C\left(\sum_{k\neq i}\lambda_i|\omega_k(a_i)|^2\varepsilon_{ki}^2 + \varepsilon_{ki}^2\varepsilon_{k\ell}^4 + \sum_{k\neq\ell}\lambda_\ell|\omega_k(a_\ell)|^2\varepsilon_{k\ell}^2 + \varepsilon_{k\ell}^2\varepsilon_{kt}^4\right)$$

$$+C\lambda_\ell\left(\sum\varepsilon_{\ell m}(\sum\varepsilon_{kt})^2 + \sum\varepsilon_{k\ell}^2\varepsilon_{kt}^4\right)$$

$$+C\lambda_i\left(\sum\varepsilon_{ki}^2\varepsilon_{k\ell}^4 + \sum\varepsilon_{im}^4(\sum\varepsilon_{st})^2\right) + C\lambda_k(\sum\varepsilon_{mk}^2\varepsilon_{kt}^4 + \sum\varepsilon_{mk}^4\varepsilon_{st}^2).$$

(3) $\varepsilon_{ks} \int_{\Omega^c} |\partial\varphi_i| \delta_\ell^4 \delta_s \leq C(\varepsilon_{ks}(\sqrt{\lambda_s}\varepsilon_{st}\sqrt{\lambda_i}\varepsilon_{i\ell}^2) + \sqrt{\lambda_i\lambda_s}\varepsilon_{im}^2\varepsilon_{mt}($ for $\lambda_m \geq \lambda_s) + \lambda_s^{1/4}\sqrt{\lambda_i}\varepsilon_{\ell i}^2\lambda_\ell^{1/4}\varepsilon_{\ell t} + \lambda_s^{1/4}(\lambda_i\varepsilon_{it})^3 + (\lambda_i^{1/4}\varepsilon_{it})^{5/2}(\lambda_s^{1/4}\varepsilon_{sq})^{3/2})$

(4) $\varepsilon_{ks} \int_{\Omega^c} |\partial\varphi_i| \delta_i^4 \delta_s \leq \varepsilon_{ks}\sqrt{\lambda_i} \int_{\Omega^c} \delta_i^6 \delta_s \leq \varepsilon_{ks}\sqrt{\lambda_i\lambda_s}\varepsilon_{it}^3($ for $\lambda_s \leq \lambda_s) + \varepsilon_{ks}\lambda_i(\sum_{\lambda_s \geq \lambda_i} \varepsilon_{sj}\varepsilon_{js}^2)($ for $\lambda_s \geq \lambda_i)$

Observation. All inequalities can be transformed into o(upperbound) i.e. an ε can be inserted as a coefficient of the upperbound. For 1 and 2 there are at various spots in the upperbound an $\varepsilon_{it}^{1-\delta}, \delta > 0$ as small as we wish which can be inserted.

Proof.   We start with 3. We split Ω^c into two pieces, Ω^{1c} and Ω^{2c} adapted to δ_s. In $\Omega^{1c}, \lambda_s|x - x_s| \geq \frac{c}{\varepsilon_{st}}$ for some $t \neq s$. In Ω^{2c}, for some $m \neq s$ with $\lambda_m \geq C\lambda_s(\lambda_m \gg \lambda_s), \lambda_s|x - x_s| \leq \frac{c}{\varepsilon_{ms}}$ and $\lambda_m|x - x_m| \leq \frac{1}{8\varepsilon_{ms}}$ so that, in Ω^{2c},

$$\delta_s \leq \delta_m$$
$$|x - x_m| = o(|x_s - s_m|)$$
$$|x - x_s|'' \sim'' |x_m - x_s|.$$

We have

$$\varepsilon_{ks} \int_{\Omega^{1c}} |\partial\varphi_i| \delta_\ell^4 \delta_s \leq C\sqrt{\lambda_i}\varepsilon_{ks}\sqrt{\lambda_s}\varepsilon_{st} \int_{\Omega^{1c}} \delta_i^2\delta_\ell^4 \leq C\left(\sqrt{\lambda_s}\varepsilon_{ks}\varepsilon_{st}\right)\left(\sqrt{\lambda_i}\varepsilon_{i\ell}^2\right). \qquad \square$$

In Ω^{2c}, since $\lambda_m \gg \lambda_s$,

$$\frac{1}{c\varepsilon_{mt}} \leq \lambda_m|x - x_m| \leq \frac{c}{\varepsilon_{ms}} \quad \text{for some } t \neq m.$$

We claim and we will prove later that:

$$\int_{\Omega_c \cap \{|x-x_i| \geq c|x_i-x_\ell|\}} \delta_i^2\delta_\ell^4 \leq C\varepsilon_{i\ell}^2\varepsilon_{\ell t}^2(\text{or } C\varepsilon_{ij}^2\varepsilon_{jt}^2 \text{ with } \lambda_j \geq \lambda_\ell)$$

and

$$\int_{\Omega_c \cap \{|x-x_i| \leq c|x_i-x_\ell|\}} \delta_i^2\delta_\ell^4 \leq \frac{C\varepsilon_{i\ell}^2}{\lambda_\ell|x_i - x_\ell|}.$$

In order to estimate $\varepsilon_{ks}\sqrt{\lambda_i} \int_{\Omega_c^2} \delta_i^2\delta_\ell^4\delta_s$, we will distinguish between two cases: either $\lambda_\ell \leq \lambda_m$ and $\delta_\ell \leq \delta_m$.

Then,

$$\varepsilon_{ks}\sqrt{\lambda_i}\int_{\Omega_c^2\cap\{|x-x_i|\geq c|x_i-x_m|\}}\delta_i^2\delta_m^4\delta_s \leq C\varepsilon_{ks}\sqrt{\lambda_i\lambda_s}\varepsilon_{im}^2\varepsilon_{mt}^2$$

$$\varepsilon_{ks}\sqrt{\lambda_i}\int_{\Omega_c^2\{|x-x_i|\leq c|x_i-x_m|\}}\delta_i^2\delta_m^4\delta_s \leq C\frac{\varepsilon_{ks}\sqrt{\lambda_i\lambda_s}\varepsilon_{im}^2}{\lambda_m|x_i-x_m|}.$$

In $\Omega_c^2 \cap \{|x - x_i| \leq c|x_i - x_m|\}$,

$$|x - x_m|\text{``}\sim''|x_i - x_m|$$

so that

$$\lambda_m|x_i - x_m| \geq \lambda_m|x - x_m| \geq \frac{1}{c\varepsilon_{mt}}$$

and

$$\varepsilon_{ks}\sqrt{\lambda_i}\int_{\Omega_c^2\cap\{|x-x_i|\leq c|x_i-x_m|\}}\delta_i^2\delta_m^4\delta_s \leq C\varepsilon_{ks}\sqrt{\lambda_i\lambda_s}\varepsilon_{im}^2\varepsilon_{mt}.$$

Or $\lambda_\ell \geq \lambda_m \geq \lambda_s$

$$\varepsilon_{ks}\sqrt{\lambda_i}\int_{\Omega_c^2\cap\{|x-x_i|\geq c|x_i-x_\ell|\}}\delta_i^2\delta_\ell^4\delta_s \leq C\varepsilon_{ks}\sqrt{\lambda_i\lambda_s}\varepsilon_{\ell i}^2\varepsilon_{\ell t}^2$$

$$\varepsilon_{ks}\sqrt{\lambda_i}\int_{\Omega_c^2\cap\{|x-x_i|\leq c|x_i-x_\ell|\}}\delta_i^2\delta_\ell^4\delta_s \leq C\varepsilon_{ks}\lambda_s^{1/4}\sqrt{\lambda_i}\varepsilon_{\ell i}^2\lambda_m^{1/4}\frac{1}{\lambda_\ell|x_i-x_\ell|}.$$

We then observe that, in $\Omega_c^2 \cap \{|x - x_i| \leq c|x_i - x_\ell|\}$,

$$|x_i - x_\ell|\text{``}\sim''|x - x_\ell|$$

so that

$$\lambda_\ell|x_i - x_\ell| \geq c\lambda_\ell|x - x_\ell|.$$

Then, either

$$\lambda_\ell|x - x_\ell| \geq \frac{c}{\varepsilon_{\ell t}}$$

and

$$\varepsilon_{ks}\sqrt{\lambda_i}\int_{\Omega_c^2\cap\{|x-x_i|\le c|x_i-x_\ell|}\delta_i^2\delta_\ell^4\delta_s \le C\varepsilon_{ks}\lambda_s^{1/4}\sqrt{\lambda_i}\varepsilon_{\ell i}^2\lambda_m^{1/4}\varepsilon_{\ell t}$$

or

$$\lambda_j|x-x_j| \le \frac{1}{8\varepsilon_{\ell j}} \text{ for some } \lambda_j > \lambda_\ell.$$

We may then replace δ_ℓ by δ_j and proceed as above.
Lastly, assume that $\lambda_\ell \le \lambda_m$ and $\delta_m \le \delta_\ell$ in a part of Ω^{2c}.
Then,

$$\delta_s \le \delta_m \le \delta_\ell.$$

$$\varepsilon_{ks}\sqrt{\lambda_i}\delta_i^2\delta_\ell^4\delta_s \le \varepsilon_{ks}\lambda_s^{1/4}\sqrt{\lambda_i}\lambda_\ell^{1/4}\delta_i^2\delta_\ell^4$$

so that

$$\varepsilon_{ks}\sqrt{\lambda_i}\int_{\Omega_c^2\cap\{|x-x_i|\ge c|x_i-x_\ell|\}}\delta_i^2\delta_\ell^4\delta_s$$

$$\le \varepsilon_{ks}\lambda_s^{1/4}\sqrt{\lambda_i}\lambda_\ell^{1/4}\int_{\Omega_c^2\cap\{|x-x_i|\ge c|x_i-x_\ell|\}}\delta_i^2\delta_\ell^4 \le C\varepsilon_{ks}\lambda_s^{1/4}\sqrt{\lambda_i}\varepsilon_{i\ell}^2\lambda_\ell^{1/4}\varepsilon_{\ell t}^2$$

and

$$\varepsilon_{ks}\sqrt{\lambda_i}\int_{\Omega_c^2\cap\{|x-x_i|\}}\delta_i^2\delta_\ell^4\delta_s \le C\varepsilon_{ks}\lambda_s^{1/4}\sqrt{\lambda_i}\lambda_\ell^{1/4}\frac{\varepsilon_{i\ell}^2}{\lambda_\ell|x_i-x_\ell|}.$$

In $\Omega_c^2 = \{|x-x_i| \le c|x_i-x_\ell|\}, |x-x_\ell|$ and $|x_i-x_\ell|$ are of the same order so that

$$\lambda_\ell|x_i-x_\ell| \ge c\lambda_\ell|x-x_\ell|.$$

Again, either

$$\lambda_\ell|x-x_\ell| \ge \frac{c}{\varepsilon_{\ell t}}$$

and we upperbound with $C\varepsilon_{ks}\lambda_s^{1/4}\sqrt{\lambda_i}\varepsilon_{i\ell}\lambda_\ell^{1/4}\varepsilon_{\ell t}$ or there exist some j, with $\lambda_j \ge \lambda_\ell, j \ne \ell$ and

$$\lambda_j|x-x_j| \le \frac{c}{\varepsilon_{j\ell}}.$$

Then,

$$\delta_\ell \leq \delta_j$$

and the induction proceeds.

If, at any point, we end up with $\ell = i$, then either $\lambda_s \leq \lambda_i$ and

$$\sqrt{\lambda_i}\varepsilon_{ks}\int_{\Omega^c}\delta_i^6\delta_s \leq C\sqrt{\lambda_i}\sqrt{\lambda_s}\varepsilon_{ks}\varepsilon_{it}^3 \leq C\left(\lambda_i^{1/4}\varepsilon_{it}\right)^3\lambda_s^{1/4}\varepsilon_{ks}$$

and if $\lambda_s \geq \lambda_i$,

$$\sqrt{\lambda_i}\varepsilon_{ks}\int_{\Omega^c}\delta_i^6\delta_s \leq \lambda_i\varepsilon_{ks}\int_{\Omega^c}\delta_i^5\delta_s \leq C\lambda_i\varepsilon_{it}^{5/2}\varepsilon_{ks}\varepsilon_{sq}^{1/2}$$

$$\leq C\left(\lambda_i^{1/4}\varepsilon_{it}\right)^{5/2}\left(\lambda_s^{1/4}\varepsilon_{sq}\right)^{3/2}.$$

We now prove our claim about $\int_{\Omega_c}\delta_i^2\delta_\ell^4$.

If $|x - x_i| \geq c|x_i - x_\ell|$, then

$$\int_{\Omega_c\cap\{|x-x_i|\geq c|x_i-x_\ell|\}}\delta_i^2\delta_\ell^4 \leq \frac{C}{\lambda_i|x_i - x_\ell|^2}\int_{\lambda_\ell|x-x_\ell|\geq\frac{c}{\varepsilon_{\ell t}}}\delta_\ell^4 \leq C\varepsilon_{i\ell}^2\varepsilon_{\ell t}^2$$

unless $\delta_\ell \leq \delta_j$ for a part of the domain of integration, with $\lambda_j \geq \lambda_\ell, j \neq \ell$ and $\lambda_j|x - x_j| \leq \frac{1}{8\varepsilon_\ell}$ in which case we start an induction.

Or, we integrate on $\Omega^c \cap \{|x - x_i| \leq c|x_i - x_\ell|\}$. Then,

$$\int_{\Omega^c\cap\{|x-x_i|\leq c|x_i-x_\ell|\}}\delta_i^2\delta_\ell^4 \leq \frac{1}{\lambda_\ell^2|x_i - x_\ell|^4}\int_{|x-x_i|\leq c|x_i-x_\ell|}\delta_i^2$$

$$\leq \frac{1}{\lambda_\ell^2|x_i - x_\ell|^4}\cdot\frac{1}{\lambda_i^2}\cdot\lambda_i|x_i - x_\ell| \leq \frac{C}{\lambda_i\lambda_\ell^2|x_i - x_\ell|^3}$$

$$\leq \frac{C\varepsilon_{i\ell}^2}{\lambda_\ell|x_i - x_\ell|}.$$

as claimed.

Proof of 1.

$$\int_{\Omega_\ell}|\partial\varphi_i|\delta_\ell^4|v_\ell| \leq C\sqrt{\lambda_i}\int_{\Omega_\ell}\delta_i^2\delta_\ell^4|v_\ell| \leq C\sqrt{\lambda_i}|v_\ell|_{L^6}\left(\int_{\Omega_\ell}\delta_i^{12/5}\delta_\ell^{24/5}\right)^{5/6}$$

$$\leq C\sqrt{\lambda_i}|v_\ell|_{L^6}\left(\left(\int_{\Omega_\ell \cap \{|x-x_i|\leq c|x_i-x_\ell|\}} \delta_i^{12/5}\delta_\ell^{24/5}\right)^{5/6} + \frac{1}{\lambda_i|x_i-x_\ell|^2}\left(\int \delta_\ell^{24/5}\right)^{5/6}\right)$$

$$= (1) + (2).$$

Clearly,

$$\leq \frac{c\sqrt{\lambda_i}|v_\ell|_{L^6}}{\lambda_\ell\lambda_i|x_i-x_\ell|^2}\sqrt{\lambda_\ell} \leq C\sqrt{\lambda_\ell}|v_\ell|_{L^6}\sqrt{\lambda_i}\varepsilon_{i\ell}^2 \tag{1.24}$$

and

$$\leq \frac{C\sqrt{\lambda_i}|v_\ell|_{L^6}}{\lambda_\ell^2|x_i-x_\ell|^4}\left(\int_{|x_i-x_\ell|\leq c|x_i-x_\rho}\delta_i^{12/5}\right)^{5/6}$$

$$= \frac{C_1\sqrt{\lambda_i}|v_\ell|_{L^6}}{\lambda_\ell^2|x_i-x_\ell|^4}\cdot\frac{1}{\lambda_i^{3/2}}(\lambda_i|x-x_i|)^{(3-\frac{12}{5})\cdot\frac{5}{6}} \tag{1.25}$$

$$= \frac{C_1\sqrt{\lambda_i}|v_\ell|_{L^6}}{\lambda_\ell^2|x_i-x_\ell|^4}\cdot\frac{1}{\lambda_i^{3/2}}(\lambda_i|x-x_i|)^{1/2} = \frac{C_1\sqrt{\lambda_i}\sqrt{\lambda_\ell}|v|_{L^6}}{\lambda_i\lambda_\ell|x_i-x_\ell|^2\lambda_\ell^{3/2}|x_i-x_\ell|^{3/2}}$$

$$= \frac{C_1\sqrt{\lambda_i}\varepsilon_{i\ell}^2\sqrt{\lambda_\ell}|v_\ell|_{L^6}}{\lambda_\ell^{3/2}|x_i-x_\ell|^{3/2}}.$$

On the other hand,

$$\int_{\Omega_\ell}|\partial\varphi_i|\delta_\ell^4|h_\ell| \leq C|h_\ell^0|_\infty\sqrt{\lambda_i}\int_{\Omega_\ell}\delta_i^2\delta_\ell^4 + C\sum_{\lambda_s\geq\lambda_\ell s\neq\ell}\int_{\Omega_\ell}|\partial\varphi_i|\delta_\ell^4|h_{\ell,s}^1|_\infty\delta_s$$

$$\leq c\sqrt{\lambda_\ell}\left(\frac{|h_\ell^0|_\infty}{\sqrt{\lambda_\ell}}\right)\sqrt{\lambda_i}\varepsilon_{i\ell}^2 + C\sum_{\lambda_s\geq\lambda_\ell s\neq\ell}\left(\int_{\Omega_\ell}|\partial\varphi_i|\delta_\ell^4\delta_s\right)|h_{\ell,s}^1|_\infty.$$

$$\square$$

We estimate $C\sum\limits_{\lambda_s\geq\lambda_\ell s\neq\ell}\left(\int|\partial\varphi_i|\delta_\ell^4\delta_s\right)|h_{\ell,s}^1|_\infty$ below, after starting the proof of 2. of Lemma 57.

Proof of 2. For $s\neq\ell,\lambda_s\geq\lambda_\ell$

$$\int_{\Omega_\ell}|\partial\varphi_i|\delta_s^4|v_\ell| \leq C\sqrt{\lambda_i\lambda_s}\varepsilon_{s\ell}\int_{\Omega_\ell}\delta_i^2\delta_s^3|v_\ell| \leq C\sqrt{\lambda_i\lambda_s}\varepsilon_{i\ell}\varepsilon_{s\ell}^{3/2+1}|v|_{H^1}$$

$$\leq C\left(\sqrt{\lambda_i}\varepsilon_{i\ell}|v_\ell|_{H^1}\right)\sqrt{\lambda_s}\varepsilon_{s\ell}^{5/2}.$$

On the other hand,

$$\int_{\Omega_\ell} |\partial\varphi_i|\delta_s^4|h_\ell| \leq C\sqrt{\lambda_i\lambda_s}\varepsilon_{s\ell}|h_\ell^0|_\infty \int_{\Omega_\ell} \delta_i^2\delta_s^3 + C \sum_{\lambda_t\geq\lambda_\ell t\neq\ell} \left(\int_{\Omega_\ell} |\partial\varphi_i|\delta_s^4\delta_t\right)|h_{\ell,t}^1|_\infty$$

$$\leq C\sqrt{\lambda_i\lambda_s}\varepsilon_{s\ell}|h_\ell^0|_\infty \left(\int_{\Omega_\ell} \delta_i^6\right)^{1/3}$$

$$\times \left(\int_{\Omega_\ell} \delta_s^{9/2}\right)^{2/3} + C \sum_{\lambda_t\geq\lambda_\ell t\neq\ell} \left(\int_{\Omega_\ell} |\partial\varphi_i|\delta_s^4\delta_t\right)|h_{\ell,t}^1|_\infty$$

$$\leq C\sqrt{\lambda_i\lambda_s}\varepsilon_{i\ell}\frac{|h_\ell^0|_\infty}{\sqrt{\lambda_\ell}}\varepsilon_{s\ell}^2 + C \sum_{\lambda_t\geq\lambda_\ell t\neq\ell} \left(\int_{\Omega_\ell} |\partial\varphi_i|\delta_s^4\delta_t\right)|h_{\ell,t}^1|_\infty$$

$$\leq C\left(\sqrt{\lambda_i}\varepsilon_{i\ell}\frac{|h_\ell^0|_\infty}{\sqrt{\lambda_\ell}}\right)\sqrt{\lambda_s}\varepsilon_{s\ell}^2 + C \sum_{\lambda_t\geq\lambda_\ell t\neq\ell} \left(\int_{\Omega_\ell} |\partial\varphi_i|\delta_s^4\delta_t\right)|h_{\ell,t}^1|_\infty.$$

$$\square$$

The two terms $C \sum_{\lambda_s\geq\lambda_\ell s\neq\ell} |h_{\ell,s}^1|_\infty \int_{\Omega_\ell} |\partial\varphi_i|\delta_\ell^4\delta_s$ and $C \sum_{\lambda_t\geq\lambda_\ell t\neq\ell} |h_{\ell,t}^1|_\infty$ $\times \int_{\Omega_\ell} |\partial\varphi_i|\delta_s^4\delta_t$ are estimated now. Recall that $|\partial\varphi_i| \leq C\left(\lambda_i|\omega_i^\infty|\delta_i^\ell + \delta_i^3\right)$. We estimate first

$$\sqrt{\lambda_i} \int_{\Omega_\ell} \delta_i^2\delta_\ell^4\delta_s, \; \sqrt{\lambda_i} \int_{\Omega_\ell} \delta_i^2\delta_s^4\delta_t.$$

Since we know that $\delta_s \leq C\delta_\ell$ on Ω_ℓ (Lemma 68 below), we need only to estimate $\sqrt{\lambda_i} \int_{\Omega_\ell} \delta_i^2\delta_\ell^4\delta_s$. We divide Ω_ℓ into three subdomains, one where $|x-a_s| \geq c|a_s-a_\ell|$. On the domain, labeled Ω_ℓ^1 we have

$$\sqrt{\lambda_i} \int_{\Omega_\ell^1} \delta_i^2\delta_\ell^4\delta_s \leq \frac{C\sqrt{\lambda_I}}{\sqrt{\lambda_s}|a_s-a_\ell|} \int_{\Omega_\ell^1} \delta_i^2\delta_\ell^4 \leq C\sqrt{\lambda_i}\varepsilon_{i\ell}^2\sqrt{\lambda_\ell}\,\varepsilon_{\ell s}.$$

On Ω_ℓ^2, we have $|x-a_s| \leq c|a_s-a_\ell|, c$ small, so that $|x-a_\ell|$ and $|a_\ell-a_s|$ are of the same order and $|x-a_i| \geq c|a_i-a_\ell|$. We then have, using 3 of Lemma 13:

$$\sqrt{\lambda_i} \int_{\Omega_\ell^2} \delta_i^2\delta_\ell^4\delta_s \leq \frac{C\sqrt{\lambda_i}}{\lambda_i|a_i-a_\ell|^2} \cdot \frac{\varepsilon_{\ell s}}{\sqrt{\lambda_\ell}} \leq C\sqrt{\lambda_i}\varepsilon_{i\ell}^2\sqrt{\lambda_\ell}\,\varepsilon_{\ell s}.$$

On $\Omega_\ell^3, |x-a_\ell|$ and $|a_\ell-a_s|$ are of the same order and $|x-a_\ell|$ and $|a_i-a_\ell|$ are of the same order, $|x-a_i| \leq c(a_i-a_\ell)$. We then upperbound

$$\sqrt{\lambda_i} \int_{\Omega_\ell^3} \delta_i^2\delta_\ell^4\delta_s \leq \frac{C\sqrt{\lambda_i}}{\lambda_\ell^2|a_i-a_\ell|^3|a_\ell-a_s|} \int \delta_i^2\delta_s.$$

Assuming that $i \neq s$, we have

$$\int_{\Omega_\ell^3} \delta_i^2 \delta_s \leq \frac{C}{\lambda_i |a_i - a_s|^2} \int_{\{|x-a_s| \leq \frac{1}{10}|a_i-a_s|\} \cap \Omega_\ell^3} \delta_s$$

$$+ \frac{C}{\sqrt{\lambda_s}|a_i - a_s|} \int_{|x-a_i| \leq \frac{1}{10}|a_i-a_s|} \delta_i^2 + \frac{C}{\lambda_i \sqrt{\lambda_s}} \int_{3|a_\ell-a_s| \geq r \geq 10|a_i-a_s|} \frac{r^2 dr}{r^3}$$

$$+ \frac{C}{\lambda_i \sqrt{\lambda_s}|a_i - a_s|^3} \int_{r \leq 10|a_i-a_s|} r^2 dr \leq \frac{C}{\lambda_i \sqrt{\lambda_s}} \left(1 + \text{Log } \frac{|a_\ell - a_s|}{|a_i - a_s|} \right).$$

and, since $|a_\ell - a_s|$ and $|a_\ell - a_i|$ are of the same order:

$$\sqrt{\lambda_i} \int_{\Omega_\ell^3} \delta_i^2 \delta_\ell^4 \delta_s \leq \frac{C\sqrt{\lambda_i}}{\lambda_i \lambda_\ell^2 |a_i - a_\ell|^3} \sqrt{\lambda_\ell} \varepsilon_{\ell s} \left(1 + \text{Log } \frac{|a_i - a_\ell|}{|a_i - a_s|} \right)$$

$$\leq \frac{C}{\sqrt{\lambda_i}|a_i - a_s|} \cdot \frac{1}{\lambda_\ell^2 |a_i - a_\ell|^2} \frac{|a_i - a_s|}{|a_i - a_\ell|} \left(1 + \text{Log } \frac{|a_i - a_\ell|}{|a_i - a_s|} \right) \cdot \sqrt{\lambda_\ell} \varepsilon_{\ell s}.$$

Since we may assume that $\frac{|a_i-a_s|}{|a_i-a_\ell|} \leq C$ (otherwise the estimate is straightforward), we find

$$\sqrt{\lambda_i} \int_{\Omega_\ell^3} \delta_i^2 \delta_\ell^4 \delta_s \leq \frac{C}{\sqrt{\lambda_i}|a_i - a_s|} \cdot \frac{1}{\lambda_\ell^2 |a_i - a_\ell|^2} \cdot \sqrt{\lambda_\ell} \varepsilon_{\ell s}.$$

Thus,

$$|\omega_i^\infty| \sqrt{\lambda_i} \int_{\Omega_\ell^3} \delta_i^2 \delta_4^2 \delta_s |h_{\ell,s}^1|_\infty \leq \frac{C|\omega_i^\infty|}{\sqrt{\lambda_i}|a_i - a_s|} \cdot \sqrt{\lambda_\ell} \varepsilon_{\ell s} |h_{\ell,s}^1|_\infty \cdot \frac{1}{\lambda_\ell^2 |a_i - a_\ell|^2}.$$

On the other hand, observing that $\delta_i^3 \leq \sqrt{\lambda_i} \delta_i^2$ and using some estimates from above,

$$\int_{\Omega_\ell^3} \delta_i^3 \delta_4^4 \delta_s \leq \frac{C}{\lambda_\ell^2 |a_\ell - a_s|^4} \left(\frac{1}{\sqrt{\lambda_s} \lambda_i} + \frac{1}{\sqrt{\lambda_s} \lambda_i^{3/2}} \int_{10|a_i-a_s| \leq r \leq 3|a_\ell-a_s|} \frac{r^2 dr}{r^4} \right)$$

$$\leq \frac{C}{\lambda_\ell^2 |a_\ell - a_s|^4} \left(\frac{\sqrt{\lambda_i}}{\sqrt{\lambda_s} \lambda_i} + \frac{1}{\lambda_i^{3/2} \sqrt{\lambda_s}} \text{Log } \frac{\lambda_i |a_\ell - a_i|}{\lambda_i |a_i - a_s|} \right)$$

$$\leq \frac{C}{\sqrt{\lambda_i} \lambda_\ell^2 |a_\ell - a_i|^3} \sqrt{\lambda_\ell} \varepsilon_{\ell s} \left(1 + \frac{1}{\sqrt{\lambda_i}} \text{Log } \lambda_i |a_\ell - a_i| \right).$$

We may assume that $|a_\ell - a_i| \leq 1$, using the proper chart. Thus, since $|\partial\varphi_i| \leq C(|\omega_i^\infty|\sqrt{\lambda_i}\delta_i^2 + \delta_i^3)$, for $i \neq s$

$$\int |\partial\varphi_i|\delta_\ell^4 \delta_s \leq C\left(\sqrt{\lambda_i}\varepsilon_{i\ell}^2\sqrt{\lambda_\ell}\varepsilon_{\ell s} + \frac{|\omega_i^\infty|}{\sqrt{\lambda_i}|a_i - a_s|}\sqrt{\lambda_\ell}\varepsilon_{\ell s}\right)$$

and

$$|h_{\ell,s}^1|_\infty \int \partial\varphi_i\delta_\ell^4 \delta_s \leq \frac{C}{\lambda_\ell|a_\ell - a_s|}\left(\frac{|\omega_i^\infty|^2}{\lambda_i|a_i - a_s|^2} + \lambda_\ell\varepsilon_{\ell s}^2|h_{\ell,s}^1|_\infty^2 + \lambda_i\varepsilon_{i\ell}^4\right).$$

For $s = i$, the estimate is more involved. We have

$$|h_{\ell,i}^1|_\infty \int_{\Omega_\ell^3} |\partial\varphi_i||\omega_\ell^4|\delta_i \leq \frac{C}{\lambda_\ell^2|a_i - a_\ell|^4}|h_{\ell,i}^1|_\infty\left(|\omega_i^\infty|\sqrt{\lambda_i}\int_{\Omega_\ell^3} \delta_i^3 + \int_{\Omega_\ell^3} \delta_i^4\right)$$

$\frac{C|h_{\ell,i}^1|_\infty}{\lambda_\ell^2|a_i-a_\ell|^4}\int_{\Omega_\ell^3}\delta_i^4$ is easy to handle and provides the same contribution than before.

On the other hand (we assume that $\lambda_i \gg \lambda_\ell$, otherwise the estimate is straightforward)

$$\frac{C|h_{\ell,i}^1|_\infty}{\lambda_\ell^2|a_i - a_\ell|^4}|\omega_i^\infty|\sqrt{\lambda_i}\int_{\Omega_\ell^3}\delta_i^3 \leq \frac{C|\omega_i^\infty||h_{\ell,i}^1|_\infty}{\lambda_i\lambda_\ell^2|a_i - a_\ell|^4}\int_{\frac{1}{\lambda_i\varepsilon_{i\ell}}\leq r\leq c|a_i-a_\ell|}\frac{r^2 dr}{r^3}$$

$$\leq \frac{C|\omega_i^\infty||h_{\ell,i}^1|_\infty}{\lambda_i\lambda_\ell^2|a_i - a_\ell|^4}\text{Log}\sqrt{\frac{\lambda_i}{\lambda_\ell}} \leq C_1\left(\frac{|\omega_i^\infty|}{\sqrt{\lambda_i}}\varepsilon_{i\ell}|h_{\ell,i}^1|_\infty\,\text{Log}\,\frac{\lambda_i}{\lambda_\ell}\right)\frac{1}{\lambda_\ell^{3/2}|a_i - a_\ell|^3}$$

$$\leq \left(C_1|\omega_i^\infty|\varepsilon_{i\ell}\cdot\varepsilon_{i\ell}|h_{\ell,i}^1|_\infty\,\text{Log}\left(\frac{\lambda_i}{\lambda_\ell}\right)\right)\frac{1}{\lambda_\ell|a_i - a_\ell|^2}.$$

We consider

$$\varepsilon_{i,\ell}|h_{\ell,i}^1|_\infty = \sqrt{\lambda_i}\varepsilon_{i\ell}\int_{B_i}|f| + \int_{\Omega_\ell}|f|\delta_i$$

and we try to estimate directly $\sqrt{\lambda_i}\varepsilon_{i\ell}\int_{B_i}|f| + \int_{\Omega_\ell}|f|\delta_i$. f is made of three types of terms:

1 - terms which are $O\left(\sum_{k\neq j}|\omega_k|^4|\omega_j|\right)$

2 - terms involving $v = v_i + h_i$ in B_i, $v = v_\ell + h_\ell$ in Ω_ℓ.

3 - projection terms

We denote f_1 the contribution of f of the terms of type 1 and 2. Thus, f reads

$$f = f_1 + \sum A_k e_k$$

where e_k could be $\omega_k, \lambda_k \frac{\partial \omega_k}{\partial \lambda_k}, \frac{\partial \omega_k}{\partial \sigma_k}, \frac{1}{\lambda_k} \frac{\partial \omega_k}{\partial a_k}$. A_k is somewhat complicated because it involves the inverse of the matrix of the scalar-product in the bases $\{\omega_i, \dots\}$ but it is essentially $\int f e_k$.

We know that

$$\int_{B_i} |e_i| \le \frac{C}{\sqrt{\lambda_i}}, \int_{B_i} |e_k| \le C \frac{\varepsilon_{ik}^2}{\sqrt{\lambda_i}} \text{ for } k \ne i.$$

Thus,

$$\sqrt{\lambda_i} \varepsilon_{i\ell} \int_{B_i} |f| + \int_{\Omega_\ell} |f| \delta_i \le C \left(\sqrt{\lambda_i} \varepsilon_{i\ell} \int_{B_i} |f_1| + \int_{\Omega_\ell} |f_1| \delta_i \right.$$

$$\left. + \varepsilon_{i\ell} \left(|A_i| + \sum_{k \ne i} \varepsilon_{ik}^2 |A_k| \right) + \sum_k |A_k| \int_{\Omega_\ell} |e_k| \delta_i \right)$$

$$\le C \left(\sqrt{\varepsilon_{i\ell}} \left(\int_{B_i} |f_1|^{6/5} \right)^{5/6} + \int_{\Omega_\ell} |f_1| \delta_i + \varepsilon_{i\ell} \left(|A_i| + \sum_{k \ne i} \varepsilon_{ik}^2 |A_k| \right) \right.$$

$$\left. + \sum_k |A_k| \int_{\Omega_\ell} |e_k| \delta_i \right).$$

In fact, $|A_i|$ stands for several terms which correspond to the various e_i's $\left(\omega_i, \frac{\partial \omega_i}{\lambda_i \partial a_i} \text{ etc.} \right)$ with the same index i.

We use then the fact that $|v - \theta_h| \le C_h \sum \delta_k$ and the fact (see Lemma 68 below) that $\delta_k \le C \delta_m$ on Ω_m to derive

$$\sqrt{\varepsilon_{i\ell}} \left(\int_{B_i} |f_1|^{6/5} \right)^{5/6} + \int_{\Omega_\ell} |f_1| \delta_i$$

$$\le C \sum_{k \ne m} \left(\sqrt{\varepsilon_{i\ell}} \left(\int_{B_i} |\omega_k|^{24/5} |\omega_m|^{6/5} \right)^{5/6} + \int_{\Omega_\ell} |\omega_k|^4 |\omega_m| \delta_i \right)$$

$$+C\sqrt{\varepsilon_{i\ell}}\left(\int_{B_i}\delta_i^{24/5}(|v_i|+|h_i|)^{6/5}\right)^{5/6}+C\int_{\Omega_\ell}\delta_\ell^4(|v_\ell|+|h_\ell|)\delta_i+C_h(\sum\varepsilon_{st})^h.$$

Considering also $\int_{\omega_\ell}|e_k|\delta_i$, either $k=\ell$ and this is $O(\varepsilon_{i\ell})$ or $k=i$ and this is $O(\varepsilon_{i\ell}^3)$ or $k\neq i, k\neq\ell$, then $\delta_k\leq C\delta_\ell$ in Ω_ℓ (Lemma 68 below) and

$$\int_{\Omega_\ell}|e_k|\delta_i\leq\int_{\Omega_\ell}\delta_k^4\delta_\ell\delta_i\leq C\varepsilon_{ik}\varepsilon_{k\ell}.$$

Thus,

$$\sqrt{\lambda_i}\varepsilon_{i\ell}\int_{B_i}|f|+\int_{\Omega_\ell}|f|\delta_i\leq C\left(\varepsilon_{i\ell}\left(|A_i|+\sum_{k\neq i}\varepsilon_{ik}^2|A_k|+|A_\ell|\right)+\right.$$

$$\left.\sum_{\substack{k\neq i\\k\neq\ell}}\varepsilon_{ik}\varepsilon_{k\ell}|A_k|+\sum_{k\neq m}\left(\sqrt{\varepsilon_{i\ell}}\left(\int_{B_i}|\omega_k|^{24/5}|\omega_m|^{6/5}\right)^{5/6}+\int_{\Omega_\ell}|\omega_k|^4|\omega_m|\delta_i\right)\right)$$

$$+\sqrt{\varepsilon_{i\ell}}\left(\int_{B_i}\delta_i^{24/5}(|v_i|+|h_i|)^{6/5}\right)^{5/6}+\int_{\Omega_\ell}\delta_\ell^4(|v_\ell|+|h_\ell|)\delta_i+C_h\left(\sum\varepsilon_{st}^h\right).$$

$$=B.$$

We thus have

$$\frac{C|h_{\ell,i}^1|_\infty|\omega_i^\infty|}{\lambda_\ell^2|a_i-a_\ell^4|}\sqrt{\lambda_i}\int_{\Omega_\ell^3}\delta_i^3\leq(C_1|\omega_i^\infty|\varepsilon_{i\ell}B)\ \mathrm{Log}\ \frac{\lambda_i}{\lambda_\ell}\times\frac{1}{\lambda_\ell|a_i-a_\ell|^2}.$$

The main issue is to see that $\mathrm{Log}\frac{\lambda_i}{\lambda_\ell}$, which is less that $C\ \mathrm{Log}\ \varepsilon_{i\ell}^{-1}$ can be absorbed into B while enough room is left to provide an upperbound of the usual type.

We track down the terms and find $(\lambda_\ell\leq\lambda_i)$

$$\frac{C|h_{\ell,i}^1|_\infty}{\lambda_\ell^2|a_i-a_\ell|^4}|\omega_i^\infty|\sqrt{\lambda_i}\int_{\Omega_\ell^3}\delta_i^3\leq\frac{C_1}{\lambda_\ell^2|a_i-a_\ell|^2}\left(\sqrt{\lambda_\ell}|\omega_i^\infty|\varepsilon_{i\ell}\right)$$

$$\times\left(\varepsilon_{i\ell}\ \mathrm{Log}\ \varepsilon_{i\ell}^{-1}\sqrt{\lambda_\ell}|v_\ell|_{H_0^1}+\frac{\sqrt{\lambda_\ell}}{\sqrt{\lambda_i}}\sqrt{\varepsilon_{i\ell}}\ \mathrm{Log}\ \varepsilon_{i\ell}^{-1}\left[\sqrt{\lambda_i}|v_i|_{H_0^1}+\sqrt{\lambda_i}\left(\frac{|h_i^0|_\infty}{\sqrt{\lambda_i}}\right.\right.\right.$$

$$+ \sum_{\substack{\lambda_s \geq \lambda_i \\ s \neq i}} |h^1_{i,s}|_\infty \varepsilon_{is} + \varepsilon^2_{im} \sum \varepsilon_{st} \Bigg]$$

$$+ \varepsilon_{i\ell} \operatorname{Log} \varepsilon_{i\ell}^{-1} \sqrt{\lambda_\ell} \left[\frac{|h^0_\ell|_\infty}{\sqrt{\lambda_\ell}} + \sum_{\subset \lambda_s \geq \lambda_i s \neq i} \varepsilon_{s\ell} |h^1_{\ell,s}|_\infty + \varepsilon^2_{\ell m} \sum \varepsilon_{st} \right]$$

$$+ \sqrt{\varepsilon_{i\ell}} \operatorname{Log} \varepsilon_{i\ell}^{-1} \sqrt{\frac{\lambda_\ell}{\lambda_i}} \left[\sqrt{\lambda_i} \left(\sum_{k \neq i} |\omega_k(a_i)| \varepsilon_{ki} + \varepsilon_{ki} \varepsilon^2_{i\ell} \right) \right]$$

$$+ \varepsilon_{i\ell} \operatorname{Log} \varepsilon_{i\ell}^{-1} \sqrt{\lambda_\ell} \left(\sum_{k \neq \ell} |\omega_k(a_\ell)| \varepsilon_{k\ell} + \varepsilon_{k\ell} \varepsilon^2_{kt} \right)$$

$$+ \varepsilon_{i\ell} \operatorname{Log} \varepsilon_{i\ell}^{-1} \left[\varepsilon_{k\ell} \frac{\sqrt{\lambda_\ell}}{\sqrt{\lambda_i}} \sqrt{\lambda_i} |A_i| + \left(\sum_{k \neq i} \varepsilon_{ik} \sqrt{\lambda_i} \varepsilon_{ik} |A_k| \right) \frac{\sqrt{\lambda_\ell}}{\sqrt{\lambda_i}} + \sqrt{\lambda_\ell} |A_\ell| \right]$$

$$+ \sum_{\substack{k \neq i \\ k \neq \ell \\ \lambda_\ell \leq \lambda_k}} \operatorname{Log} \varepsilon_{i\ell}^{-1} \varepsilon_{ik} \varepsilon_{k\ell} \sqrt{\frac{\lambda_\ell}{\lambda_k}} \sqrt{\lambda_k} |A_k| + \sum_{\substack{k \neq i \\ k \neq \ell \\ \lambda_k \leq \lambda_\ell}} \left(\varepsilon_{ik} \operatorname{Log} \frac{\sqrt{\lambda_i}}{\sqrt{\lambda_\ell}} \right) \sqrt{\lambda_\ell} \varepsilon_{k\ell} |A_k| \Bigg) .$$

Observe that

$$\varepsilon_{ik} \varepsilon_{k\ell} \leq C \varepsilon_{i\ell}.$$

Hence

$$\operatorname{Log} \varepsilon_{i\ell}^{-1} \varepsilon_{i\ell} \varepsilon_{k\ell} \sqrt{\frac{\lambda_\ell}{\lambda_k}} \leq C \varepsilon_{i\ell} \operatorname{Log} \varepsilon_{i\ell}^{-1} \text{ if } \lambda_k \geq \lambda_\ell.$$

If $\lambda_k \leq \lambda_\ell$,

$$\varepsilon_{ik} \operatorname{Log} \frac{\sqrt{\lambda_i}}{\sqrt{\lambda_\ell}} \leq \varepsilon_{ik} \operatorname{Log} \frac{\sqrt{\lambda_i}}{\sqrt{\lambda_k}} \leq C \varepsilon_{ik} \operatorname{Log} \frac{\varepsilon_{ik}^{-1}}{\lambda_k |a_i - a_k|} \leq C_1 \varepsilon_{ik} \operatorname{Log} \varepsilon_{ik}^{-1}.$$

Observe also that

$$|A_m| \leq C \left(\int_{\Omega_m} |f_1| \delta_m + \sum \varepsilon_{ms} \int_{\Omega_s} |f_1| \delta_s \right).$$

The estimate on $\int_{\Omega_m} |f_1|\delta_m$ follows in the same way the estimates of Theorem 1 were derived.

Our estimate follows.

Proof of Lemma 43.

Considering $\int_{\widetilde{\Omega}_c} \delta_\ell^5 \delta_i$, either we are on the part of $\widetilde{\Omega}_c$ where $|x - x_i| \geq c|x_\ell - x_i|$. We call this part $\widetilde{\Omega}_c^1$. Then,

$$\int_{\widetilde{\Omega}_c^1} \delta_\ell^5 \delta_i \leq \frac{C}{\sqrt{\lambda_i}|x_\ell - x_i|} \int_{\widetilde{\Omega}_\ell^c} \delta_\ell^5 \leq C\varepsilon_{i\ell}\varepsilon_{\ell s}^2.$$

On the remainder $\widetilde{\Omega}_c^2$ of $\widetilde{\Omega}_c$,

$$|x - x_i| \leq c|x_\ell - x_i|$$

and

$$|x - x_\ell| \geq (1 - c)|x_\ell - x_i|.$$

Thus,

$$\int_{\widetilde{\Omega}_c^2} \delta_\ell^5 \delta_i \leq \frac{C}{\lambda_\ell^{5/2}|x_\ell - x_i|^5} \frac{1}{\sqrt{\lambda_i}} \int_{r \leq c|x_\ell - x_i|} rdr \leq \frac{C\varepsilon_{i\ell}}{\lambda_\ell^2|x_i - x_\ell|^2}.$$

Since $\widetilde{\Omega}_c^2$ is non empty and $\widetilde{\Omega}_c^2 \subset \widetilde{\Omega}_\ell^c$, either at a given $x \in \widetilde{\Omega}_c^2$

$$\lambda_\ell|x - x_\ell| \geq \frac{c_1}{\varepsilon_{\ell s}}$$

or

$$\lambda_\ell|x - x_j| \leq \frac{c_2}{\varepsilon_{\ell j}} \text{ for some } \lambda_j > \lambda_\ell.$$

In the second case,

$$\delta_\ell \leq C\delta_j$$

at such an x and we can replace their contribution to $\int \delta_\ell^5 \delta_i$ by $\int \delta_j^5 \delta_i$.

Since $\lambda_j > \lambda_i$, the induction will eventually stabilize.

Thus, we may assume that we are in the domain where

$$\lambda_\ell|x - x_\ell| \geq \frac{c_1}{\varepsilon_{\ell s}}.$$

On the other hand,

$$|x - x_\ell| \leq |x - x_i| + |x_i - x_\ell| \leq (1 + c)|x_i - x_\ell| \text{ in } \Omega_c^2.$$

Hence,

$$\lambda_\ell |x_i - x_\ell| \geq \frac{c_3}{\varepsilon_{\ell s}}$$

and

$$\int_{\widetilde{\Omega}_c^2} \delta_\ell^5 \delta_i \leq C\varepsilon_{i\ell}\varepsilon_{\ell s}^2.$$

$\square$

Proof of Lemma 44.

We split, according to the estimate of Lemma 12 and according to (1.21) and (1.22), h_ℓ into h_ℓ^0 and h_ℓ^1. We split Ω_ℓ in two parts, Ω_ℓ^1 and Ω_ℓ^2.

In Ω_ℓ^1,

$$|x - x_i| \geq c|x_i - x_\ell| \quad \text{and}$$

$$\int_{\Omega_\ell^1} \delta_i |\bar{v}|^5 \leq \frac{C}{\sqrt{\lambda_i}|x_i - x_\ell|} \int_{\Omega_\ell^1} |\bar{v}|^5 \leq \frac{C\varepsilon_{i\ell}}{\sqrt{\varepsilon_{\ell t}}} \left(\int_{\Omega_\ell} |v_\ell|^6 \right)^{5/6}$$

$$+\frac{C}{\sqrt{\lambda_i}|x_i - x_\ell|} \left(\sum_k \int_{\widetilde{\Omega}_\ell^1} \delta_k^4 |h_\ell^0| + \int_{\Omega_\ell - \widetilde{\Omega}_\ell} \delta_i |h_\ell|^5 + \int_{\widetilde{\Omega}_\ell^1} \delta_i h_\ell^4 \inf(|h_\ell|, |h_\ell^1|) \right)$$

$$+C_h \sum \varepsilon_{st}^h \leq \frac{C\varepsilon_{i\ell}}{\sqrt{\varepsilon_{\ell t}}} |v_\ell|_{H_0^1}^5 + C\sqrt{\lambda_\ell}\varepsilon_{i\ell} \sum_{\lambda_k \geq \lambda_\ell} \int_{\widetilde{\Omega}_\ell^1} \delta_k^4 |h_\ell^0|$$

$$+ \int_{\widetilde{\Omega}_\ell^1} \delta_i h_\ell^4 \inf(|h_\ell|, |h_\ell^1|) + \int_{\Omega_\ell - \widetilde{\Omega}_\ell} \delta_i |h_\ell|^5 + C_h \left(\sum \varepsilon_{st} \right)^h .$$

Observe now that, using Lemma 68 and which we prove below ($\delta_k \leq C\delta_\ell$ in $\widetilde{\Omega}_\ell$).

$$\sqrt{\lambda_\ell}\varepsilon_{i\ell} \int_{\widetilde{\Omega}_\ell} \delta_k^4 |h_\ell^0| + \int_{\widetilde{\Omega}_\ell} \delta_i h_\ell^4 \inf(|h_\ell|, |h_\ell^1|) + \int_{\Omega_\ell - \widetilde{\Omega}_\ell} \delta_i |h_\ell|^5$$

$$\leq C\frac{\varepsilon_{i\ell}|h_\ell^0|_\infty}{\sqrt{\lambda_\ell}} + \sum_{\substack{\lambda_s \geq \lambda_\ell \\ s \neq \ell}} |h_{\ell,s}^1|_\infty \int_{\widetilde{\Omega}_\ell} \delta_i \delta_\ell^4 \delta_s + C \int_{\Omega_\ell - \widetilde{\Omega}_\ell} \delta_i \sum_{\substack{\lambda_k \geq \lambda_\ell \\ k \neq \ell}} \delta_k^5$$

$$+C\int_{\Omega_\ell-\tilde{\Omega}_\ell}\delta_i\delta_\ell^5+C_h\left(\sum\varepsilon_{st}^h\right)\le C\varepsilon_{i\ell}\frac{|h_\ell^0|_\infty}{\sqrt{\lambda_\ell}}+\left(\sum_{\substack{\lambda_s\ge\lambda_\ell\\s\ne\ell}}\varepsilon_{s\ell}|h_{\ell,s}^1|_\infty\right)\varepsilon_{i\ell}$$

$$+C\left(\sum_{\substack{\lambda_k\ge\lambda_\ell\\k\ne\ell}}\varepsilon_{ik}\varepsilon_{k\ell}^2+\sum\varepsilon_{i\ell}\varepsilon_{\ell s}^2\right)+C_h(\sum\varepsilon_{st})^h.$$

Indeed, the function $|h_\ell|$ up to a function θ_ℓ with $|\theta_\ell|_{H^1}\le C_h(\sum\varepsilon_{st})^h$ is using Proposition 2, either bounded by $C\delta_\ell$ or by $C\sum_{\substack{\lambda_k\ge\lambda_\ell\\k\ne\ell}}\delta_k$.

In Ω_ℓ^2,

$$|x-x_i|\le c|x_i-x_\ell|.$$

Then,

$$\int_{\Omega_\ell^2}\delta_i|\bar{v}|^5\le C\sum_k\int_{\Omega_\ell^2}\delta_i\delta_k^4|\bar{v}_\ell|+C\int_{\Omega_\ell}\delta_i|h_\ell|^5+C\varepsilon_{st}^h$$

$$\le C\sum_{\lambda_k\ge\lambda_\ell}\int_{\Omega_\ell^2}\delta_i\delta_k^4|\bar{v}_\ell|+C\int_{\Omega_\ell}\delta_i|h_\ell|^5+C\sum\varepsilon_{st}^h$$

$\int_{\Omega_\ell}\delta_i|h_\ell|^5$ has been estimated above.

Considering now $\int_{\Omega_\ell^2}\delta_i\delta_k^4|\bar{v}_\ell|$ with $\lambda_k\ge\lambda_\ell$, either $\lambda_k>\lambda_i$. Then,

$$\int_{\Omega_\ell^2}\delta_i\delta_k^4|\bar{v}_\ell|\le\left(\int\delta_i^2\delta_k^4\right)^{1/2}\left(\int_{\Omega_\ell}\delta_k^4\bar{v}_\ell^2\right)^{1/2}\le C\varepsilon_{ik}\varepsilon_{k\ell}|\bar{v}_\ell|_{H_0^1}\le C_1\varepsilon_{i\ell}|\bar{v}_\ell|_{H_0^1}.$$

If $k=\ell$, we derive the same upperbound (in both cases, it is $o(\varepsilon_{i\ell}|v_\ell|_{H_0^1})$).

If $\lambda_\ell\le\lambda_k\le\lambda_i$, then either

$$|x_i-x_k|\ge c|x_i-x_\ell|$$

and

$$\int_{\Omega_\ell}\delta_i\delta_k^4|\bar{v}_\ell|\le C\varepsilon_{ik}|\bar{v}_\ell|_{H_0^1}\le C_1\varepsilon_{i\ell}|\bar{v}_\ell|_{H_0^1}$$

or

$$|x_i-x_k|\le c|x_i-x_\ell|.$$

Thus,

$$|x_k - x_\ell| \geq |x_i - x_k| - |x_k - x_i| \geq (1 - c)|x_i - x_\ell|.$$

Since $\lambda_k > \lambda_\ell$, on Ω_ℓ,

$$\lambda_k|x - x_k| \geq \frac{c}{\varepsilon_{k\ell}}$$

which rereads

$$\delta_k(x) \leq \frac{C}{\sqrt{\lambda_\ell}|x_k - x_\ell|}.$$

Then,

$$\int\limits_{\substack{|x-x_i|\leq c|x_i-x_\ell| \\ \Omega_\ell}} \delta_i \delta_k^4 |v_\ell| \leq \frac{C}{\lambda_\ell^2 |x_k - x_\ell|^4} \int\limits_{\substack{|x-x_i|\leq c|x_i-x_\ell| \\ \Omega_\ell}} \delta_i |\bar{v}_\ell|$$

$$\leq \frac{C_1}{\sqrt{\lambda_i}\lambda_\ell^2|x_\ell - x_i|^4} \int\limits_{r\leq c|x_i-x_\ell|} \frac{|\bar{v}_\ell|}{r} \leq \frac{C_1|\bar{v}_\ell|_{H_0^1}}{\sqrt{\lambda_i}\lambda_\ell^2|x_i - x_\ell|^4} \left(\int\limits_{r\leq c|x_i-x_\ell|} \frac{r^2 dr}{r^{6/5}} \right)^{5/6}$$

$$\leq \frac{C_1|\bar{v}_\ell|_{H_0^1}}{\sqrt{\lambda_i}\lambda_\ell^3|x_i - x_\ell|^4}|x_i - x_\ell|^{3/2} \leq C_1'\varepsilon_{i\ell}|\bar{v}_\ell|_{H_0^1}.$$

$$\square$$

Again, it is easy to replace this bound with $0(\varepsilon_{i\ell}|\bar{v}_\ell|_{H_0^1})$.

Proof of Lemma 45. The estimate follows closely the one completed in Lemma 43 about $\int_{\Omega_c} \delta_i\delta_\ell^5$. We split $\int\limits_{\Omega_\ell\cap\Omega_i^c} \delta_i\delta_k^5$ in three pieces. The first one is $\int\limits_{(\Omega_\ell\cap\Omega_i^c)^1} \delta_i\delta_k^5$ where x satisfies $|x - x_i| \geq c|x_i - x_\ell|$ in $(\Omega_\ell \cap \Omega_i^c)^1$.

This yields

$$\int\limits_{(\Omega_\ell\cap\Omega_i^c)^1} \delta_i\delta_k^5 \leq \frac{C}{\sqrt{\lambda_i}|x_i - x_\ell|} \int\limits_{(\Omega_\ell\cap\Omega_i^c)^1} \delta_k^5 \leq \frac{C}{\sqrt{\lambda_i}\sqrt{\lambda_k}|x_i - x_\ell|} \int\limits_{r\geq\frac{1}{\varepsilon_{k\ell}}} \frac{r^2 dr}{(1 + r^2)^{5/2}}$$

$$\leq C\varepsilon_{i\ell}\varepsilon_{k\ell}^2 \text{ since } \lambda_k \geq \lambda_\ell.$$

The second piece is $\int\limits_{(\Omega_\ell\cap\Omega_i^c)^2} \delta_i\delta_k^5$ where x in $(\Omega_\ell \cap \Omega_i^c)^2$ satisfies $|x - x_i| \leq c|x_i - x_\ell|$.

We split it into two pieces: a first contribution on $(\Omega_\ell \cap \Omega_i^c)_1^2$ where $|x - x_i| \geq c|x_i - x_\ell|$. For this contribution, we derive as above:

$$\int_{(\Omega_\ell \cap \Omega_i^c)_1^2} \delta_i \delta_k^5 \leq C\varepsilon_{ik}\varepsilon_{k\ell}^2 \leq C_1 \varepsilon_{i\ell}\varepsilon_{k\ell}.$$

The remainder is $\displaystyle\int_{(\Omega_\ell \cap \Omega_i^c)_2^2} \delta_i \delta_k^5$.

On $(\Omega_\ell \cap \Omega_i^c)_2^2$, $|x - x_i| \leq |x_i - x_k|$; so that

$$(1 - c)|x_i - x_k| \leq |x - x_k| \leq (1 + c)|x_i - x_k|.$$

Since x is in Ω_ℓ with $\lambda_k \geq \lambda_\ell, k \neq \ell$:

$$\lambda_k |x - x_k| \geq \frac{1}{8\varepsilon_{k\ell}}.$$

Hence

$$c\lambda_k |x_i - x_k| \geq \sqrt{\lambda_k \lambda_\ell}|x_k - x_\ell|$$

and

$$\frac{c}{\sqrt{\lambda_k}|x - x_k|} \leq \frac{1}{\sqrt{\lambda_k}|x_i - x_k|} \leq \frac{c}{\sqrt{\lambda_\ell}|x_k - x_\ell|}$$

so that

$$\delta_k(x_k, \lambda_k)(x) \leq \frac{C}{\sqrt{\lambda_k}|x_i - x_k|}$$

and

$$\int_{(\Omega_\ell \cap \Omega_i^c)_2^2} \delta_i \delta_k^5 \leq \frac{C}{\lambda_k^{5/2}|x_i - x_k|^5} \int_{|x-x_i| \leq c|x_i - x_k|} \delta_i \leq \frac{C}{\sqrt{\lambda_i}\lambda_k^{5/2}|x_i - x_k|^5}$$

$$\times \int_{r \leq c|x_i - x_k|} r\,dr \leq \frac{C_1}{\sqrt{\lambda_i}\sqrt{\lambda_k}} \cdot \frac{1}{\lambda_k^2|x_i - x_k|^2} \leq \frac{C\varepsilon_{ik}}{\lambda_k \lambda_\ell|x_k - x_\ell|^2} \leq C_3\varepsilon_{ik}\varepsilon_{k\ell}^2.$$

$$\square$$

Proof of Lemma 46. The estimate is straightforward for $k = \ell$, so we assume that $k \neq \ell$.

102   *Recent Progress in Conformal Geometry*

Let

$$\Omega_\ell^1 = \Omega_\ell \cap \{|x - x_i| \geq c|x_i - x_\ell|\}$$
$$\Omega_\ell^2 = \Omega_\ell \cap \{|x - x_i| \geq c|x_i - x_k|\}$$
$$\Omega_\ell^3 = \Omega_\ell \cap \{|x - x_i| \leq c|x_i - x_k|\}$$
$$\subset \Omega_\ell \cap \{(1 - c)|x_i - x_k| \leq |x - x_k| \leq (1 + c)|x_i - x_\ell|\}.$$

If Ω_ℓ^3 is not empty, then since $\lambda_k \geq \lambda_\ell, k \neq \ell$,

$$\lambda_k|x - x_k| \geq \frac{c}{\varepsilon_{k\ell}} \quad \text{for some } x \in \Omega_\ell^3$$

and therefore

$$\lambda_k|x_i - x_k| \geq \frac{c_1}{\varepsilon_{k\ell}},$$

an estimate which we will use on Ω_ℓ^3. $\square$

We compute:

$$\int_{\Omega_\ell^1} \delta_i \delta_k^4 \leq \frac{C}{\sqrt{\lambda_i}|x_i - x_\ell|} \times \frac{1}{\lambda_k} \leq \frac{C_1 \varepsilon_{i\ell}}{\sqrt{\lambda_\ell}}$$

$$\int_{\Omega_\ell^2} \delta_i \delta_k^4 \leq \frac{C}{\sqrt{\lambda_i}|x_i - x_k|} \int_{\lambda_k r \geq \frac{c}{\varepsilon_{k\ell}}} \frac{\lambda_k^2 r^2 dr}{(1 + \lambda_k^2 r^2)^2} \leq \frac{C \varepsilon_{k\ell}}{\lambda_k \sqrt{\lambda_i}|x_i - x_k|}$$

$$\leq \frac{C_1 \varepsilon_{ik} \varepsilon_{k\ell}}{\sqrt{\lambda_k}} \leq \frac{C \varepsilon_{i\ell}}{\sqrt{\lambda_\ell}}$$

$$\int_{\Omega_\ell^3} \delta_i \delta_k^4 \leq \frac{C}{\lambda_k^2 |x_i - x_k|^4} \int_{r \leq c|x_i - x_k|} \frac{r^2 dr}{\sqrt{\lambda_i} r} \leq \frac{C|x_i - x_k|^2}{\sqrt{\lambda_i} \lambda_k^2 |x_i - x_k|^4}$$

$$\leq \frac{C \varepsilon_{ik}}{\sqrt{\lambda_k} \lambda_k |x_i - x_k|} \leq \frac{C \varepsilon_{ik} \varepsilon_{k\ell}}{\sqrt{\lambda_k}} \leq \frac{c \varepsilon_{i\ell}}{\sqrt{\lambda_\ell}}.$$

     $\square$

Proof of Lemma 47. The proof of (ii) follows closely the proof of (i). First we have, using Proposition 2

$$\sqrt{\lambda_i} \varepsilon_{ij} \int_{B_j} |\bar{v}_j|^5 \leq C\sqrt{\lambda_j} \varepsilon_{ij} \int_{B_j} \sum_{\lambda_k \geq \lambda_j} \delta_k^4 |\bar{v}_j| + C_h \left(\sum \varepsilon_{st}\right)^h \sqrt{\varepsilon_{ij}}$$

$$\leq C \varepsilon_{ij} |\bar{v}_j|_{H_0^1} + C_h \left(\sum \varepsilon_{st}\right)^h \sqrt{\varepsilon_{ij}}.$$

Then,

$$\sqrt{\lambda_j}\varepsilon_{ij}\int_{B_j}|\bar{h}_j|^5 \leq C\sqrt{\lambda_j}\varepsilon_{ij}\int_{\tilde{B}_j}\sum_{\lambda_k\geq\lambda_j}\delta_k^4|\bar{h}_j^0| + C\sqrt{\lambda_j}\varepsilon_{ij}$$

$$\times\left(\int_{\tilde{B}_j}\sum_{\substack{\lambda_k\geq\lambda_j\\\lambda_s\geq\lambda_j}}\delta_k^4\delta_s\right)|\bar{h}_{j,s}^1|_\infty + C_h\left(\sum\varepsilon_{st}\right)^h\sqrt{\varepsilon_{ij}} + \sqrt{\lambda_j}\varepsilon_{ij}\int_{B_j-\tilde{B}_j}\sum_{\lambda_k\geq\lambda_j}\delta_k^5$$

$$\leq C\varepsilon_{ij}\frac{|\bar{h}_j^0|_\infty}{\sqrt{\lambda_j}} + C\varepsilon_{ij}\sum\varepsilon_{sj}|\bar{h}_{j,s}^1|_\infty + C_h\left(\sum\varepsilon_{st}\right)^h\sqrt{\varepsilon_{ij}}$$

$$+ C\varepsilon_{ij}\left(\sum_{\lambda_j\geq\lambda_j}\varepsilon_{kj}^{5/2} + \varepsilon_{jt}^{5/2}\right).$$

The claim follows. The proof of (ii) is identical. $\square$

Proof of Lemma 48. We split the integral between an integral on Ω_ℓ and an integral on Ω_ℓ^c. We have using Proposition 2

$$\int_{\Omega_\ell^c}|\bar{v}|^5|L^{-1}e_\ell| \leq C\sum_{\substack{k\neq m\\m\neq\ell}}\int_{\Omega_\ell^c\cap\Omega_m}\delta_k^5\delta_\ell + C_h\left(\sum\varepsilon_{st}\right)^h\sqrt{\varepsilon_{\ell q}}$$

$$+ C\sum_{m\neq\ell}\int_{\Omega_\ell^c\cap\Omega_m}\delta_\ell|\bar{v}|^5 \leq C\sum\varepsilon_{\ell k}\varepsilon_{ks}^2 + C_h\left(\sum\varepsilon_{st}\right)^h\sqrt{\varepsilon_{\ell q}}$$

$$+ \sum\varepsilon_{\ell m}\left(|\bar{v}_m|_{H_0^1} + \frac{|h_m^0|_\infty}{\sqrt{\lambda_m}} + \sum_{\substack{\lambda_s\geq\lambda_m\\s\neq m}}\varepsilon_{sm}|h_{m,s}^1|_\infty\right) + C\left(\sum\varepsilon_{\ell k}\varepsilon_{ks}^2\right).$$

On the other hand,

$$\int_{\Omega_\ell}|\bar{v}_\ell + \bar{h}_\ell|^5|L^{-1}e_\ell| \leq C|\bar{v}_\ell|_{H_0^1}^2 + C\left(\int_{\tilde{\Omega}_\ell}\sum_{\lambda_k\geq\lambda_\ell}\delta_k^3|L^{-1}e_\ell||\bar{h}_\ell|\right)|\bar{h}_\ell^0|_\infty$$

$$+ C\int_{\tilde{\Omega}_\ell}\sum_{\substack{\lambda_k\geq\lambda_\ell\\\lambda_s\geq\lambda_\ell\\s\neq\ell}}\delta_k^4\delta_s|L^{-1}e_\ell||h_{\ell,s}^1|_\infty + C_h\left(\sum\varepsilon_{st}\right)^h + C\sum\varepsilon_{\ell k}^3$$

$$\leq C \left(|\bar{v}_\ell|^2_{H^1_0} + \frac{|h^0_\ell|_\infty}{\sqrt{\lambda_\ell}} |h_\ell|_{L^6} + |h^1_{\ell,s}|_\infty \sum_{\substack{\lambda_k \geq \lambda_\ell \\ \lambda_s \geq \lambda_\ell \\ s \neq \ell}} \left(\int \delta_k^4 \delta_s^2 \right)^{1/2} \left(\int \delta_k^4 \delta_\ell^2 \right)^{1/2} \right.$$

$$\left. + C_h \left(\sum \varepsilon_{st} \right)^h + \sum \varepsilon_{\ell k}^3 \right) \leq C \left(|\bar{v}_\ell|^2_{H^1_0} + o \left(\frac{|h^0_\ell|_\infty}{\sqrt{\lambda_\ell}} \right) \right.$$

$$\left. + \sum \left(\varepsilon_{\ell k} \varepsilon_{ks} |h^1_{\ell,s}|_\infty + \varepsilon_{\ell,s} |h^1_{\ell,s}|_\infty \right) + C \sum \varepsilon_{\ell k}^3 + C_h \left(\sum \varepsilon_{st} \right)^h \right).$$

$\square$

Proof of Lemma 49. Using Lemma 43, we have:

$$\int \omega_s^4 |k^*| |L^{-1} e_\ell| \leq C \sum \int_{\Omega_c} \delta_k^5 \delta_\ell + C_h \sum \varepsilon_{st}^h \leq \sum \varepsilon_{\ell k} \varepsilon_{ks}^2 + C_h \sum \varepsilon_{st}^h.$$

$\square$

Proof of Lemma 50. This is a modification of Lemma 49, using the fact that $|\bar{v}_\ell + \bar{h}_\ell - \theta_\ell| \leq C \sum \delta_k$ with $|\theta_\ell|_{H^1_0} \leq C_h \left(\sum \varepsilon_{st} \right)^h$.

This allows to upperbound

$$\int_{\Omega_\ell - \tilde{\Omega}_\ell} |\omega_\ell|^3 |v_\ell + h_\ell|^2 |L^{-1} e_\ell|$$

by

$$C \varepsilon_{\ell m}^2 \varepsilon_{k\ell} + C_h \left(\sum \varepsilon_{st} \right)^h.$$

$\square$

Proof of Lemma 51. We first estimate $O \left(\sqrt{\lambda_j} \varepsilon_{ij} \int_{B_j} |\bar{v}|^5 \right)$. Using Lemmas 12, and Proposition 2, we upperbound this quantity with:

$$O \left(\varepsilon_{ij} \sqrt{\lambda_j} \right) \int_{\tilde{B}_j} \sum \delta_k^4 \left(|\bar{v}| + |\bar{h}^o_j| + \sum_s |\bar{h}^1_{j,s}| \delta_s \right)$$

$$+ O \left(\varepsilon_{ij} \sqrt{\lambda_j} \right) \int_{B_j - \tilde{B}_j} \sum \delta_k^5 + O \left(\varepsilon_{ij} \sqrt{\lambda_j} \right) C_h \left(\sum \varepsilon_{st} \right)^h |\tilde{B}_j|^{1/6}.$$

We may retain among the δ_k's only those with $\lambda_k \geq C \lambda_j$.

We then derive

$$O\left(\varepsilon_{ij}\sqrt{\lambda_j}\right)\int_{B_j}|\bar{v}|^5 \le O(\varepsilon_{ij})|\bar{v}_j|_{H_0^1}$$

$$+ O\left(\varepsilon_{ij}\sqrt{\lambda_j}\right)\int_{\tilde{B}_j}\sum_{\lambda_k \ge \lambda_j}\delta_k^4\left(|\bar{h}_j^o| + |\bar{h}_{j,s}^1|\delta_s\right)$$

$$+ O\left(\varepsilon_{ij}\sqrt{\lambda_j}\right)\int_{B_j - \tilde{B}_j}\sum_{\lambda_k \ge \lambda_j}\delta_k^5 + O\left(\sqrt{\varepsilon_{ij}}\right)C_h\left(\sum\varepsilon_{st}\right)^h$$

$$\le O(\varepsilon_{ij})\left(|\bar{v}_j|_{H_0^1} + \frac{|\bar{h}_j^o|_\infty}{\sqrt{\lambda_j}} + \sum_{\substack{\lambda_s \ge \lambda_j \\ s \ne j}}\varepsilon_{js}|\bar{h}_{j,s}^1|_\infty\right)$$

$$+ \varepsilon_{ij}O\left(\sum\varepsilon_{j\ell}^2\right) + O\left(\sqrt{\varepsilon_{ij}}\right)C_h\left(\sum\varepsilon_{st}\right)^h. \qquad\square$$

Proof of Lemma 52. The first term has been estimated in Lemma 43, after the use of Proposition 2.

For the second term, we use again Proposition 2. Up to a term of order $C_h\left(\sum\varepsilon_{st}\right)^h$, this term is then upperbounded by (on $\Omega_j, \lambda_k|x - x_k| \ge \frac{c}{\varepsilon_{kj}}$ if $\lambda_k \ge C\lambda_j, k \ne j$):

$$\sum_k \sqrt{\lambda_j}O(\varepsilon_{ij})\int_{(B_j - \tilde{B}_j)\cap\Omega_k^c}\delta_k^5$$

$$\le C\sum_{\lambda_k \ge \lambda_j}\sqrt{\lambda_j}O(\varepsilon_{ij})\int_{(B_j - \tilde{B}_j)\cap\Omega_k^c}\delta_k^5 \le C\sum\varepsilon_{ij}\varepsilon_{js}^2. \qquad\square$$

Proof of Lemma 53. Using our observations, we derive the following upperbounds:

$$\int_{\Omega_\ell}\omega_k^4|\omega_s||\partial\omega_\ell| \le (|\underline{\omega}_k(\underline{a}_\ell)| + |\underline{\omega}_\ell(\underline{a}_k)|)\varepsilon_{k\ell} + \varepsilon_{k\ell}^3 \text{ if } \lambda_\ell \le \lambda_k$$

$$\le |\underline{\omega}_k(\underline{a}_\ell)|\varepsilon_{k\ell} + \varepsilon_{k\ell}^3 \text{ if } \lambda_k \le \lambda_\ell.$$

If $\lambda_\ell \le \lambda_k$, we upperbound

$$\int_{\Omega_k}\omega_k^4|\omega_s||\partial\omega_\ell| \le \left(\int_{\Omega_k}|\omega_k|^{24/5}|\partial\omega_\ell|^{6/5}\right)^{5/6}$$

$$\le C\left((|\underline{\omega}_k(\underline{a}_\ell)| + |\underline{\omega}_\ell(\underline{a}_k)|)\varepsilon_{k\ell} + \varepsilon_{k\ell}^3\right).$$

If $\lambda_k \leq \lambda_\ell$,

$$\int_{\Omega_k} \omega_k^4 |\omega_s||\partial \omega_\ell| \leq \left(\int_{\Omega_k} \omega_k^4 \omega_s^2 \right)^{1/2} \left(\int_{\Omega_k} \omega_k^4 |\partial \omega_\ell|^2 \right)^{1/2}$$

$$\leq C \varepsilon_{k\ell} \left(\int_{\Omega_k} \omega_k^4 \omega_s^2 \right)^{1/2}.$$

$\square$

We then distinguish between the case $\lambda_s \geq \lambda_k$: we then have

$$\int_{\Omega_k} \omega_k^4 |\omega_s||\partial \omega_\ell| \leq C \varepsilon_{k\ell} \left(\left(|\underline{\omega}_k(\underline{a}_s)| + |\underline{\omega}_s(\underline{a}_k)| \right) \varepsilon_{ks} + \varepsilon_{ks}^3 \right)$$

and the case $\lambda_k \geq \lambda_s$, which we study below.

Considering now

$$\int_{\Omega_k^c \cap \Omega_\ell^c} \omega_k^4 |\omega_s||\partial \omega_\ell|,$$

either $|\omega_s(x)| \leq |\omega_k(x)|$. For those x's, by Lemma 45, we upperbound the contribution with

$$\int_{\Omega_k^c \cap \Omega_\ell^c} |\omega_k|^5 |\partial \omega_\ell| \leq C \sum_{\lambda_j \geq \lambda_k} \varepsilon_{\ell j} \varepsilon_{jm}^2.$$

Or $|\omega_k(x)| \leq |\omega_s(x)|$. For those x's, we upperbound the contribution with

$$\int_{\Omega_s} \omega_s^4 |\omega_k||\partial \omega_\ell| + \int_{\Omega_s^c \cap \Omega_\ell^c} |\omega_s|^5 |\partial \omega_\ell| \leq C \left(\int_{\Omega_s} \omega_s^4 \omega_k^2 \right)^{1/2} \varepsilon_{s\ell} + \sum_{\lambda_j \geq \lambda_s} \varepsilon_{j\ell} \varepsilon_{jm}^2.$$

Again, if $\lambda_k \geq \lambda_s$, we can upperbound $\left(\int_{\Omega_s} \omega_s^4 \omega_k^2 \right)^{1/2} \varepsilon_{s\ell}$ with

$$C \varepsilon_{s\ell} \left(\left(|\underline{\omega}_s(\underline{a}_k)| + |\underline{\omega}_k(\underline{a}_s)| \right) \varepsilon_{sk} + \varepsilon_{sk}^2 \right).$$

If $\lambda_s \geq \lambda_k$, we have to control $\left(\int_{\Omega_s} \omega_s^4 \omega_k^2 \right)^{1/2}$ and this is similar to $\left(\int_{\Omega_k} \omega_k^4 \omega_s^2 \right)^{1/2}$ with $\lambda_k \geq \lambda_s$.

We thus consider $\int_{\Omega_k} \omega_k^4 \underline{\omega}_s^2$ with $\lambda_k \geq \lambda_s$. Since such quantities are conformally invariant, we deconcentrate ω_s so that $\bar{\lambda}_s = 1, \omega_s$ becomes $\underline{\omega}_s, \lambda_k$ becomes $\bar{\lambda}_k$ with $\bar{\lambda}_k \sim c\varepsilon_{ks}^{-2}$. ω_k is concentrated at $\underline{a}_k (= 0)$ and we write:

$$\int_{\Omega_k} \omega_k^4 \underline{\omega}_s^2 \leq \underline{\omega}_s^2(\underline{a}_k) \int_{|x|\leq 1} \omega_k^4 + C \int_{|x|\leq 1} r^2 \omega_k^4 + \int_{|x|\geq 1} \omega_k^4 \underline{\omega}_s^2.$$

We can think of all this decomposition as occuring on the sphere S^3 and we then have, since $|\omega_k| \leq \frac{C}{\sqrt{\bar{\lambda}_k}}$ for $|x| \geq 1$,

$$\int_{|x|\geq 1} \omega_k^4 \underline{\omega}_s^2 \leq \frac{C}{\bar{\lambda}_k^2} = O(\varepsilon_{ks}^4)$$

$$\int_{|x|\leq 1} \omega_k^4 = O\left(\frac{1}{\bar{\lambda}_k}\right) = O\left(\varepsilon_{ks}^2\right)$$

$$\int_{|x|\leq 1} r^2 \omega_k^4 \leq C \int_{r\leq 1} \frac{r^2 \bar{\lambda}_k^2 r^2 dr}{\left(1+\bar{\lambda}_k^2 r^2\right)^2} \leq \frac{C\bar{\lambda}_k}{\bar{\lambda}_k^3} = \frac{C}{\bar{\lambda}_k^2} = O(\varepsilon_{ks}^4)$$

so that

$$\left(\int_{\Omega_k} \omega_k^4 \omega_s^2\right)^{1/2} \leq C\left(|\underline{\omega}_s|(\underline{a}_k)\varepsilon_{ks} + \varepsilon_{ks}^2\right)$$

and, if $\lambda_k \geq \lambda_s$,

$$\int_{\Omega_k} \omega_k^4 |\omega_s||\partial\omega_\ell| \leq C\varepsilon_{k\ell}\left(|\underline{\omega}_s|(\underline{a}_k)\varepsilon_{ks} + \varepsilon_{ks}^2\right)$$

while, if $\lambda_s \geq \lambda_k$,

$$\int_{\Omega_s} \omega_s^4 |\omega_k||\partial\omega_\ell| \leq C\varepsilon_{s\ell}\left(|\underline{\omega}_k|(\underline{a}_s)\varepsilon_{ks} + \varepsilon_{ks}^2\right).$$

$\square$

1.10　Proof of the Morse Lemma at Infinity

1.10.1　*Decomposition in groups, gradient and L^∞-estimates on $\bar{v}$, proof of the Morse Lemma at infinity*

1.10.2　*Content of Part III*

Parts I and II have clearly indicated that the main obstacle to the derivation of a general Morse Lemma at infinity relies on the fact that some masses, which interact very little, could be concentrated at points very close to one another. The remainder terms are then impossible to control.

If this occurs, $\sum \alpha_i \omega_i$ can be split in several clusters, denoted for simplicity as I and II. The masses in each cluster are very close to one another compared to the masses of I with respect to the masses of II (the distance of the cluster I to the cluster II is of the order of $|a|$). We then have two clusters and their optimal $\bar{v}$'s:

$$\sum_{i \in I} \alpha_i \omega_i + \bar{v}_I, \quad \sum_{i \in II} \alpha_i \omega_i + \bar{v}_{II}$$

and a remainder $\bar{v} - (\bar{v}_I + \bar{v}_{II})$.

In Part III we complete major improvements of the estimates of Part II.

On one hand, using the uniqueness feature of the implicit function theorem applied to $\bar{v}$, we get rid of all the remainder terms $C_h(\sum \varepsilon_{st})^h$ involved in the pointwise estimates on $\bar{v}$ in Part II.

On the other hand, we derive a much better estimate, for every x of $\mathbb{R}^3$, of $\int \frac{\omega_k^4 |\omega_\ell| + |\omega_\ell|^4 |\omega_k|}{|x-y|}$ of the type $C((|\omega_\ell^\infty| + |\omega_\ell(a_k)| \sqrt{\lambda_\ell} |a_k - a_\ell| + \varepsilon_{\ell k})\delta_k + (|\omega_k^\infty| + |\omega_k(a_\ell)| \sqrt{\lambda_k} |a_k - a_\ell| + \varepsilon_{\ell k})\delta_\ell)\varepsilon_{\ell k}$.

We also show that $|\bar{v} - (\bar{v}_I + \bar{v}_{II})|$ is pointwise estimated by such an expression but with $\ell \in I$ and $k \in II$ (Proposition 3).

Another meaningful estimate is derived on ∇v_I in $\Omega_{II}($ and ∇v_{II} in $\Omega_I)$ in terms of

$$\frac{C}{|a|} \left(\sum_{\substack{(k,j) \in I \\ k \neq j}} \left(|\omega_j^\infty| + |\omega_j(a_k)| \sqrt{\lambda_j} |a_k - a_j| + \varepsilon_{kj} \right) \varepsilon_{kj} \delta_k \right).$$

All these improvements concur and yield a better estimate on $|(v - (v_I + v_{II}))|_{H_0^1}$ than the one provided in Theorem 1 of Part II. The new estimate has the definite advantage that instead of splitting the contribution in the upperbound of the ω_j's of each cluster ($C \sum |\omega_j^\infty|$ etc.), it combines them in expressions such as $\sum_{j\in I}\alpha_j\omega_j(a_k)$ i.e. $\sum_{j\in I}\alpha_j\omega_j$ behaves as a single function $\omega_j, j\in I$.

We then complete the expansion and derive the estimates needed in order to get rid of the remainder terms. The new fact is that, instead of estimating the variation of J as we translate each w_j independently, we now translate each cluster altogether; the various relative positions of the masses ω_j's in each cluster are unchanged.

Using the improvements described above and a tricky way to write $\partial J(\sum_{i\in I}\alpha_i\omega_i + \bar{v})\cdot\partial_I(\sum\alpha_i\omega_i + \bar{v}) - \partial_I$ is the derivative with respect to translation of the first cluster - we derive a key lemma, Lemma 67 which shows that all remainder terms in our expansion are now $O\left(\frac{\varepsilon_{ij}^3}{|a|}\right)$ (for the derivative with respect to the translation of the point of concentration of each cluster). We indicate then how to turn $O\left(\frac{\varepsilon_{ij}^3}{|a|}\right)$ into $o\left(\frac{\varepsilon_{ij}^3}{|a|}\right)$ and we derive the general Morse Lemma at infinity under (A1).

1.10.3 *Basic conformally invariant estimates*

This section is devoted to the derivation of new estimates on $\bar{v}, v_I, v_{II}$ and their derivatives. In previous works e.g. [Bahri 1989], when considering only positive functions or even in [Bahri 2001] for the case of two masses, we could work without such estimates because, at first and second order in the expansion of J, $\bar{v}$ intervened only through its projection $Q(\bar{v})$ on $\mathrm{Span}\{\omega_i, \frac{\partial\omega_i}{\partial\sigma}, \frac{\partial\omega_i}{\partial a_i}$ etc.$\}$ which was relatively easily estimated. The higher order $\bar{v}$-terms were dominated by the principal part of the expansion (of J and of ∂J).

This does not happen anymore here and we have to estimate $\bar{v}, v_I, v_{II}, \nabla v_I, \nabla v_{II}, \bar{v} - (v_I + v_{II})$ appropriately (in appropriate regions).

The new estimates which we will prove read as follows:

Proposition 3

(i) $|\bar{v}(x)| \leq C\sum(|\omega_k^\infty| + |\omega_k(a_\ell)|(\sqrt{\lambda_k}|a_k - a_\ell|) + \varepsilon_{k\ell})\varepsilon_{k\ell}\delta_\ell(x)$

(ii) $|v_I(x)| \leq C\sum_{k,\ell\in I}(|\omega_k^\infty| + |\omega_k(a_\ell)|(\sqrt{\lambda_k}|a_k - a_\ell|) + \varepsilon_{k\ell})\varepsilon_{k\ell}\delta_\ell$

(iii) $|v_{II}(x)| \leq C\sum_{k,\ell\in II}(|\omega_k^\infty| + |\omega_k(a_\ell)|(\sqrt{\lambda_k}|a_k - a_\ell|) + \varepsilon_{k\ell})\varepsilon_{k\ell}\delta_\ell$

(iv) $\displaystyle |v - (v_I + v_{II})| \leq \sum_{(k,\ell)\in(I,II)}\left[\left(\left(\sum|\omega_k^\infty| + \varepsilon_{k\ell}\right)\varepsilon_{k\ell} + \frac{\sum|\omega_k(a_\ell)|}{\sqrt{\lambda_\ell}}\right)\delta_\ell\right.$

$$+ \left.\left(\left(\sum|\omega_\ell^\infty| + \varepsilon_{k\ell}\right)\varepsilon_{k\ell} + \frac{\sum|\omega_\ell(a_k)|}{\sqrt{\lambda_k}}\right)\delta_k\right] + \sum_{\substack{(i,j)\in(I,II)\\ or\ (II,I)}}O\left(\sum\varepsilon_{\ell m}\right)\varepsilon_{ij}\delta_i.$$

Proposition 4

(i) *Assume that y is in $\widetilde{\Omega}_{II}$, a smaller version of Ω_{II} such that $d(\Omega_I, \partial\widetilde{\Omega}_{II}) \geq c|a|$. Then, for $y \in \widetilde{\Omega}_{II}$,*

$$|\nabla v_I(y)| \leq \frac{C}{|a|}\left(\sum_{\substack{k,j\in I \\ k\neq j}}(|\omega_j^\infty| + |\omega_j(a_k)|\sqrt{\lambda_j}|a_k - a_j| + \varepsilon_{kj})\varepsilon_{kj}\delta_k\right)$$

(ii) *Assume that y is in $\widetilde{\Omega}_I$, a smaller version of Ω_I such that $d(\Omega_{II}, \partial\widetilde{\Omega}_I) \geq c|a|$. Then, for $y \in \widetilde{\Omega}_I$,*

$$|\nabla v_{II}(y)| \leq \frac{C}{|a|}\left(\sum_{\substack{k,j\in II \\ k\neq j}}(|\omega_j^\infty| + |\omega_j(a_k)|\sqrt{\lambda_j}|a_k - a_j| + \varepsilon_{kj})\varepsilon_{kj}\delta_k\right).$$

(iii) *For all $y \in S^3, |\nabla v_I| \leq C\sum_{i\in I}\sqrt{\lambda_i}\delta_i^2$.*

We will establish these estimates in several steps.
In a first step, we assume that

$$\lambda_k > \lambda_\ell, \Omega_k = \{x \in \mathbb{R}^3 s.t. \lambda_k|x - a_k| < \frac{1}{8\varepsilon_{k\ell}}\};$$

$$\Omega_\ell = \{x \in \mathbb{R}^3 s.t. \lambda_\ell|x - a_\ell| < \frac{1}{8\varepsilon_{k\ell}}, \lambda_k|x - a_k| \geq \frac{1}{\varepsilon_{k\ell}}\}.$$

We start with:

Lemma 58   *Assume that $\lambda_k \geq 8\lambda_\ell$.*
     Then

$$\int_{\Omega_k} \frac{\omega_k^4|\omega_\ell| + |\omega_k|\omega_\ell^4}{|x - y|} \leq C\left(|\omega_\ell^\infty| + |\omega_\ell(a_k)|\sqrt{\lambda_\ell}|a_k - a_\ell| + \varepsilon_{\ell k}\right)\varepsilon_{k\ell}\delta_k$$

and

Lemma 59

$$\varepsilon_{k\ell}\delta_k \leq C\delta_\ell.$$

Proof of Lemma 58. Considering

$$(1) = \int_{\Omega_k} \frac{\omega_k^4|\omega_\ell|}{|x - y|}, \text{we have since } \lambda_k \geq 8\lambda_\ell$$

$$|w_\ell(x)| \le C\left(|w_\ell(a_k)|\sqrt{\lambda_\ell}|a_k - a_\ell| + \left(\frac{|w_\ell^\infty|}{|a_k - a_\ell|} + \frac{1}{\lambda_\ell|a_k - a_\ell|^2}\right)|x - a_k|\right)$$

$$\times \frac{1}{\sqrt{\lambda_\ell}|a_k - a_\ell|} \quad \text{for } x \in \Omega_k$$

so that since $|x - a_k| \le C/\lambda_k \varepsilon_{\ell k}$

$$(1) \le C\left(|w_\ell(a_k)|\sqrt{\lambda_\ell}|a_k - a_\ell| + |w_\ell^\infty| + \varepsilon_{\ell k}\right)\frac{1}{\sqrt{\lambda_\ell}|a_k - a_\ell|}\int_{\Omega_k}\frac{w_k^4}{|x - y|}.$$

$\square$

Observe now that

$$\int_{\Omega_k}\frac{w_k^4}{|x - y|} \le \sqrt{\lambda_k}\int_{\Omega_k\cap\{\lambda_k|x-y|\ge 1\}} w_k^4\delta_k(y)$$

$$+ \int_{\Omega_k\cap\{|x-y|\le\frac{1}{\lambda_k}\}}\frac{w_k^4}{|x - y|} \quad \text{if } \lambda_k|y - a_k| \text{ is large}$$

$$\le \frac{C\delta_k}{\sqrt{\lambda_k}} + \frac{C}{\lambda_k^2|a_k - y|^4}\times\frac{1}{\lambda_k^2} \le \frac{C}{\lambda_k|y - a_k|}.$$

Thus,

$$(1) \le C\left(|w_\ell^\infty| + |w_\ell(a_k)|\sqrt{\lambda_\ell}|a_k - a_\ell| + \varepsilon_{\ell k}\right)\varepsilon_{k\ell}\delta_k \quad \text{if } \lambda_k|y - a_k| \text{ is large.}$$

If $\lambda_k|y - a_k|$ is bounded by C, then

$$\int_{\Omega_k}\frac{w_k^4}{|x - y|} \le C\sqrt{\lambda_k}\int_{\lambda_k|x-y|\ge 100C} w_k^4\delta_k(y)$$

$$+ \int_{\lambda_k|x-y|\le 100C}\frac{w_k^4}{|x - y|} \le \frac{C\delta_k}{\sqrt{\lambda_k}} + C \le C\frac{\delta_k(y)}{\sqrt{\lambda_k}}$$

and the estimate above extends.

We now consider

$$(2) = \int_{\Omega_k}\frac{|w_k|w_\ell^4}{|x - y|}$$

$$\le C\left(\frac{w_\ell^{\infty 4}}{\lambda_\ell^2|a_k - a_\ell|^4} + (|w_\ell(a_k)|\sqrt{\lambda_\ell}|a_k - a_\ell|)^4 + \varepsilon_{k\ell}^4\right)\int_{\Omega_k}\frac{|w_k|}{|x - y|}$$

and we subdivide between the case

a. $\lambda_k|y - a_k| \le C.$

Then,

$$\int_{\Omega_k} \frac{|\omega_k|}{|x-y|} \le C\sqrt{\lambda_k} \int_{|x-y|\le \frac{100C}{\lambda_k}} \frac{1}{|x-y|} + C\sqrt{\lambda_k} \int_{\Omega_k} \delta_k^2(x)dx$$

$$\le \frac{C}{\lambda_k^{3/2}} + \frac{C\sqrt{\lambda_k}}{\lambda_k^2} \int_{r\le \frac{1}{\varepsilon_{\ell k}}} \frac{r^2 dr}{1+r^2} \le \frac{\bar{C}}{\lambda_k^{3/2}} \cdot \frac{1}{\varepsilon_{k\ell}}.$$

Thus,

$$(2) \le C \left(\omega_\ell^{\infty^4} + (|\omega_\ell(a_k)|\sqrt{\lambda_\ell}|a_k - a_\ell|)^4 + \varepsilon_{k\ell}^4 \right) \varepsilon_{k\ell}\delta_k.$$

b. $\lambda_k|y - a_k| \ge C.$

Then,

$$\int_{\Omega_k} \frac{|\omega_k|}{|x-y|} \le C \left(\frac{1}{|y-a_k|} \int_{|x-a_k|\le \frac{C}{100\lambda_k}} \delta_k \right.$$

$$+ \frac{1}{\sqrt{\lambda_k}|y-a_k|} \int_{|x-y|\le \frac{C}{100\lambda_k}} \frac{1}{|x-y|} + \frac{1}{\sqrt{\lambda_k}} \int_{\substack{|x-a_k|\ge \frac{C}{100\lambda_k} \\ |x-y|\ge \frac{C}{100\lambda_k} \\ x\in\Omega_k}} \left. \frac{1}{|x-a_k||y-x|} \right).$$

Observe that

$$\text{Max} \left(|x - a_k|, |y - x| \right) \ge \frac{1}{2}|y - a_k|$$

and that

$$\text{Min} \left(|x - a_k|, |y - x| \right) \le \frac{\bar{C}}{\lambda_k\varepsilon_{k\ell}} \quad \text{for } x \in \Omega_k.$$

Thus,

$$\int_{\Omega_k} \frac{|\omega_k|}{|x-y|} \le \frac{C}{\lambda_k^{5/2}|y-a_k|} + \frac{C}{\sqrt{\lambda_k}|y-a_k|} \int_{\frac{\bar{C}}{\lambda_k\varepsilon_{k\ell}}\ge r\ge \frac{C}{100\lambda_k}} rdr$$

$$\leq \frac{C}{\lambda_k^{5/2}|y - a_k|} + \frac{C}{\lambda_k^2 \varepsilon_{k\ell}^2} \cdot \frac{1}{\sqrt{\lambda_k}|y - a_k|}$$

and

$$(2) \leq C\delta_k \left(|\omega_\ell^\infty|^4 + (|\omega_\ell(a_k)|\sqrt{\lambda_\ell}|a_k - a_\ell|)^4 + \varepsilon_{k\ell}^4\right) \varepsilon_{k\ell}.$$

Combining the estimates on (1.5) and (1.2), we derive Lemma 58.

Proof of Lemma 59. We are assuming that

$$\varepsilon_{k\ell} \sim \frac{C}{\sqrt{\lambda_k \lambda_\ell}|a_k - a_\ell|}$$

so that

$$\varepsilon_{k\ell}\delta_k \sim \frac{C}{\sqrt{\lambda_k \lambda_\ell}|a_k - a_\ell|} \cdot \frac{\sqrt{\lambda_k}}{(1 + \lambda_k^2|x - a_k|^2)^{1/2}}$$
$$= \frac{C}{\sqrt{\lambda_\ell}|a_k - a_\ell|(1 + \lambda_k^2|x - a_k|^2)^{1/2}}.$$

Our inequality is thus equivalent to:

$$(1 + \lambda_\ell^2|x - a_\ell|^2)^{1/2} \leq C\lambda_\ell|a_k - a_\ell|(1 + \lambda_k^2|x - a_k|^2)^{1/2}$$

which clearly holds if $|x - a_\ell| \leq C_1|a_k - a_\ell|$. If $|x - a_\ell|$ is very large with respect to $|a_k - a_\ell|$, then it is of the same order than $|x - a_k|$ and the inequality becomes

$$\lambda_\ell|x - a_k| \leq C\lambda_\ell\lambda_k|x - a_k||a_k - a_\ell|$$

i.e. $\lambda_\ell|a_k - a_\ell|$ large, which holds. $\qquad\square$

Next, we extend Lemma 58 to Ω_ℓ and establish:

Lemma 60

$$\int_{\Omega_\ell} \frac{\omega_k^4|\omega_\ell| + \omega_\ell^4|\omega_k|}{|x - y|} \leq C \left(|\omega_k^\infty| + |\omega_k(a_\ell)|\sqrt{\lambda_k}|a_k - a_\ell| + \varepsilon_{k\ell}\right) \varepsilon_{k\ell}\delta_\ell.$$

Proof. If $\lambda_\ell \geq 8\lambda_k$, Lemma 60 follows readily from Lemma 58. If $\lambda_k \geq 8\lambda_\ell$, $\lambda_k|x - a_k| \geq \varepsilon_{\ell k}^{-1}$ for $x \in \Omega_\ell$ and the estimate is straightforward. If λ_k and λ_ℓ are equivalent, the estimate is also straightforward. $\qquad\square$

Lemma 58 and Lemma 60 combine and with a slight generalization, we derive the estimate:

Lemma 61

$$\int \frac{\omega_k^4|\omega_\ell| + \omega_\ell^4|\omega_k|}{|x-y|} \leq C\left(\left(|\omega_\ell|^\infty + |\omega_\ell(a_k)|\sqrt{\lambda_\ell}|a_k - a_\ell| + \varepsilon_{\ell k}\right)\delta_k\right.$$

$$\left. + \left(|\omega_k^\infty| + |\omega_k(a_\ell)|\sqrt{\lambda_k}|a_k - a_\ell| + \varepsilon_{\ell k}\right)\delta_\ell\right)\varepsilon_{\ell k}.$$

We now establish (i), (ii), (iii) of Proposition 3. The proof of (iv) requires a few further remarks and we will be completing later, in a section devoted to $v - (v_I + v_{II})$.

Proof of (i), (ii), (iii) of Proposition 3. $\bar{v}$ satisfies

$$Q_{F^\perp}\left(J'\left(\sum \alpha_j \omega_j + \bar{v}\right)\right) = 0. \tag{1.26}$$

Picking up a large constant C_1 (much larger than $C\sum|\omega_j^\infty|$), we modify (1.26) as follows; we write it as

$$Q^* L\left(J'\left(\sum \alpha_j \omega_j\right)\right) + \Delta \bar{v} + Q^*\left(5\left(\sum \alpha_j \omega_j\right)^4 \bar{v} + \sum_{m=2}^{5} f_m \bar{v}^m\right) = 0 \tag{1.27}$$

which we transform into

$$-\Delta\tilde{v} = Q^* L\left(J'\left(\sum \alpha_j \omega_j\right)\right) - Q^*\left(5\left(\sum \alpha_j \omega_j\right)^4 \tilde{v}\chi_{U\Omega_i}\right.$$

$$+ \sum_{m=2}^{5} f_m \inf\left(C_1 \sum \varepsilon_{ij}\delta_i, |\tilde{v}|\right)^{m-1} sgn(\tilde{v}^{m-1})\tilde{v}$$

$$\left. + 5\left(\sum \alpha_j \omega_j\right)^4 \chi_{(U\Omega_i)^c} \times sgn\tilde{v} \times \inf\left(|\tilde{v}|, C_1 \sum \varepsilon_{ij}\delta_i\right)\right) = \tilde{f} \tag{1.28}$$

(1.28) is only a slight modification of (1.27) and one can easily see that the inversion theorem applies to it yielding a unique solution $\tilde{v}$ in $F^\perp$.

If we are able to prove on $\tilde{v}$ the estimate

$$|\tilde{v}(x)| \leq C_1 \sum \varepsilon_{ij}\delta_i \tag{1.29}$$

then $\tilde{v} = \bar{v}$ and $\bar{v}$ satisfies (1.28). In order to prove (1.29), we derive from (1.28)

$$\tilde{v} = c\int \frac{\tilde{f}}{|x-y|} \tag{1.30}$$

$\tilde{v}$ splits also in pieces $\tilde{v}_i, \tilde{h}_i$ just as $\bar{v}$ did. Theorem 1 will extend readily with an improvement: there is no contribution of $C_h\left(\sum \varepsilon_{st}\right)^h$ because we can use (1.29) in $\left(U\widetilde{\Omega}_i\right)^c$. The constant C_1 is then multiplied by $o(1)$ we derive the estimate

$$|\tilde{v}_i|_{H_0^1} + \frac{|\tilde{h}_i^0|}{\sqrt{\lambda_i}} + \sum_{\substack{\lambda_s \geq \lambda_i \\ s \neq i}} \varepsilon_{is}|\tilde{h}_{is}^1| \leq C \sum \varepsilon_{ij} \qquad (1.31)$$

where C does not depend on C_1 because

$$\inf\left(|\tilde{v}|, C_1 \sum \varepsilon_{ij}\delta_i\right) \leq |\tilde{v}|. \qquad (1.32)$$

In order to prove (1.31) - (1.32), we first prove that $\tilde{v}$ satisfies (i) of Proposition 2, page 72 with $\theta_h = 0$. Indeed, using (1.30) in (1.28) and forgetting a first step the projection operator Q^*, we find

$$\begin{aligned}
|\tilde{v}| \leq C &\left(\int \sum_{k \neq j} \frac{\omega_j^4|\omega_k| + \omega_k^4|\omega_j|}{|x-y|} \right.\\[2mm]
&+ \sum_i \left(|\tilde{v}_i|_{H_0^1} \int_{\Omega_i} \sum_j \frac{\delta_j^5}{|x-y|} + |\tilde{h}_i^0|_\infty \int_{\Omega_i} \sum_k \frac{\delta_k^4}{|x-y|} \right.\\[2mm]
&+ \sum_{\substack{s \\ \lambda_s \geq \lambda_i \\ s \neq i}} |\tilde{h}_i^s|_\infty \int_{\Omega_i} \sum_k \frac{\delta_k^4 \delta_s}{|x-y|} \left. \right) + \sum_2^5 C_1^{m-1} \varepsilon_{ik}^{m-1} \left(|\tilde{v}_i|_{H_0^1} \int_{\Omega_i} \frac{\delta_k^5}{|x-y|} \right.\\[2mm]
&+ |\tilde{h}_i^0|_\infty \int_{\Omega_i} \frac{\delta_k^4}{|x-y|} + \sum_{\substack{s \\ \lambda_s \geq \lambda_i \\ s \neq i}} |\tilde{h}_i^s|_\infty \int_{\Omega_i} \frac{\delta_k^4 \delta_s}{|x-y|} \left. \right)\\[2mm]
&+ \sum_j \int_{(U\Omega_\ell)^c} \frac{C_1 \delta_j^4 \sum \varepsilon_{ik}\delta_k + C_1^5 \sum \varepsilon_{ik}^5 \delta_k^5}{|x-y|} \right)
\end{aligned}$$

$$(1.33)$$

C is independent of C_1.
We will prove later that

$$\int_{\Omega_i} \frac{\delta_j^4}{|x-y|} \leq \frac{c\delta_i}{\sqrt{\lambda_i}}, \quad \int_{\Omega_i} \frac{\delta_j^5}{|x-y|} \leq C\delta_i \qquad (1.34)$$

$$\int\limits_{(U\Omega_\ell)^c} \frac{\delta_i^5}{|x-y|} = o(\delta_i), \quad \int_{\Omega_i} \frac{\delta_j^4 \delta_s}{|x-y|} = O(\varepsilon_{is})\delta_i \text{ for } \lambda_s \underset{s\neq i}{\geq} \lambda_i \qquad (1.35)$$

(i) of Proposition 2 follows. Theorem 1 thereby follows for $\tilde{v}$, (iii) of Proposition 2 is not needed because (1.28) is a modified version of (1.26) and the nonlinearity involve $\inf(|\tilde{v}|, C_1 \sum \varepsilon_{k\ell}\delta_\ell)$ in $(\cup\Omega_j)^c$.

Then, after tracking down the contribution of Q^*, (1.33) implies that

$$|\tilde{v}(x)| \leq C \sum \varepsilon_{ij}\delta_i + o\left(C_1 \sum \varepsilon_{ij}\delta_i\right) \qquad (1.36)$$

(1.29) follows and $\tilde{v} = \bar{v}$.

Plugging then in (1.28) the estimates on $|\tilde{v}|_{H_0^1}$ and $\frac{|h_i|_\infty}{\sqrt{\lambda_i}}$ provided by Theorem 1 (without $C_h(\sum \varepsilon_{st})^h$, we got rid of this contribution), using the estimates on $\int_{\Omega_i} \frac{\delta_k^4}{|x-y|}, \int_{(U\Omega_\ell)^c} \frac{\delta_j^5}{|x-y|}$ and using also the estimate (Lemma 59)

$$\varepsilon_{ik}\delta_k \leq C\delta_i$$

we derive that

$$|\tilde{v}(x)| = |\bar{v}(x)| \leq$$

$$C\left(\sum(|\omega_k^\infty| + |\omega_k(a_\ell)|\sqrt{\lambda_k}|a_k - a_\ell| + \varepsilon_{k\ell})\varepsilon_{k\ell}\delta_\ell(x) + \sum_{i\neq\ell} \varepsilon_{ki}\varepsilon_{k\ell}\delta_\ell\right).$$
$$(1.37)$$

The extra-term $\sum \varepsilon_{ik}\varepsilon_{k\ell}\delta_\ell$ is due to $\int_{(U\Omega_\ell)^c} \frac{\delta_j^4 \sum \varepsilon_{ik}\delta_k}{|x-y|}$ (see Lemma 40 $k \neq j$).

Bootstrapping (1.37) into (1.28), we get a better estimate with $\sum_{i\neq\ell} \varepsilon_{ki}\varepsilon_{k\ell}\delta_\ell$ replaced by $\varepsilon_{ik}\varepsilon_{k\ell}\varepsilon_{k_1\ell_1}\delta_{\ell_1}$ with $i \neq \ell, i \neq \ell_1, k, \ell \neq \ell_1, k_1 = k$ or ℓ (otherwise we are done). Thus ε_{ij} and $\varepsilon_{k\ell}$ belong to different pairs, ε_{ij} and $\varepsilon_{k_1\ell_1}$ belong to different pairs as well as $\varepsilon_{k\ell}$ and $\varepsilon_{k_1\ell_1}$. Since there is a finite number of pairs, we end up with a term of the type $\varepsilon_{k\ell}^2\delta_\ell$ and we are done.

(i), (ii), (iii) of Proposition 3 follow.

□

We now establish (i) and (ii) of Proposition 4.

Proof of Proposition 4. We write

$$v_I(y) = \int \frac{f_I}{|x-y|}.$$

We assume that $y \in \widetilde{\Omega}_{II}$ and split

$$\nabla v_I(y) = \int f_I O\left(\frac{1}{|x-y|^2}\right)$$

$$= \int_{|x-y| \geq c|a|} f_I O\left(\frac{1}{|x-y|^2}\right) + \int_{|x-y| \leq c|a|} f_I O\left(\frac{1}{|x-y|^2}\right).$$

We can easily estimate the first term and we derive

$$\left| \int_{|x-y| \geq c|a|} f_I O\left(\frac{1}{|x-y|^2}\right) \right| = O\left(\frac{1}{|a|}\right) \int \frac{|f_I|}{|x-y|}.$$

The second term yields

$$\left| \int_{|x-y| \leq c|a|} f_I O\left(\frac{1}{|x-y|^2}\right) \right| \leq C|a| \operatorname*{Sup}_{|x-y| \leq c|a|} |f_I(x)|.$$

In f_I, all terms are explicit except those involving v_I.

For those, we can use (ii) of Proposition 3. Thus, since $y \in \widetilde{\Omega}_{II}$,

$$|a| \operatorname*{Sup}_{|x-y| \leq c|a|} |f_I(x)| \leq \frac{C}{|a|} \cdot |a|^2 \left(\sum_{k,\ell \in I} \left(|\omega_k^\infty| + \frac{1}{\lambda_k |a|} + \varepsilon_{k\ell} \right)^4 \right.$$

$$\times \left(|\omega_\ell^\infty| + \frac{1}{\lambda_\ell |a|} + \varepsilon_{k\ell} \right) \left(\varepsilon_{k\ell} \frac{1}{\sqrt{\lambda_k}|a|} + \frac{1}{\lambda_k^2 \sqrt{\lambda_\ell}|a|^5} \right) \right)$$

$$\leq \frac{C}{|a|} \left(\sum_{k,\ell \in I} \left(|\omega_k^\infty| + \frac{1}{\lambda_k |a|} + \varepsilon_{k\ell} \right)^4 \right.$$

$$\times \left(|\omega_\ell^\infty| + \frac{1}{\lambda_\ell |a|} + \varepsilon_{k\ell} \right) \times \left(\frac{\varepsilon_{k\ell}}{\sqrt{\lambda_k}|a|} + \frac{1}{\sqrt{\lambda_k}\sqrt{\lambda_\ell}|a|} \cdot \frac{1}{\lambda_k |a|} \cdot \frac{1}{\sqrt{\lambda_k}|a|} \right) \right).$$

Combining the two estimates, using (1.28) and (1.31) to estimate $\int \frac{|f_I|}{|x-y|}$ as already done above, using Lemma 61, we derive the estimate on ∇v_I, hence also on ∇v_{II}. $\qquad\square$

We prove now (1.34), (1.35).

We start with $\int_{\Omega_i} \frac{\delta_j^4}{|x-y|}$. In Lemma 68, we establish that $\delta_j \leq C\delta_i$ on $\Omega_i, \forall j$.

Thus,

$$\int_{\Omega_i} \frac{\delta_j^4}{|x-y|} \le C \int_{\Omega_i} \frac{\delta_i^4}{|x-y|} \le \frac{C\delta_i}{\sqrt{\lambda_i}}. \tag{1.38}$$

We now consider($\widetilde{\Omega}_j$ are smaller versions of Ω_j).

$$\int_{(U\Omega_\ell)^c} \frac{\delta_j^5}{|x-y|} \le C\sqrt{\lambda_j} \sum_{\lambda_\ell < \lambda_j} \int_{(U\Omega_\ell)^c} \frac{\delta_j^4}{|x-y|} + C \sum_{\lambda_\ell > \lambda_j} \int_{\widetilde{\Omega}_j^c} \frac{\delta_j^4 \delta_\ell}{|x-y|}$$

$$\le C \sum \varepsilon_{j\ell} \delta_j. \tag{1.39}$$

Finally, considering $\int_{\Omega_i} \frac{\delta_k^4 \delta_s}{|x-y|}$ and using Lemma 68, we upperbound it by $\int_{\Omega_i} \frac{\delta_i^4 \delta_s}{|x-y|}$. We have, if $\lambda_i |y - x_i|$ is large

$$\int_{\Omega_i} \frac{\delta_i^4 \delta_s}{|x-y|} \le \sqrt{\lambda_i} \int_{\Omega_i} \delta_i^4 \delta_s \delta_{y,\lambda_i} + \int_{\Omega_i \cap \{|x-y| \le \frac{1}{\lambda_i}\}} \frac{\delta_i^4 \delta_s}{|x-y|} \tag{1.40}$$

$$\le \frac{c}{\sqrt{\lambda_i}|y-x|}\varepsilon_{is} + \int_{\Omega_i \cap \{|x-y| \le \frac{1}{\lambda_i}\}} \frac{\delta_i^4 \delta_s}{|x-y|} \le C\varepsilon_{is}\delta_i + \int_{\Omega_i \cap \{|x-y| \le \frac{1}{\lambda_i}\}} \frac{\delta_i^4 \delta_s}{|x-y|}.$$

If $\lambda_i |y - x_i|$ is bounded, then

$$\int_{\Omega_i} \frac{\delta_i^4 \delta_s}{|x-y|} \le C\sqrt{\lambda_i} \int_{\Omega_i} \delta_i^5 \delta_s + \int_{\Omega_i \cap \{|x-y| \le \frac{C}{\lambda_i}\}} \frac{\delta_i^4 \delta_s}{|x-y|}$$

$$\le C\varepsilon_{is}\delta_i + \int_{\Omega_i \cap \{|x-y| \le \frac{C}{\lambda_i}\}} \frac{\delta_i^4 \delta_s}{|x-y|}.$$

If $|y - x_s| \ge c|x_s - x_i|$, we write

$$\int_{\Omega_i \cap \{|x-y| \le \frac{1}{\lambda_i}\}} \frac{\delta_i^4 \delta_s}{|x-y|} \le \frac{C\delta_i^4(y)}{\sqrt{\lambda_s}} \int_{|x-y| \le \frac{C}{\lambda_i}} \frac{1}{|x - x_s||x - y|}$$

$$\le \frac{C\delta_i^4(y)}{\sqrt{\lambda_s}|y - x_s|\lambda_i^2} \le C_1 \varepsilon_{si}\delta_i.$$

If $|y - x_s| = o(|x_s - x_i|)$, we write

$$\int_{\Omega_i \cap \{|x-y| \leq \frac{1}{\lambda_i}\}} \frac{\delta_i^4 \delta_s}{|x-y|} \leq \frac{c\delta_i^4(y)}{\sqrt{\lambda_s}} \int_{r \leq \frac{1}{\lambda_i}} dr$$

$$\leq \frac{C}{\lambda_i^3 \sqrt{\lambda_s}|y - x_i|^4} \leq C_1 \varepsilon_{si} \delta_i.$$

Summing up, we derive our claim. (i) and (ii) of Proposition 4 follow.

We now establish (iii).

We observe that ∇v_I satisfies:

$$-\Delta \nabla v_I = \nabla f_I$$

so that

$$|\nabla v_I(y)| \leq C \int \frac{|f_I|}{|x - y|^2}.$$

Using then Proposition 3, (ii) and Lemma 59, we derive that

$$|\nabla v_I(y)| \leq C \sum_{k \neq \ell} \int \frac{\delta_k^4 \delta_\ell}{|x - y|^2}.$$

If $\lambda_k |y - a_k| \geq 1$, we split $\int \frac{\delta_k^4 \delta_\ell}{|x-y|^2}$ in two pieces:

$$(I) = \int_{|x-y| \geq c|y-a_k|} \frac{\delta_k^4 \delta_\ell}{|x - y|^2}, \quad (II) = \int_{|x-y| \leq c|y-a_k|} \frac{\delta_k^4 \delta_\ell}{|x - y|^2}$$

c is a small constant.

We derive easily:

$$(I) \leq \frac{C}{|y - a_k|^2} \frac{\varepsilon_{k\ell}}{\sqrt{\lambda_k}} \leq C\varepsilon_{k\ell} \sqrt{\lambda_k} \delta_k^2.$$

For (II), we observe that, with c small, $|x - a_k|$ and $|y - a_k|$ are of the same order on the domain of integration of (II). Thus

$$(II) \leq \frac{C\lambda_k^2}{(1 + \lambda_k^2|y - a_k|^2)^2} \left(\int_{\substack{|x-y| \leq c|y-a_k| \\ |x-a_\ell| \geq c|a_k-a_\ell|}} \frac{\delta_\ell}{|x - y|^2} + \int_{\substack{|x-y| \leq c|y-a_k| \\ |x-a_\ell| \leq |a_k-a_\ell|}} \frac{\delta_\ell}{|x - y|^2} \right)$$

$$= (III) + (IV).$$

(III) is easily upperbounded by

$$(III) \leq \frac{C\lambda_k^2}{(1 + \lambda_k^2|y - a_k|^2)^2} \frac{1}{\sqrt{\lambda_\ell}|a_k - a_\ell|}|y - a_k| \leq C\varepsilon_{k\ell}\sqrt{\lambda_k}\delta_k^2.$$

We split the domain of integration of (IV) into two parts:

On the first one, $|x - y| \leq c|y - a_\ell| + \frac{1}{\lambda_\ell}$ so that $|x - a_\ell|$ and $|y - a_\ell|$ are of the same order unless $\lambda_\ell|y - a_\ell| \leq C$. This contribution is upperbounded by

$$\frac{C\lambda_k^2}{(1 + \lambda_k^2|y - a_k|^2)^2} \times \frac{\sqrt{\lambda_\ell}}{(1 + \lambda_\ell^2|y - a_\ell|^2)^{1/2}} \times \left(|y - a_\ell| + \frac{1}{\lambda_\ell}\right).$$

Since $|x - y| \leq c|y - a_k|$, $|x - a_k|$ and $|y - a_k|$ are of the same order. Since $|x - a_\ell| \leq c|a_k - a_\ell|$, $|y - a_k|$ is of the order of $|a_k - a_\ell|$.

We thus derive the upperbound

$$\frac{C\lambda_k^2}{\lambda_k^2|a_k - a_\ell|^2} \times \frac{1}{\sqrt{\lambda_\ell}} \times \frac{1}{1 + \lambda_k^2|y - a_k|^2} \leq C\varepsilon_{k\ell}\sqrt{\lambda_k}\delta_k^2.$$

On the second part of the domain of integration of IV, we use the upperbound $(|x - y| \geq c|y - a_\ell| + \frac{1}{\lambda_\ell})$

$$\frac{C\lambda_k^2}{(1 + \lambda_k^2|y - a_k|^2)^2} \left(\frac{\lambda_\ell^2}{\lambda_\ell^2|y - a_\ell|^2 + 1}\right) \int\limits_{|x - a_\ell| \leq c|a_k - a_\ell|} \delta_\ell$$

$$\leq \frac{C\lambda_k^2}{(1 + \lambda_k^2|y - a_k|^2)^2} \times \lambda_\ell\delta_\ell^2 \times \frac{|a_k - a_\ell|^2}{\sqrt{\lambda_\ell}}.$$

As seen above, $|y - a_k|$ is of the order of $|a_k - a_\ell|$ and this is upperbounded by

$$\frac{C\sqrt{\lambda_\ell}\delta_\ell^2}{(1 + \lambda_k^2|a_k - a_\ell|^2)} \leq C\sqrt{\lambda_\ell}\delta_\ell^2.$$

This concludes the proof if $\lambda_k|y - a_k| \geq 1$.

If $\lambda_k|y - a_k| \leq 1$, we split:

$$\int \frac{\delta_k^4\delta_\ell}{|x - y|^2} \leq \int\limits_{|x - y| \leq \frac{1}{\lambda_k}} \frac{\delta_k^4\delta_\ell}{|x - y|^2} + \int\limits_{|x - y| \geq \frac{1}{\lambda_k}} \frac{\delta_k^4\delta_\ell}{|x - y|^2}.$$

On one hand:

$$\int\limits_{|x - y| \geq \frac{1}{\lambda_k}} \delta_k^4\delta_\ell \leq C\lambda_k^2 \frac{\varepsilon_{k\ell}}{\sqrt{\lambda_k}} = C\lambda_k^{3/2}\varepsilon_{k\ell} \leq C_1\varepsilon_{k\ell}\sqrt{\lambda_k}\delta_k^2.$$

On the other hand, if $|x - y| \leq \frac{1}{\lambda_k}$, then

$$\left| |x - a_\ell| - |a_k - a_\ell| \right| \leq |x - a_k| = O\left(\frac{1}{\lambda_k}\right).$$

Thus, since $\lambda_k |a_k - a_\ell|$ is very large, $|x - a_\ell|$ is of the order of $|a_k - a_\ell|$ and

$$\int_{|x-y| \leq \frac{1}{\lambda_k}} \frac{\delta_k^4 \delta_\ell}{|x - y|^2} \leq \frac{C\lambda_k^2}{\sqrt{\lambda_\ell} |a_k - a_\ell|} \times \frac{1}{\lambda_k} \leq C \varepsilon_{k\ell} \delta_k^2 \cdot \sqrt{\lambda_k}$$

(iii) follows.

1.10.4   *Estimates on $v - (v_I + v_{II})$*

We want in this section to show how our estimates on v, v_I, v_{II} extend to $v - (v_I + v_{II})$. The equation satisfied by $v - (v_I + v_{II})$ is basically the same than the one satisfied by the other quantities but it has an additional splitting between group I and group II which makes it more specific. We want to use this splitting to derive better estimates. This requires some care.

The equation satisfied by v is

$$-\Delta v = Q^* \left(\left(\sum \alpha_i \omega_i + v \right)^5 \right)$$

while v_I and v_{II} satisfy

$$-\Delta v_I = Q_I^* \left(\left(\sum_{i \in I} \alpha_i \omega_i + v_I \right)^5 \right)$$

$$-\Delta v_{II} = Q_{II}^* \left(\left(\sum_{i \in I} \alpha_i \omega_i + v_{II} \right)^5 \right).$$

Let us forget in a first step the presence of Q^*, Q_I^*, Q_{II}^* and have a look at the equation satisfied by $v - (v_I + v_{II})$. Since we are not taking care of Q^*, Q_I^*, Q_{II}^*, we will leave an additional term U for this contribution. We thus have

$$- \Delta(v - (v_I + v_{II})) =$$

$$\left(\sum \alpha_i \omega_i + v \right)^5 - \left(\sum_{i \in I} \alpha_i \omega_i + v_I \right)^5 - \left(\sum_{i \in II} \alpha_i \omega_i + v_{II} \right)^5 + U$$

$$= O \left(\left| \sum_{i \in II} \alpha_i \omega_i \right| \left| \sum_{i \in I} \alpha_i \omega_i \right|^4 + \left| \sum_{i \in I} \alpha_i \omega_i \right| \left| \sum_{i \in II} \alpha_i \omega_i \right|^4 \right)$$

$$+ 5 \left(\sum \alpha_i \omega_i \right)^4 (v - (v_I + v_{II}))$$

$$+ \sum_{\mu=1}^{3} O \left(\left| \sum_{i \in II} \alpha_i \omega_i \right| \left| \sum_{i \in I} \alpha_i \omega_i \right|^{\mu-1} + \left| \sum_{i \in II} \alpha_i \omega_i \right|^{\mu} \right) |v_I|^{4-\mu}$$

$$+ \sum_{\mu=1}^{3} O \left(\left| \sum_{i \in II} \alpha_i \omega_i \right|^{\mu-1} \left| \sum_{i \in I} \alpha_i \omega_i \right| + \left| \sum_{i \in I} \alpha_i \omega_i \right|^{\mu} \right) |v_{II}|^{4-\mu}$$

$$+ \sum_{\theta=2}^{5} f_\theta (v - (v_I + v_{II}))^\theta + U.$$

We view it as

$$-\Delta(v - (v_I + v_{II})) - 5 \left(\sum \alpha_i \omega_i \right)^4 (v - (v_I + v_{II})) - \sum_{\theta=2}^{5} f_\theta (v - (v_I + v_{II}))^\theta =$$

$$O \left(\left| \sum_{i \in II} \alpha_i \omega_i \right| \left| \sum_{i \in I} \alpha_i \omega_i \right|^4 + \left| \sum_{i \in I} \alpha_i \omega_i \right| \left| \sum_{i \in II} \alpha_i \omega_i \right|^4 \right)$$

$$+ \sum_{\mu=1}^{3} O \left(\left(\left| \sum_{i \in II} \alpha_i \omega_i \right| \left| \sum_{i \in I} \alpha_i \omega_i \right|^{\mu-1} + \left| \sum_{i \in II} \alpha_i \omega_i \right|^{\mu} \right) |v_I|^{4-\mu} \right.$$

$$\left. + \left(\left| \sum_{i \in II} \alpha_i \omega_i \right|^{\mu-1} \left| \sum_{i \in I} \alpha_i \omega_i \right| + \left| \sum_{i \in I} \alpha_i \omega_i \right|^{\mu} \right) |v_{II}|^{4-\mu} \right) + U.$$

We split $\mathbb{R}^3$ or S^3 as usual in its different pieces Ω_i, adjusted to the ω_i's and $\Omega^c = (U\Omega_i)^c$.

Accordingly, $v - (v_I + v_{II})$ is split in each Ω_ℓ into an H_0^1-piece $(v - (v_I + v_{II}))_\ell$ and a harmonic piece h_ℓ. The non-linear operator $-\Delta u - 5(\sum \alpha_i \omega_i)^4 u - \sum_{\theta=2}^{5} f_\theta u^\theta$ is the same than the one used for $\bar{v}$ and the splitting gives rise to a matrix relating $|(v-(v_i+v_{II}))\ell|_{H_0^1}$, $\frac{|h_\ell^0|_\infty}{\sqrt{\lambda_\ell}}$, $\sum\limits_{\substack{\lambda_s \geq \lambda_{\ell s} \\ s \neq \ell}} \varepsilon_{\ell s} \frac{|h_\ell^s|_\infty}{\sqrt{\lambda_\ell}}$ just

as for $\bar{v}$. We thus need only to keep track of the terms which do not involve $(v-(v_I+v_{II}))\ell$, h_ℓ. These terms all come either from

$$g = O\left(|\sum_{i\in II}\alpha_i\omega_i||\sum_{i\in I}\alpha_i\omega_i|^4 + |\sum_{i\in I}\alpha_i\omega_i||\sum_{i\in II}\alpha_i\omega_i|^4\right)$$

$$+\sum_{\mu=1}^{3} O\left(\left(|\sum_{i\in II}\alpha_i\omega_i||\sum_{i\in I}\alpha_i\omega_i|^{\mu-1} + |\sum_{i\in II}\alpha_i\omega_i|^\mu\right)|v_I|^{4-\mu}+\right.$$

$$\left.\left(|\sum_{i\in II}\alpha_i\omega_i|^{\mu-1}|\sum_{i\in I}\alpha_i\omega_i| + |\sum_{i\in I}\alpha_i\omega_i|^\mu\right)|v_{II}|^{4-\mu}\right)$$

and its projections or from the contribution on $v - (v_I + v_{II})$ in Ω^c. The contribution of g is controlled by:

$$\int_{\Omega_i^c} g\delta_i + \frac{1}{\sqrt{\lambda_i}\rho_i}\int_{\Omega_i} g + \int_{\Omega_i} g\delta_j + \sqrt{\lambda_j}\varepsilon_{ij}\int_{B_j} g.$$

Indeed, if we replace g by f of (1.1) in Section 1.1, the above quantity is related to h_i (see Lemma 12) and upperbounds what we denoted $\frac{|h_i^0|_\infty}{\sqrt{\lambda_i}}$ + $\sum\limits_{\substack{\lambda_s \geq \lambda_i \\ s \neq i}} \varepsilon_{is}|h_i^s|_\infty$. These quantities, in turn, are involved in the upperbound

of $|v_i|_{H_0^1}$, together with $\left(\int_{\Omega_i} g^{6/5}\right)^{5/6} \cdot |v_i|_{H_0^1}$ (or $|(v-(v_I+v_{II}))i|_{H_0^1}$ here) is involved, together with a bound on v (or $v - (v_I+v_{II})$ here) in Ω^c in the derivation of the bound on

$$\int_{\Omega_i^c} |f|\delta_i + \frac{1}{\sqrt{\lambda_i}\rho_i}\int_{\Omega_i} |f| + \int_{\Omega_i} |f|\delta_j + \sqrt{\lambda_j}\varepsilon_{ij}\int_{B_j} |f|.$$

Due to the form of the matrix involved in this estimate, see (3), estimates on $|(v-(v_I+v_{II}))\ell|_{H_0^1}$ on $\frac{|h_\ell^0|_\infty}{\sqrt{\lambda_\ell}}$ + $\sum\limits_{\substack{\lambda_s \geq \lambda_\ell \\ s \neq \ell}} \varepsilon_{\ell s}|h_\ell^s|_\infty$ are easily derived from the following:

Proposition 5

(i) $\int_{\Omega_i^c} g\delta_i + \frac{1}{\sqrt{\lambda_i\rho_i}} \int_{\Omega_i} g + \int_{\Omega_i} g\delta_j \leq$

$$C\left[\sum_{\substack{k\neq i \\ k\in I \\ \text{or } k\in II}} \left|\sum_{\substack{j\in II \\ \text{or} \\ j\in I}} \alpha_j\omega_j(a_k)\right|\frac{\varepsilon_{ik}}{\sqrt{\lambda_k}} + \left|\sum_{\substack{j\in I \ \text{if } i\in II \\ j\in II \ \text{if } i\in I}} \alpha_j\omega_j(a_i)\right|\frac{\sum \varepsilon_{im}}{\sqrt{\lambda_i}}\right]$$

$$+ O\left(\sum_{j\neq i} \frac{|\omega_j^\infty|\varepsilon_{ij}}{\lambda_i|a|} + \sum \varepsilon_{ij} \sum_{\substack{(j,m)\in \\ (I,II) \ \text{or} \ (II,I)}} \varepsilon_{jm}^2\right).$$

(ii)

$$\left(\int_{\Omega_i} g^{6/5}\right)^{5/6} \leq \frac{C}{\sqrt{\lambda_i}}\left|\sum_{\substack{j\in I \ \text{if } i\in II \\ j\in II \ \text{if } i\in I}} \alpha_j\omega_j(a_i)\right|$$

$$+ O\left(\sum_{j\neq i} \frac{|\omega_j^\infty|\varepsilon_{ij}}{\lambda_i|a|} + \sum \varepsilon_{ij} \sum_{\substack{(j,m)\in \\ (I,II) \ \text{or} \ (II,I)}} \varepsilon_{jm}^2\right).$$

(iii)

$$|v - (v_I + v_{II})|(x) \leq C \sum_{\substack{(i,j)\in \\ (I,II) \ \text{or} \ (II,I)}} \left(|\omega_j^\infty| + \varepsilon_{ij} + \frac{1}{\lambda_j|a|}\right)\varepsilon_{ij}\delta_i.$$

We will derive from Proposition 5 the following theorem.

Theorem 2 $\left|(v - (v_I + v_{II}))_i\right|_{H_0^1} + \frac{|h_i^0|_\infty}{\sqrt{\lambda_i}} + \sum_{\substack{\lambda_s\geq\lambda_i \\ s\neq i}} \varepsilon_{is}|h_i^s|_\infty$

$$\leq C\left[\sum_{\substack{k\neq i \\ k\in I \\ \text{or } k\in II}} \left|\sum_{\substack{j\in II \\ \text{or} \\ j\in I}} \alpha_j\omega_j(a_k)\right|\frac{\varepsilon_{ik}}{\sqrt{\lambda_k}} + \frac{1}{\sqrt{\lambda_i}}\left|\sum_{\substack{j\in I \ \text{if } i\in II \\ j\in II \ \text{if } i\in I}} \alpha_j\omega_j(a_i)\right|\right]$$

$$+ O\left(\sum_{j \neq i} \frac{|\omega_j^\infty|\varepsilon_{ij}}{\lambda_i |a|} + \sum \varepsilon_{ij} \sum_{\substack{(j,m)\in \\ (I,II) \text{ or } (II,I)}} \varepsilon_{jm}^2 \right).$$

Proof of Proposition 5.

Proof of (i). The only contribution which is not related to an Ω_ℓ (more accurately to a smaller version of $\Omega_\ell, \widetilde{\Omega}_\ell$, since we need to leave some room in the estimate of the harmonic part h_ℓ) comes from $\int_{\Omega_i^c} g\delta_i$ and is of the type $\int_{\Omega^c} g\delta_i$. $\qquad\square$

g contains products of δ_k's with $k \in I$ with δ_m's with $m \in II$. Both terms are present. Assume i is in II for example. Consider

$$\sum \int_{\Omega^c} \delta_i(\delta_k^4 \delta_m + \delta_m^4 \delta_k).$$

We can always upperbound one of δ_k or δ_m by $\frac{C}{\sqrt{\lambda_k}|a|}$ or $\frac{C}{\sqrt{\lambda_m}|a|}$, or δ_i by $\frac{C}{\sqrt{\lambda_i}|a|}$.

Thus, using Lemma 69,

$$\int_{\Omega^c} \delta_i \delta_m^4 \delta_k \leq C \sum_{\substack{(s,t)\in \\ (I,II) \text{ or } (II,I)}} \varepsilon_{is}\varepsilon_{st}^2.$$

We now estimate, for $k \neq i$, $\int_{\Omega_k} g\delta_i$. Assume $k \in I$ for example. Then, since $\delta_j \leq \delta_k$ on Ω_k,

$$\int_{\Omega_k} g\delta_i \leq C \left| \sum \alpha_j \omega_j(a_k) \right| \int_{\Omega_k} \delta_k^4 \delta_i + C \sum_{j\in II} \int_{\Omega_k} \frac{|x - a_k|}{\sqrt{\lambda_j}|a|^2} \delta_k^4 \delta_i$$

$$+ C \sum_{\substack{(\ell,m)\in \\ (I,I)}} \varepsilon_{\ell m} \int_{\Omega_k} \delta_m \delta_i \delta_k^3 \sum_{s\in II} \delta_s + C \sum_{\substack{(\ell,m)\in \\ (II,II)}} \varepsilon_{\ell m} \int_{\Omega_k} \delta_k^3 \delta_i \delta_m \sum_{s\in II} \delta_s$$

$$\leq C \left| \sum_{j\in II} \alpha_j \omega_j(a_k) \right| \frac{\varepsilon_{ki}}{\sqrt{\lambda_k}} + C \sum_{j\in II} \frac{1}{\sqrt{\lambda_j}\sqrt{\lambda_k}|a|^2} \int_{\Omega_k} \delta_k^3 \delta_i$$

$$+ C \sum_{\substack{(\ell,m)\in \\ (I,I)}} \varepsilon_{\ell m} \sum_{s\in II} \frac{1}{\sqrt{\lambda_s}|a|} \int_{\Omega_k} \delta_m \delta_k^3 \delta_i + C \sum_{\substack{(\ell,m)\in \\ (II,II)}} \varepsilon_{\ell m} \sum_{s\in II} \frac{1}{\sqrt{\lambda_s}|a|} \int_{\Omega_k} \delta_k^3 \delta_i \delta_m.$$

Observe that

$$\int_{\Omega_k} \delta_m \delta_i \delta_k^3 \le C \int_{\Omega_k} \delta_k^4 \delta_i \le C \frac{\varepsilon_{ki}}{\sqrt{\lambda_k}}$$

and

$$\int_{\Omega_k} \delta_k^3 \delta_i \le \int_{\Omega_k \cap \{|x-a_i| \ge 10|a_i-a_k|\}} + \int_{\Omega_k \cap \{c|a_i-a_k| \le |x-a_i| \le 10|a_i-a_k|\}}$$

$$+ \int_{\Omega_k \cap \{|x-a_i| \le c|a_i-a_k|\}} \le \frac{C}{\sqrt{\lambda_i} \lambda_k^{3/2}} \frac{1}{|a_i - a_k|} + \frac{C}{\sqrt{\lambda_i}|a_i - a_k|}$$

$$\times \frac{1}{\lambda_k^{3/2}} \ \mathrm{Log} \ (|a_i - a_k|\lambda_k) + \frac{C}{\lambda_k^{3/2}|a_i - a_k|^3} \frac{1}{\sqrt{\lambda_i}} |a_i - a_k|^2$$

$$\le \frac{C\varepsilon_{ik}}{\lambda_k}(1 + \ \mathrm{Log} \ (\lambda_k|a_i - a_k|)).$$

Thus,

$$\int_{\Omega_k} g\delta_i \le C \left| \sum_{j \in II} \alpha_j \omega_j(a_k) \right| \frac{\varepsilon_{ik}}{\sqrt{\lambda_k}}$$

$$+ C \sum_{j \in II} \frac{1}{\sqrt{\lambda_j}\sqrt{\lambda_k}|a|} \cdot \frac{\varepsilon_{ik}}{\lambda_k|a|}(1 + \ \mathrm{Log} \ (\lambda_k|a_i - a_k|))$$

$$+ C \sum \varepsilon_{\ell m} \sum_{s \in II} \frac{1}{\sqrt{\lambda_s}\sqrt{\lambda_k}|a|} \varepsilon_{ki}.$$

This estimate adapts easily if $k \in II$. Let us now estimate $\frac{1}{\rho_i \sqrt{\lambda_i}} \int_{\Omega_i} g = \sqrt{\lambda_i} \ \mathrm{Max} \ \varepsilon_{i\ell} \int_{\Omega_i} g$.

Assume that $i \in II$ for example. We then have (again $\delta_s \le C\delta_i$ in Ω_i).

$$\sqrt{\lambda_i}\varepsilon_{im} \int_{\Omega_i} g \le C\sqrt{\lambda_i}\varepsilon_{im} \left| \sum_{j \in I} \alpha_j \omega_j(a_i) \right| \int_{\Omega_i} \delta_i^4$$

$$+ C\sqrt{\lambda_i}\varepsilon_{im} \sum_{j \in I} \frac{1}{\sqrt{\lambda_j}|a|} \int_{\Omega_i} |x-a_i|\delta_i^4 + C\sqrt{\lambda_i}\varepsilon_{im} \left(\sum \varepsilon_{ts} \right) \sum_{j \in I} \frac{1}{\sqrt{\lambda_j}|a|} \int_{\Omega_i} \delta_i^4$$

$$\leq \frac{\varepsilon_{im}}{\sqrt{\lambda_i}}\left|\sum_{j\in I}\alpha_j\omega_j(a_i)\right| + C\varepsilon_{im}\sum_{j\in I}\frac{1}{\sqrt{\lambda_i}\sqrt{\lambda_j}|a|}\int_{\Omega_i}\delta_i^3$$

$$+ C\varepsilon_{im}\sum\varepsilon_{ts}\sum_{j\in I}\frac{1}{\sqrt{\lambda_i}\sqrt{\lambda_j}|a|} + C\varepsilon_{im}\sum\varepsilon_{ts}\sum_{j\in I}\frac{1}{\sqrt{\lambda_i}\sqrt{\lambda_j}|a|}$$

$$\leq \frac{\varepsilon_{im}}{\sqrt{\lambda_i}}\left|\sum_{j\in I}\alpha_j\omega_j(a_i)\right| + C\varepsilon_{im}\sum_{j\in I}\frac{1}{\sqrt{\lambda_i}\sqrt{\lambda_j}|a|}\frac{1}{\lambda_i^{3/2}}\mathrm{Log}(\lambda_i|a|)$$

$$+ C\varepsilon_{im}\sum\varepsilon_{ts}\sum_{j\in I}\frac{1}{\sqrt{\lambda_i}\sqrt{\lambda_j}|a|}.$$

For $\sqrt{\lambda_j}\varepsilon_{ij}\int_{B_j}g$, the estimate is identical to the one completed above. As we multiply by $\sqrt{\lambda_i}$, the upperbound takes away this multiplicative factor which cancels in a natural way with $\sqrt{\lambda_j}$ in the denominator ($\lambda_j \geq \lambda_i$). We estimate now $\int_{\Omega_i} g\delta_j$ and $\left(\int_{\Omega_i} g^{6/5}\right)^{5/6}$.

Recalling that we are assuming that $i \in II$, we upperbound $\int_{\Omega_i} g\delta_j$ as follows

$$\int_{\Omega_i} g\delta_j \leq C\left|\sum_{m\in I}\alpha_m\omega_m(a_i)\right|\int_{\Omega_i}\delta_i^4\delta_j + C\sum_{m\in I}\frac{1}{\sqrt{\lambda_m}|a|}\int_{\Omega_i}|x-a_i|\delta_i^4\delta_j$$

$$+ C\left(\sum\varepsilon_{\ell s}\right)\sum_{m\in I}\frac{1}{\sqrt{\lambda_m}|a|}\int_{\Omega_i}\delta_i^4\delta_j$$

$$\leq C\left(\left|\sum_{m\in I}\alpha_m\omega_m(a_i)\right| + \left(\sum\varepsilon_{\ell s}\right)\sum_{m\in I}\frac{1}{\sqrt{\lambda_m}|a|}\right)$$

$$\times\frac{\varepsilon_{ij}}{\sqrt{\lambda_i}} + C\sum_{m\in I}\frac{1}{\sqrt{\lambda_m}\sqrt{\lambda_i}|a|}\int_{\Omega_i}\delta_i^3\delta_j$$

$$\leq C\left(\left|\sum_{m\in I}\alpha_m\omega_m(a_i)\right| + \left(\sum\varepsilon_{\ell s}\right)\sum_{m\in I}\frac{1}{\sqrt{\lambda_m}|a|}\right)\frac{\varepsilon_{ij}}{\sqrt{\lambda_i}}$$

$$+ C\sum_{m\in I}\frac{1}{\sqrt{\lambda_m}\sqrt{\lambda_i}|a|}\frac{\varepsilon_{ij}}{\lambda_i}\,\mathrm{Log}\left(\lambda_i|a_i-a_j|\right).$$

We finally estimate $\left(\int_{\Omega_i} g^{6/5}\right)^{5/6}$ which we upperbound as follows using the finer structure of g:

$$\left(\int_{\Omega_i} g^{6/5}\right)^{5/6} \le \frac{C}{\sqrt{\lambda_i}} \left| \sum_{j\in I} \alpha_j \omega_j(a_i) \right|$$

$$+ C \sum_{j\in I} \frac{1}{\sqrt{\lambda_j}|a|^2} \left(\int_{\Omega_i} |x - a_i|^{6/5} \delta_i^{24/5}\right)^{5/6}$$

$$+ C \left(\sum \varepsilon_{\ell s}\right) \sum_{j\in I} \frac{1}{\sqrt{\lambda_j}|a|} \left(\int_{\Omega_i} \delta_i^{24/5}\right)^{5/6} \le \frac{C}{\sqrt{\lambda_i}} \left| \sum_{j\in I} \alpha_j \omega_j(a_i) \right|$$

$$+ C \sum_{j\in I} \frac{1}{\sqrt{\lambda_i}\sqrt{\lambda_j}\lambda_i|a|^2} + C \left(\sum \varepsilon_{\ell s}\right) \sum_{j\in I} \frac{1}{\sqrt{\lambda_i \lambda_j}|a|}$$

U contains two type of terms: on one hand it contains terms which we are expecting, that is Q^* of the nonlinearly involved in the equation satisfied by $v - (v_I + v_{II})$. On the other hand it contains $(Q^* - Q_I^*)\left(\left(\sum_{i\in I} \alpha_i \omega_i + v_I\right)^5\right), (Q^* - Q_{II}^*)(\sum_{i\in II} \alpha_i \omega_i + v_{II})^5$. Without changing the equation much we may apply the operator Q^* to it again. U then contains

$$Q^*(\Delta v_I), Q^*(\Delta v_{II}), Q^*(Q^* - Q_I^*)\left(\left(\sum_{i\in I} \alpha_i \omega_i + v_I\right)\right)^5 \quad \text{etc.}$$

We claim that all these terms are

$$\sum_{\substack{(i,j)\in(I,II) \\ \text{or } (II,I)}} O\left(\sum \varepsilon_{\ell m}\right) \varepsilon_{ij} \delta_i.$$

Indeed,

$$Q^* = Q^* Q_{II}^* = Q^* Q_I^*$$

so that

$$Q^*(\Delta v_I) = Q^* Q_{II}^*(\Delta v_I), Q^*(\Delta v_{II}) = Q^* Q_{II}^*(\Delta v_I)$$

$$Q^*\left((Q^* - Q_I^*)\left(\left(\sum_{i\in I} \alpha_i \omega_i + v_I\right)^5\right)\right) = Q^*(Q^* - Q_I^*)Q_I^*\left(\left(\sum_{I\in I} \alpha_i \omega_i + v_I\right)^5\right)$$

$$= Q^* Q_{II}^* Q_I^* \left(\left(\sum_{i \in I} \alpha_i \omega_i + v_I \right)^5 \right) - Q^* Q_{II}^* Q_I^* \left(\left(\sum_{i \in I} \alpha_i \omega_i + v_I \right)^5 \right) = 0 \text{ etc.}$$

It is easy to see that $Q^* Q_{II}^*(\Delta v_I) = \Delta Q^* Q_{II}^*(v_I)$ satisfies the estimate as well as $Q^* Q_I^*(\Delta v_{II}) = \Delta Q^* Q_I^*(v_{II})$.

We can carry out then the estimates established on $\bar{v}, \bar{v}_I, \bar{v}_{II}$.

All the remainder terms are as stated in Proposition 5 and Theorem 2 but for the contribution of $\sum \alpha_j \omega_j - \sum \alpha_j \omega_j(a_i)$ in the expressions which we study. We have bounded above this quantity by $O(|x - a_i|) \sum \frac{1}{\sqrt{\lambda_j |a|}}$ in our computations. Because of this term, the remainder which we found is $O\left(\frac{\varepsilon_{ij}}{\lambda_i |a|} \right)$ and terms of this type instead of being $(\varepsilon_{ij} \sum \varepsilon_{\ell m}^2)$. This can be improved as follows: using the expansion of $\omega_j, O\left(\frac{|x - a_i|}{\sqrt{\lambda_j |a|}} \right)$ is in fact $O\left(\frac{|\omega_j^\infty| |x - a_i|}{\sqrt{\lambda_j |a|}} + \frac{1}{\lambda_j \sqrt{\lambda_j |a|}} |x - a_i| \right)$. Then the contribution of this term becomes, after a straightforward computation, $O\left(\frac{|\omega_j^\infty| \varepsilon_{ij}}{\lambda_i |a|} + \varepsilon_{ij}^3 \right)$.

(iv) of Proposition 3 is easily derived in the same way it was completed for $\bar{v}$. Instead of $\displaystyle\sum_{\substack{(i,j) \in (I,II) \\ \text{or } (II,I)}} \left(\sum \varepsilon_{\ell m} \right) \varepsilon_{ij} \delta_i$, one could easily write

$$\sum_{\substack{(i,j) \in (I,II) \\ \text{or } (II,I)}} \left(\sum \varepsilon_{i\ell} \right) \varepsilon_{ij} \delta_i.$$

(iii) of Proposition 5 follows.

1.10.5 *The expansion*

We expand (all computations are completed up to the same multiplicative constant).

$$\partial J \left(\sum \alpha_i \omega_i + v \right) \cdot \partial_I \left(\sum_{i \in I} \alpha_i \omega_i + v \right)$$

$$= \partial J \left(\sum_{i \in I} \alpha_i \omega_i + v_I + \sum_{i \in II} \alpha_i \omega_i + v_{II} + v - (v_I + v_{II}) \right)$$

$$\cdot \partial_I \left(\sum_{i \in I} \alpha_i \omega_i + v_I \right) + \partial J \left(\sum \alpha_i \omega_i + v \right) \cdot \partial_I (v - v_I)$$

$$= \partial^2 J \left(\sum_{i\in I} \alpha_i \omega_i + v_I \right) \cdot \left(\sum_{i\in II} \alpha_i \omega_i + v - v_I \right) \cdot \partial \left(\sum_{i\in I} \alpha_i \omega_i + v_I \right)$$

$$+ \sum_{\theta=1}^{5} L_\theta \left(\sum_{i\in I} \alpha_i \omega_i + v_{II} \right) \cdot (v - (v_I + v_{II}))^\theta \cdot \partial \left(\sum_{i\in I} \alpha_i \omega_i + v_I \right)$$

$$+ \partial J \left(\sum_{i\in II} \alpha_i \omega_i + v_{II} \right) \cdot \partial \left(\sum_{i\in II} \alpha_i \omega_i + v_I \right) - c \int \nabla \left(\sum_{i\in II} \alpha_i \omega_i + v_{II} \right)$$

$$\cdot \nabla \partial \left(\sum_{i\in I} \alpha_i \omega_i + v_I \right) + \partial J(\sum \alpha_i \omega_i + v) \cdot \partial_I (v - v_I) + O_2$$

$$\cdot \partial \left(\sum_{i\in I} \alpha_i \omega_i + v_I \right) = (1).$$

By O_2 we indicate terms which are at least first order in $\sum_{i\in I} \alpha_i \omega_i + v_I$ and at least second order in $\omega_{II}, v_{II}, \sum_{i\in II} \alpha_i \omega_i + v_{II}$ or $v - (v_I + v_{II})$ (combined).

Observe that

$$\partial J \left(\sum_{i\in I} \alpha_i \omega_i + v_I \right) \cdot \partial_{II}(v - v_I) = 0$$

Thus (1) reads:

$$(1) = -\partial J \left(\sum_{i\in I} \alpha_i \omega_i + v_I \right) \cdot \partial_{II} \left(\sum_{i\in II} \alpha_i \omega_i \right)$$

$$+ \left(\partial J(\sum \alpha_i \omega_i + v) - \partial J \left(\sum_{i\in II} \alpha_i \omega_i + v_{II} \right) \right) \cdot \partial_I (v - v_I)$$

$$+ \partial J(\sum_{i\in II} \alpha_i \omega_i + v_{II}) \cdot \partial \left(\sum_{i\in I} \alpha_i \omega_i + v_I \right)$$

$$- \int \nabla \left(\sum_{i\in II} \alpha_i \omega_i + v_{II} \right) \partial \left(\sum_{i\in I} \alpha_i \omega_i + v_I \right)$$

$$+ \sum_{\theta=1}^{5} L_\theta \left(\sum_{i\in II} \alpha_i \omega_i + v_{II} \right) \cdot (v - (v_I + v_{II}))^\theta \cdot \partial \left(\sum_{i\in I} \alpha_i \omega_i + v_I \right)$$

$$+ O_2 \cdot \partial \left(\sum_{i \in I} \alpha_i \omega_i + v_I \right).$$

Remark 1 *We check that we are not missing any quadratic term in v.*

Tracking down $\int \nabla v \nabla \partial_I v$, which is the only quadratic term in v in $\partial J(\sum \alpha_i \omega_i + v) \cdot \partial_I v$, we find $\int \nabla(v - v_I)\nabla \partial_I v_I$ coming from $\partial^2 J(\sum_{i \in I} \alpha_i \omega_i + v_I) \cdot \left(\sum_{i \in II} \alpha_i \omega_i + v - v_I \right) \cdot \partial \left(\sum_{i \in I} \alpha_i \omega_i + v_I \right)$, $\int \nabla v_{II} \nabla \partial v_I$ coming from $\partial J\left(\sum_{i \in II} \alpha_i \omega_i + v_{II} \right) \cdot \partial v_I$ and $\int \nabla v \nabla \partial_I (v - v_I)$ coming from $\partial J(\sum \alpha_i \omega_I + v) \cdot \partial_I(v - v_I)$. Since $\int \nabla v_I \nabla \partial v_I = 0$, we find $\int \nabla v \nabla \partial_I v$. There is no gradient term in L_1.

Lemma 62

$$\partial J \left(\sum_{i \in II} \alpha_i \omega_i + v_{II} \right) \cdot \partial v_I = O\left(\frac{1}{|a|} \right)$$

$$\times \left(\sum_{(i,j) \in (I,I)} \varepsilon_{ij} \right) \times \left(\sum_{(m,\ell) \in (II,II)} \varepsilon_{m\ell} \right) \times \left(\sum_{(s,t) \in (I,II)} \varepsilon_{st} \right).$$

Proof. We first observe that

$$\partial J \left(\sum_{i \in II} \alpha_i \omega_i + v_{II} \right) \cdot \partial v_I = \partial J \left(\sum_{i \in II} \alpha_i \omega_i + v_{II} \right) \cdot Q_{II}(\partial v_I).$$

Let θ_{II} be one of the functions among $\omega_{II}, \frac{\partial}{\partial \sigma}\omega_{II}, \lambda_{II}\frac{\partial}{\partial \lambda_{II}}\omega_{II}, \frac{1}{\lambda_{II}}\frac{\partial \omega_{II}}{\partial a_{II}}$.
We estimate

$$\int \nabla \partial v_I \nabla \theta_{II} = -\int \Delta \theta_{II} \partial v_I = -\int_{\Omega_I} \Delta \theta_{II} \partial v_I - \int_{\Omega_{II}} \Delta \theta_{II} \partial v_I$$

$$= \int_{\partial \Omega_I} \Delta \theta_{II} v_I + \int_{\Omega_I} \partial \Delta \theta_{II} v_I - \int_{\Omega_{II}} \Delta \theta_{II} \partial v_I.$$

Using the estimate on ∂v_I provided in Proposition 4, we derive:

$$\int \nabla \partial v_I \nabla \theta_{II} = O\left(\frac{1}{|a|} \right) \left(\sum_{(i,j) \in (I,I)} \varepsilon_{ij} \right) \left(\sum_{(m,\ell) \in (I,II)} \varepsilon_{m\ell} \right).$$

On the other hand, a standard calculation shows that

$$\partial J\left(\sum_{i\in II}\alpha_i\omega_i + v_{II}\right)\cdot\theta_{II} = O\left(\sum_{(m,\ell)\in(II,II)}\varepsilon_{m\ell}\right).$$

$\square$

The result follows.

Remark 2

$$\int_{\Omega_i}\partial\Delta\theta_{II}v_I = -\partial\left(\int_{\Omega_I}\Delta\theta_{II}v_I\right) = -\int_{\Omega_I}\Delta\theta_{II}\partial v_I - \int_{\partial\Omega_I}\Delta\theta_{II}v_I$$

Lemma 63

$$\partial J\left(\sum_{i\in II}\alpha_i\omega_i + v_{II}\right)\cdot\partial\left(\sum_{i\in I}\alpha_i\omega_i\right)$$

$$= \partial J\left(\sum_{i\in II}\alpha_i\omega_i + v_{II}\right)\cdot Q_{II}\left(\partial\left(\sum_{i\in II}\alpha_i\omega_i\right)\right) = \partial J\left(\sum_{i\in II}\alpha_i\omega_i\right)$$

$$\cdot Q_{II}\left(\partial\left(\sum_{i\in I}\alpha_i\omega_i\right)\right) + O\left(\frac{1}{|a|}\right)\left(\sum_{(i,j)\in(I,II)}\varepsilon_{ij}\right)\left(\sum_{(\ell,s)\in(II,II)}\varepsilon_{\ell s}^2\right).$$

Proof.   We estimate

$$\left(\partial J\left(\sum_{i\in II}\alpha_i\omega_i + v_{II}\right) - \partial J\left(\sum_{i\in II}\alpha_i\omega_i\right)\right)\cdot\theta_{II}$$

$$= \partial^2 J\left(\sum_{i\in II}\alpha_i\omega_i\right)\cdot v_{II}\cdot\theta_{II} + O(|v_{II}|_H^2).$$

Because v_{II} and θ_{II} are orthogonal,

$$\partial^2 J\left(\sum_{i\in II}\alpha_i\omega_i\right)\cdot v_{II}\cdot\theta_{II} + O(|v_{II}|_H^2) = O\left(\sum_{(i,j)\in(II,II)}\varepsilon_{ij}^2\right).$$

Lemma 64

$$\partial J\left(\sum_{i\in I}\alpha_i\omega_i + v_I\right)\cdot\partial_{II}\left(\sum_{i\in II}\alpha_i\omega_i\right) = \partial J\left(\sum_{i\in I}\alpha_i\omega_i\right)$$

$$\cdot Q_I \left(\partial_{II} \left(\sum_{i \in II} \alpha_i \omega_i \right) \right) + O \left(\frac{1}{|a|} \right) \left(\sum_{(i,j) \in (I,II)} \varepsilon_{ij} \right) \left(\sum_{(\ell,s) \in (I,I)} \varepsilon_{\ell s}^2 \right). \qquad \square$$

Proof. same than above $\qquad \square$

Lemma 65 *Let $\theta \geq 2, \mu + \mu_1 \geq 2, \mu_1 \geq 1, \theta + \mu + \mu_1 = 6$. Then*

$$\left| \int \partial \left(\sum_{i \in I} \alpha_i \omega_i + v_I \right)^\theta \omega_{II}^\mu v_{II}^{\mu_1} \right| \leq \frac{C}{|a|} \left(\sum_{\substack{(k,\ell) \in (I,II) \\ \text{or } (I,I)}} \varepsilon_{k\ell} \right) \left(\sum_{\substack{(s,t) \in \\ (I,II)}} \varepsilon_{st}^2 \right).$$

Proof. We split $\int \partial \left(\sum_{i \in I} \alpha_i \omega_i + v_I \right)^\theta \omega_{II}^\mu v_{II}^{\mu_1}$ into

$$\int_{\Omega_I} \partial \left(\sum_{i \in I} \alpha_i \omega_i + v_I \right)^\theta \omega_{II}^\mu v_{II}^{\mu_1} + \int_{\Omega_{II}} \partial \left(\sum_{i \in I} \alpha_i \omega_i + v_I \right)^\theta \omega_{II}^\mu v_{II}^{\mu_1}.$$

The second integrand is easily upperbounded using Proposition 4 by

$$C \frac{1}{|a|} \sum_{\substack{(k,\ell) \\ \in (II,II)}} \varepsilon_{k\ell} \int_{\Omega_{II}} \delta_I^\theta \delta_{II}^{\mu + \mu_1} \leq \frac{C}{|a|} \left(\sum_{\substack{(k,\ell) \\ \in (II,II)}} \varepsilon_{k\ell} \right) \sum_{\substack{(s,t) \in \\ (I,II)}} \varepsilon_{st}^2.$$

For the first integrand we complete an integration by parts and the result follows. $\qquad \square$

Lemma 66 *Let $i_0 \in I$, $\partial \varphi_{I_0} = \frac{\partial \varphi_{I_0}}{\partial a_{i_0}} = -\nabla \varphi_{i_0}$.*

$$\left| \int \nabla(v - (v_I + v_{II})) \nabla \partial \varphi_{i_0} \right|$$

$$\leq C \sqrt{\lambda_{i_0}} \left(\sqrt{\lambda_{i_0}} (|(v - (v_I + v_{II}))_{i_0}|_{H_0^1} + \frac{|h_{i_0}^0|_\infty}{\sqrt{\lambda_{i_0}}} \right.$$

$$\left. + \sum_{\substack{\lambda_s \geq \lambda_{i_0} \\ s \neq i_0}} \varepsilon_{i_0 s} |h_{i_0}^s|_\infty + \sum_j \sqrt{\lambda_{i_0}} \left(\sum \varepsilon_{i_0 s}^3 + \sum_{\lambda_t \geq \lambda_{i_0}} \varepsilon_{i_0 t} \varepsilon_{ts}^2 \right) \right).$$

Proof. Integrating by parts, we find

$$\int \nabla(v - (v_I + v_{II})) \nabla \partial \varphi_{i_0} = - \int (v - (v_I + v_{II})) \Delta \partial \varphi_{i_0}.$$

Clearly

$$\left| \int_{\Omega_{i_0}} (v - (v_I + v_{II})) \Delta \partial \varphi_{i_0} \right|$$

$$\leq C \lambda_{i_0} \left(|(v - (v_I + v_{II}))_{i_0}|_{H_0^1} + \frac{|h_{i_0}|_\infty}{\sqrt{\lambda_{i_0}}} + \sum_{\substack{\lambda_s \geq \lambda_{i_0} \\ s \neq i_0}} \varepsilon_{i_0 s} |h_{i_0}^s|_\infty \right).$$

On the other hand,

$$\left| \int_{\Omega_{i_0}^c} (v - (v_I + v_{II})) \Delta \partial \varphi_{i_0} \right|$$

$$\leq C \lambda_{i_0} \sum_{\subset (i,j) \in (I,II) \text{ or } (II,I)} \left(|\omega_j^\infty| + \frac{1}{\lambda_j |a|} \right) \varepsilon_{ij} \int_{\Omega_{i_0}^c} (\delta_{i_0}^5 \delta_i + \delta_{i_0}^5 \delta_j)$$

Observe now that

$$\int_{\Omega_{i_0}^c} \delta_{i_0}^s \delta_i \leq C \varepsilon_{i_0 s}^3 \text{ if } i = i_0, \leq \int_{\Omega_{i_0}^c \cap \Omega_i^c} \delta_{i_0}^5 \delta_i + \int_{\Omega_{i_0}^c \cap \Omega_i} \delta_{i_0}^5 \delta_i$$

$$\leq C \sum_{\lambda_t \geq \lambda_{i_0}} \varepsilon_{it} \varepsilon_{ts}^2 + C \varepsilon_{i_0 s}^3.$$

Thus

$$\left| \int_{\Omega_{i_0}^c} (v - (v_I + v_{II})) \Delta \partial \varphi_{i_0} \right| \leq C \sqrt{\lambda_{i_0}} \left(\sqrt{\lambda_{i_0}} \varepsilon_{ij} \varepsilon_{i_0 s}^3 + \varepsilon_{ij} \sqrt{\lambda_{i_0}} \sum_{\lambda_j \geq \lambda_{i_0}} \varepsilon_{it} \varepsilon_{ts}^2 \right).$$

$\square$

Observe that

$$Q_I(v - v_I) = 0.$$

Thus

$$(\partial_I Q_I)(v - v_I) = -Q_I(\partial_I(v - v_I)).$$

We then have

Lemma 66′

$$\partial J\left(\sum_{i \in II} \alpha_i \omega_i + v_{II} \right) \cdot Q_I \partial_I(v - v_I - v_{II})$$

$$= -\partial J\left(\sum_{i\in II}\alpha_i\omega_i + v_{II}\right)\cdot Q_{II}\left((\partial_I Q_I)(v - v_I - v_{II})\right)$$

$$-\partial J\left(\sum_{i\in II}\alpha_i\omega_i + v_{II}\right)\cdot Q_{II}(\partial_I Q_I(v_{II}))$$

$$= O\left(\frac{1}{|a|}\right)\left(\sum_{(i,j)\in(I,II)}\varepsilon_{ij}\right)\left(\sum_{(s,t)\in(II,II)}\varepsilon_{st}^2\right)$$

$$+\sum_{i_0\in I}\lambda_{i_0}\left(|(v-(v_I+v_{II}))_{i_0}|_{H_0^1}^2 + \frac{|h_{i_0}^0|_\infty^2}{\lambda_{i_0}} + \sum_{\substack{\lambda_s\geq\lambda_{i_0}\\ s\neq i_o}}\varepsilon_{i_0 s}^2|h_{i_0}^s|_\infty^2\right).$$

Proof.   Observe that

$$\partial J\left(\sum_{i\in II}\alpha_i\omega_i + v_{II}\right)\cdot\theta_{II} = O\left(\sum_{\substack{(s,t)\in\\ (II,II)}}\varepsilon_{st}\right).$$

On the other hand, for $i_0\in I$;

$$\int\nabla\theta_{II}\nabla\theta_{i_o} = O\left(\sum_{j\in II}\varepsilon_{i_0 j}\right),\quad \int\nabla\theta_{II}\nabla\partial\theta_{i_0} = O\left(\frac{1}{|a|}\sum_{j\in II}\varepsilon_{i_0 j}\right).$$

Using Lemma 66 and extensions of it

$$-\int\Delta\partial\theta_{i_0}(v - v_I - v_{II})$$

$$\leq C\sqrt{\lambda_{i_0}}\left(\sqrt{\lambda_{i_o}}\left(|(v-(v_I+v_{II}))_{i_0}|_{H_0^1} + \frac{|h_{i_0}^0|_\infty}{\sqrt{\lambda_{i_0}}}\right.\right.$$

$$\left.\left.+\sum_{\substack{\lambda_s\geq\lambda_{i_0}\\ s\neq i_0}}\varepsilon_{i_0 s}|h_{i_0}^s|_\infty + \sum_j\sqrt{\lambda_{i_0}}\left(\varepsilon_{i_0 s}^3 + \sum_{\lambda_t\geq\lambda_{i_0}}\varepsilon_{it}\varepsilon_{ts}^2\right)\varepsilon_{i_0 s}\right)\right).$$

On the other hand,

$$-\int \Delta \partial \theta_I v_{II} = O\left(\sum_{\substack{(i,j)\in \\ (I,II)}} \frac{\varepsilon_{ij}^2}{|a|} \right).$$

Observe that each $-\int \Delta \partial \theta_{i_0}(v - (v_I + v_{II}))$ is multiplied by $\int \nabla\theta_{II}\nabla\theta_{i_0}$ and by $\partial J\left(\sum_{i\in II} \alpha_i \omega_i + v_{II} \right) \cdot \theta_{II}$. The result follows $\qquad \square$

Lemma 67

(1)

$$\int \left(\left| \sum_{i\in I} \alpha_i \omega_i + v_I \right| + \left| \sum_{i\in II} \alpha_i \omega_i + v_{II} \right| + |v - (v_I + v_{II})| \right)^2$$

$$\left(\left| \sum_{i\in II} \alpha_i \omega_i + v_{II} \right| + |v - (v_I + v_{II})| \right) \cdot$$

$$\cdot |v - (v_I + v_{II})| \cdot \left| \partial \left(\sum_{i\in I} \alpha_i \omega_i + v_I \right)^2 \right|$$

$$\leq C \left(\sum_{\substack{(i,j)\in \\ (I,II)}} \frac{\varepsilon_{ij}^3}{|a|} + \sum_{\ell\in I} \lambda_\ell \left(|(v - (v_I + v_{II}))_\ell|_{H_0^1}^2 + \frac{|h_\ell^0|_\infty^2}{\lambda_\ell} \right. \right.$$

$$\left. + \sum_{\substack{\lambda_s \geq \lambda_\ell \\ s\neq \ell}} \varepsilon_{s\ell}^2 |h_\ell^s|_\infty^2 \right) + \sum_{\ell\in I} \left(\sum_{i\in II} \alpha_i \omega_i(a_\ell) \right)^2 + \sum_{\substack{i\in II \\ \ell\in I}} \frac{\omega_i^{\infty^2} \varepsilon_{i\ell}^2}{\lambda_\ell |a_i - a_\ell|^2} \right).$$

(2)

$$\int \left(\left| \sum_{i\in I} \alpha_i \omega_i + v_I \right| + \left| \sum_{i\in II} \alpha_i \omega_i + v_{II} \right| + |v - (v_I + v_{II})| \right)^2$$

$$\left(\left| \sum_{i\in II} \alpha_i \omega_i + v_{II} \right| + |v - (v_I + v_{II})| \right)^2$$

$$\cdot |v - (v_I + v_{II})| \cdot \left| \partial \left(\sum_{i\in I} \alpha_i \omega_i + v_I \right) \right|$$

$$\leq C \left(\sum_{\substack{(i,j)\in \\ (I,II)}} \frac{\varepsilon_{ij}^3}{|a|} + \sum_{\ell\in I} \lambda_\ell \left(|(v - (v_I + v_{II}))_\ell|_{H_0^1}^2 + \frac{|h_\ell^0|_\infty^2}{\lambda_\ell} + \sum_{\substack{\lambda_s \geq \lambda_\ell \\ s\neq\ell}} \varepsilon_{s\ell}^2 |h_\ell^s|_\infty^2 \right) \right.$$

$$\left. + \sum_{\ell\in I} \left(\sum_{i\in II} \alpha_i \omega_i(a_\ell) \right)^2 + \sum_{\substack{i\in II \\ \ell\in I}} \frac{\omega_i^{\infty\,2} \varepsilon_{i\ell}^2}{\lambda_\ell |a_i - a_\ell|^2} \right).$$

(3)

$$\left(\bar{\partial}\omega_I \text{ is either } \frac{1}{\lambda_{i_0}} \frac{\partial\omega_{i_0}}{\partial a_{i_0}} \text{ or } \lambda_{i_0} \frac{\partial\omega_{i_0}}{\partial \lambda_{i_0}} \text{ or } \frac{\partial\omega_{i_0}}{\partial \sigma_{i_0}} \text{ with } i_0 \in I \right)$$

$$\int \left(\left(\sum_{i\in I} \alpha_i\omega_i + v_I \right)^4 \left| \sum_{i\in II} \alpha_i\omega_i + v_{II} \right| \right.$$

$$+ \left(\sum_{i\in II} \alpha_i\omega_i + v_{II} \right)^4 \left| \sum_{i\in I} \alpha_i\omega_i + v_I \right|$$

$$+ \left(\sum_{i\in I} \alpha_i\omega_i + v_I \right)^4 |v - (v_I + v_{II})| + |v - (v_I + v_{II})|^4 \left| \sum_{i\in I} \alpha_i\omega_i + v_I \right|$$

$$\left. + |v - (v_I + v_{II})|^5 \right) |\bar{\partial}\omega_I|$$

$$\leq C \frac{1}{\sqrt{\lambda_{i_0}}} \left(\sum_{\substack{\ell\in I \\ \ell\neq i_0}} \frac{\sqrt{\lambda_\ell}}{\lambda_\ell |a_\ell - a_{i_0}|} \left[|(v - (v_I + v_{II}))_\ell|_{H_0^1} \right. \right.$$

$$+ \sum_{\substack{\lambda_s \geq \lambda_\ell \\ s\neq\ell}} |h_\ell^s|_\infty \varepsilon_{\ell s} + \frac{|h_\ell^0|_\infty}{\sqrt{\lambda_\ell}} + \frac{\left| \sum_{i\in II} \alpha_i\omega_i(a_\ell) \right|}{\sqrt{\lambda_\ell}}$$

$$\left. \left. + \sum_{(i,j)\in(II,II)} \varepsilon_{ij}\varepsilon_{i\ell} + \frac{\sqrt{\varepsilon_{\ell i}}}{\lambda_\ell |a|} \right] + \sum_{\ell\in II} \frac{\varepsilon_{i\ell}^{3/2}}{\sqrt{\lambda_\ell} |a|} + \frac{1}{|a|} \sum \frac{\varepsilon_{ij}^2}{\sqrt{\lambda_i}} \right)$$

$$+C\sum\frac{\varepsilon_{i_0 i}\varepsilon_{is}}{\sqrt{\lambda_i\lambda_j}|a|}+C\left(\left|\frac{\sum\limits_{i\in II}\alpha_i\omega_i(a_{i_0})}{\sqrt{\lambda_{i_0}}}\right|+|(v-(I+v_{II}))|_{H_0^1}\right.$$

$$\left.+\sum_{\substack{\lambda_s\geq\lambda_{i_0}\\ s\neq i_0}}\varepsilon_{i_0 s}|h_{i_0}^s|_\infty+\frac{|h_{i_0}^0|_\infty}{\sqrt{\lambda_{i_0}}}+\sum_{i\in II}\varepsilon_{ii_0}^2\right).$$

Proof.   □

Proof of (1). The contribution on Ω_{II} of the integral is easily upper-bounded by

$$\frac{1}{|a|}\sum_{(i,j)\in(I,II)}\varepsilon_{ij}^3.$$

Indeed, in Ω_{II},

$$\left|\partial\left(\sum_{i\in I}\alpha_i\omega_i+v_I\right)\right|\leq C\frac{1}{|a|}\sum_{i\in I}\delta_i$$

and

$$|v-(v_I-v_{II})|\leq C\sum_{(i,j)\in(I,II)}\varepsilon_{ij}(\delta_i+\delta_j).$$

The claim then follows from the standard estimate

$$\sum_{(i,j)\in(I,II)}\left(\int\delta_j^4\delta_i^2+\int_{\Omega_{II}}\delta_i^6\right)\leq C\sum_{\substack{(i,j)\in\\(I,II)}}\varepsilon_{ij}^2.$$

We turn now to the contribution on Ω_I which we split between the contribution on each Ω_I^ℓ and the contribution on $(U\Omega_I^\ell)^c\cap\Omega_I$:

Let us first consider the contribution on $(U\Omega_I^\ell)^c\cap\Omega_I$. We observe that in Ω_I:

$$|v-(v_I+v_{II})|(x)+|\sum_{i\in II}\alpha_i\omega_i+v_{II}|(x)\leq C\sum_{\substack{(i,j)\in\\(I,II)}}\left(|\omega_j^\infty|+\frac{1}{\lambda_j|a|}+\varepsilon_{ij}\right)\varepsilon_{ij}\sqrt{\lambda_i}.$$

Indeed, for $j\in II$, in Ω_I

$$|\delta_j(x)|\leq\frac{C}{\sqrt{\lambda_j}|a|}\leq C\varepsilon_{ij}\sqrt{\lambda_i}\quad\text{if }i\in I.$$

Also,

$$|\partial \omega_i| \le C\delta_i^3(\lambda_i|x - a_i|).$$

Let us first estimate the contribution of $\partial \omega_i, i \in I$, ignoring the contribution of ∂v_I.

Thus,

$$\int_{(U\Omega_I^\ell)^c \cap \Omega_I} \le C \sum_{(i,j)\in(I,II)} \left(|\omega_j^\infty|^2 + \frac{1}{\lambda_j^2|a|^2} + \varepsilon_{ij}^2\right) \lambda_i\varepsilon_{ij}^2$$

$$\times \int_{(U\Omega_I^\ell)^c \cap \Omega_I} \sum_{\substack{i,j,\ell\in I \\ m\in II}} \delta_i^3(\lambda_i|x - a_i|)\delta_j(\delta_\ell^2 + \delta_m^2).$$

Observe that

$$\left(\int_{(U\Omega_I^\ell)^c \cap \Omega_I} \delta_i^6(\lambda_i r)^2 r^2 dr\right)^{1/2}$$

$$\le C\left(\int_{r \ge \frac{c}{\varepsilon_{is}}} \frac{r^4 dr}{(1+r^2)^3} + \sum_{\substack{k\neq i \\ \lambda_k > \lambda_i}} \int_{\tilde{\Omega}_k^c \cap \Omega_i^c} \delta_k^6\lambda_i^2(|a_i - a_k|^2 + |x - a_k|^2)\right)^{1/2}.$$

Indeed, in $(U\Omega_I^\ell)^c$, either

$$\lambda_i|x - a_i| \ge \frac{c}{\varepsilon_{is}},$$

or, for some k such that $\lambda_k > \lambda_i$,

$$\frac{c}{\varepsilon_{km}} \le \lambda_k|x - a_k| \le \frac{1}{\varepsilon_{ki}}.$$

Then,

$$|x - a_i| \le |a_i - a_k| + |x - a_k|$$

and

$$\delta_i \le C\delta_k.$$

Thus,

$$\left(\int_{(U\Omega_I^\ell)^c\cap\Omega_I}\delta_i^6(\lambda_i r)^2 r^2 dr\right)^{1/2}\leq C\sum_{\lambda_k\geq\lambda_i}\sqrt{\varepsilon_{ks}}.$$

Summing, the contribution on $(U\Omega_I^\ell)^c\cap\Omega_I$ is bounded by

$$\int_{(U\Omega_I^\ell)^c\cap\Omega_I}\leq\frac{C}{|a|}\left(\sum_{j\in II}\frac{1}{\lambda_j|a|}\left(\omega_j^{\infty2}+\frac{1}{\lambda_j^2|a|^2}+\varepsilon_{ij}^2\right)\right)\sum\varepsilon_{ks}^2.$$

since

$$\left(\int_{(U\Omega_I^\ell)^c\cap\Omega_I}\delta_j^2(\delta_\ell^2+\delta_m^2)^2\right)^{1/2}\leq C\sum\varepsilon_{ks}^{3/2}.$$

This does not quite fit into the expression of the remainder as we are seeking it. We need to improve this estimate. We observe that there is a factor equal to $v-(v_I+v_{II})$ in the expression which we want to upperbound. Above, we replaced it by $|v-(v_I+v_{II})|(x)+|\sum_{i\in II}\alpha_i\omega_i+v_{II}|(x)$.

$|v-(v_I+v_{II})|$ is less in Ω_I than $C\varepsilon_{ij}\left(\frac{1}{\sqrt{\lambda_j|a|}}+\delta_i\right),i,j\in I,II$. For $\frac{\varepsilon_{ij}}{\sqrt{\lambda_j|a|}}$, the estimate is straightforward since this factor comes out of the integral entirely. The factor $\lambda_i\varepsilon_{ij}^2$ is then recomposed, but the expression under the integral has one power of δ less. The estimate follows for this part of $v-(v_I+v_{II})$.

For $\varepsilon_{ij}\delta_i$, we repeat our initial argument, only that now a factor equal to $\sqrt{\lambda_i}$ is missing. In the end, we derive an extra δ_i^2 in

$$\left(\int_{(U\Omega_I^\ell)^c\cap\Omega_I}\delta_i^2\delta_j^2(\delta_\ell^2+\delta_m^2)^2\right)^{1/2}.$$

$\left(\int_{\Omega^c}\delta_i^\delta\right)^{1/2}$ is less than $C\sqrt{\lambda_i}\varepsilon_{it}^{5/2}$. Our upperbound becomes:

$$\sqrt{\lambda_i}\varepsilon_{ij}^2\left(\sqrt{\lambda_k}\varepsilon_{kt}^{5/2}\right)\sqrt{\varepsilon_{mt}},\quad i,j\text{ belong to }I,II.$$

These two groups have a large interaction. Discussing various cases according to the fact that λ_j is the smallest concentration in its group or not and using the fact that the interaction of I and II is largest or equivalent to the largest, we derive the result.

We now consider the contribution on each Ω_I^ℓ (or a smaller version $\widetilde{\Omega}_I^\ell$).
We expand

$$\left(\sum_{i\in II}\alpha_i\omega_i + v_{II}\right)(x) = \sum_{i\in II}\alpha_i\omega_i(a_\ell) + \sum_{i\in II}\alpha_i(\omega_i(x) - \omega_i(a_\ell)) + v_{II}(x)$$

$$= \sum_{i\in II}\alpha_i\omega_i(a_\ell) + O\left(\frac{1}{|a|}\sum_{i\in II}\delta_i|x - a_\ell|\right) + O\left(\sum_{(i,j)\in(II,II)}\varepsilon_{ij}\delta_i\right).$$

$$(1.41)$$

We then have, with $j \in I, j_1, k_1 \in I, \ell_1 \in II$

$$\int_{\Omega_I^\ell} |\partial\omega_j||v - (v_I + v_{II})|\left(|v - (v_I + v_{II})| + |\sum_{i\in II}\alpha_i\omega_i + v_{II}|\right)\delta_{j_1}(\delta_{k_1}^2 + \delta_{\ell_1}^2)$$

$$\leq C\int_{\Omega_I^\ell}\sqrt{\lambda_j}\delta_j^2|v - (v_I + v_{II})|\left(|v - (v_I + v_{II})| + |\sum_{i\in II}\alpha_i\omega_i + v_{II}|\right)$$

$$\times \delta_{j_1}(\delta_{k_1}^2 + \delta_{\ell_1}^2)$$

$$\leq C\int_{\Omega_I^\ell}\sqrt{\lambda_j}\delta_j^2|v - (v_I + v_{II})|\left(|v - (v_I + v_{II})| + |\sum_{i\in II}\alpha_i\omega_i + v_{II}|\right)$$

$$\times (\delta_{j_1}^3 + \delta_{k_1}^3 + \delta_{\ell_1}^3).$$

We have already used the claim (which we reestablish below) that on Ω_I^ℓ

$$\delta_m \leq C\delta_\ell \quad \forall m.$$

Thus,

$$\int_{\Omega_I^\ell} \leq C\int_{\Omega_I^\ell}\sqrt{\lambda_j}\delta_j^2|v - (v_I + v_{II})|(|v - (v_I + v_{II})| + |\sum_{i\in II}\alpha_i\omega_i + v_{II}|)\delta_\ell^3.$$

We first consider $\int_{\Omega_I^\ell}\sqrt{\lambda_j}\delta_j^2|v - (v_I + v_{II})|^2\delta_\ell^3$.

We split $v - (v_I + v_{II})$ in Ω_I^ℓ into $(v - (v_I + v_{II}))_\ell + h_\ell^0 + \sum_{\substack{\lambda_s\geq\lambda_\ell\\s\neq\ell}} h_\ell^s\delta_s$ and

we have

$$\int_{\Omega_I^\ell} \leq C\sqrt{\lambda_j}\int_{\Omega_I^\ell}\delta_j^2|(v - (v_I + v_{II}))_\ell|^2\delta_\ell^3 + C\sqrt{\lambda_j}|h_\ell^0|_\infty^2\int_{\Omega_I^\ell}\delta_j^2\delta_\ell^3$$

$$+ \sum_s C\sqrt{\lambda_j}|h_\ell^s|_\infty^2\int_{\Omega_j^\ell}\delta_j^2\delta_s^2\delta_\ell^3.$$

Observe that

$$C\sqrt{\lambda_j}\int_{\Omega_I^\ell}\delta_j^2|(v-(v_I+v_{II}))_\ell|^2\delta_\ell^3$$

$$\leq C\sqrt{\lambda_j}\left(\int_{\Omega_I^\ell}\delta_j^6\right)^{1/3}|(v-(v_I+v_{II}))_\ell|_{H_0^1}^2\times\sqrt{\lambda_\ell}.$$

For $j\neq\ell$, this yields

$$C\sqrt{\lambda_j}\varepsilon_{j\ell}\sqrt{\lambda_\ell}|(v-(v_I+v_{II}))_\ell|_{H_0^1}^2\leq\lambda_\ell|v-(v_I+v_{II})|_{H_0^1}^2.$$

For $j=\ell$, the same estimate holds readily.

Observe also that $(\delta_j\leq C\delta_\ell$ on $\Omega_I^\ell)$

$$\sqrt{\lambda_j}\int_{\Omega_I^\ell}\delta_j^2\delta_\ell^3\leq C\sqrt{\lambda_j}\int_{\Omega_I^\ell}\delta_j\delta_\ell^4\leq\frac{C\sqrt{\lambda_j}\varepsilon_{j\ell}}{\sqrt{\lambda_\ell}}\leq C_1.\ \text{(See Lemma 70)}$$

This also works for $j=\ell$.

By Lemma 70 (below),

$$\sqrt{\lambda_j}\int_{\Omega_I^\ell}\delta_j^2\delta_s^2\delta_\ell^3=o(\lambda_\ell\varepsilon_{s\ell}^2)\quad\left(O(\lambda_\ell\varepsilon_{s\ell}^2)\text{ if }j=\ell\right).$$

Thus

$$\int_{\Omega_I^\ell}\leq C(\lambda_\ell|(v-(v_I+v_{II}))_\ell|_{H_0^1}^2+|h_\ell^0|_\infty^2+\lambda_\ell\sum_{C\lambda_s\geq\lambda_\ell s\neq\ell}\varepsilon_{s\ell}^2|h_\ell^2|_\infty^2).$$

We now consider

$$\int_{\Omega_I^\ell}\sqrt{\lambda_j}\delta_j^2|v-(v_I+v_{II})|\left|\sum_{i\in II}\alpha_i\omega_i+v_{II}\right|\delta_\ell^3,$$

which we upperbound using (1.41) by

$$C\left(\sum_{(i,m)\in(II,II)}\left(|\omega_m^\infty|+\frac{1}{\lambda_m|a_m-a_i|}\right)\frac{\varepsilon_{im}}{\sqrt{\lambda_i}|a|}+\left|\sum_{i\in II}\alpha_i\omega_i(a_\ell)\right|\right)$$

$$\times\int_{\Omega_I^\ell}\sqrt{\lambda_j}\delta_j^2|v-(v_I+v_{II})|\delta_\ell^3+\sum_{i\in II}\sqrt{\lambda_j}\left(C\frac{|\omega_i^\infty|}{\sqrt{\lambda_i}|a|^2}+\frac{C}{\lambda_i^{3/2}|a|^2}\right)$$

$$\times\int_{\Omega_I^\ell}\delta_j^2|v-(v_I+v_{II})||x-a_\ell|\delta_\ell^3=(I)+(II).$$

Observe that

$$\sqrt{\lambda_\ell}|x - a_\ell|\delta_\ell \le C$$

so that

$$(II) \le C \sum_{i \in II} \frac{\sqrt{\lambda_j}}{\sqrt{\lambda_\ell}} \left(C\frac{|\omega_i^\infty|}{\sqrt{\lambda_i}|a|^2} + \frac{C}{\lambda_i^{3/2}|a|^2} \right) \int_{\Omega_I^\ell} \delta_j^2 |v - (v_I + v_{II})| \delta_\ell^2$$

$$\le C \sum_{i \in II} \sqrt{\lambda_j} \left(\frac{|\omega_i^\infty|\varepsilon_{i\ell}}{|a|} + \frac{C\varepsilon_{i\ell}}{\lambda_i|a|^2} \right) \left(|h_\ell^0|_\infty \int_{\Omega_I^\ell} \delta_j^2 \delta_\ell^2 + \sum_{\substack{\lambda_s \ge \lambda_\ell \\ s \ne \ell}} |h_\ell^s|_\infty \int_{\Omega_\ell} \delta_j^2 \delta_s \delta_\ell^2 \right.$$

$$\left. + |(v - (v_I + v_{II}))_\ell|_{H_0^1} \left(\int_{\Omega_I^\ell} \delta_j^{12/5} \delta_\ell^{12/5} \right)^{5/6} \right).$$

For $j \ne \ell$, we have

$$\int_{\Omega_I^\ell} \delta_j^2 \delta_s \delta_\ell^2 \le \left(\int_{\Omega_I^\ell} \delta_s^2 \delta_\ell^4 \right)^{1/2} \left(\int_{\Omega_I^\ell} \delta_j^4 \right)^{1/2} \le C\varepsilon_{s\ell} \left(\frac{1}{\lambda_j} \int_{r \ge \frac{1}{\varepsilon_{\ell j}}} \frac{r^2 dr}{(1 + r^2)^2} \right)^{1/2}$$

$$\le C\varepsilon_{s\ell}\sqrt{\frac{\varepsilon_{\ell j}}{\lambda_j}} \text{ if } \lambda_j \ge \lambda_\ell, \le C\varepsilon_{s\ell} \left(\int_{\Omega_I^\ell} \delta_j^6 \right)^{1/3} \times |\Omega_I^\ell|^{1/6} \le \frac{C\varepsilon_{s\ell}\varepsilon_{j\ell}}{\sqrt{\lambda_\ell} \text{ Max } \varepsilon_{\ell m}}$$

$$\le C\varepsilon_{s\ell}\sqrt{\frac{\varepsilon_{\ell j}}{\lambda_\ell}} \text{ if } \lambda_j \le \lambda_\ell.$$

For $j = \ell$, we have

$$\int_{\Omega_I^\ell} \delta_\ell^4 \delta_s \le C\frac{\varepsilon_{\ell s}}{\sqrt{\lambda_\ell}}.$$

Combining, we derive that

$$\frac{\sqrt{\lambda_j}\varepsilon_{i\ell}}{|a|} |h_\ell^s|_\infty \int_{\Omega_I^\ell} \delta_j^2 \delta_s \delta_\ell^2 \le C\sqrt{\lambda_\ell}\varepsilon_{\ell s}|h_\ell^s|_\infty \times \frac{\varepsilon_{i\ell}}{|a|} \left(\frac{\sqrt{\varepsilon_{\ell j}}}{\sqrt{\lambda_\ell}} + \frac{1}{\sqrt{\lambda_\ell}} \right).$$

We estimate now, using the above computations,

$$\int_{\Omega_I^\ell} \delta_j^2 \delta_\ell^2 \le \left(\int_{\Omega_I^\ell} \delta_j^4 \right)^{1/2} \left(\int_{\Omega_I^\ell} \delta_\ell^4 \right)^{1/2} \le C\sqrt{\frac{\varepsilon_{\ell j}}{\lambda_j}} \times \frac{1}{\sqrt{\lambda_\ell}} \quad \text{for } j \ne \ell$$

$$\le \frac{C}{\lambda_\ell} \qquad \text{for } j = \ell.$$

Thus,

$$\frac{\sqrt{\lambda_j}\,\varepsilon_{i\ell}}{|a|}|h_\ell^0|_\infty \int_{\Omega_I^\ell} \delta_j^2 \delta_\ell^2 \le C|h_\ell^0|_\infty \cdot \frac{\varepsilon_{i\ell}}{|a|}\left(\frac{\sqrt{\varepsilon_{\ell j}}}{\sqrt{\lambda_\ell}} + \frac{1}{\sqrt{\lambda_\ell}}\right).$$

We then consider

$$\left(\int_{\Omega_I^\ell} \delta_j^{12/5}\delta_\ell^{12/5}\right)^{5/6} \le \left(\int_{\Omega_I^\ell} \delta_j^3\delta_\ell^3\right)^{2/3}|\Omega_I^\ell|^{1/6} \le C\frac{\varepsilon_{j\ell}^2\,\log^{2/3}\varepsilon_{\ell j}^{-1}}{\sqrt{\lambda_j\varepsilon_{\ell j}}}\quad\text{for } j \ne \ell$$

$$\left(\int_{\Omega_I^\ell} \delta_\ell^{24/5}\right)^{5/6} \le \frac{C}{\sqrt{\lambda_\ell}}\qquad\qquad\qquad\qquad\text{for } j = \ell.$$

We thus derive

$$\frac{\sqrt{\lambda_j}\,\varepsilon_{i\ell}}{|a|}|(v-(v_I+v_{II}))|_{H_0^1}\left(\int_{\Omega_I^\ell} \delta_j^{12/5}\delta_\ell^{12/5}\right)^{5/6} \le C\sqrt{\lambda_\ell}|(v-(v_I+v_{II}))\ell|_{H_0^1}$$

$$\times \frac{\varepsilon_{i\ell}}{\sqrt{\lambda_\ell}|a|}\left(\sqrt{\varepsilon_{\ell j}}\,\log^{2/3}\varepsilon_{\ell j}^{-1} + 1\right).$$

This, using the factor $|\omega_i^\infty|+\frac{C}{\lambda_i|a|}$, $i \in II$, takes care of (II). We now estimate (I).

We split

$$\sqrt{\lambda_j}\int_{\Omega_I^\ell} \delta_j^2|v-(v_I+v_{II})|\delta_\ell^3$$

$$\le C\sqrt{\lambda_j}\left(|h_\ell^0|_\infty \int_{\Omega_I^\ell}\delta_j^2\delta_\ell^3 + \sum_{\substack{\lambda_s \ge \lambda_\ell\\ s\ne\ell}}|h_\ell^s|_\infty \int_{\Omega_I^\ell}\delta_j^2\delta_s\delta_\ell^3\right.$$

$$\left. + |(v-(v_I+v_{II}))\ell|_{H_0^1}\left(\int_{\Omega_I^\ell}\delta_j^{12/5}\delta_\ell^{18/5}\right)^{5/6}\right).$$

Observe that $(\delta_j \le c\delta_\ell)$

$$\sqrt{\lambda_j}\int_{\Omega_I^\ell}\delta_j^2\delta_\ell^3 \le \sqrt{\lambda_j}\int_{\Omega_I^\ell}\delta_\ell^4\delta_j \le \frac{C\sqrt{\lambda_j}\varepsilon_{\ell j}}{\sqrt{\lambda_\ell}} \le C_1$$

$$\sqrt{\lambda_j}\int_{\Omega_I^\ell}\delta_j^2\delta_s\delta_\ell^3 \le \sqrt{\lambda_j}\left(\int_{\Omega_I^\ell}\delta_j^6\right)^{1/3}\left(\int_{\Omega_I^\ell}\delta_s^{3/2}\delta_\ell^{9/2}\right)$$

$$\leq C\sqrt{\lambda_j}\varepsilon_{\ell j}\left(\int_{\Omega_I^\ell}(\delta_s^2\delta_\ell^4)^{3/4}\delta_\ell^{3/2}\right)^{2/3}\leq C\sqrt{\lambda_j}\varepsilon_{\ell j}\left(\int_{\omega_I^\ell}\delta_s^2\delta_\ell^4\right)^{1/2}$$

$$\leq C\sqrt{\lambda_j}\varepsilon_{\ell j}\varepsilon_{\ell s}\leq C\sqrt{\lambda_\ell}\varepsilon_{s\ell}\qquad(\text{ it works also for }j=\ell)$$

$$\sqrt{\lambda_j}\left(\int_{\Omega_I^\ell}\delta_j^{12/5}\delta_\ell^{18/5}\right)^{5/6}\leq C\sqrt{\lambda_j}\left(\int_{\Omega_I^\ell}\delta_j^2\delta_\ell^4\right)^{5/6}\leq C\sqrt{\lambda_j}\varepsilon_{\ell j}^{5/3}\leq C\sqrt{\lambda_\ell}.$$

Combining, we derive that

$$C\left(\sum_{\substack{(i,m)\\\in(II,II)}}\left(|\omega_m^\infty|+\frac{1}{\lambda_m|a_i-a_m|}\right)\frac{\varepsilon_{im}}{\sqrt{\lambda_i}|a|}\right.$$

$$\left.+\left|\sum_{i\in II}\alpha_i\omega_i(a_\ell)\right|\right)\int_{\Omega_I^\ell}\sqrt{\lambda_j}\delta_j^2|v-(v_I+v_{II})|\delta_\ell^3$$

$$\leq C\left(\sum_{\substack{(i,m)\\\in(II,II)}}\left(|\omega_m^\infty|+\frac{1}{\lambda_m|a_i-a_m|}\right)\frac{\varepsilon_{im}}{\sqrt{\lambda_i}|a|}+\left|\sum_{i\in II}\alpha_i\omega_i(a_\ell)\right|\right)$$

$$\times\left(|h_\ell^0|^\infty+\sum_{\substack{\lambda_s\geq\lambda_\ell\\s\neq\ell}}\sqrt{\lambda_\ell}\varepsilon_{s\ell}|h_\ell^s|_\infty+\sqrt{\lambda_\ell}|(v-(v_I+v_{II}))_\ell|_{H_0^1}\right).$$

$$\qquad\qquad\qquad\qquad\qquad\qquad\qquad\qquad\qquad\qquad\qquad\qquad\qquad\square$$

1) follows for $\partial\omega_i$.

The above proof extends almost verbatim to ∂v_I using (iii) of Proposition 4 since the main estimate we have used in the proof of 1) for $\partial\omega_I$ was

$$|\partial\omega_I|\leq C\sum_{i\in I}\sqrt{\lambda_i}\delta_i^2$$

and ∂v_I satisfies the same estimate.

However, in the beginning of the proof of 1) when estimating the contribution on $(U\Omega_I^\ell)^c\cap\Omega_I$, we have used

$$|\partial\omega_i|\leq C\delta_i^3(\lambda_i|x-a_i|).$$

This estimate is not verified by ∂v_I and we have to modify the argument. In the worst case, using Proposition 2 and 4, we have to estimate

$$(1) = \sum_s \sum_{i \in I} \sqrt{\lambda_i} \int_{\Omega^c_{\ell,I} \cap \Omega_I} \delta_i^2 \left(\sum_{\substack{j \in II \\ (k,m) \in (I,II)}} \delta_j \varepsilon_{km} \delta_k \right) \delta_s^3$$

$$\leq C \sum_s \sum_{i \in I} \sum_{\substack{j \in II \\ (k,m) \in (I,II)}} \frac{\sqrt{\lambda_i} \varepsilon_{km}}{\sqrt{\lambda_j} |a|} \int_{\Omega^c_{\ell,I} \cap \Omega_I} \delta_i^2 \delta_k \delta_s^3$$

$$\leq C \sum_s \sum_{i \in I} \sum_{\substack{j \in II \\ (k,m) \in (I,II)}} \frac{\sqrt{\lambda_i} \varepsilon_{km}}{\sqrt{\lambda_j} |a|} \varepsilon_{s\ell}^{3/2} \left(\int_{\Omega^c_{\ell,I} \cap \Omega_I} \delta_i^4 \delta_k^2 \right)^{1/2}.$$

We now distinguish two cases:

(1) $\lambda_i \leq \lambda_k$.

     We then upperbound by

$$\frac{C}{\sqrt{\lambda_j} \sqrt{\lambda_m} |a|^2} \varepsilon_{s\ell}^{3/2} \varepsilon_{ik} = o\left(\sum \frac{\varepsilon_{st}^3}{|a|} \right).$$

(2) $\lambda_i \geq \lambda_k$.

We then have, after splitting the domain of integration between $|x - a_i| \geq c|a_i - a_k|$, c small, and its complement:

$$\int_{\Omega^c_{\ell,I}} \delta_i^4 \delta_k^2 \leq \frac{C}{\lambda_k |a_i - a_k|^2} \int_{\Omega^c_{\ell,I}} \delta_i^4 + \frac{C}{\lambda_i^2 |a_i - a_k|^4} \times \frac{\lambda_k |a_i - a_k|}{\lambda_k^2}$$

$$\leq \frac{C}{\lambda_k |a_i - a_k|^2} \sum_{\lambda_m \geq \lambda_i} \frac{\varepsilon_{ms}}{\lambda_m} + \frac{C}{\lambda_i^2 \lambda_k |a_i - a_k|^3}.$$

This yields on (1) the upperbound:

$$(1) \leq C \sum_s \sum_{i \in I} \sum_{\substack{j \in II \\ (k,m) \in (I,II)}} \frac{\varepsilon_{s\ell}^{3/2}}{\sqrt{\lambda_m} \sqrt{\lambda_j} |a|^2} \left(\frac{\varepsilon_{nt}}{\lambda_i \lambda_k |a_i - a_k|^2} + \frac{1}{\lambda_k \lambda_i^2 |a_i - a_k|^3} \right)$$

$$= o\left(\frac{\sum \varepsilon_{st}^3}{|a|} \right).$$

$$(1.42)$$

Proof of 3). We now turn to estimate

$$\int \left[\left(\sum_{i \in I} \alpha_i \omega_i + v_I \right)^4 \left(\left| \sum_{i \in II} \alpha_i \omega_i + v_{II} \right| + \left| v - (v_I + v_{II}) \right| \right) \right.$$

$$\left. + \left(\left(\sum_{i \in II} \alpha_i \omega_i + v_{II} \right)^4 + \left| v - (v_I + v_{II}) \right|^4 \right) \times \left| \sum_{i \in I} \alpha_i \omega_i + v_I \right| \right] |\bar{\partial} \omega_{i_0}|.$$

Here $\bar{\partial} \omega_{i_0}$ is either ω_{i_0} or $\lambda_{i_0} \frac{\partial \omega_{i_0}}{\partial a_{i_0}}$ or $\frac{1}{\lambda_{i_0}} \frac{\partial \omega_{i_0}}{\partial a_{i_0}}$ or $\frac{\partial \omega_{i_0}}{\partial \sigma_{i_0}}$. If a derivative is taken on ω_{i_0}, it is tamed.

We have

$$\int_{(U\Omega_\ell)^c \cap \Omega_{II}} \leq \frac{C}{\sqrt{\lambda_{i_0}} |a|} \sum \frac{\varepsilon_{ij}^2}{\sqrt{\lambda_i}}.$$

$$\int_{(U\Omega_\ell)^c \cap \Omega_I} \leq C \left(\sum_{j \in II} \frac{1}{\sqrt{\lambda_j} |a|} \right) \sum \int_{\Omega^c} \delta_i^4 \delta_{i_0}.$$

We now claim that

$$\sum \int_{\Omega^c} \delta_i^4 \delta_{i_0} \leq C \left(\sum \frac{\varepsilon_{i_0 i} \varepsilon_{is}}{\sqrt{\lambda_i}} + \sum \frac{\varepsilon_{is}^2}{\sqrt{\lambda_{i_0}}} \right).$$

We will establish this claim later. Thus,

$$\int_{(U\Omega_\ell)^c \cap \Omega_I)} \leq C \left(\sum_{j \in II} \frac{1}{\sqrt{\lambda_j} |a|} \right) \left(\sum \frac{\varepsilon_{i_0 i} \varepsilon_{is}}{\sqrt{\lambda_i}} + \sum \frac{\varepsilon_{is}^2}{\lambda_{i_0}} \right).$$

We are left with the contribution in $U\Omega_\ell$.

The contribution on $(U\Omega_\ell) \cap \Omega_{II}$ is upperbounded as follows

$$\int_{(U\Omega_\ell) \cap \Omega_{II}} \leq \frac{C}{\sqrt{\lambda_{i_0}} |a|} \sum_{\ell \in II} \left\{ \left(\sum_{i \in I} \frac{1}{\sqrt{\lambda_i} |a|} \right)^4 \left(\sum \int_{\Omega_\ell \cap \Omega_{II}} \delta_j^6 \right)^{1/6} \right.$$

$$\times \frac{1}{(\lambda_\ell \max \varepsilon_{\ell m})^{5/2}} + \sum_{i \in I} \frac{1}{\sqrt{\lambda_i} |a|} \frac{1}{\lambda_\ell} \Bigg\} \leq \sum_{\ell \in II} \frac{C}{\sqrt{\lambda_{i_0}} \sqrt{\lambda_\ell} |a|} \left(\sum_{i \in I} \frac{\varepsilon_{i\ell}^4}{\max \varepsilon_{\ell m}^{5/2}} \right)$$

$$\leq C \sum_{\ell \in II} \frac{C}{\sqrt{\lambda_{i_0}} \sqrt{\lambda_\ell} |a|} \cdot \varepsilon_{i\ell}^{3/2}$$

Using the fact that $\delta_j \leq C\delta_\ell$ on Ω_ℓ, we upperbound the contribution on $\Omega_\ell \cap \Omega_I (\ell \in I)$ as follows

$$\int_{\Omega_\ell \cap \Omega_I} \leq C \int_{\Omega_\ell \cap \Omega_I} \delta_\ell^4 |\partial \omega_{i_0}| \left(\left| \sum_{i \in II} \alpha_i \omega_i + v_{II} \right| + |v - (v_I + v_{II})| \right)$$

$$\leq C \left(\int \delta_\ell^4 \delta_{i_0}^2 \right)^{1/2} \left(\int_{\Omega_\ell \cap \Omega_I} \delta_\ell^4 \left(\left| \sum_{i \in II} \alpha_i \omega_i + v_{II} \right|^2 + |v - (v_I + v_{II})|^2 \right) \right)^{1/2}$$

$$\leq C\varepsilon_{\ell i_0} \left(\int_{\Omega_\ell \cap \Omega_I} \delta_\ell^4 \left(\left| \sum_{i \in II} \alpha_i \omega_i + v_{II} \right|^2 + |v - (v_I + v_{II})|^2 \right) \right)^{1/2}$$

$$\leq C \frac{1}{\lambda_\ell |a_\ell - a_{i_0}|} \frac{\sqrt{\lambda_\ell}}{\sqrt{\lambda_{i_0}}} \left(|(v - (v_I + v_{II}))\ell|_{H_0^1} + \sum_{\substack{\lambda_s \geq \lambda_\ell \\ s \neq \ell}} |h_\ell^s|_\infty \varepsilon_{\ell s} + \frac{|h_\ell^0|_\infty}{\sqrt{\lambda_\ell}} \right.$$

$$\left. + \frac{| \sum_{i \in II} \alpha_i \omega_i(a_\ell)|}{\sqrt{\lambda_\ell}} + \sum_{(i,j) \in (II,II)} \varepsilon_{ij}\varepsilon_{i\ell} + \frac{1}{|a|} \sum_{i \in II} \left(\int_{\Omega_\ell \cap \Omega_I} \delta_\ell^4 \delta_i^2 |x - a_\ell|^2 \right)^{1/2} \right).$$

Observe now that

$$\left(\int_{\Omega_\ell \cap \Omega_I} \delta_\ell^4 \delta_i^2 |x - a_\ell|^2 \right)^{1/2} \leq \frac{1}{\sqrt{\lambda_\ell}} \left(\int_{\Omega_\ell \cap \Omega_I} \delta_\ell^2 \delta_i^2 \right)^{1/2} \leq \frac{C}{\sqrt{\lambda_\ell}\sqrt{\lambda_i}|a|} \left(\int_{\Omega_\ell} \delta_\ell^2 \right)^{1/2}$$

$$\leq \frac{C\varepsilon_{\ell i}}{\lambda_\ell} \times \frac{1}{\sqrt{\text{Max } \varepsilon_{\ell m}}} \leq C \frac{\sqrt{\varepsilon_{\ell i}}}{\lambda_\ell}$$

for $i \in II$.

    Thus,

$$\int_{\Omega_\ell \cap \Omega_I} \leq C \sqrt{\frac{\lambda_\ell}{\lambda_{i_0}}} \times \frac{1}{\lambda_\ell |a_\ell - a_{i_0}|} \left(|(v - (v_I + v_{II}))\ell|_{H_0^1} + \sum_{\substack{\lambda_s \geq \lambda_\ell \\ s \neq \ell}} |h_\ell^s|_\infty \varepsilon_{\ell s} \right.$$

$$\left. + \frac{|h_\ell^0|_\infty}{\sqrt{\lambda_\ell}} + \frac{| \sum_{i \in II} \alpha_i \omega_i(a_\ell)|}{\sqrt{\lambda_\ell}} + \sum_{(i,j) \in (II,II)} \varepsilon_{ij}\varepsilon_{i\ell} + \frac{\sqrt{\varepsilon_{\ell i}}}{\lambda_\ell |a|} \right).$$

We also have an additional term which comes from the integration over $\Omega_{i_0} \cap \Omega_I$.

$$\int_{\Omega_{i_0} \cap \Omega_I} \leq C \left(\frac{|\sum_{i \in II} \alpha_i \omega_i(a_{i_0})|}{\sqrt{\lambda_{i_0}}} + \sum_{i \in II} \varepsilon_{i i_0}^2 + |(v - (v_I + v_{II}))_{i_0}|_{H_0^1} \right.$$

$$\left. + \sum_{\substack{\lambda_s \geq \lambda_{i_0} \\ s \neq i_0}} \varepsilon_{i_0 s} |h_{i_0}^s|_\infty + \frac{|h_{i_0}^0|_\infty}{\sqrt{\lambda_{i_0}}} \right).$$

$\square$

Proof of 2). In Ω_I, 2) follows from 1). In Ω_{II}, 2) is either immediate or follows closely 3) in Ω_{II} because, in Ω_{II},

$$|\partial \omega_{i_0}| \leq \frac{C}{|a|} \delta_{i_0} \quad \text{for } i_0 \in I.$$

Thus, we obtain an extra factor $\frac{1}{|a|}$ with respect to the estimate in 3). $\square$

Observation. Let us remove from the expression in 3) the term

$$\int \left(\sum_{i \in I} \alpha_i \omega_i + v_I \right)^4 \left| \sum_{i \in II} \alpha_i \omega_i + v_{II} \right| |\bar{\partial} \omega_I| \quad \text{and replace it by}$$

$$\int \left| \sum_{i \in I} \alpha_i \omega_i + v_I \right|^3 \left| \sum_{i \in II} \alpha_i \omega_i + v_{II} \right|^2 |\bar{\partial} \omega_I|.$$

The estimate then improves because $\dfrac{|\sum_{i \in II} \alpha_i \omega_i(a_{i_0})|}{\sqrt{\lambda_{i_0}}}$ is replaced by its square.

The term contributing

$$C \left(|(v - (v_I + v_{II}))_{i_0}|_{H_0^1} + \sum \varepsilon_{i_0 s} |h_{i_0}^s|_\infty + \frac{|h_s^0|_\infty}{\sqrt{\lambda_{i_0}}} \right)$$

in 3) of Lemma 67 comes from

$$\int_{\Omega_{i_0}} \left(\sum_{i \in I} \alpha_i \omega_i \right)^4 (v - (v_I + v_{II})) \bar{\partial} \omega_{i_0},$$

more precisely from

$$\int_{\Omega_{i_0}} \omega_{i_0}^4 (v - (v_I + v_{II})) \bar{\partial}\omega_{i_0}.$$

All the other terms will yield a contribution which is small o of the same quantity. Up to an additional small o term, we find

$$\int_{\mathbb{R}^3} \omega_{i_0}^4 (v - (v_I + v_{II})) \bar{\partial}\omega_{i_0}.$$

Using our orthogonality conditions, $\int \omega_{i_0}^4 \bar{\partial}\omega_{i_0}(v-v_I) = 0$ so that the above quantity reduces to

$$-\int_{\mathbb{R}^3} \omega_{i_0}^4 v_{II} \bar{\partial}\omega_{i_0} = 0 \left(\sum_{(\ell,j)\in(II,II)} \left(|\omega_j^\infty| + \frac{1}{\lambda_j |a_\ell - a_j|} + \varepsilon_{\ell j} \right) \right) \varepsilon_{\ell j} \varepsilon_{i_0 \ell}$$

$$= \frac{1}{\sqrt{\lambda_{i_0}}} o \left(\frac{|\omega_j^\infty|}{\sqrt{\lambda_j} |a_{i_0} - a_\ell|} + \sum \frac{\varepsilon_{ij}^3}{\sqrt{\lambda_j} |a_i - a_j|} \right).$$

$\square$

Lemma 68   *Let $j \neq \ell$. There exists C such that $\delta_j \leq C\delta_\ell$ on Ω_ℓ.*

Proof.   This is already established if $\lambda_j \leq \lambda_\ell$.
     Assume $\lambda_j \geq \lambda_\ell$. Then,

$$\lambda_j |x - a_j| \geq \frac{1}{\varepsilon_{j\ell}} \sim \sqrt{\lambda_j \lambda_\ell} |a_j - a_\ell| \text{ on } \Omega_\ell$$

i.e.

$$\frac{1}{\sqrt{\lambda_\ell} |a_j - a_\ell|} \geq \frac{1}{\sqrt{\lambda_j} |x - a_j|}.$$

This yields the result if $|x - a_j| \leq 10|a_j - a_\ell|$. Otherwise, $|x - a_j|$ and $|x - a_\ell|$ are of the same order and the inequality is immediate. $\square$

We now establish our claim.

Lemma 69

$$\sum \int_{\Omega^c} \delta_i^4 \delta_{i_0} \leq C \left(\sum \frac{\varepsilon_{i_0 i} \varepsilon_{is}}{\sqrt{\lambda_i}} + \sum \frac{\varepsilon_{is}^2}{\sqrt{\lambda_{i_0}}} \right).$$

Proof.   We split the integral into two pieces

$$\int_{\Omega^c\cap\{|x-a_{i_0}|\geq c|a_i-a_{i_0}|\}} \leq \frac{C}{\sqrt{\lambda_{i_0}}|a_i-a_{i_0}|}\int_{\Omega^c}\delta_i^4 \leq \frac{C\varepsilon_{is}^2}{\lambda_i\sqrt{\lambda_{i_0}}|a_i-a_{i_0}|} \leq \frac{C\varepsilon_{is}^2}{\sqrt{\lambda_{i_0}}}$$

and

$$\int_{\Omega^c\cap\{|x-a_i|\geq c_1|a_i-a_{i_0}|\}\cap\{|x-a_{i_0}|\leq c|a_i-a_{i_0}|\}}$$

$$\leq \frac{C}{\lambda_i^2|a_i-a_{i_0}|^4}\int_{\Omega^c\cap\{|x-a_{i_0}|\leq c|a_i-a_{i_0}|\}}\delta_{i_0}$$

$$\leq \frac{C}{\lambda_i^2|a_i-a_{i_0}|^4}\times\frac{1}{\lambda_{i_0}^{5/2}}\times\lambda_{i_0}^2|a_i-a_{i_0}|^2 \leq \frac{C\varepsilon_{i_0 i}}{\sqrt{\lambda_i}\lambda_i|a_i-a_{i_0}|}.$$

Observe now that if the domain of integration is not empty,

$$c_1|a_i-a_{i_0}| \leq |x-a_i| \leq \frac{1}{c_1}|a_i-a_{i_0}|$$

for every x in this domain. Thus

$$\lambda_i|a_i-a_{i_0}| \geq c_1\lambda_i|x-a_i| \geq \frac{1}{8\varepsilon_{is}}$$

for some s, unless x is in (an extended version) of an $\tilde{\Omega}_m\cap\Omega^c$ for some m such that $\lambda_m \geq \lambda_i, m\neq i$. On such a domain

$$\delta_i \leq C\delta_m$$

and we can start an induction where $\int_{\Omega^c}\delta_i^4\delta_{i_0}$ is replaced by $\int_{\Omega^c}\delta_m^4\delta_{i_0}$. $\square$

Lemma 70   *i) Assume that $\lambda_s \geq \lambda_\ell, s\neq\ell$. Assume that $j\neq\ell, \lambda_j\geq\lambda_\ell$. Then,*

$$\sqrt{\lambda_j}\int_{\Omega_\ell}\delta_j^2\delta_s^2\delta_\ell^3 \leq C\sqrt{\lambda_j}\varepsilon_{j\ell}\sqrt{\lambda_\ell}\varepsilon_{s\ell}^2 = o(\lambda_\ell\varepsilon_{s\ell}^2).$$

ii) If $j=\ell$, then

$$\sqrt{\lambda_\ell}\int_{\Omega_\ell}\delta_\ell^5\delta_s^2 \leq C\lambda_\ell\varepsilon_{s\ell}^2.$$

This estimate extends to $j\neq\ell, \lambda_j\leq\lambda_\ell, \sqrt{\lambda_j}\int_{\Omega_\ell}\delta_j^2\delta_\ell^3\delta_s^2$.

Proof of i). We first estimate the contribution on $\Omega_\ell^1 = \Omega_\ell \cap \{|x - a_s| \geq \frac{1}{100}|a_\ell - a_s|\}$.

We find

$$\sqrt{\lambda_j} \int_{\Omega_\ell^1} \delta_j^2 \delta_s^3 \delta_\ell^3 \leq \frac{C\sqrt{\lambda_j}}{\lambda_s|a_s - a_\ell|^2} \int_{\Omega_\ell^1} \delta_j^2 \delta_\ell^3.$$

We split Ω_ℓ^1 into

$$\tilde{\Omega}_\ell^0 = \Omega_\ell^1 \cap \{|x - a_\ell| \leq 10|a_j - a_\ell|\}, \Omega_\ell^1 \cap \{|x - a_\ell| \geq 10|a_j - a_\ell|\} = \tilde{\Omega}_\ell^1.$$

In Ω_ℓ^1,

$$\frac{1}{2} \leq \frac{|x - a_j|}{|x - a_\ell|} \leq 2$$

and

$$\delta_j^2 \delta_\ell^2 \leq \frac{C}{\lambda_\ell \lambda_j} \frac{1}{|x - a_j|^4}$$

so that

$$\int_{\tilde{\Omega}_\ell^1} \delta_j^2 \delta_\ell^3 \leq \sqrt{\lambda_\ell} \int_{\tilde{\Omega}_\ell} \delta_j^2 \delta_\ell^2 \leq \frac{C\sqrt{\lambda_\ell}}{\lambda_\ell \lambda_j} \int_{\tilde{\Omega}_\ell^1} \frac{1}{|x - a_j|^4}$$

$$\leq \frac{C\sqrt{\lambda_\ell}}{\lambda_\ell} \mathrm{Max}_{\Omega_\ell} \frac{1}{\lambda_j|x - a_j|} \leq \frac{C\sqrt{\lambda_\ell}}{\lambda_\ell} \varepsilon_{\ell j}.$$

So that

$$\frac{C\sqrt{\lambda_j}}{\lambda_s|a_s - a_\ell|^2} \int_{\tilde{\Omega}_\ell} \delta_j^2 \delta_\ell^3 \leq C\sqrt{\lambda_\ell}\sqrt{\lambda_j}\varepsilon_{\ell j}\varepsilon_{s\ell}^2 = o\left(\lambda_\ell \varepsilon_{s\ell}^2\right).$$

Considering now $\int_{\tilde{\Omega}_\ell^0}$, we have

$$\int_{\tilde{\Omega}_\ell^0} \leq \frac{C\sqrt{\lambda_j}}{\lambda_s|a_s - a_\ell|^2} \left(\frac{1}{\lambda_j|a_j - a_\ell|^2} \int_{\{|x-a_\ell|\leq 10|a_j-a_\ell|\}} \delta_\ell^3 + \frac{1}{\lambda_\ell^{3/2}|a_j - a_\ell|^3} \right.$$

$$\left. \times \int_{\{|x-a_j|\leq 11|a_j-a_\ell|\}} \delta_j^2 + \int_{\tilde{\Omega}_\ell^1} \delta_\ell^3 \delta_j^2 \right).$$

Indeed, if $|x - a_j| \geq 11|a_j - a_\ell|$, then $|x - a_\ell| \geq 10|a_j - a_\ell|$ and we are in $\tilde{\Omega}_\ell^1$.

The contribution on $\widetilde{\Omega}_\ell^1$ has been estimated.
We obtain

$$\int_{\widetilde{\Omega}_\ell^0} \leq \frac{C\sqrt{\lambda_j}}{\lambda_s|a_s - a_\ell|^2}\left(\frac{1}{\lambda_j|a_j - a_\ell|^2} \times \frac{1}{\lambda_\ell^{3/2}} \times \text{Log}\,(\lambda_\ell|a_j - a_\ell|)\right.$$

$$\left. + \frac{1}{\lambda_\ell^{3/2}|a_j - a_\ell|^3}\cdot\frac{1}{\lambda_j}|a_j - a_\ell|\right)o(\lambda_\ell\varepsilon_{\ell s}^2) \leq o(\lambda_\ell\varepsilon_{\ell s}^2).$$

Next, we consider

$$\Omega_\ell^2 = \Omega_\ell \cap \{|x - a_s| \leq \frac{1}{100}|a_\ell - a_s|\} \subset \Omega_\ell \cap \{|x - a_\ell| \geq \frac{1}{2}|a_\ell - a_s|\}.$$

We have

$$\sqrt{\lambda_j}\int_{\Omega_\ell^2}\delta_j^2\delta_s^2\delta_\ell^3 \leq \frac{C\sqrt{\lambda_j}}{\lambda_\ell^{3/2}|a_\ell - a_s|^3}\int_{\Omega_\ell^2}\delta_j^2\delta_s^2.$$

We then distinguish two cases
$1.\lambda_j|a_j - a_s| \geq c\varepsilon_{\ell j}^{-1}$
Arguing as above

$$\int\delta_j^2\delta_s^2 \leq \frac{C}{\lambda_s|a_s - a_j|^2}\int_{|x-a_j|\leq 10|a_s-a_j|}\delta_j^2 + \frac{C}{\lambda_s|a_s - a_j|^2}\int_{|x-a_s|\leq 10|a_s-a_j|}$$

$$+\frac{C}{\lambda_j\lambda_s}\int_{\Omega_\ell}\frac{1}{|x - a_j|^4} \leq \frac{C_1}{\lambda_s\lambda_j|a_s - a_j|} + \frac{C}{\lambda_j\lambda_s}\,\text{Max}_{\Omega_\ell}\frac{1}{|x - a_j|}.$$

In Ω_ℓ,

$$\lambda_j|x - a_j| \geq c\varepsilon_{\ell j}^{-1}$$

so that

$$\sqrt{\lambda_j}\int_{\Omega_\ell^2}\delta_j^2\delta_s^2\delta_\ell^3 \leq \frac{C\sqrt{\lambda_j}}{\lambda_\ell^{3/2}|a_\ell - a_s|^3}\left(\frac{1}{\lambda_s\lambda_j|a_s - a_j|} + \frac{\varepsilon_{\ell j}}{\lambda_s}\right)$$

$$\leq C\sqrt{\lambda_j}\varepsilon_{\ell j}\sqrt{\lambda_\ell}\varepsilon_{s\ell}^2 = o(\lambda_\ell\varepsilon_{s\ell}^2).$$

$2.\lambda_j|a_j - a_s| = o(\varepsilon_{\ell j}^{-1})$; then $|a_j - a_s| = o(|a_\ell - a_j|)$.
For any x in Ω_ℓ, $\lambda_j|x - a_j| \geq c\varepsilon_{\ell j}^{-1}$ so that $|x - a_j| \geq 100|a_j - a_s|$ and
$|x - a_s| \geq 99|a_j - a_s|$. Thus,

$$\frac{1}{2} \leq \frac{|x - a_j|}{|x - a_s|} \leq 2.$$

Then,

$$\sqrt{\lambda_j} \int_{\Omega_\ell^2} \delta_j^2 \delta_s^2 \delta_\ell^3 \le \frac{C\sqrt{\lambda_j}}{\lambda_\ell^{3/2}|a_\ell - a_s|^3} \int_{\Omega_\ell^2} \delta_j^2 \delta_s^2$$

$$\le \frac{C\sqrt{\lambda_j}}{\lambda_\ell^{3/2}|a_\ell - a_s|^3} \times \frac{1}{\lambda_j \lambda_s} \int_{\Omega_\ell^2} \frac{1}{|x - a_j|^4}$$

$$\le \frac{C\sqrt{\lambda_j}}{\lambda_\ell^{3/2}|a_\ell - a_s|^3} \times \frac{1}{\lambda_s} \underset{\Omega_\ell}{\text{Max}} \frac{1}{\lambda_j|x - a_j|} \le \frac{C\sqrt{\lambda_j}}{\lambda_\ell^{3/2}|a_\ell - a_s|^3} \times \frac{1}{\lambda_s}\varepsilon_{\ell j} \le o(\lambda_\ell \varepsilon_{\ell s}^2)$$

(i) follows. $\qquad\qquad\qquad\qquad\qquad\qquad\qquad\qquad\qquad\qquad\qquad\qquad\square$

Proof of (ii).

$$\sqrt{\lambda_\ell} \int_{\Omega_\ell} \delta_\ell^4 \delta_s^2 \le \lambda_\ell \int_{\Omega_\ell} \delta_\ell^4 \delta_s^2 \le C\lambda_\ell \varepsilon_{\ell s}^2.$$

If $\lambda_j \le \lambda_\ell, \delta_j \le C\delta_\ell$ in Ω_ℓ and the estimate follows readily. $\qquad\square$

1.10.6   *The coefficient in front of $\varepsilon_{k\ell}\delta_\ell$*

In all the estimates derived on $v, v_I, v_{II}, v - (v_I + v_{II})$, we use a coefficient in front of the main term of the estimate which varies greatly. In Theorem 1, this coefficient is ω_k^∞ and the estimate reads:

$$|v_\ell|_{H_0^1} + \frac{|h_\ell^0|_\infty}{\sqrt{\lambda_\ell}} + \sum_{\lambda_s \ge \lambda_\ell} \varepsilon_{s\ell}|h_\ell^s|_\infty \le \sum (\omega_k^\infty|\varepsilon_{\ell k} + \ldots.$$

In Proposition 3, this coefficient changes into.

$$|\bar{v}(x)| \le C \sum (|\omega_k^\infty| + |\omega_k(a_k)|(\sqrt{\lambda_k}|a_k - a_\ell|) + \varepsilon_{k\ell})\varepsilon_{k\ell}\delta_\ell(x)$$

etc. (see Theorem 2 in particular).

    This coefficient is obviously close to ω_k^∞ but we want to make it more precise here as it has a direct effect on our Morse Lemma at infinity as well as an effect on the remainder terms found in all estimates of Theorem 1, Propositions 3, 4, Theorem 2 *etc.* The value of this coefficient involves two masses ω_k and ω_ℓ. We rescale ω_k into $\bar{\omega}_k$ which has concentration 1 i.e. is in its standard form. ω_ℓ then, after this rescaling, is concentrated at a point which we denote $\bar{a}_\ell^k$ and we have:

Proposition 6 $(|\omega_k^\infty| + |\omega_k(a_\ell)|(\sqrt{\lambda_k}|a_k - a_\ell|)|)\, \varepsilon_{k\ell}$ *can be replaced in Theorem 1, Proposition 3, Theorem 2 by* $|\alpha_k \bar{\omega}_k(\bar{a}_\ell^k)|\frac{\sqrt{\lambda_k}}{\sqrt{\lambda_\ell}}.$

In Proposition 5 and Theorem 2, the combinations $\sum \alpha_j \omega_j(a_k)$ can be replaced with $\sum \alpha_j \bar{\omega}_j(\bar{a}_k^j)\sqrt{\lambda_j}$.

The remainder terms in Theorem 2 becomes third order in $\sum \varepsilon_{ks}^3$.

Proof. The proof for the L^∞-estimates is similar to the proof in the other cases i.e. to the proof for results such as Theorems 1 and 2.

Let us show how to estimate expressions such as $\int_{\Omega_i} \left(\sum_{j \neq i} \alpha_j \omega_j \right) \omega_i^4 \bar{v}_i$

or, setting:

$$f_t = \omega_t \omega_j^3 \bar{\partial} \omega_i, \bar{\partial} = \frac{\partial}{\partial \sigma_i} \text{ or } \frac{1}{\lambda_i} \frac{\partial}{\partial a_i} \text{ or } \lambda_i \frac{\partial}{\partial \lambda_i}$$

how to estimate $\int_{\Omega_j} \omega_s f_t$. We will see in each case how the expression $\omega_s(\bar{a}_j^s)\sqrt{\lambda_s}$ emerges and we will estimate the remainder term.

We start with $\int_{\Omega_j} \omega_s f_t$. We rescale ω_s to a concentration 1. ω_j then concentrates around a new point $\bar{a}_j^s$ and we have:

$$\int_{\Omega_j} \omega_s f_t = \int_{\tilde{\Omega}_j^s} \bar{\omega}_s \tilde{f}_t = \bar{\omega}_s(\bar{a}_j^s)\tilde{f}_t + O\left(\int_{\tilde{\Omega}_j^s} |x - \bar{a}_j^s||\tilde{f}_t| \right)$$

$$= \bar{\omega}_s(\bar{a}_j^s)\sqrt{\lambda_s} \int_{\Omega_j} f_t + O\left(\int_{\Omega_j} |x - \bar{a}_j^s||\tilde{f}_t| \right).$$

For $t \neq j$,

$$\int_{\tilde{\Omega}_j^s} |x - \bar{a}_j^s||\tilde{f}_t| = o\left(\int_{\tilde{\Omega}_j^s} |x - \bar{a}_j^s|\tilde{\delta}_j^4 \tilde{\delta}_i \right)$$

$$= o\left(\varepsilon_{ij} \left(\int_{\tilde{\Omega}_j^s} |x - \bar{a}_j^s|^2 \tilde{\delta}_j^4 \right)^{1/2} \right).$$

Let $\tilde{\lambda}_j$ be the new concentration of $\tilde{\delta}_j$. Then $\frac{1}{\sqrt{\tilde{\lambda}_j}} \sim c\varepsilon_{js}$ and

$$\left(\int_{\tilde{\Omega}_j^s} |x - \bar{a}_j^s|^2 \tilde{\delta}_j^4 \right)^{1/2} \leq \frac{C}{\tilde{\lambda}_j^{3/2}} \left(\int_{\tilde{\lambda}_j \tilde{\Omega}_j^s} dr \right)^{1/2} \leq \frac{C}{\tilde{\lambda}_j^{3/2}} \times \frac{1}{\sqrt{\text{Max } \varepsilon_{jm}}} \leq C\varepsilon_{js}^{5/2}.$$

Thus,

$$\int_{\tilde{\Omega}_j^s} |x - \bar{a}_j^s||\tilde{f}_t| = 0(\varepsilon_{ij}\varepsilon_{js}^{5/2}).$$

On the other hand,

$$\int_{\Omega_j} |f_t| \le C \int_{\Omega_j} \delta_t \delta_j^3 \delta_i \le C\varepsilon_{ij} \left(\int_{\Omega_j} \delta_t^2 \delta_j^2 \right)^{1/2} \le C \frac{\varepsilon_{ij}}{\sqrt{\lambda_j}}.$$

Thus,

$$\int_{\Omega_j} \left(\sum_{s \ne j} \alpha_s \omega_s \right)^2 \omega_j^3 \bar\partial \omega_i = O(\varepsilon_{ij}) \left(\left| \sum_{s \ne j} \alpha_s \frac{\bar\omega_s(\bar a_j^s)\sqrt{\lambda_s}}{\sqrt{\lambda_j}} \right| + O(\varepsilon_{js}^{5/2}) \right).$$

This estimate extends readily, for $j \ne i$, to ($\delta_s \le c\delta_j$ in Ω_j)

$$\int_{\Omega_j} \left(\sum_{s \ne j} \alpha_s \omega_s \right)^\gamma \omega_j^{5-\gamma} \bar\partial \omega_i \quad \text{for } \gamma \ge 3.$$

We can also estimate directly the contribution on Ω_j^c since

$$\int_{\Omega_j^c} \delta_j^5 \delta_i = O(\varepsilon_{ij} \sum \varepsilon_{jt}^2).$$

We consider now the case when $j = t$ and estimate, for $j \ne i$:

$$\int_{\Omega_j} \left(\sum_{s \ne j} \alpha_s \omega_s \right) \omega_j^4 \bar\partial \omega_i.$$

Arguing as above, this is:

$$0\left(\left| \sum_{s \ne j} \alpha_s \bar\omega_s(\bar a_j^s) \frac{\sqrt{\lambda_s}}{\sqrt{\lambda_j}} \right| \varepsilon_{ij} \right) + 0\left(\sum_s \int_{\tilde\Omega_j^s} |x - \bar a_j^s||\tilde g| \right)$$

with $\tilde g = \tilde\omega_j^4 \bar\partial \tilde\omega_i$. This remainder yields as above $0(\sum \varepsilon_{ij}\varepsilon_{js}^{5/2})$.
    For $j = i$, we consider

$$\int_{\Omega_j} \left(\sum_{s \ne i} \alpha_s \omega_s \right)^2 \omega_i^3 \bar\partial \omega_i = O\left(\frac{|\sum \alpha_j \bar\omega_j(\bar a_i^j)\sqrt{\lambda_j}|}{\sqrt{\lambda_i}} \varepsilon_{it} + \int_{\tilde\Omega_i^j} |x - \bar a_i^j||\tilde\delta_i^4 \tilde\delta_t| \right)$$

$$= O\left(\sum \frac{|\sum \alpha_j \bar\omega_j(\bar a_i^j)\sqrt{\lambda_j}|}{\sqrt{\lambda_i}} \varepsilon_{it} + \varepsilon_{it} \sum \varepsilon_{is}^{5/2} \right).$$

This estimate extends to

$$\sum_{\gamma \geq 3} \int_{\Omega_j} \left| \left(\sum_{s \neq j} \alpha_s \omega_s \right)^{\gamma} \omega_j^{5-\gamma} \frac{\partial \omega_i}{\partial \sigma_i} \right|.$$

We also consider

$$\int_{\Omega_i} \left(\sum_{s \neq i} \alpha_s \omega_s \right) \omega_i^4 \bar{\partial} \omega_i$$

$$= \sum_{s \neq i} \alpha_s \bar{\omega}_s(\bar{a}_i^s) \sqrt{\lambda_s} \int_{\Omega_i} \omega_i^4 \bar{\partial} \omega_i + O \left(\int_{\tilde{\Omega}_i^s} |x - \bar{a}_i^s| \tilde{\delta}_i^5 \right)$$

$$= \left(\sum_{s \neq i} \alpha_s \bar{\omega}_s(\bar{a}_i^s) \sqrt{\lambda_s} \right) \int_{\mathbb{R}^3} \omega_i^4 \bar{\partial} \omega_i + 0 \left(\sum_{\lambda_s \geq \lambda_i} \varepsilon_{is} \varepsilon_{st}^2 \right) + O \left(\frac{1}{\tilde{\lambda}_j^{3/2}} \right)$$

$$= c \sum_{s \neq i} \alpha_s \bar{\omega}_s(\bar{a}_i^s) \frac{\sqrt{\lambda_s}}{\sqrt{\lambda_i}} \bar{\partial} \omega_i^{\infty} + O \left(\sum_{\lambda_s \geq \lambda_i} \varepsilon_{is} \varepsilon_{st}^2 \right).$$

We used again the fact that $\frac{1}{\sqrt{\tilde{\lambda}_j}} \sim c \varepsilon_{ij}$.

For $\int_{\Omega_i} \left(\sum_{j \neq i} \alpha_j \omega_j \right) \omega_i^4 \bar{v}_i$, we rescale again and derive

$$\left(\sum_{j \neq i} \alpha_j \bar{\omega}_j(\bar{a}_i^j) \sqrt{\lambda_j} \right) \int_{\Omega_i} \omega_i^4 \bar{v}_i + O \left(\sum \int_{\Omega_i^j} |x - \bar{a}_i^j| \tilde{\omega}_i^4 |\tilde{v}_i| \right)$$

which we upperbound by

$$C | \sum_{j \neq i} \frac{\alpha_j \bar{\omega}_j(\bar{a}_i^j) \sqrt{\lambda_j}}{\sqrt{\lambda_i}} \|\bar{v}_i\|_{H_0^1} + |\bar{v}_i|_{H_0^1} \sum O \left(\left(\int_{\Omega_i^j} |x - \bar{a}_i^j|^{6/5} |\tilde{\omega}_i|^{24/5} \right)^{5/6} \right)$$

$$\leq C \left(| \sum_{j \neq i} \frac{\alpha_j \bar{\omega}_j(\bar{a}_i^j) \sqrt{\lambda_j}}{\sqrt{\lambda_i}} | + \sum_j \frac{1}{\tilde{\lambda}_i^{3/2}} \right) |\bar{v}_i|_{H_0^1}$$

$$\leq C \left(| \sum_{j \neq i} \alpha_j \frac{\bar{\omega}_j(\bar{a}_i^j) \sqrt{\lambda_j}}{\sqrt{\lambda_i}} | + \sum \varepsilon_{ij}^3 \right) |\bar{v}_i|_{H_0^1}.$$

These arguments repeat and repeat again. They can also be adapted to derive the L^∞-estimates of Proposition 3, 4, 5 *etc.* We need only to use the precise expressions which we have rather than bounds such as

$$\int \frac{|\sum\limits_{i\neq j} \alpha_i \omega_i |\omega_j^4 + |\sum\limits_{i\neq j}\alpha_i\omega_i|^4|\omega_j|}{|x-y|}$$ since our rescaling are completed for each ω_i.

Though tedious, the checking of these estimates is straightforward. $\square$

The next question which arises is to compare $\bar\omega_j(\bar a_i^j)$ with $\bar\omega_j^\infty$ and with $\omega_j(a_i)$ which are the expressions used previously.

As we use (instead of rescaling) the expansion of ω_j arounf a_i in the above estimate, we typically derive expressions such as

$$\omega_j(a_i)\int_{\Omega_i} f_t + \int_{\Omega_i} L(x,\omega_j)(x-a_i)f_t$$

where $L(x,\omega_j)$ is some nonlinear operator depending on ω_j.

ω_j expands as:

$$\omega_j = \frac{c\omega_j^\infty\sqrt{\lambda_j}}{(1+\lambda_j^2|x-a_j|^2)^{1/2}} + c\frac{\sqrt{\lambda_j}\lambda_j T(x-a_j)}{(1+\lambda_j^2|x-a_j|^2)^{3/2}} + \text{h.o} \tag{1.43}$$

Let

$$\gamma_j^i = \omega_j(a_i) - \frac{c\omega_j^\infty\sqrt{\lambda_j}}{(1+\lambda_j^2|a_i-a_j|^2)^{1/2}}. \tag{1.44}$$

Clearly

$$\gamma_i^j = O\left(\frac{1}{\lambda_j^{3/2}|a_i-a_j|^2}\right). \tag{1.45}$$

On the other hand, $L(x,\omega_j)$ is obtained through the differential of ω_j. Thus,

$$L(x,\omega_j) = O\left(\frac{|\omega_j^\infty|}{\sqrt{\lambda_j}|a_i-a_j|^2}\right) + O\left(\frac{1}{\lambda_j^{3/2}|a_i-a_j|^3}\right)$$

for $x\in\Omega_i$ if $\lambda_i\geq\delta\lambda_j(|x-a_j|\sim c|a_i-a_j|$ in $\Omega_i)$ or if $|x-a_j|\geq c|a_i-a_j|$,

$$L(x,\omega_j) = O\left(\frac{|\omega_j^\infty|}{\sqrt{\lambda_j}|a_i-a_j|^2}\right) + O\left(\frac{1}{\lambda_j^{3/2}|a_i-a_j|^2}\right)$$

otherwise.

Observe that

$$\int_{\Omega_i}|x-a_i|f_t = \frac{1}{\sqrt{\lambda_i}}\int_{\Omega_i}\delta_i^4 = O\left(\frac{1}{\lambda_i^{3/2}}\right)$$

$$\int_{\Omega_i} \frac{1}{\sqrt{\lambda_j}|x-a_j|} |x-a_i||f_t| = O\left(\int_{\Omega_i} \delta_j |x-a_i|\delta_i^5\right)$$

$$= O\left(\frac{1}{\sqrt{\lambda_i}} \int_{\Omega_i} \delta_j \delta_i^4\right) = O\left(\frac{\varepsilon_{ij}}{\lambda_i}\right)$$

$$\int_{\Omega_i} \frac{1}{\lambda_j|x-a_j|^2} |x-a_i||f_t| = O\left(\int_{\Omega_i} \frac{\delta_j^2 \delta_i^4}{\sqrt{\lambda_i}}\right) = O\left(\frac{\varepsilon_{ij}^2}{\sqrt{\lambda_i}}\right).$$

Thus,

Proposition 7

$$\omega_j(a_i) \int_{\Omega_i} f_t + \int_{\Omega_i} L(x,\omega_j)(x-a_i)f_t =$$

$$O\left(|\omega_j^\infty|\varepsilon_{ij} + \frac{|\gamma_i^j|}{\sqrt{\lambda_i}}\right) + O\left(\varepsilon_{ij}^3\right).$$

Observation.

The conclusion is that there are two ways to obtain in all our estimates in Theorems 1, 2, Proposition 3, 5, a remainder equal $O(\varepsilon_{ij}^3)$ after the $|\omega_j^\infty|\varepsilon_{ij}$ term: either we rescale at each step and we get a linear expression

$$\left|\sum_{j\neq i} \alpha_j \bar{\omega}_j(\bar{a}_i^j) \frac{\sqrt{\lambda_j}}{\sqrt{\lambda_i}}\right| + O(\varepsilon_{ij}^3).$$

Or we do not rescale, we just expand around a_i as we did in most of this paper. We also can then regain some linearity with $\sum \alpha_j \omega_j^\infty$ but we have an additional term which is $O\left(\sum |\omega_j^\infty|\varepsilon_{ij}\right)$ and another additional term which comes from $\frac{|\gamma_j^i|}{\sqrt{\lambda_i}}$.

The next section has two aims: derive an equation for $\frac{\partial}{\partial \sigma_i}$ on one hand, establish an estimate on $\sum |\gamma_i^j|$ on the other hand.

1.10.7 *The σ_i-equation, the estimate on $\sum |\gamma_i^j|$*

Proposition 8 *We may assume that, for each i*

(1)

$$u_i = \sum_{j \neq i} \alpha_j \bar{\omega}_j(\bar{a}_i^j) \frac{\sqrt{\lambda_j}}{\sqrt{\lambda_i}}$$

$$= \varepsilon_{i\ell} O \left(\sum_{j \neq \ell} \frac{|\omega_j^\infty| \varepsilon_{j\ell}}{\lambda_\ell |a_j - a_\ell|} + \varepsilon_{j\ell} \sum \varepsilon_{mt}^2 \right) + O \left(\sum_{\lambda_s \geq \lambda_i} \varepsilon_{is} \varepsilon_{st}^2 \right).$$

(2)

$$u_i \partial_{\sigma_i} \omega_i^\infty = c \sum_{j \neq i} \bar{\omega}_j^\infty \varepsilon_{ij}(1 + o(1)) \partial_{\sigma_i} \omega_i^\infty + O \left(\sum \frac{\varepsilon_{ij}^{3/2}}{\sqrt{\lambda_i} |a_i - a_j|^{1/2}} \right)$$

$$= \varepsilon_{i\ell} 0 \left(\sum_{j \neq \ell} \frac{|\omega_j^\infty| \varepsilon_{j\ell}}{\lambda_\ell |a_j - a_\ell|} + \varepsilon_{j\ell} \sum \varepsilon_{mt}^2 \right) + O \left(\sum_{\lambda_s \geq \lambda_i} \varepsilon_{is} \varepsilon_{st}^2 \right).$$

Observation.

The second equation is a linear dependency equation between vectors of $\mathbb{R}^3$. If σ_i is a rotation which does not change ω_i^∞ (a rotation around the polar axis of S^3), we still have an equation. We will use this fact later.

Proof. We expand

$$J' \left(\sum \alpha_j \omega_j + v \right) \cdot \frac{\partial \omega_i}{\partial \sigma_i}$$

$$= J' \left(\sum \alpha_j \omega_j + v \right) \cdot \frac{\partial \omega_i}{\partial \sigma_i} = J' \left(\sum \alpha_j \omega_j \right) \cdot \frac{\partial \omega_i}{\partial \sigma_i}$$

$$+ J'' \left(\sum \alpha_j \omega_j \right) \cdot v \cdot \frac{\partial \omega_i}{\partial \sigma_i} + O \left(\int \left| \sum \alpha_j \omega_j \right|^3 v^2 \times \left| \frac{\partial \omega_i}{\partial \sigma_i} \right| + |v|^5 \times \left| \frac{\partial \omega_i}{\partial \sigma_i} \right| \right).$$

Observe that, because of the orthogonality conditions satisfied by v,

$$J'' \left(\sum \alpha_j \omega_j \right) \cdot v \cdot \frac{\partial \omega_i}{\partial \sigma_i} = O \left(\sum_{\gamma=1}^4 \int \left(\sum_{j \neq i} \alpha_j \omega_j \right)^\gamma \omega_i^{4-\gamma} \times v \times \frac{\partial \omega_i}{\partial \sigma_i} \right).$$

Let

$$f = O \left(\sum_{\gamma=1}^4 \int \left(\sum_{j \neq i} \alpha_j \omega_j \right)^\gamma \omega_i^{4-\gamma} \times v \times \frac{\partial \omega_i}{\partial \sigma_i} + O \left(\left| \sum \alpha_j \omega_j \right| v^2 \right) + O(|v|^5) \right)$$

$$= \qquad f_1 \qquad + f_2 \qquad f_3.$$

We decompose $\int f \frac{\partial \omega_i}{\partial \sigma_i}$ into $\sum \int_{\Omega_\ell} f \frac{\partial \omega_i}{\partial \sigma_i} + \int_{(\cup \Omega_\ell)^c} f \frac{\partial \omega_i}{\partial \sigma_i}$. In each Ω_ℓ, we think of $\sum \alpha_j \omega_j$ as made of two pieces, $\alpha_\ell \omega_\ell$ and $\sum_{j \neq \ell} \alpha_j \omega_j$. There are optimal $\bar{v}$'s, $\bar{v}_\ell$, and $\bar{v}'_\ell$ related to this decomposition into groups just as we have v_I and v_{II} in the case of two combinations. $\bar{v}_\ell$ is related to ω_ℓ and is therefore equal to zero since ω_ℓ is a solution of the Yamabe problem. $\bar{v}'_\ell$ satisfies the estimate (after our remarks above).

$$|\bar{v}'_\ell| \leq \sum_{\substack{j,k \neq \ell \\ j \neq k}} \left(|\omega_k^\infty| + \frac{1}{\lambda_k |a_k - a_j|} + \varepsilon_{jk} \right) \varepsilon_{jk} \delta_j$$

and (Theorem 2)

$$|(\bar{v} - (\bar{v}_\ell + \bar{v}'_\ell))|_{H_0^1} + \frac{|h_\ell^0|_\infty}{\sqrt{\lambda_\ell}} + \sum_{\lambda_s > \lambda_\ell} \varepsilon_{s\ell} |h_\ell^s|_\infty$$

$$\leq C \left(\left| \sum_{j \neq \ell} \frac{\sqrt{\lambda_j}}{\sqrt{\lambda_\ell}} \alpha_j \bar{\omega}_j(\bar{a}_j^i) \right| + \sum_{k \neq \ell} \left| \sum_{j \neq \ell} \sqrt{\lambda_j} \alpha_j \bar{\omega}_j(\bar{a}_k^j) \right| \frac{\varepsilon_{\ell k}}{\sqrt{\lambda_k}} \right.$$

$$\left. + \sum_{j \neq \ell} |\sqrt{\lambda_\ell} \bar{\omega}_\ell(\bar{a}_j^\ell)| \frac{\varepsilon_{j\ell}}{\sqrt{\lambda_j}} \right) + O \left(\sum \frac{|\omega_j^\infty| \varepsilon_{j\ell}}{\lambda_j |a_j - a_\ell|} \right) + O \left(\sum \varepsilon_{j\ell} \sum \varepsilon_{mt}^2 \right).$$

We then derive that (we show below how to estimate $\int_{\Omega_i} \left(\sum_{j \neq i} \alpha_j \omega_j \right) \cdot \omega_i^4 \bar{v}$ and expressions of the same type)

$$\left| \int f \frac{\partial \omega_i}{\partial \sigma_i} \right| \leq C \left(\sum_{i \neq \ell} \varepsilon_{i\ell} \left| \sum_{j \neq \ell} \sqrt{\lambda_j} \frac{\alpha_j \bar{\omega}_j(\bar{a}_\ell^j)}{\sqrt{\lambda_\ell}} \right| + \right.$$

$$C \sum_{m \neq i} \left| \sum_{j \neq i} \frac{\sqrt{\lambda_j}}{\sqrt{\lambda_i}} \alpha_j \bar{\omega}_j(\bar{a}_i^j) \right| \varepsilon_{im} + \sum \varepsilon_{i\ell} \left(\frac{|\omega_j^\infty| \varepsilon_{j\ell}}{\lambda_\ell |a_j - a_\ell|} + \sum \varepsilon_{j\ell} \left(\sum \varepsilon_{mt}^2 \right) \right).$$

This estimate is transparent for $\gamma \geq 2$.

On the other hand,

$$J' \left(\sum \alpha_j \omega_j \right) \cdot \frac{\partial \omega_i}{\partial \sigma_i} = -c \int \left(\left(\sum \alpha_j \omega_j \right)^5 - \sum \alpha_j^5 \omega_j^5 \right) \frac{\partial \omega_i}{\partial \sigma_i}$$

$$= -c\sum_j \int_{\Omega_j} \left(\left(\sum \alpha_s\omega_s\right)^5 - \alpha_j\omega_j^5\right)\frac{\partial\omega_i}{\partial\sigma_i} + O\left(\sum \int_{\Omega_j^c} \delta_j^5\delta_i\right)$$

$$= \sum_j\sum_{\gamma\geq 3} \int_{\Omega_j} O\left(\left|\left(\sum_{s\neq j}\alpha_s\omega_s\right)^\gamma \omega_j^{5-\gamma}\frac{\partial\omega_i}{\partial\sigma_i}\right|\right) + O\left(\sum \int_{\Omega_j^c} \delta_j^5\delta_i\right)$$

$$- c\sum_j 15\alpha_j^2 \int_{\Omega_j}\left(\sum_{s\neq j}\alpha_s\omega_s\right)^2 \omega_j^3\frac{\partial\omega_i}{\partial\sigma_i} - 5c\sum_j \alpha_j\int_{\Omega_j}\left(\sum_{s\neq j}\alpha_s\omega_s\right)^4 \omega_j^4\frac{\partial\omega_i}{\partial\sigma_i}.$$

Our estimates from here are a repetition of the remarks made in the previous section. $\int_{\Omega_j}\left(\sum_{s\neq j}\alpha_s\omega_s\right)^2\omega_j^3\frac{\partial\omega_i}{\partial\sigma_i}$ is estimated for $j\neq i$ as follows:

$$\int_{\Omega_j}\left(\sum_{s\neq j}\alpha_s\omega_s\right)^2 \omega_j^3\frac{\partial\omega_i}{\partial\sigma_i} = \sum_{t\neq j}\alpha_t\int_{\Omega_j}\omega_t\omega_j^3\frac{\partial\omega_i}{\partial\sigma_i}.$$

Let

$$f_t = \omega_t\omega_j^3\frac{\partial\omega_i}{\partial\sigma_i}.$$

We consider $\int \omega_s f_t$ and rescale ω_s to a concentration 1.

ω_j then concentrates around a new part $\bar{a}_j^s$ and we have:

$$\int_{\Omega_j}\omega_s f_t = \int_{\tilde{\Omega}_j}\bar{\omega}_s\tilde{f}_t = \bar{\omega}_s(\bar{a}_j^s)\int_{\tilde{\Omega}_j}\tilde{f}_t + O\left(\int_{\tilde{\Omega}_j}|x - \bar{a}_j^s||\tilde{f}_t|\right).$$

Thus,

$$\int_{\Omega_j}\left(\sum_{s\neq j}\alpha_s\omega_s\right)^2 \omega_j^3\frac{\partial\omega_i}{\partial\sigma_i} = \left(\sum_{s\neq j}\alpha_s\bar{\omega}_s(\bar{a}_j^s)\sqrt{\lambda_s}\right) O\left(\int_{\Omega_j}|f_t|\right)$$

$$+ O\left(\int_{\tilde{\Omega}_j^s}|x - \bar{a}_j^s||\tilde{f}_t|\right).$$

Clearly, because $t \neq j$.

$$\int_{\tilde{\Omega}_j^s} |x - \bar{a}_j^s| |\tilde{f}_t| = o\left(\int_{\tilde{\Omega}_j^s} |x - \bar{a}_j^s| \tilde{\omega}_j^4 \tilde{\delta}_i\right) = o\left(\varepsilon_{ij} \left(\int_{\tilde{\Omega}_j^s} |x - \bar{a}_j^s|^2 \tilde{\omega}_j^4\right)^{1/2}\right).$$

Let $\tilde{\lambda}_j$ be the new concentration of $\tilde{\omega}_j$. Then, $\frac{1}{\sqrt{\tilde{\lambda}_j}} \sim c\varepsilon_{js}$ and:

$$\left(\int_{\tilde{\Omega}_j^s} |x - \bar{a}_j^s|^2 \tilde{\omega}_j^4\right)^{1/2} \leq \frac{C}{\tilde{\lambda}_j^{3/2}} \left(\int_{\tilde{\lambda}_j \tilde{\Omega}_j^s} dr\right)^{1/2} \leq \frac{C}{\tilde{\lambda}_j^{3/2}} \times \frac{1}{\sqrt{\text{Max } \varepsilon_{jm}}} \leq C\varepsilon_{js}^{5/2}.$$

Thus,

$$\left|\int_{\tilde{\Omega}_j^s} |x - \bar{a}_j^s| |\tilde{f}_t|\right| = o\left(\varepsilon_{ij} \varepsilon_{js}^{5/2}\right).$$

On the other hand,

$$\int_{\Omega_j} |f_t| \leq C \int_{\Omega_j} \delta_t \delta_j^3 \delta_i \leq C\varepsilon_{ij} \left(\int_{\Omega_j} \delta_t^2 \delta_j^2\right)^{1/2} \leq C\frac{\varepsilon_{ij}}{\sqrt{\lambda_j}}.$$

Thus, for $j \neq i$:

$$\int_{\Omega_j} \left(\sum_{s \neq j} \alpha_s \omega_s\right)^2 \omega_j^3 \frac{\partial \omega_i}{\partial \sigma_i} = O(\varepsilon_{ij}) \left(\left|\sum_{s \neq j} \alpha_s \frac{\bar{\omega}_s(\bar{a}_j^s)\sqrt{\lambda_s}}{\sqrt{\lambda_j}}\right| + o\left(\varepsilon_{js}^{5/2}\right)\right).$$

The estimate extends readily, for $j \neq i$, to ($\delta_s \leq C\delta_j$ in Ω_j).

$$\int_{\Omega_j} \left(\sum_{s \neq j} \alpha_s \omega_s\right)^\gamma \omega_j^{5-\gamma} \frac{\partial \omega_i}{\partial \sigma_i} \quad \text{for } \gamma \geq 3.$$

A direct estimate shows that:

$$\int_{\Omega_j^c} \delta_j^5 \delta_i = O\left(\varepsilon_{ij} \sum \varepsilon_{jt}^2\right).$$

In order to finish the study of our expansion for $j \neq i$, we consider

$$\int_{\Omega_j} \left(\sum_{s \neq j} \alpha_s \omega_s\right) \omega_j^4 \frac{\partial \omega_i}{\partial \sigma_i} = O\left(\left|\sum_{s \neq j} \alpha_s \frac{\bar{\omega}_s(\bar{a}_j^s)\sqrt{\lambda_s}}{\sqrt{\lambda_j}}\right| \varepsilon_{ij}\right)$$

$$+ O\left(\sum_s \int_{\tilde{\Omega}_j^s} |x - \bar{a}_j^s| |\tilde{g}| \right).$$

$|\tilde{g}|$ is equal to $\tilde{\omega}_j^4 \frac{\partial \omega_i}{\partial \sigma_i}$. We have estimated such quantities above.

We find

$$\int_{\Omega_j} \left(\sum_{s \neq j} \alpha_s \omega_s \right) \omega_j^4 \frac{\partial \omega_i}{\partial \sigma_i} = O\left(\left| \sum_{s \neq j} \frac{\alpha_s \bar{\omega}_s(\bar{a}_j^s) \sqrt{\lambda_s}}{\sqrt{\lambda_j}} \right| \varepsilon_{ij} \right) + O\left(\sum \varepsilon_{ij} \varepsilon_{js}^{5/2} \right).$$

For $j = i$, the estimates change.

We have since $|\frac{\partial \omega_i}{\partial \sigma_i}| \leq C \delta_i$

$$\int_{\Omega_j} \left(\sum_{s \neq i} \alpha_s \omega_s \right)^2 \omega_i^3 \frac{\partial \omega_i}{\partial \sigma_i} = O\left(\frac{|\sum \alpha_j \bar{\omega}_j(\bar{a}_i^j) \sqrt{\lambda_j}|}{\sqrt{\lambda_i}} \varepsilon_{it} + \int_{\tilde{\Omega}_i^j} |x - \bar{a}_i^j| \tilde{\delta}_i^4 \tilde{\delta}_t \right)$$

$$= O\left(\sum \left| \frac{\sum \alpha_j \bar{\omega}_j(\bar{a}_i^j) \sqrt{\lambda_j}}{\sqrt{\lambda_i}} \right| \varepsilon_{it} + \varepsilon_{it} \sum \varepsilon_{is}^{5/2} \right).$$

This estimate extends to $\sum_{\gamma \geq 3} \int_{\Omega_j} O\left(\left| \left(\sum_{s \neq j} \alpha_s \omega_s \right)^{\gamma} \omega_j^{5-\gamma} \frac{\partial \omega_i}{\partial \sigma_i} \right| \right)$.

We are left with $\int_{\Omega_i} \left(\sum_{s \neq i} \alpha_s \omega_s \right) \omega_i^4 \frac{\partial \omega_i}{\partial \sigma_i} = \int_{\mathbb{R}^3} \left(\sum_{s \neq i} \alpha_s \omega_s \right) \omega_i^4 \frac{\partial \omega_i}{\partial \sigma_i} +$

$O\left(\varepsilon_{is} \sum \varepsilon_{it}^2 \right).$

We split $\int_{\mathbb{R}^3} \left(\sum_{s \neq i} \alpha_s \omega_s \right) \omega_i^4 \frac{\partial \omega_i}{\partial \sigma_i}$ into two pieces. The first piece is

$$(I) = \sum_{\lambda_s > \lambda_i} \alpha_s \int \omega_s \omega_i^4 \frac{\partial \omega_i}{\partial \sigma_i}$$

and the second one is

$$(II) = \sum_{\substack{\lambda_s \leq \lambda_i \\ s \neq i}} \alpha_s \int \omega_s \omega_i^4 \frac{\partial \omega_i}{\partial \sigma_i}.$$

We rewrite (II) as

$$(II) = \sum_{\substack{\lambda_s \leq \lambda_i \\ s \neq i}} \frac{\alpha_s}{5} \frac{\partial}{\partial \sigma_i} \int \omega_i \omega_s^5.$$

We expand (I) as

$$\sum_{\lambda_s > \lambda_i} \alpha_s \int \omega_s \omega_i^4 \frac{\partial \omega_i}{\partial \sigma_i} = \sum_{\lambda_s > \lambda_i} \alpha_s \bar\omega_s(\bar a_i^s) \sqrt{\lambda_s} \int_{\mathbb{R}^3} \omega_i^4 \frac{\partial \omega_i}{\partial \sigma_i}$$

$$+ O\left(\int_{\mathbb{R}^3} |x - \bar a_i^s| \tilde\delta_i^5 \right) = \sum_{\lambda_s > \lambda_i} \alpha_s \bar\omega_s(\bar a_i^s) \frac{\sqrt{\lambda_s}}{\sqrt{\lambda_i}} \partial_{\sigma_i} \omega_i^\infty$$

$$+ O\left(\frac{1}{\tilde\lambda_i^{3/2}} \right) = \sum_{\lambda_s > \lambda_i} \alpha_s \bar\omega_s(\bar a_i^s) \sqrt{\frac{\lambda_s}{\lambda_i}} \frac{\partial_{\sigma_i} \omega_i^\infty}{5} + 0\left(\sum \varepsilon_{is}^3 \right)$$

$$= \sum_{\lambda_s > \lambda_i} \left(\alpha_s \bar\omega_s \varepsilon_{is}(1 + o(1)) + O\left(\frac{1}{\lambda_s^{3/2} \sqrt{\lambda_i}} \frac{1}{|a_i - a_s|^2} \right) \right) \frac{\partial_{\sigma_i} \omega_i^\infty}{5} + O\left(\sum \varepsilon_{is}^3 \right)$$

$$= \sum_{\lambda_s > \lambda_i} \left(\frac{\alpha_s \bar\omega_s^\infty}{5 \sqrt{\lambda_s} |a_i - a_s|} (1 + o(1)) \frac{\partial \omega_i^\infty}{\partial \sigma_i} + O\left(\frac{\varepsilon_{is}^{3/2}}{|a_i - a_s|^{1/2}} \right) \right) \times \frac{1}{\sqrt{\lambda_i}}.$$

On the other hand, using the expansion of ω_i:

(II)

$$= \sum_{\substack{\lambda_s \leq \lambda_i \\ s \neq i}} \frac{\alpha_s}{5} \frac{\partial}{\partial \sigma_i} \left(c \omega_i^\infty \int \delta_i \omega_s^5 + O\left(\frac{1}{\lambda_i^{3/2} |a_s - a_i|^2} \right) \frac{1}{\sqrt{\lambda_s}} + O\left(\sum_{\substack{\lambda_s \leq \lambda_i \\ s \neq i}} \varepsilon_{is}^3 \right) \right).$$

We can—a simple argument shows this—differentiate with respect to σ_i and we derive:

(II)

$$= \sum_{\substack{\lambda_s \leq \lambda_i \\ s \neq i}} \left\{ c \frac{\alpha_s}{5} \left(\bar\omega_s^\infty \varepsilon_{is} + O(\varepsilon_{is}^3) \right) \frac{\partial \omega_i^\infty}{\partial \sigma_i} + O\left(\frac{1}{\sqrt{\lambda_s} \lambda_i^{3/2} |a_s - a_i|^2} + \sum_{\substack{\lambda_s \leq \lambda_i \\ s \neq i}} \varepsilon_{is}^3 \right) \right\}$$

$$= c \sum_{\substack{\lambda_s \leq \lambda_i \\ s \neq i}} \left\{ \left(\frac{\alpha_s}{5} \frac{\bar\omega_s^\infty}{\sqrt{\lambda_s} |a_i - a_s|} (1 + o(1)) \frac{\partial \omega_i^\infty}{\partial \sigma_i} + O\left(\frac{\varepsilon_{is}^{3/2}}{|a_i - a_s|^{1/2}} \right) \right) \right\} \times \frac{1}{\sqrt{\lambda_i}}.$$

We thus have established that

$$\int_{\Omega_i} \left(\sum_{s \neq i} \alpha_s \omega_s \right) \omega_i^4 \frac{\partial \omega_i}{\partial \sigma_i}$$

$$= \frac{c}{\sqrt{\lambda_i}} \left(\sum_{s \neq i} \frac{\bar{\omega}_s^\infty}{\sqrt{\lambda_s}|a_i - a_s|}(1 + o(1)) \frac{\partial \omega_i^\infty}{\partial \sigma_i} + O\left(\sum \frac{\varepsilon_{is}^{3/2}}{|a_i - a_s|^{1/2}} \right) \right).$$

On the other hand, we may write:

$$\int_{\Omega_i} \left(\sum_{s \neq i} \alpha_s \omega_s \right) \omega_i^4 \frac{\partial \omega_i}{\partial \sigma_i}$$

$$= \sum_{s \neq i} \alpha_s \bar{\omega}_s(\bar{a}_i^s) \sqrt{\lambda_s} \int_{\Omega_i} \omega_i^4 \frac{\partial \omega_i}{\partial \sigma_i} + O\left(\int_{\tilde{\Omega}_i^j} |x - \bar{a}_i^s| \tilde{\delta}_i^5 \right)$$

$$= \left(\sum_{s \neq i} \alpha_s \bar{\omega}_s(\bar{a}_i^s) \sqrt{\lambda_s} \right) \left(\int_{\mathbb{R}^3} \omega_i^4 \frac{\partial \omega_i}{\partial \sigma_i} + O\left(\sum_{\lambda_s \geq \lambda_i} \frac{\varepsilon_{st}^2}{\sqrt{\lambda_s}} \right) \right) + O\left(\frac{1}{\tilde{\lambda}_i^{3/2}} \right)$$

$$= c \sum_{s \neq i} \alpha_s \bar{\omega}_s(\bar{a}_i^s) \sqrt{\frac{\lambda_s}{\lambda_i}} \partial_{\sigma_i} \omega_i^\infty + O\left(\sum_{\lambda_s \geq \lambda_i} \varepsilon_{st}^2 \varepsilon_{is} \right).$$

We used again above the fact that

$$\frac{1}{\sqrt{\tilde{\lambda}_j}} \sim c\varepsilon_{ij}.$$

The techniques used above apply as well to $\int_{\Omega_i} \left(\sum_{j \neq i} \alpha_j \omega_j \right) \omega_i^4 \bar{v}$ after splitting $\bar{v}$ into $\bar{v}_i + \bar{h}_i^0 + \sum_{\lambda_s \geq \lambda_i} \delta_s \bar{h}_i^s$, rescaling *etc.*

Summing up and denoting

$$u_i = \sum_{s \neq i} \alpha_j \omega_j(\bar{a}_i^j) \frac{\sqrt{\lambda_j}}{\sqrt{\lambda_i}},$$

our equation reads

$$J'\left(\sum \alpha_j \omega_j + v \right) \cdot \frac{\partial \omega_i}{\partial \sigma_i} = c \left(1 + O\left(\sum \varepsilon_{it} \right) \right) u_i \partial_{\sigma_i} \omega_i^\infty + \sum O(\varepsilon_{ij} u_j)$$

$$+ \varepsilon_{i\ell} O \left(\sum \frac{|\omega_j^\infty| \varepsilon_{j\ell}}{\lambda_\ell |a_j - a_\ell|} + \sum \varepsilon_{j\ell} \sum \varepsilon_{mt}^2 \right) + O \left(\sum_{\lambda_s \geq \lambda_i} \varepsilon_{st}^3 \right)$$

$$= c \sum_{s \neq i} \bar{\omega}_s^\infty \varepsilon_{is} (1 + o(1)) \partial_{\sigma_i} \omega_i^\infty + O \left(\sum \frac{\varepsilon_{is}^{3/2}}{\sqrt{\lambda_i} |a_i - a_s|^{1/2}} \right).$$

On the other hand, since the σ_i are compact variables, we may assume that

$$J' \left(\sum \alpha_j \omega_j + \bar{v} \right) \cdot \left(\alpha_i \frac{\partial \omega_i}{\partial \sigma_i} + \frac{\partial \bar{v}}{\partial \sigma_i} \right) = o(\varepsilon_{ij}^3).$$

Thus

$$J' \left(\sum \alpha_j \omega_j + \bar{v} \right) \cdot \frac{\partial \omega_i}{\partial \sigma_i} = -J' \left(\sum \alpha_j \omega_j + \bar{v} \right) \frac{\partial \bar{v}}{\partial \sigma_i} + o(\varepsilon_{ij}^3).$$

Using the fact that $J'(u + \bar{v})$ is orthogonal to the v-space, we can write

$$J'(u + \bar{v}) \cdot \frac{\partial \bar{v}}{\partial \sigma_i} = \sum A_i^j \left(J'(u + \bar{v}) \cdot \frac{\partial \omega_i}{\partial \sigma_i^j} \right) + B_i \left(J'(u + \bar{v}) \cdot \lambda_i \frac{\partial \omega_i}{\partial \lambda_i} \right)$$

$$+ \sum C_i^j J'(u + \bar{v}) \cdot \frac{1}{\lambda_i} \frac{\partial \omega_i}{\partial a_i^j}.$$

Thus,

$$\left(1 + \frac{A_i}{\alpha_i} \right) J'(u + \bar{v}) \cdot \frac{\partial \bar{v}}{\partial \sigma_i} = \frac{A_i}{\alpha_i} \left(J'(u + \bar{v}) \cdot (\alpha_i \frac{\partial \omega_i}{\partial \sigma_i} + \frac{\partial \bar{v}}{\partial \sigma_i}) \right)$$

$$+ B_i J'(u + \bar{v}) \cdot \lambda_i \frac{\partial \omega_i}{\partial \lambda_i} + \sum C_i^j J'(u + \bar{v}) \cdot \frac{1}{\lambda_i} \frac{\partial \omega_i}{\partial a_i^j}.$$

It is easy to see that

$$|A_i| + |B_i| + \sum |C_i^j| \leq C \int |\bar{v}| \delta_i^5 \leq C \left(\int_{\Omega_i} |\bar{v}| \delta_i^5 + \sum_j \int_{\Omega_i^c} \delta_i^5 \delta_j \right).$$

As above, we decompose $\bar{v}$ into $\bar{v} - \bar{v}'_i$ where $\bar{v}'_i$ is the optimal $\bar{v}$ related to the decomposition $\sum\limits_{j \neq i} \alpha_j \omega_j$. We know that:

$$|\bar{v}'_i| \leq \sum_{\substack{j,k \neq i \\ j \neq k}} \varepsilon_{jk} \delta_j$$

so that

$$\int |\bar{v}_i'|\delta_i^5 = O(\varepsilon_{ij}\varepsilon_{jk})$$

$$= o(\varepsilon_{ij}\varepsilon_{jk}) \quad \text{when the } \omega_k^\infty = o(1)$$

and we know that (Theorem 2)

$$\left|(\bar{v} - \bar{v}_i')_i\right|_{H_0^1} + \frac{|h_i^0|_\infty}{\sqrt{\lambda_i}} + \sum_{\lambda_s \geq \lambda_i} \varepsilon_{si}|h_i^s|_\infty$$

$$\leq C \left(|u_i| + \sum_{k \neq \ell} \varepsilon_{ik} u_k + O\left(\sum |\omega_j^\infty| \frac{\varepsilon_{ij}}{\lambda_i |a_i - a_j|} + O\left(\sum \varepsilon_{ij} \left(\sum \varepsilon_{mt}^2 \right) \right) \right) \right).$$

Thus,

$$\int_{\Omega_i^c} |\bar{v}|\delta_i^5 + \int_{\Omega_i} |\bar{v}|\delta_i^5 = O\left(|u_i| + \sum_{k \neq \ell} \varepsilon_{ik}|u_k| + \sum \frac{|\omega_j^\infty|}{\lambda_i |a_i - a_j|} \varepsilon_{ij} + \sum \varepsilon_{mt}^2 \right).$$

On the other hand, since

$$\left| \lambda_i \frac{\partial \omega_i}{\partial \lambda_i} \right| + \frac{1}{\lambda_i} \left| \frac{\partial \omega_i}{\partial a_i^j} \right| \leq C \left(|\omega_i^\infty|\delta_i + \frac{\delta_i^2}{\sqrt{\lambda_i}} \right),$$

$$\left| J'(u + \bar{v}) \cdot \lambda_i \frac{\partial \omega_i}{\partial \lambda_i} \right| + \left| J'(u + \bar{v}) \cdot \frac{1}{\lambda_i} \frac{\partial \omega_i}{\partial a_i^j} \right|$$

$$\leq C \left(\sum_\ell |\omega_i^\infty|\varepsilon_{i\ell} + \sum_{\substack{\ell \neq i \\ s \neq \ell}} \int \delta_\ell^4 \delta_s \frac{\delta_i^2}{\sqrt{\lambda_i}} \right.$$

$$\left. + \sum_{\ell \neq i} \int (\delta_\ell^4 + \delta_i^3|\omega_\ell|) |\bar{v}| \frac{\delta_i^2}{\sqrt{\lambda_i}} + \int |\bar{v}|^5 \frac{\delta_i^2}{\sqrt{\lambda_i}} \right).$$

We know that

$$\int \delta_\ell^4 \delta_s \frac{\delta_i^2}{\sqrt{\lambda_i}} \leq \int \delta_\ell^4 \delta_i^2 + \int \delta_s^4 \delta_i^2 \leq C(\varepsilon_{\ell i}^2 + \varepsilon_{si}^2) \text{ if } \lambda_\ell \text{ or } \lambda_s \leq \lambda_i$$

$$\int \delta_\ell^4 \delta_s \frac{\delta_i^2}{\sqrt{\lambda_i}} \leq \int \delta_\ell^4 \delta_s \delta_i \leq C\varepsilon_{\ell s}\varepsilon_{\ell i} \text{ if } \lambda_s \text{ and } \lambda_\ell \geq \lambda_i$$

$$\sum_{\ell \neq i} \int \delta_\ell^4 |\bar{v}| \frac{\delta_i^2}{\sqrt{\lambda_i}} \leq C \left(\sum_{\substack{\ell \neq i \\ s \neq \ell}} \int \delta_\ell^4 \delta_s \frac{\delta_i^2}{\sqrt{\lambda_i}} \right)$$

$$\int \delta_i^3 |\omega_\ell| \frac{\delta_i^2}{\sqrt{\lambda_i}} |\bar{v}| \leq C \int_{\Omega_i} \delta_i^4 \delta_\ell |\bar{v}| + C \int_{\Omega_i^c} \delta_i^4 \delta_\ell |\bar{v}|$$

$$\leq C \left(\varepsilon_{i\ell}(|\bar{v}_i|_{H_0^1} + \frac{|\bar{h}_i^0|_\infty}{\sqrt{\lambda_i}} + \sum_{\lambda_s > \lambda_i} \varepsilon_{is}|\bar{h}_i^s|_\infty) + \sum \varepsilon_{im}^2 \right) \text{ for } \ell \neq i$$

$$\int |\bar{v}|^5 \frac{\delta_i^2}{\sqrt{\lambda_i}} \leq C \left(\sum \varepsilon_{ki}^5 + \sum_{\ell \neq i} \delta_\ell^4 |\bar{v}| \frac{\delta_i^2}{\sqrt{\lambda_i}} \right)$$

$|\bar{v}|_{H_0^1} + \frac{|\bar{h}_i^0|_\infty}{\sqrt{\lambda_i}} + \sum\limits_{\lambda_s > \lambda_i} \varepsilon_{is}|\bar{h}_i^s|_\infty$ is bounded above by ε_{it}.

Combining, we derive that:

$$c \left(1 + O \left(\sum \varepsilon_{it} \right) \right) u_i \partial_{\sigma_i} \omega_i^\infty + O \left(\sum \varepsilon_{i\ell} u_j \right)$$

$$= \varepsilon_{i\ell} O \left(\sum \frac{|\omega_j^\infty| \varepsilon_{j\ell}}{\lambda_\ell |a_j - a_\ell|} + \sum \varepsilon_{j\ell} \sum \varepsilon_{mt}^2 \right) + \sum_{\lambda_s \geq \lambda_i} O(\varepsilon_{st}^2 \varepsilon_{is}).$$

If $\omega_i^\infty = o(1)$, $\nabla \omega_i^\infty$ is non zero (an assumption) and there exists σ_i such that $\partial_{\sigma_i} \omega_i^\infty$ is away from zero (we complete a rigid rotation of $\bar{\omega}_i$, then reconcentrate it around the same point a_i, with the same concentration λ_i). These equations take a matrical form. We derive that:

$$u_i = \varepsilon_{i\ell} O \left(\sum \frac{\omega_j^\infty \varepsilon_{j\ell}}{\lambda_\ell |a_j - a_\ell|} + \sum \varepsilon_{j\ell} \sum \varepsilon_{mt}^2 \right) + O \left(\sum_{\lambda_s \geq \lambda_i} \varepsilon_{is} \varepsilon_{st}^2 \right)$$

and

$$u_i \partial_{\sigma_i} \omega_i^\infty = \varepsilon_{i\ell} O \left(\sum \frac{|\omega_j^\infty| \varepsilon_{j\ell}}{\lambda_\ell |a_j - a_\ell|} + \sum \varepsilon_{j\ell} \sum \varepsilon_{mt}^2 \right) + O \left(\sum_{\lambda_s \geq \lambda_i} \varepsilon_{is} \varepsilon_{st}^2 \right).$$

$\square$

We now easily derive from Proposition 8 and similar computations the following identity:

$$\sqrt{\lambda_i}(1 + o(1))\omega_j^\infty \varepsilon_{ij} + \gamma_i^\gamma = \bar{\omega}_j(\bar{a}_i^j)\sqrt{\lambda_j} + O\left(\sqrt{\lambda_i} \sum_{\lambda_s \geq \lambda_i} \varepsilon_{is}\varepsilon_{st}^2\right).$$

Using (1) of Proposition 8, we derive:

$$\sum_{j \neq i} \alpha_j \gamma_i^j = -\sum \frac{\alpha_j \omega_j^\infty}{\sqrt{\lambda_j}|a_i - a_j|}(1 + o(1)) + O\left(\sqrt{\lambda_i} \sum_{\lambda_s \geq \lambda_i} \varepsilon_{is}\varepsilon_{st}^2\right). \quad (1.46)$$

We then have the following Lemma:

Let

$$|A| = \left|\gamma_i^j \gamma_i^k\right| \quad \text{for } j \neq k.$$

Lemma 71   *(a) If $|a_i - a_j|$ and $|a_i - a_k|$ are comparable,*

$$|A| \leq C\frac{\varepsilon_{jk}^3}{|a_j - a_k|}$$

(b) If $\lambda_i \leq C \operatorname{Sup}(\lambda_j, \lambda_k)$ e.g. $\lambda_i \leq C\lambda_i \leq C\lambda_k$, then

$$|A| \leq o(\gamma_i^{j2}) + C\frac{\varepsilon_{ik}^3}{|a_i - a_k|}$$

(c) If $|a_i - a_j| = o(|a_i - a_k|)$ and $\lambda_k \geq C\lambda_j$,

$$|A| \leq o(\gamma_i^{j2}) + C\frac{\varepsilon_{jk}^3}{|a_j - a_k|}$$

(d) If $|a_i - a_j| = o(|a_i - a_k|)$, $\lambda_k + \lambda_j = o(\lambda_i)$, $\lambda_k = o(\lambda_j)$,

$$|A| \leq C|\gamma_i^j \gamma_j^k| + C\frac{\varepsilon_{jk}^3}{|a_i - a_j|}$$

Proof.
    Proof of (a) Since $|a_i - a_j|$ and $|a_i - a_k|$ are comparable,

$$|a_j - a_k| \leq C|a_i - a_j|$$

$$|a_j - a_k| \leq C|a_i - a_k|$$

so that

$$|A| = |\gamma_i^j \gamma_i^k| \le C \times \frac{1}{\lambda_j^{3/2}\lambda_k^{3/2}} \times \frac{1}{|a_k - a_j|^4} \le C' \frac{\varepsilon_{jk}^3}{|a_k - a_j|}.$$

Proof of (b) Assume $\lambda_i \le C\lambda_k$. Then,

$$|A| \le C\gamma_i^j \times \frac{1}{\lambda_i^{3/4}\lambda_k^{3/4}|a_i - a_k|^2} \le o(\gamma_i^{j2}) + C\frac{\varepsilon_{ik}^3}{|a_i - a_k|}.$$

Proof of (c) If $|a_i - a_j| = o(|a_i - a_k|)$, then $|a_j - a_k|$ and $|a_i - a_k|$ are some of the same order. Thus, if $\lambda_k \ge C\lambda_j$,

$$|A| \le C\gamma_i^j \times \frac{\varepsilon_{jk}^{3/2}}{|a_j - a_k|^{1/2}} \le o(\gamma_i^{j2}) + C\frac{\varepsilon_{jk}^3}{|a_j - a_k|}.$$

Proof of (d) By definition,

$$\gamma_i^k = \omega_k(a_i) - \frac{c}{(1 + \lambda_k^2|a_i - a_k|^2)^{1/2}} = \omega_k(a_j) - c\frac{\omega_k^\infty \sqrt{\lambda_k}}{(1 + \lambda_k^2|a_j - a_k|^2)^{1/2}}$$

$$+ O\left(\frac{1}{\lambda_k^{3/2}|a_j - a_k|^3}\right)|a_i - a_j| = \gamma_j^k + O\left(\frac{1}{\lambda_k^{3/2}|a_j - a_k|^3}\right)|a_i - a_j|.$$

Thus,

$$|A| = |\gamma_i^j \gamma_i^k| \le |\gamma_j^k \gamma_i^j| + \frac{C}{\lambda_k^{3/2}\lambda_j^{3/2}|a_j - a_k|^3} \times \frac{|a_i - a_j|}{|a_i - a_j|^2}$$

$$\le |\gamma_j^k \gamma_i^j| + C\frac{\varepsilon_{jk}^3}{|a_i - a_j|}.$$

$\square$

We now introduce a definition:

Definition 1 A configuration $\sum \alpha_i \omega_i + \bar{v}$ is **well-distributed** if $\forall i, j, k$ pairwise distinct, with $\lambda_k + \lambda_j = o(\lambda_i), \lambda_k = o(\lambda_j)$, then $\varepsilon_{kj} \le C\varepsilon_{ij}$ if $|a_i - a_j| = o(|a_i - a_k|)$

and we have

Proposition 9   *If a configuration $\sum \alpha_i \omega_i + \bar{v}$ is **well-distributed**, then*

$$|\gamma_i^k| \leq C \sum_{\substack{k \\ \lambda_k \leq \lambda_i}} \left| \sum \frac{\omega_j^\infty}{\sqrt{\lambda_j}|a_k - a_j|}(1 + o(1)) \right| + C \sum_{\lambda_j \leq C\lambda_i} \frac{\varepsilon_{jk}^{3/2}}{|a_j - a_k|^{1/2}}$$

for each i and ℓ.

Proof.   If a configuration is well distributed, we may replace in (d) of Lemma 71 $\frac{\varepsilon_{jk}^3}{|a_i - a_j|}$ by $\frac{\varepsilon_{ij}^3}{|a_i - a_j|}$.

We consider then all the relations (1.46) for $i \leq \bar{i}$ where $\bar{i}$ is a fixed index. We square all these relations and we add them in a weighted combination, with much larger weights on lower i, these weights decreasing as i increases to $\bar{i}$. (b) of Lemma 71 allows us to replace all products for a given i with $\lambda_i \leq C$ Sup (λ_j, λ_k) by $o(\gamma_i^{j2}) + C \frac{\varepsilon_{ik}^3}{|a_i - a_k|} \cdot o(\gamma_i^{j2})$ is of course absorbed into $\alpha_j^2 \gamma_i^{j2}$. We are left with j, k such that Sup $(\lambda_j, \lambda_k) = o(\lambda_i)$.

We then use (c) and also (d) since $C\gamma_i^j \gamma_j^k = o(\gamma_i^{j2}) + O(\gamma_j^{k2})$ and $O(\gamma_j^{k2})$ can be aborbed into $M\gamma_j^{k2}$ provided by the same relation for $i = j$ now, squared and weighted appropriately. Finally, if $|a_i - a_j|$ and $|a_i - a_k|$ are comparable, we use (a). $\square$

Assuming now that we have a general configuration which is not necessarily well-distributed, we prove:

Lemma 72 $\frac{|\gamma_i^j|}{\sqrt{\lambda_i}} \leq C\varepsilon_{ij}^2$.

Proof. $\frac{\gamma_i^j}{\sqrt{\lambda_i}}$ is invariant by all conformal deformations which do not change ω_i^∞ and a_i because $\frac{\gamma_i^j}{\sqrt{\lambda_i}}$ can be identified as $c \int (\delta_j - \delta_j(a_i))\omega_i^5$ from the expansion of ω_j and of $\int \omega_j \omega_i^5$.

We thus dilate ω_i around a_i, on S^3, until λ_i and λ_j are about of the same order. With the new $a_i, a_j, \lambda_i, \lambda_j$ (computed in any chart we please, as long as $|a_i - a_j| \leq 10$ for example), we have:

$$\frac{\gamma_i^j}{\sqrt{\lambda_i}} = O\left(\frac{1}{\sqrt{\lambda_i}\lambda_j^{3/2}|a_i - a_j|^2}\right) = O(\varepsilon_{ij}^2).$$

$\square$

Both sides are conformally invariant. Lemma 72 follows.

Corollary 1 *In the estimate of* $|(v - (v_I + v_{II}))|_{H_0^1} + \frac{|h_i^0|_\infty}{\sqrt{\lambda_i}} + \sum\limits_{\lambda_s \geq \lambda_i} \varepsilon_{is}$
$|h_i^s|_\infty$ *carried out in Theorem 2,* $\sum\limits_{\substack{k \neq i \\ k \in I \ or \ k \in II}} |\sum\limits_{\substack{j \in II \\ or \\ j \in I}} \alpha_j \omega_j(a_k)| \frac{\varepsilon_{ik}}{\sqrt{\lambda_k}}$

can be replaced by $\sum\limits_{\substack{k \neq i \\ k \in I \ or \ k \in II}} |\sum\limits_{\substack{j \in II \\ or \\ j \in I}} \alpha_j \omega_j^\infty| + O(\sum \varepsilon_{ik} \sum \varepsilon_{jk}^2).$

We introduce now a new definition:

Definition 2   A partition of a configuration in **groups** is the data of a packing of the various masses into at least two distinct groups, $G_1, \ldots, G_s$ such that $|a_i - a_j| = o(|a_i - a_s|)$ if i, j belong to the same group and s is in a different group.

Given a partition, each group G_m has a lowest concentration λ_{i_m}. The interaction of G_m and $G_s, m \neq s$, is $\varepsilon_{i_m i_s}$.

A partition of a configuration is groups is **well-distributed** if there exists $C_1, C_2 > 0$ such that for every m, j, k pairwise distinct

$$d(G_m, G_j) \leq \frac{1}{C_1} d(G_m, G_k) \Rightarrow \varepsilon_{i_m i_k} \leq C_2 \varepsilon_{i_m i_j}.$$

Observation.   If a configuration is well-distributed, then it is obviously well-distributed as a partition, with groups reduced to single masses.

We then have:

Lemma 73   *Let $p \in G_m, j$ and $k \in G_\ell, G_q$ with $\ell, q \neq m$.*
If $|a_p - a_j| = o(|a_p - a_k|)$, ℓ is different from q.
If in addition, $\lambda_j + \lambda_k = o(\lambda_p), \lambda_k = o(\lambda_j)$, then

$$|\gamma_p^j \gamma_p^k| \leq C|\gamma_p^j \gamma_j^k| + C \frac{\varepsilon_{i_m i_\ell}^3}{|a_{i_m} - a_{i_\ell}|}.$$

Proof.   If j and k are in the same group, then $|a_p - a_j|$ and $|a_p - a_k|$ are of the same order. If in addition $\lambda_j + \lambda_k = o(\lambda_p), \lambda_k = o(\lambda_j)$ then, using (d) of Lemma 71:

$$|\gamma_p^j \gamma_p^k| \leq C \gamma_p^j \gamma_j^k| + C \frac{\varepsilon_{jk}^3}{|a_p - a_j|}.$$

Since $|a_p - a_j| = o(|a_p - a_k|), |a_p - a_j| = o(|a_j - a_k|)$ and

$$d(G_\ell, G_m) \leq \frac{1}{C_1} d(G_\ell, G_q)$$

so that

$$\varepsilon_{jk} \leq C\varepsilon_{i_\ell i_q} \leq CC_2\varepsilon_{i_\ell i_m}$$

while

$$|a_p - a_j| \sim |a_{i_m} - a_{i_\ell}|.$$

$\square$

We also observe that:

Lemma 74 *If $j, k, \in G_m$ with $\lambda_j \leq \lambda_k$, then*

$$|\gamma_i^k| = o\left(\frac{\varepsilon_{jk}^{3/2}}{|a_j - a_k|^{1/2}}\right) \quad \text{for } i \in G_\ell, \ell \neq m.$$

Proof.

$$|\gamma_i^k| \leq \frac{C}{\lambda_k^{3/2}|a_i - a_k|^2} \leq \frac{C'}{(\lambda_j\lambda_k)^{3/4}|a_i - a_k|^2} = o\left(\frac{\varepsilon_{jk}^{3/2}}{|a_j - a_k|^{1/2}}\right).$$

$\square$

And

Lemma 75 *If $i, j, \in G_m$ and $\lambda_j \geq \lambda_i$, then*

$$|\gamma_i^j| \leq C\frac{\varepsilon_{ij}^{3/2}}{|a_i - a_j|^{1/2}}.$$

Proof. Straightforward. $\square$

1.10.8 *The system of equations corresponding to the variations of the points a_i*

We turn now to the system of equations corresponding to the variation of the points a_i. We move all the points of the same group together. We assume that

$$(H) \qquad \sum_{\substack{i \text{ under motion} \\ j \text{ not moving}}} |\gamma_j^i| = O\left(\sum \frac{\varepsilon_{mt}^3}{|a_m - a_t|} + \sum \frac{\omega_k^{\infty 2}}{\lambda_k|a_k - a_\ell|^2}\right)$$

(H) is satisfied, without any restriction on i and j, for well-distributed configuration by Proposition 9.

Under (H), we use Lemmas 62, 63, 64, 65, and Lemma 67 (1) and (2).

The remainder is then controlled, after Lemmas 62–67 and is, when all ω_i^∞'s are $o(1)$, which we are assuming since this is the most difficult case,

$$o\left(\sum \frac{\varepsilon_{mt}^3}{|a_m - a_t|} + \sum \frac{\omega_k^{\infty 2}}{\lambda_k |a_k - a_\ell|^2}\right).$$

In fact, the remainder is

$$o\left(\sum_{\substack{(i,j)\in \\ \text{different} \\ \text{groups}}} \frac{\varepsilon_{ij}^3}{|a_i - a_j|}\right) + R$$

where R are all remainders in (1) and (2) of Lemma 67.

We find (we take for simplicity the case of two groups):

$$-\partial J\left(\sum_{i\in I} \alpha_i \omega_i + v_I\right) \cdot \partial_{II}\left(\sum_{i\in II} \alpha_i \omega_i\right)$$

$$+\partial J\left(\sum_{i\in II} \alpha_i \omega_i + v_{II}\right) \cdot \partial_I\left(\sum_{i\in I} \alpha_i \omega_i + v_I\right)$$

$$-\int \nabla\left(\sum_{i\in I} \alpha_i \omega_i + v_{II}\right) \nabla \partial_I\left(\sum_{i\in I} \alpha_i \omega_i + v_I\right)$$

$$= o\left(\sum \frac{\varepsilon_{mt}^3}{|a_m - a_t|} + \sum \frac{\omega_k^{\infty 2}}{\lambda_k |a_k - a_\ell|^2}\right).$$

We estimate

$$\int \nabla\left(\sum_{i\in II} \alpha_i \omega_i + v_{II}\right) \nabla \partial\left(\sum_{i\in I} \alpha_i \omega_i + v_I\right)$$

which we split into

$$\int \nabla\left(\sum_{i\in II} \alpha_i \omega_i\right) \nabla \partial\left(\sum_{i\in I} \alpha_i \omega_i\right) + \int \left(\sum_{i\in II} \alpha_i \omega_i^5\right) \partial v_I$$

$$+5\int \left(\sum_{i\in I} \alpha_i \omega_i^4\right) \partial \omega_i v_{II} + \int \nabla v_{II} \nabla \partial v_I = \int \left(\sum_{i\in II} \alpha_i \omega_i^5\right) \partial\left(\sum_{i\in I} \alpha_i \omega_i\right)$$

$$+\int \left(\sum_{i\in II} \alpha_i \omega_i^5\right) \partial v_I - \int \left(\sum_{i\in I} \alpha_i \omega_i^5\right) \partial v_{II} + \int \nabla v_{II} \nabla \partial v_I.$$

The term $\int \left(\sum_{i \in II} \alpha_i \omega_i^5 \right) \partial \left(\sum_{i \in I} \alpha_i \omega_i \right)$ is the main term of the expansion. $\int \nabla v_{II} \nabla \partial v_I$ reads as $- \int \Delta v_{II} \partial v_I$. Using the estimates of Proposition 4 on ∂v_I, the equation satisfied by $-\Delta v_{II}$ and the estimate on v_{II}, it is easy to see that this expression is $o \left(\frac{\varepsilon_{ij}^3}{|a_j - a_j|} \right)$ in fact $o \left(\sum \frac{\varepsilon_{ij}^3}{|a|} \right)$ when all the ω_i^∞'s are small.

The terms $\int \omega_i^5 \partial v_I, i \in II$, and $\int \omega_i^5 \partial v_{II}, i \in I$, are of the same type. We focus on $\int \omega_i^5 \partial v_I$ for $i \in II$.

We split it as follows:

$$\int \omega_i^5 \partial v_I = \int_{\Omega_{II}} \omega_i^5 \partial v_I + \int_{\Omega_{II}^c} \omega_i^5 \partial v_I.$$

We rewrite

$$\int_{\Omega_{II}^c} \omega_i^5 \partial v_I = \int_{\partial \Omega_{II}^c} \omega_i^5 v_I - 5 \int_{\Omega_{II}^c} \omega_i^4 \partial \omega_i v_I.$$

On the other hand,

$$\int_{\Omega_{II}} \omega_i^5 \partial v_I = \partial v_I(a_i) \int_{\Omega_{II}} \omega_i^5 + O \left(\underset{x \in \Omega_{II}}{\mathrm{Sup}} |\nabla \left(\nabla(v_I) \right)| \right) \int_{\Omega_{II}} |\omega_i|^5 |x - a_i|.$$

We have in Ω_{II} an estimate on ∇v_I and we can derive a similar estimate on $|\nabla(\nabla v_I)|$: each time we take an additional derivative, we divide by $\frac{1}{|a|}$.

Thus,

$$\left| \partial v_I(a_i) \int_{\Omega_{II}} \omega_i^5 \right|$$

$$\leq \frac{C}{|a|} \sum_{(\ell,m) \in (I,I)} \left(|\omega_\ell^\infty| + \varepsilon_{\ell m} + \frac{1}{\lambda_\ell |a_\ell - a_m|} \right) \varepsilon_{\ell m} \frac{1}{\sqrt{\lambda_m} |a_m - a_i|}$$

$$\times \left| \int_{\mathbb{R}^3} \omega_i^5 + O \left(\frac{1}{\sqrt{\lambda_i}} \sum \varepsilon_{ij}^2 \right) \right| \leq o \left(\sum \frac{|\omega_i^\infty|}{\sqrt{\lambda_i} |a_m - a_i|} \right.$$

$$\times \left(\frac{|\omega_\ell^\infty| \varepsilon_{\ell m}}{\sqrt{\lambda_m} |a|} + \frac{\varepsilon_{\ell m}^2}{|a| \sqrt{\lambda_m}} + \frac{\varepsilon_{\ell m}^2}{|a| \sqrt{\lambda_\ell}} \right) \right) + o \left(\sum \frac{\varepsilon_{\ell j}^3}{|a|} \right)$$

$$= o \left(\sum \frac{\omega_j^{\infty 2}}{\lambda_j |a_m - a_j|^2} + \sum \frac{\varepsilon_{\ell j}^3}{|a_\ell - a_j|} \right).$$

A similar estimate is readily available for $\int_{\partial \Omega_{II}^c} \omega_i^5 v_I - 5 \int_{\Omega_{II}^c} \omega_i^4 \partial \omega_i v_I$.

Considering $\int_{\Omega_{II}} |\omega_i|^5 |x - a_i|$, we have:

$$\int_{\Omega_{II}} |\omega_i|^5 |x - a_i| \le \frac{C}{\lambda_i^{3/2}}$$

so that

$$\operatorname*{Sup}_{x \in \Omega_{II}} |\nabla(\nabla v_I)| \int_{\Omega_{II}} |\omega_i|^5 |x - a_i|$$

$$\le \frac{C}{|a|^2} \cdot \frac{1}{\lambda_i^{3/2}} \sum_{\substack{(\ell,m) \in \\ (I,I)}} \left(|\omega_\ell^\infty| + \varepsilon_{\ell m} + \frac{1}{\lambda_\ell |a_m - a_\ell|} \right) \times \varepsilon_{\ell m} \times \frac{1}{\sqrt{\lambda_m}|a_m - a_i|}.$$

Observe that since we may assume that $|a_m - a_j|$ is of the order of $|a|$ (when there are more than two groups, we improve easily the estimate on $v_I, \nabla v_i, \nabla(\nabla v_I)$):

$$\frac{|\omega_\ell^\infty| \varepsilon_{\ell m}}{\lambda_i^{3/2} |a|^2 \sqrt{\lambda_m} |a_m - a_i|} \le \frac{C|\omega_\ell^\infty}{\sqrt{\lambda_\ell} |a_\ell - a_i|} \times \frac{1}{\lambda_m |a_m - a_\ell|} \times \frac{1}{\lambda_i^{3/2} |a|^2}.$$

Observe that $i \in II$ which has a basic concentration larger than the basic concentration of I. Thus, this expression can be upper bounded by

$$\frac{|\omega_\ell^\infty| \varepsilon_{\ell_o i}^3 \times \varepsilon_{\ell m}}{|a|} = o \left(\sum_{\substack{(i,j) \in \\ \text{different groups}}} \frac{\varepsilon_{ij}^3}{|a|} \right)$$

if ω_ℓ^∞ is $o(1)$.

We also have

$$\frac{\varepsilon_{\ell m}^2}{\lambda_i^{3/2} |a|^2 \sqrt{\lambda_m} |a_m - a_i|} \le o \left(\sum \frac{\varepsilon_{jt}^3}{|a|} \right)$$

$$\frac{\varepsilon_{\ell m}}{\lambda_i^{3/2} |a|^2 \lambda_\ell |a_m - a_\ell|} \times \frac{1}{\sqrt{\lambda_m} |a_m - a_i|}$$

$$\le \frac{C \varepsilon_{\ell m}^2}{\lambda_i |a|^2 \sqrt{\lambda_\ell} \sqrt{\lambda_i} |a_m - a_i|} \le \frac{C_1 \varepsilon_{\ell m}^2}{\lambda_i |a|^2 \sqrt{\lambda_\ell} \sqrt{\lambda_i} |a_\ell - a_i|}$$

$$= o \left(\sum \frac{\varepsilon_{jt}^3}{|a|} \right).$$

Combining, we derive that

$$\underset{x\in\Omega_{II}}{\mathrm{Sup}}\ |\nabla(\nabla v_I)|\int_{\Omega_{II}}|\omega_i|^5|x-a_i|\le o\left(\sum\frac{\varepsilon_{jt}^3}{|a_j-a_t|}+\sum\frac{\omega_\ell^{\infty 2}}{\lambda_\ell|a_\ell-a_m|^2}\right)$$

and thus that

$$\int\nabla\left(\sum_{i\in II}\alpha_i\omega_i+v_{II}\right)\nabla\partial\left(\sum_{i\in I}\alpha_i\omega_i+v_I\right)$$

$$=\int\left(\sum_{i\in II}\alpha_i\omega_i^5\right)\partial\left(\sum_{i\in I}\alpha_i\omega_i\right)+o\left(\sum\frac{\varepsilon_{jt}^3}{|a_j-a_t|}+\sum\frac{\omega_\ell^{\infty 2}}{\lambda_\ell|a_\ell-a_m|^2}\right).$$

The above estimates extend to the study of

$$\partial J\left(\sum_{i\in I}\alpha_i\omega_i\right)\cdot Q_I\left(\partial_{II}\left(\sum_{i\in II}\alpha_i\omega_i\right)\right).$$

Indeed, denoting φ_I one of $\omega_i,\frac{1}{\lambda_i}\frac{\partial\omega_i}{\partial a_i},\frac{\partial\omega_i}{\partial\sigma_i}$ for $i\in I$, the above quantity is essentially controlled by

$$\left|\partial J\left(\sum_{i\in I}\alpha_i\omega_i\right)\cdot\varphi_I\right|\left|\int\nabla\varphi_I\nabla\partial_{II}\left(\sum_{i\in II}\alpha_i\omega_i\right)\right|.$$

The projection operator is slightly more complicated, but these are the essential terms.

Thus, we have:

$$\sum_{\substack{i\in II\\j\in I}}\alpha_i\alpha_j\int\omega_i^5\partial_I\omega_j=o\left(\sum\frac{\varepsilon_{mt}^3}{|a_m-a_t|}+\sum\frac{\omega_k^{\infty 2}}{\lambda_k|a_k-a_\ell|^2}\right).$$

We know that (see [Bahri 2001])

$$\int\omega_i^5\omega_j=c\omega_j(a_i)\frac{\omega_i^\infty}{\sqrt{\lambda_i}}-c_{ij}\varepsilon_{ij}^3+o\left(\sum\varepsilon_{ij}^3+(\omega_i^{\infty 2}+\omega_j(a_i)^2\lambda_j|a_i-a_j|^2)\varepsilon_{ij}^2\right)$$

$$=c\omega_i(a_j)\frac{\omega_j^\infty}{\sqrt{\lambda_j}}-\tilde{c}_{ij}\varepsilon_{ij}^3+o\left(\sum\varepsilon_{ij}^3+(\omega_j^{\infty 2}+\omega_i(a_j)^2\lambda_i|a_i-a_j|^2)\varepsilon_{ij}^2\right)$$

and we can establish as well that (λ_{i_0} is the smallest concentration of group I)

$$\int w_i^5 \partial_I w_j = -c\frac{w_i^\infty}{\sqrt{\lambda_i}}\frac{\partial}{\partial a_i}w_j(a_i) - \frac{\partial}{\partial a_i}(c_{ij}\varepsilon_{ij}^3)$$

$$+ o\left(\sum \frac{\varepsilon_{\ell t}^3}{|a|} + \sum \frac{w_\ell^{\infty 2}}{\lambda_\ell |a_\ell - a_t|^2} + \sum_{\substack{i\in I \\ j\in II}} \frac{\lambda_{io}}{\lambda_i}\gamma_i^{j2}\right).$$

Thus using Lemmas 71–75:

$$\int w_i^5 \partial_I w_j = -c\frac{w_i^\infty}{\sqrt{\lambda_i}}\frac{\partial}{\partial a_i}w_j(a_i) - \frac{\partial}{\partial a_i}(c_{ij}\varepsilon_{ij}^3) + o\left(\sum \frac{\varepsilon_{\ell t}^3}{|a|} + \sum \frac{w_\ell^{\infty 2}}{\lambda_\ell |a_\ell - a_t|^2}\right)$$

or using the other expression.

$$\int w_j^5 \partial_{II} w_i = c\frac{w_j^\infty}{\sqrt{\lambda_j}}\frac{\partial}{\partial a_j}w_i(a_j) - \frac{\partial}{\partial a_j}(\tilde{c}_{ij}\varepsilon_{ij}^3) + o\left(\sum \frac{\varepsilon_{\ell t}^3}{|a|} + \sum \frac{w_\ell^{\infty 2}}{\lambda_\ell |a_\ell - a_t|^2}\right).$$

If $\lambda_j \geq \lambda_i$ we use the first expression and if $\lambda_i > \lambda_j$ we use the second expression.

$w_j(a_i)$ reads as:

$$w_j(a_i) = \frac{w_j^\infty}{\sqrt{\lambda_j}|a_i - a_j|} + O\left(\frac{\left|D\bar{\omega}_j\left(\frac{a_i - a_j}{|a_i - a_j|}\right)\right|}{\lambda_j^{3/2}|a_i - a_j|^2}\right).$$

$$\frac{\partial}{\partial a_i}w_j(a_i) = -\frac{w_j^\infty(a_i - a_j)}{\sqrt{\lambda_j}|a_i - a_j|^3} + O\left(\frac{1}{\lambda_j^{3/2}|a_i - a_j|^3}\right).$$

Observe that for $\lambda_j \geq \lambda_i$

$$\frac{|w_i^\infty|}{\sqrt{\lambda_i}\lambda_j^{3/2}|a_i - a_j|^3} = O\left(\frac{w_i^{\infty 2}}{\lambda_i|a_i - a_j|^2} + \frac{1}{\lambda_j^3|a_i - a_j|^4}\right)$$

$$= O\left(\frac{w_i^{\infty 2}}{\lambda_i|a_i - a_j|^2} + \frac{\varepsilon_{ij}^3}{|a_i - a_j|}\right).$$

We want to transform this large O into a small o.

We observe that since $|a_i - a_{i_0}| = o(|a_i - a_j|)$ if i, i_o belong to the same group while i and j are not in the same group:

$$\frac{\partial}{\partial a_i}w_j(a_i) = \frac{\partial}{\partial a_i}w_j(a_i)\Big|_{a_i = a_{io}} + o\left(\frac{|w_j^\infty|}{\sqrt{\lambda_j}|a_i - a_j|^2} + \frac{1}{\lambda_j^{3/2}|a_i - a_j|^3}\right)$$

so that

$$\sum_{\lambda_i < \lambda_j} \alpha_i \left(\omega_i^\infty \frac{\partial}{\partial a_i} \omega_j(a_i) \right) \frac{1}{\sqrt{\lambda_i}} = - \sum_{\lambda_i < \lambda_j} \alpha_i \frac{\omega_i^\infty \omega_j^\infty (a_i - a_j)}{\sqrt{\lambda_i}\sqrt{\lambda_j}|a - a_j|^3}$$

$$+ \sum_{\lambda_i < \lambda_j} \alpha_i \frac{\omega_i^\infty}{\sqrt{\lambda_i}} \frac{\partial}{\partial a_i} \left(\omega_j(a_i) - \frac{\omega_j^\infty}{\sqrt{\lambda_j}|a_i - a_j|} \right) \Big|_{a_i = a_{i_0}}$$

$$+ o \left(\sum \frac{\omega_k^{\infty 2}}{\lambda_k |a_k - a_i|^2} + \sum \frac{\varepsilon_{mt}^3}{|a_m - a_t|} \right).$$

Observe that now for $\lambda_i < c\lambda_j (i \in I, j \in II)$

$$\sum_{\lambda_i < c\lambda_j} \alpha_i \frac{\omega_i^\infty}{\sqrt{\lambda_i}} \frac{\partial}{\partial a_i} \left(\omega_j(a_i) - \frac{\omega_j^\infty}{\sqrt{\lambda_j}|a_i - a_j|} \right)$$

$$= o \left(\sum \frac{\omega_i^{\infty 2}}{\lambda_i |a_i - a_j|^2} + \frac{\varepsilon_{ij}^3}{|a_i - a_j|} \right).$$

We thus find

$$\sum_{\substack{i \in II \\ j \in I}} \alpha_i \alpha_j \int \omega_i^5 \partial_I \omega_j = - \sum_{\substack{i \in II \\ j \in I}} \alpha_i \alpha_j \frac{\omega_i^\infty \omega_j^\infty}{\sqrt{\lambda_i}\sqrt{\lambda_j}} \frac{a_i - a_j}{|a_i - a_j|^3}$$

$$+ \sum_{c\lambda_j < \lambda_i < \frac{1}{c}\lambda_j} \alpha_j \frac{\alpha_j \omega_j^\infty}{\sqrt{\lambda_j}} \frac{\partial}{\partial a_j} \left(\omega_i(a_j) - \frac{\omega_i^\infty}{\sqrt{\lambda_i}|a_i - a_j|} \right) \Big|_{a_j = a_{j_0}}$$

$$- \sum_{\lambda_i \leq \lambda_j} \alpha_i \alpha_j \frac{\partial}{\partial a_i} (c_{ij}\varepsilon_{ij}^3) - \sum_{\lambda_i > \lambda_j} \alpha_i \alpha_j \frac{\partial}{\partial a_j} (\tilde{c}_{ij}\varepsilon_{ij}^3)$$

$$+ o \left(\sum \frac{\omega_i^{\infty 2}}{\lambda_i |a_i - a_j|^2} + \frac{\varepsilon_{ij}^3}{|a_i - a_j|} \right).$$

We know that

$$\omega_i(a_j) = \frac{\omega_i^\infty}{\sqrt{\lambda_i}|a_i - a_j|} + \nabla \omega_i^\infty \left(\frac{a_i - a_j}{\lambda_i^{3/2}|a_i - a_j|^3} \right) + h.o.$$

Thus,

$$\frac{\partial}{\partial a_j} \left(\omega_i(a_j) - \frac{\omega_i^\infty}{\sqrt{\lambda_i}|a_i - a_j|} \right)$$

$$= -\frac{1}{\lambda_i^{3/2}} \frac{\nabla \omega_i^\infty}{|a_i - a_j|^3} - \frac{3(a_i - a_j)}{\lambda_i^{3/2} |a_i - a_j|^4} \nabla \omega_i^\infty \left(\frac{a_i - a_j}{|a_i - a_j|} \right) + h.o$$

and

$$\sum_{i \in II} \alpha_i \alpha_j \int \omega_i^3 \partial_I \omega_j = -\sum_{\substack{i \in II \\ j \in I}} \alpha_i \alpha_j \frac{\omega_i^\infty \omega_j^\infty}{\sqrt{\lambda_i \lambda_j}} \frac{a_i - a_j}{|a_i - a_j|^3} - \sum_{\lambda_i \leq \lambda_j} \alpha_i \alpha_j \frac{\partial}{\partial a_i} (c_{ij} \varepsilon_{ij}^3)$$

$$- \sum_{\lambda_i > \lambda_j} \alpha_i \alpha_j \frac{\partial}{\partial a_j} (\tilde{c}_{ij} \varepsilon_{ij}^3) + \sum_{c\lambda_j < \lambda_i < \frac{\lambda_j}{c}} 3\alpha_i \alpha_j \frac{\omega_j^\infty}{\sqrt{\lambda_j} |a_i - a_j|} \frac{a_i - a_j}{|a_i - a_j|}$$

$$\times \frac{\nabla \omega_i^\infty \left(\frac{a_i - a_j}{|a_i - a_j|} \right)}{\lambda_i^{3/2} |a_i - a_j|^2} + \frac{\nabla \omega_i^\infty}{\lambda_i^{3/2}} \sum_{c\lambda_j < \lambda_i < \frac{\lambda_j}{c}} \frac{\alpha_i \alpha_j \omega_j^\infty}{\sqrt{\lambda_j} |a_i - a_j|} \times \frac{1}{|a_i - a_j|^2}$$

$$+ o \left(\sum \frac{\omega_j^{\infty 2}}{\lambda_j |a_i - a_j|^2} + \sum \frac{\varepsilon_{ij}^3}{|a_i - a_j|} \right)$$

$$= o \left(\sum \frac{\omega_j^{\infty 2}}{\lambda_j |a_i - a_j|^2} + \sum \frac{\varepsilon_{ij}^3}{|a_i - a_j|} \right).$$

$$(1.47)$$

Observation. The remainder are in fact $o \left(\displaystyle\sum_{\substack{(i,j) \in \\ \text{different groups}}} \frac{\varepsilon_{ij}^3}{|a_i - a_j|} \right)$ or are
found among the remainders in the estimates of Lemma 67.

(1.47) has been derived for the motion of groups relative one to the other under (H).

We need, in order to derive our Morse Lemma at infinity, a complete and coherent set of equations. (1.47) provides part of these equations, the remainder being given by Proposition 8. Our argument for the Morse Lemma at infinity is based on Conjectures 3, 3′ (see the Introduction) and requires therefore that we pass to the limit, getting rid of the remainders. This process is tricky because the vectorial equations of Proposition 8 are about single masses, while (1.47) is about groups, furthermore it holds only under (H).

We thus have to pause briefly in our computations, to show how to derive equations similar to (1.47) for single masses and explain how these equations, using Conjectures 3, 3′, provide us with bounds after a contradiction argument.

These bounds read:

$$\sum \frac{\omega_i^{\infty 2}}{\lambda_i |a_i - a_j|} \leq C \sum_{\substack{(i,j)\in \\ \text{different} \\ \text{groups}}} \frac{\varepsilon_{ij}^3}{|a_i - a_j|}. \tag{K}$$

We start with:

Definition 3   (basic concentration): *We define the basic concentration* λ_{m_0} *of a group* G_m *to be the smallest among the concentrations of the group* G_m.

We introduce the following rule which we follow as we deform the concentrations of the points a_i:

1.10.9   *Rule about the variation of the points of concentrations of the various groups*

Given a number A, we will be moving, in a single movement, all groups having a basic concentration smaller than a given number A.

We now introduce five basic observations about Lemma 67, its remainders and our expansion above.

Five observations.

Observation 1. Coming back to the estimates of Lemma 67, we see that these estimates involve $\sum_{\ell \in I} \left(\sum_{j \in II} \alpha_j \omega_j(a_\ell) \right)^2$. These quantities involve γ_ℓ^{j2} for $j \in II$ (the non-moving group) and $\ell \in I$, the moving group. Since $\gamma_\ell^j = O\left(\frac{1}{\lambda_j^{3/2} |a_j - a_\ell|^2} \right) = O\left(\frac{1}{\lambda_{j_0}^{3/2} |a_{j_0} - a_{\ell_0}|} \right)$ and since $\lambda_{j_0} \geq \lambda_{i_0}$, we can claim that

$$\gamma_\ell^{j2} = O\left(\sum_{\substack{i \in I \\ m \in II}} \frac{\varepsilon_{im}^3}{|a_i - a_m|} \right).$$

Observation 2.   The remainder terms in Lemma 67 also contain $\sum_{\ell \in I} \lambda_\ell (|(v - (v_I + v_{II}))\ell|_{H_0^1}^2 + \ldots)$.   Coming back to Theorem 2, with i, such quantities have been estimated.   The estimate involves $\left(\sum_{j \in II} \alpha_j \omega_j(a_\ell) \right)^2$ (see Theorem 2, after squaring the estimate and multiplying by λ_ℓ). Observation 1 holds for this quantity. But we also find

$$\lambda_\ell \sum_{\substack{k \neq \ell \\ k \in I}} \left(\sum_{\substack{j \in II \\ \text{or} \\ j \in I}} \alpha_j \omega_j(a_k) \right)^2 \frac{\varepsilon_{\ell k}^2}{\lambda_k} \quad \text{which contains} \quad \frac{\lambda_\ell}{\lambda_k} \varepsilon_{\ell k}^2 \gamma_k^{j2}.$$

ℓ belongs to the moving group, while j and k alternate. If k does not belong to the moving group, then, using basic concentrations:

$$\lambda_\ell \varepsilon_{\ell k}^2 \leq C \lambda_{\ell_0} \varepsilon_{\ell_0 k}^2.$$

Using Lemma 72,

$$\lambda_\ell \varepsilon_{\ell k}^{j2} \frac{\gamma_k^{j2}}{\lambda_k} \leq C \lambda_{\ell_0} \varepsilon_{\ell_0 k}^2 \varepsilon_{kj}^4 \leq C \lambda_{\ell_0} \left(\varepsilon_{\ell_0 k}^6 + \varepsilon_{kj}^6 \right) \leq C \sum_{\substack{m \in I \\ n \in II}} o \left(\frac{\varepsilon_{mm}^3}{|a_m - a_n|} \right)$$

since $\lambda_{\ell_0} \leq C \lambda_k$.

If, on the other hand, k belongs to the moving group, then j does not. Then,

$$\lambda_\ell \varepsilon_{\ell k}^2 \frac{\gamma_k^{j2}}{\lambda_k} \leq C \lambda_\ell \varepsilon_{\ell k}^2 \varepsilon_{kj}^4 \leq C \lambda_k \varepsilon_{kj}^4 \leq C' \lambda_{k_0} \varepsilon_{k_0 j}^4.$$

Since $\lambda_{k_0} \leq C \lambda_j$,

$$\lambda_{k_0} \varepsilon_{k_0 j}^4 \leq C \sqrt{\lambda_{k_0} \lambda_j} \varepsilon_{k_0 j}^4 \leq C \frac{\varepsilon_{k_0 j}^3}{|a_{k_0} - a_j|}.$$

We thus see that, if we split our masses at each step, into two larger groups, each made of our basic groups G_m and if the first large group has basic concentrations smaller than the ones of the second large group, then we have a good control on $\frac{\lambda_\ell}{\lambda_k} \varepsilon_{\ell k}^2 \gamma_k^{j2}$ as we move, in a single movement, the concentration points of the first large group.

Observation 3. We consider here the remainder $O(\sum \varepsilon_{ij} \sum \varepsilon_{jm}^2)$ in Theorem 2. This remainder comes from $\int_{\Omega^c} \delta_i \delta_k^4 \delta_m$, with k and m belonging to different groups. Assume for simplicity that $i \neq k$. This expression, when squared and multiplied by λ_i, becomes $O(\lambda_i \varepsilon_{ik}^2 \varepsilon_{km}^2 \varepsilon_{kt}^2) \cdot \omega_i$ is in the group which moves. If ω_k is in the other group, Observation 2 extends: $\lambda_i \varepsilon_{ik}^2$ reads $O(\lambda_{i_0} \varepsilon_{i_0 k}^2)$ with $\lambda_{i_0} \leq CA$. Our expression becomes, with obvious notations and after some work, $O(\lambda_{II} \varepsilon_{I,II}^6)$ and we are done. If ω_k is in the same group, i.e. the moving group, then our remainder becomes $o(\lambda_k \varepsilon_{km}^2 \varepsilon_{kt}^2) = o(\lambda_{k_0} \varepsilon_{k_0 m}^2 \varepsilon_{kt}^2) = o(\lambda_{k_0} \varepsilon_{k_0 m_0}^2 \varepsilon_{kt}^2) =$

$$o\left(\frac{\varepsilon_{k_0 m_0}}{|a_{k_0}-a_{m_0}|}\varepsilon_{kt}\right) = o\left(\frac{\varepsilon_{k_0 m_0}}{|a_{k_0}-a_{m_0}|}\sum_{\substack{(\ell,j)\\ \in \text{ different groups}}}\varepsilon_{\ell j}^2\right).$$

This latter expression becomes, **if we assume that the packing in groups is well-distributed**, $o\left(\displaystyle\sum_{\substack{(\ell,j)\\ \in \text{ different groups}}}\frac{\varepsilon_{\ell j}^3}{|a_\ell-a_j|}\right).$ This last part of Observation

3 can be used for all remainders $O\left(\frac{\varepsilon_{ij}^3}{|a|}\right)$ which we stated in Lemma 67 in

order to see them as $O\left(\displaystyle\sum_{\substack{(\ell,j)\\ \in \text{ different groups}}}\frac{\varepsilon_{\ell j}^3}{|a_\ell-a_j|}\right).$

Observation 4. We have an additional term which we cannot control by $\displaystyle\sum_{i\in I, j\in II}\frac{\varepsilon_{ij}^3}{|a_i-a_j|}$ in Lemma 67, in the remainder. This term reads

$$\sum_{\substack{i\in II\\ \ell\in I}}\frac{\omega_i^{\infty 2}\varepsilon_{i\ell}^2}{\lambda_\ell|a_i-a_\ell|^2}.$$

It can be traced back to three terms in the proof of Proposition 5. These terms are all of the same type; as we expand $\sum_{j\in G_\ell}\alpha_j\omega_j$ around a_k,

we find first $\displaystyle\sum_{j\in G_\ell}\alpha_j\omega_j(a_k)$, then $\displaystyle\sum_{j\in G_\ell}\alpha_j\omega_j^\infty(\delta_j(x)-\delta_j(a_i))$ which provides

this term (after integration and squaring). Finally, we find a third term

which is $O\left(\displaystyle\sum_{\substack{i\in I\\ j\in II}}\frac{\varepsilon_{ij}^3}{|a_i-a_j|}\right).$ With our present techniques, we cannot claim

that the second term is $O\left(\displaystyle\sum_{\substack{i\in I\\ j\in II}}\frac{\varepsilon_{ij}^3}{|a_i-a_j|} + \sum_{\substack{\ell,m\\ \in \text{ different}\\ \text{groups}}}\left(\sum_{j\in G_\ell}\frac{\omega_j^\infty}{\sqrt{\lambda_j}|a_j-a_m|}\right)^2\right).$

Observe however that if λ_ℓ is not λ_{ℓ_0}, this term is readily

$$O\left(\sum_{\substack{i\in I\\ j\in II}}\frac{\varepsilon_{ij}^3}{|a_i-a_j|}\right), \text{ even } o\left(\sum_{\substack{i\in I\\ j\in II}}\frac{\varepsilon_{ij}^3}{|a_i-a_j|}\right).$$

When the configuration is well-distributed, the groups are reduced to sin-

gletons and this second term is right away $o\left(\displaystyle\sum_{i\neq j}\frac{\omega_i^{\infty 2}}{\lambda_i|a_i-a_j|^2}\right).$ The estimate

on this term holds on all configurations, well-distributed or not. i and j

can be taken in different groups.

Observation 5.

In the expansion completed above for $\partial J(\sum \alpha_i \omega_i + v)$.

$\partial_I \left(\sum_{i \in I} \alpha_i \omega_i + v \right)$, we have found as we were trying to estimate $\partial v_I(a_i) \int_{\Omega_{II}} \omega_i^5$ a term which was $O\left(\sum \frac{|\omega_i^\infty|}{\sqrt{\lambda_i}|a_m - a_i|} \times \frac{|\omega_\ell^\infty| \varepsilon_{\ell m}}{\sqrt{\lambda_m}|a|} \right)$. This is a term similar, to a large extent, to the expression discussed in Observation 4. One would like, for this term, to gather the contribution for all i's $\in II$, thinking of this expression as

$$\left(\sum_\ell O\left(\frac{\varepsilon_{\ell m}}{\sqrt{\lambda_m}|a|} \right) \right) \sum_{i \in II} \frac{\omega_i^\infty}{\sqrt{\lambda_i}|a_m - a_i|}.$$

This would suit more our estimates ($m \in I$). We cannot derive such an esti-

mate at this point. However, this expression is again $o\left(\sum_{\substack{i,j \in \\ \text{different} \\ \text{groups}}} \frac{\omega_i^{\infty 2}}{\lambda_i |a_i - a_j|^2} \right)$

1.10.10 *The basic parameters and the end of the expansion*

1.10.11 *Remarks on the basic parameters*

Our basic parameters are α_i, λ_i and a_i and σ_i.

σ_i are parameters of rigid motions, motions on the basic rescaled $\bar{\omega}_i$. This includes rotation around an axis but also global rotation. The global rotations, i.e. the rotations which do not preserve the polar axis of S^3 may be viewed in fact as the composition of a rotation around the polar axis and of a translation of a_i. Only that this translation occurs in the basic parameter i.e. is scaled by $\frac{1}{\lambda_i}$ and that it must be viewed, because it involves a change of the point at infinity, as a translation in a different chart than the one where the points a_i are read. Accordingly, we can express our basic equations in the following way: α_i and λ_i equations (easy to derive and yield $\sum_i \left| \frac{\partial J}{\partial \alpha_i} \right| + \lambda_i \left| \frac{\partial J}{\partial \lambda_i} \right| = o\left(\sum \frac{\omega_i^{\infty 2}}{\lambda_i |a_i - a_j|^2} + \sum \varepsilon_{ij}^3 \right)$; they can be completed uniformly on groups the estimate becoming then an estimate for $(i, j) \in$ different groups), a_i - equations derived in **two** distinct charts i.e. a chart and the chart derived after a Kelvin transform and then σ_i-equations. However, for the σ_i-equations, we can then think in terms of rotations which preserve the polar axis.

We should be able to pass to the limit in the σ_i-equations when we consider a packing in groups each one of which have a large interaction.

These more restricted equations might also, if suitably read, pass to the limit even as groups or single masses involved do not have a large interaction.

For the first conclusion, we come back to (1.47) to have a second look at the equation involved:

This expression reads

$$\bar{c} \sum_{\substack{i \in II \\ j \in I}} \alpha_i \alpha_j \frac{\omega_i^\infty \omega_j^\infty}{\sqrt{\lambda_i \lambda_j}} \frac{a_i - a_j}{|a_i - a_j|^3} - \sum_{\lambda_i \leq \lambda_j} \alpha_i \alpha_j \frac{\partial}{\partial a_i}(c_{ij}\varepsilon_{ij}^3)$$

$$- \sum_{\lambda_i > \lambda_j} \alpha_i \alpha_j \frac{\partial}{\partial a_j}(\tilde{c}_{ij}\varepsilon_{ij}^3)$$

$$-\bar{c} \sum_{\substack{c\lambda_j \leq \lambda_i \leq \frac{\lambda_j}{c} \\ i \in I, j \in II}} \frac{\alpha_i \alpha_j \omega_j^\infty}{\sqrt{\lambda_j}|a_i - a_j|} \left(\frac{\nabla \omega_i^\infty}{\lambda_i^{3/2}|a_i - a_j|^2} \right.$$

$$\left. + 3 \frac{a_i - a_j}{|a_i - a_j|} \frac{\nabla \omega_i^\infty}{\lambda_i^{3/2}|a_i - a_j|^2} \cdot \frac{a_i - a_j}{|a_i - a_j|} \right).$$

The sum for $c\lambda_i \leq \lambda_i \leq \frac{\lambda_j}{c}$ can be replaced by

$$+ \sum_{\substack{i \in I \\ j \in II}} \frac{\alpha_i \alpha_j \omega_j^\infty}{\sqrt{\lambda_j}|a_i - a_j|}(1 + o(1))$$

$$\times \left(\frac{\nabla \omega_i^\infty}{\lambda_i^{3/2}|a_{i_0} - a_{j_0}|^2} + 3 \frac{a_{i_0} - a_{j_0}}{|a_{i_0} - a_{j_0}|} \frac{\nabla \omega_i^\infty}{\lambda_i^{3/2}|a_{i_0} - a_{j_0}|^2} \cdot \frac{a_{i_0} - a_{j_0}}{|a_{i_0} - a_{j_0}|} \right)$$

and this sum might be restricted to $i = i_0$. $\frac{1}{\lambda_{i_0}^{3/2}|a_{i_0} - a_{j_0}|^2}$ is larger than

$c \sum_{\substack{i \in I \\ j \in II}} \frac{\varepsilon_{ij}^{3/2}}{|a_i - a_j|^{1/2}}$ and we thus see that, indeed, (1.47) rewritten in this way

provides us with an estimate on $\sum_{j \in II} \frac{\alpha_j \omega_j^\infty}{\sqrt{\lambda_j}|a_{i_0} - a_{j_?}|}$ similar to the one of

Proposition 8 provided that

$$- \sum_{\substack{i \in II \\ j \in I}} \alpha_i \alpha_j \frac{\omega_i^\infty \omega_j^\infty}{\sqrt{\lambda_i \lambda_j}} \frac{a_i - a_j}{|a_i - a_j|^3} = O\left(\sum_{\substack{i \in II \\ j \in I}} \frac{\varepsilon_{ij}^3}{|a_i - a_j|} \right).$$

Coming back to (1.47) under the earlier form, we find that once $i \in I, j \in II$ and λ_i and λ_j are of the same order

$$\sum_{i \in II, j \in I} \alpha_i \alpha_j \frac{w_i^\infty w_j^\infty}{\sqrt{\lambda_i \lambda_j}} \frac{a_i - a_j}{|a_i - a_j|^2}$$

$$= O\left(\sum_{i \in I, j \in II} \frac{\varepsilon_{ij}^3}{|a_i - a_j|} \right) + o\left(\sum_{i \in I, j \in II} \frac{w_i^{\infty 2}}{\lambda_i |a_i - a_j|^2} \right)$$

which is nearly the required estimate. We thus see that, indeed, the a_i equations, viewed in various charts should contain the estimates of Proposition 8. Accordingly, instead of Conjecture 2, we introduce:

Conjecture 2′   Let C_1 and C_2 be two symmetric charts of S^3, with a_i far from the poles

$$\sup_{C_i} \sup_{a_i^k} \left| \left(\cdots \frac{w_m^\infty}{\sqrt{\lambda_m}} \cdots \right) \left(\frac{\partial A}{\partial a_l^k} \right) \left(\begin{array}{c} \vdots \\ \frac{w_m^\infty}{\sqrt{\lambda_m}} \\ \vdots \end{array} \right) \right| \geq c \sum \frac{w_i^{\infty 2}}{\lambda_i |a_i - a_j|^2}.$$

The $\frac{w_i^\infty}{\sqrt{\lambda_i}}$'s should be taken in each chart to be equal to their respective values.

For the second conclusion, we use the expansion of w_i. Under a variation $\delta \sigma_i$ which preserves the polar axis, w_i^∞ is unchanged, as well as a_i. Thus,

$$\frac{\partial w_i}{\partial \sigma_i} = \frac{c(\frac{\partial}{\partial \sigma_i} D\bar{w}_i)(x - a_i)\lambda_i \sqrt{\lambda_i}}{(1 + \lambda_i^2 |x - a_i|^2)^{3/2}} + h.o.$$

When $\lambda_i |x - a_i|$ is large, we find:

$$\frac{\partial w_i}{\partial \sigma_i} = O\left(\frac{1}{\lambda_i \sqrt{\lambda_i} |x - a_i|^2} \right) + h.o.$$

We have derived in (1.2) of Proposition 8, the σ_i-equations for groups. We derive these equations here in another way which makes clear that a factor equal to $\frac{1}{\lambda_i^{3/2}}$ is present in all the terms involved in these equations. This, when applied to single masses, offers a way to pass to the limit when the concentration tend to $+\infty$ even though the total interaction of a given single mass is $o(\sum \varepsilon_{ij})$.

When we are considering a group G_ℓ, we may perform on **all** the masses of the group the same variation $\delta \sigma_i$ so that their relative positions are unchanged and their specific contribution to the functional J is untouched.

So are ω_i^∞, ω_j^∞ for each i and j. We follow for these σ_i's the rule set for the points a_i, so that the masses of the configuation $\sum \alpha_i \omega_i + v$ are split into two groups. The first group undergoes the variation $\delta \sigma_i$ while the second one does not. If we expand $\frac{\partial J}{\partial \sigma_i}$, we find that the main terms of the expansion are i

$$\sum_{(i,j)\in(I,II)} -\frac{\omega_i^\infty}{\sqrt{\lambda_i}|a_i - a_j|} \frac{\partial_{\sigma_i}((D\bar{\omega}_j + o(1))(\frac{a_i - a_j}{|a_i - a_j|}))}{\lambda_j \sqrt{\lambda_j}|a_i - a_j|} - \sum_{(i,j)\in(I,II)} (\partial_{\sigma_i} c_{ij})\varepsilon_{ij}^3.$$

The remainder is easily controlled (The $|\frac{\partial \omega_i}{\partial \sigma_i}|$ are bounded by $C\delta_i$, in fact by a much better estimate.) and we can pass to the limit after dividing by $\sum_{i\in I, j\in II} (\frac{|\omega_i^\infty|\varepsilon_{ij}}{\lambda_j |a_i - a_j|} + \varepsilon_{ij}^3)$. This provides us with a set of σ_i-equations in the limit, in this framework. As explained above, when we have to come back to single mass ω_i for which the ε_{ij}'s could be very small, the system which we find at the limit might not contain the mass ω_i because the remainder involves $o(\varepsilon_{ij} \sum \varepsilon_{js}^2)$, thus might overwhelm the principal term as we pass to the limit. There is however another way to pass to the limit under such conditions: in view of the expansion of $\frac{\partial \omega_i}{\partial \sigma_i}$ carried out above and of the principal terms of the expansion, we can multiply the whole equation by $\lambda_i^{3/2}$. If we carry our expansion so that we have $\omega_j^\infty \omega_i(a_i)\varepsilon_{ij}$ as leading terms in the expansion of J, instead of the symmetric expression with (j, i), (which we may assume), we can see easily that there is a factor $\frac{1}{\lambda_i^{3/2}}$ in all our expansions. This factor is easy to trace back: As we compute $\int g\omega_i^4 \frac{\partial \omega_i}{\partial \sigma_i} = g(a_i) \int \omega_i^4 \frac{\partial \omega_i}{\partial \sigma_i} + |\nabla g|_{L^\infty} \int O(|x - a_i|)\delta_i^5$, we recognize this factor immediately. This works for $g = \delta_\ell$ or ω_l or $\Delta^{-1} w_k^4 \omega_l$ etc. We thus see that this factor should be present everywhere, including in the remainder. This remainder includes $\partial J . \partial_{\sigma_i} v$ which reads

$$\sum \partial J . \varphi \int \nabla \partial_{\sigma_i} v \nabla \varphi$$

with $\varphi = \omega_i, \frac{1}{\lambda_i} \frac{\partial \omega_i}{\partial a_i}, \frac{\partial \omega_i}{\partial \bar{\sigma}_i}$ etc. Observe that

$$\int \nabla \partial_{\sigma_i} v \nabla \varphi = -\int \nabla v \nabla \partial_{\sigma_i} \varphi = \int v \Delta \partial_{\sigma_i} \varphi.$$

We now have L^∞ estimates on v and also on ∇v_I, ∇v_{II} etc which we can use. The factor $\frac{1}{\lambda_i^{3/2}}$ can be expected here too. Multiplying by $\lambda_i^{3/2}$, we find that our σ_i-equation take another form which allows under suitable conditions to pass to the limit even though some of the masses might have a very small total interaction.

At the end of the next section, we justify this claim.

1.10.12 *The end of the expansion and the concluding remarks*

Combining Proposition 8 and (1.47), we can conclude if we have a **well-distributed** configuration of masses such that for each single mass $\theta_i = \sum_{j\neq i} \varepsilon_{ij} \geq c \sum_{k\neq t} \varepsilon_{kt}$, since we may then warrant (H) for each motion if we follow the rule which we set above for the motion of the points. Looking in the end to the complete set of equations, we can rebuild an equation for the motion of each a_i. Combining with the σ_i-equation of Proposition 8, we conclude.

At the end of this section, we show how to pass to the limit in the σ_i-equation even though $\theta_i = \sum_{j\neq i} \varepsilon_{ij} = o\left(\sum \varepsilon_{kt}\right)$.

We can also, in view of our five observations which we made in the last section, conclude in the case of a well-distributed packing in groups under the additional condition that

$$(K) \qquad \sum_{(i,j)\in\text{different groups}} \frac{{\omega_i^\infty}^2}{\lambda_i|a_i - a_j|^2} \leq C \sum_{(i,j)\in\text{ different groups}} \frac{\varepsilon_{ij}^3}{|a_i - a_j|}.$$

We could even multiply the left hand side in (K) by $\frac{1}{\lambda_l|a_m - a_l|}$, the requirement would be sufficient in order to conclude. If the basic concentrations of all groups are comparable, we can absorb all the terms covered by Oberservation 4. We are left only with the term covered by Observation 5, i.e. we need:

$$(K') \qquad \sum_{l\in I, i\in II, m\in I} \frac{|\omega_i^\infty \omega_j^\infty|\varepsilon_{lm}}{\sqrt{\lambda_m}\sqrt{\lambda_i}|a||a_m - a_i|} \leq C \sum_{(i,j)\in\text{different group}} \frac{\varepsilon_{ij}^3}{|a_i - a_j|}.$$

The conclusions stated above require us to transform all O's in Lemma 67 into o's. This can be completed when the ω_l^∞'s are $o(1)$. We need also for this to use the $v - (v_I + v_{II})$ factor in these estimates, splitting this expression on low and high modes, using Progsition 8 on low modes and eigenvalues for large modes. We assume now that (K) and (K') do not hold.

We want a bound on $\sum \frac{{\omega_i^\infty}^2}{\lambda_i|a_i-a_j|^2}$ in terms of $\sum_{(i,j)\in\text{different groups}} \frac{\varepsilon_{ij}^3}{|a_i-a_j|}$. This depends on the derivation of σ_i-equations for single masses. We start by showing how and why we can move each single mass with a reasonable control on the remainders:

Lemma 76   *Assume that $\lambda_i < \sup \lambda_j$, we may then assume that our configuration satisfies the estimate:*

$$\lambda_i^2 \partial J \left(\sum \alpha_j \omega_j + v \right) \cdot \frac{\partial \omega_i}{\partial \lambda_i} = o \left(\sum_{j \neq i} \frac{{w_j^\infty}^2}{\lambda_j |a_j - a_i|^2} + \sum_{j \neq i} \frac{\varepsilon_{ij}^3}{|a_i - a_j|} \right).$$

Proof.   Since $\lambda_i < \sup \lambda_j$, we are free to move λ_i and thus we may assume, unless our proof of the Morse Lemma at infinity is complete for the configuration, that

$$\lambda_i^2 \partial J \left(\sum \alpha_j \omega_j + v \right) \cdot \left(\alpha_i \frac{\partial \omega_i}{\partial \lambda_i} + \frac{\partial v}{\partial \lambda_i} \right) = 0.$$

We compute, with Q the usual projection operator, $\lambda_i Q(\frac{\partial v}{\partial \lambda_i})$. We have for $\varphi = \frac{1}{\lambda_i} \frac{\partial \omega_i}{\partial \lambda_i}, \ \lambda_i \frac{\partial \omega_i}{\partial \lambda_i}, \ \frac{\partial \omega_i}{\partial \sigma_i^j}$,

$$\int \nabla \lambda_i \frac{\partial v}{\partial \lambda_i} \nabla \varphi = - \int \nabla v \nabla \lambda_i \frac{\partial \varphi}{\partial \lambda_i}.$$

Thus,

$$\lambda_i^2 \partial J \left(\sum \alpha_j \omega_j + v \right) \cdot \frac{\partial v}{\partial \lambda_i} = A_i \partial J \left(\sum \alpha_j \omega_j + v \right) \cdot \frac{\partial \omega_i}{\partial a_i}$$

$$+ B_i \partial J \left(\sum \alpha_j \omega_j + v \right) \cdot \lambda_i^2 \frac{\partial \omega_i}{\partial \lambda_i} + \sum_k C_i^k \partial J \left(\sum \alpha_j \omega_j + v \right) \cdot \lambda_i \frac{\partial \omega_i}{\partial \sigma_i^j}.$$

Thus,

$$\left(1 + \frac{B_i}{\alpha_i} \right) \lambda_i^2 \partial J \left(\sum \alpha_j \omega_j + v \right) \cdot \alpha_i \frac{\partial \omega_i}{\partial \lambda_i} = -A_i \partial J \left(\sum \alpha_j \omega_j + v \right) \cdot \frac{\partial \omega_i}{\partial a_i}$$

$$- \sum_k C_i^k \partial J \left(\sum \alpha_j \omega_j + v \right) \cdot \lambda_i \frac{\partial \omega_i}{\partial \sigma_i^j}.$$

The same computation can be carried out for

$$\partial J \left(\sum \alpha_j \omega_j + v \right) \cdot \lambda_i \left(\frac{\partial \omega_i}{\partial \sigma_i^j} + \frac{\partial v}{\partial \sigma_i^j} \right).$$

We find (after inverting a matrix)

$$\lambda_i \partial J \left(\sum \alpha_j \omega_j + v \right) \cdot \frac{\partial \omega_i}{\partial \sigma_i^j} = -\tilde{A}_i \partial J \left(\sum \alpha_j \omega_j + v \right) \cdot \frac{\partial \omega_i}{\partial a_i}$$

$$- \tilde{B}_i \partial J \left(\sum \alpha_j \omega_j + v \right) \cdot \lambda_i^2 \frac{\partial \omega_i}{\partial \lambda_i}.$$

Combining, we derive:

$$\lambda_i^2 \partial J\left(\sum \alpha_j \omega_j + v\right).\alpha_i \frac{\partial \omega_i}{\partial \lambda_i} = O(|\tilde{A}_i| + |\tilde{B}_i| + |\tilde{C}_i^k| + |A_i| + |B_i| + |C_i^k|)$$

$$\times \partial J\left(\sum \alpha_j \omega_j + v\right).\frac{\partial \omega_i}{\partial a_i}.$$

$\square$

Next, we have:

Lemma 77   *Let ω_1 be a mass in a group G_1 which is not the most concentrated mass among all masses. In a chart where a_1 is a compact variable, we may assume that*

$$\partial J\left(\sum \alpha_j \omega_j + v\right).(\partial_{a_1}\omega_1 + \partial_{a_1}v) = o\left(\sum_{j\neq 1} \frac{\omega_j^{\infty^2}}{\lambda_j |a_j - a_1|^2} + \sum_{j\neq 1}\right) \frac{\varepsilon_{1j}^3}{|a_j - a_1|}.$$

$$(1.48)$$

Furthermore, the above estimate rereads:

$$\partial J\left(\sum \alpha_j \omega_j + v\right).\partial_{a_1}\omega_1 = o\left(\sum_{j\neq 1} \frac{\omega_j^{\infty^2}}{\lambda_j |a_j - a_1|^2} + \sum_{j\neq 1} \frac{\varepsilon_{1j}^3}{|a_j - a_1|}\right). \quad (1.49)$$

Proof.   We start with the proof of (1.48). If (1.48) does not hold, we may decrease J by moving the point a_1. If for every $j \neq 1$, $|a_j - a_1|$ does not tend to zero along the deformation, the only hinderance in this decrease would come from the fact that a_1 would move away to infinity.

However, this would not happen without a decrease in J which we can estimate easily and which suffices in order to carry out the Morse Lemma at infinity.

If $|a_j - a_1|$ tends to zero for a certain j, then we use the fact that some control on the concentrations—they cannot increase beyond twice their initial value along the global deformation—is built in our deformation. $\displaystyle\sum_{j\neq 1}\left(\frac{\omega_j^{\infty^2}}{\lambda_j |a_j - a_1|^2} + \sum_{j\neq 1}\frac{\varepsilon_{1j}^3}{|a_j - a_1|}\right)$ comes eventually to dominate $\displaystyle\sum_{k\neq s}\frac{\omega_j^{\infty^2}}{\lambda_k |a_k - a_s|} + \sum \frac{\varepsilon_{1j}^3}{|a_k - a_s|}$ (assuming that a_1 and a_j are the only points colliding, otherwise the argument can be modified and generalized).

Thus, before the collision occurs, J has decreased enough so that the Morse Lemma at infinity can be carried out.

We now expand $\partial J(\sum \alpha_j \omega_j + v).(\partial_{a_1}\omega_1 + \partial_{a_1}v)$. Considering $\partial J(\sum \alpha_j \omega_j + v).\partial_{a_1}v$, we use the projection operator Q and we derive that

$$\partial J\left(\sum \alpha_j \omega_j + v\right).\partial_{a_1}v = A_1 \partial J.\partial_{a_1}\omega_1 + \sum_j B_1^j \partial J.\lambda_1 \frac{\partial \omega_1}{\partial \sigma_1^j} + C_1 \partial J.\lambda_1^2 \frac{\partial \omega_1}{\partial \lambda_1}.$$

Since ω_1 is not the most concentrated of its group, we may use Lemma 76. $\partial J.\lambda_1^2 \frac{\partial \omega_1}{\partial \lambda_1}$ is as small as we please depending on $\partial_{a_1}\omega_1$. The same conclusion can be derived for $\partial J. \frac{\partial \omega_1}{\partial \sigma_1^j}$. A_1, B_1^j, C_1 are all $o(1)$, in fact much smaller. Therefore, the estimate on $\partial J(\sum \alpha_j \omega_j + v).(\partial_{a_1}\omega_1 + \partial_{a_1}v)$ is in fact an estimate on $\partial J(\sum \alpha_j \omega_j + v).\partial_{a_1}\omega_1$. $\qquad\qquad\qquad \square$

We thus have a good control on the size of $\partial J(\partial_{a_1}\omega_1 + \partial_{a_1}v)$. But we need now to complete an expansion. As we already know, this expansion provides additional remainders and these remainders cannot be controlled if a condition of type (H) does not hold along the deformation. We thus need to move **together**, in a single movement, all the groups having a basic concentration of the same order than G_1 and all the masses of G_1 having a concentration at most λ_1. Then, (H), restricted to the appropriate indexes, hold. We still need the expansion, this expansion is not provided by Lemma 67 right away since we are also moving certain masses of G_1 relative to the other ones.

We sketch here how the estimates of Lemma 67 extend to this new framework, observe that $\partial J(\partial_{a_1}\omega_1 + \partial_{a_1}v).\partial v$ has disappeared from the estimates. We need only to estimate $\sum_{j\,\text{moving}} \partial J(\partial_{a_1}\omega_1 + \partial_{a_1}v).(\alpha_j \partial \omega_j)$. The expansion is then easier and completed as follows:

We decompose the configuration into a group I made of groups G_l where all masses (some for G_1) move and a group II where all the masses of all group do not move:

$$\sum \alpha_i \omega_i + v = \left(\sum_{i\in I} \alpha_i \omega_i + v_I\right) + \left(\sum_{i\in II} \alpha_i \omega_i + v_{II}\right) + (v - (v_I + v_{II}))$$

$$= U_1 + U_2 + v_{12}.$$

We expand ($\tilde{\omega}_1$ stands for all moving masses in I):

$$\partial J(U_1 + U_2 + v_{12}).\partial \tilde{\omega}_1 = \partial J(U_1).\partial \tilde{\omega}_1 + \partial^2 J(U_1).U_2.\partial \tilde{\omega}_1 + \partial^2 J(U_1).v_{12}.\partial \tilde{\omega}_1$$

$$- c \int U_2^5 \partial \tilde{\omega}_1 + \sum_{2 \le \mu_1 \le 4, \mu + \mu_1 = 5} c_{\mu \mu_1} \int U_1^\mu U_2^{\mu_1} \partial \tilde{\omega}_1 + R.$$

$c_{\mu\mu_1}$ are constants bounded independently of λ_i, a_i, etc. R has been estimated throught Lemma 62 untill Lemma 68, only that $\partial\omega_1$ was $\partial(\sum_{i\in I}\alpha_i\omega_i+v_I)$ in these lemmas. However, the exact expression of $\partial\omega_1$ was $\partial(\sum_{i\in I}\alpha_i\omega_i+v_I)$ was never used in the proofs of these lemmas. $\partial^2 J(\omega_1).\partial\omega_1$ is zero so that

$$\partial^2 J(U_1).v_{12}.\partial\tilde{\omega}_1 = O\left(\sum_{j\in I,j\neq 1}\int \omega_j^4|v_{12}||\partial\tilde{\omega}_1| + \int v_I^4|v_{12}||\partial\tilde{\omega}_1|\right).$$

We split the domain of integration between the various Ω_j, $j\in I$ and their complement.

For $j=1$, $\int_{\Omega_1}\omega_k^4|v_{12}||\partial\tilde{\omega}_1|$ behaves as R. For $j\neq 1$, we upperbound

$$\int_{\Omega_j}\omega_k^4|v_{12}||\partial\tilde{\omega}_1| \leq \int_{\Omega_j}\omega_k^4|v_{12}|(|\omega_1^\infty|\sqrt{\lambda_1}\delta_1^2 + \delta_1^3)$$

$$\leq C\int_{\Omega_j}\omega_j^4|v_{12}|(|\omega_1^\infty|\sqrt{\lambda_1}\delta_1^2 + \delta_1^3).$$

$|\omega_1^\infty|\int_{\Omega_j}\omega_k^4|v_{12}|\sqrt{\lambda_1}\delta_1^2$ behaves like R. For $\lambda_1\leq\lambda_j$,

$$\int_{\Omega_j}\omega_j^4|v_{12}|\sqrt{\lambda_1}\delta_1^3 \leq \frac{C}{\lambda_1^{3/2}|a_i - a_j|^3} \leq \int_{\Omega_j}\omega_j^4|v_{12}| \leq C|(v_{12}^0)_j|_\infty$$

$$\times\frac{\varepsilon_{1j}^2}{\sqrt{\lambda_1}|a_1 - a_j|} + C\sum_{\lambda_s\geq\lambda_j}\sqrt{\lambda_j}\varepsilon_{js}|(v_{12}^s)_j|_\infty\frac{\varepsilon_{1j}^2}{\sqrt{\lambda_1}|a_1 - a_j|}$$

$$+ C\sqrt{\lambda_j}|(v_{12}^0)_j|_{H_0^1}\frac{\varepsilon_{1j}^2}{\sqrt{\lambda_1}|a_1 - a_j|}.$$

For $\lambda_1\geq\lambda_j$, $|\partial\omega_1 \leq C(|\omega_1^\infty| + \varepsilon_{1j})\sqrt{\lambda_1}\delta_1^2$ and the estimate is straight forward. For $\int_{\Omega_j}v_I^4|v_{12}||\partial\omega_1|$, we use Proposition 3. We can upperbound this quantity by

$$\sum_{(m,n)\in(I,II),(q,k)\in(I,I)}\sqrt{\lambda_1}\varepsilon_{qk}^4\varepsilon_{mn}\int_{\Omega_j}\delta_q^4(\delta_m + \delta_n)\delta_1^2$$

$$\leq \sum\sqrt{\lambda_1}\sqrt{\lambda_q}\varepsilon_{qk}^4\varepsilon_{mn}\int_{\Omega_j}\delta_q^3(\delta_m + \delta_n)\delta_1^2 \leq C\sqrt{\lambda_1}\sqrt{\lambda_q}\varepsilon_{qk}^5.$$

This upperbound works for $\lambda_1 \leq \lambda_q$. If $\lambda_1 \geq \lambda_q$, we upperbound by

$$\lambda_q \varepsilon_{qk}^4 \varepsilon_{mn} \int \delta_q^2 (\delta_m + \delta_n)\delta_1^2 \leq C\lambda_q \varepsilon_{qk}^5.$$

$\sum_{2 \leq \mu_1 \leq 4, \mu + \mu_1 = 5} \int |U_1|^\mu |U_2|^{\mu_1} |\partial \omega_1|$ can be treated in a similar way. We then conclude that (all bounds in Lemma 67 can be transformed into small o's rather than large O's when the ω_i^∞'s are small as explained above):

$$\left| \partial J \left(\sum \alpha_i \omega_i + v \right) \cdot (\partial_{a_1} \tilde{\omega}_1 + \partial_{a_1} v) \right| \geq |\partial J(U_I)(\partial_{a_I} \tilde{\omega}_1 + \partial_{a_1} v)|$$

$$+ o\left(\sum \frac{\omega_j^{\infty 2}}{\lambda_j |a_i - a_j|^2} + \sum \frac{\varepsilon_{ij}^3}{|a_i - a_j|} \right)$$

$$- C \sum_{(i,j) \in G_l \times G_m, l \neq m} \frac{\varepsilon_{ij}^3}{|a_i - a_j|}.$$

$\tilde{\omega}_1$ stands for all moving masses. The set of these moving masses depends on the choice of a single mass ω_1. Summing up on all the masses ω_1 which are not the most concentrated in their groups, with appropriate scales M_i so the concentrations increase, we find

$$\sum_{\substack{\omega_i \text{ not the most concentrated} \\ \text{of their group}}} M_i \left| \partial J \left(\sum \alpha_j \omega_j + v \right) \cdot (\partial_{a_i} \tilde{\omega}_i + v) \right|$$

$$\geq \sum_l |\partial J(U_\ell)(\partial_{a_1} \omega_i + \partial_{a_1} v_l)|$$

$$+ o\left(\sum \frac{\omega_j^{\infty 2}}{\lambda_j |a_i - a_j|^2} + \sum \frac{\varepsilon_{ij}^3}{|a_i - a_j|} \right) - C \sum_{(i,j) \in G_l \times G_m, l \neq m} \frac{\varepsilon_{ij}^3}{|a_i - a_j|}.$$

U_ℓ stands for the configuration $\sum_{i \in G_l} \alpha_i \omega_i + v_l$ of the group G_l. There is no restriction on i in the right hand side since the following identity holds (translation invariance of J):

$$\partial J(U_\ell) \left(\sum_{i \in G_l} (\partial_{a_i} \omega_j + \partial_{a_i} v_l) \right) = 0.$$

Assuming that the U_l's are well distributed **and that we can derive limit** σ_i**-equations for each** ω_i **relative to** U_ℓ **i.e.** if we knew how to pass to the limit for each mass ω_i in these equations (only for U_ℓ's such that (K) does not hold; σ_i is a rotation preserving the polar axis), we would find

$$(K'') \sum_l \sum_{(i,j)\in G_l\times G_l} \left(\frac{\omega_i^{\infty^2}}{\lambda_i|a_i - a_j|^2} + \frac{\varepsilon_{ij}^3}{|a_i - a_j|} \right) \leq C \sum_{(i,j)\in G_l\times G_l, l\neq m} \frac{\varepsilon_{ij}^3}{|a_i - a_j|}$$

if

$$\sum |\partial J \left(\sum \alpha_j \omega_j + v \right) \cdot (\partial_{a_i}\omega_i + v)| \leq C \sum_{(i,j)\in G_l\times G_m, l\neq m} \frac{\varepsilon_{ij}^3}{|a_i - a_j|}$$

$$+ o \left(\sum_{(i,j)\in G_l\times G_m, l\neq m} \left(\frac{\omega_i^{\infty^2}}{\lambda_i|a_i - a_j|^2} + \frac{\varepsilon_{ij}^3}{|a_i - a_j|} \right) \right).$$

Observe that (1.47) combined with Proposition 8 and Conjecture 3 provides an estimate of the type

$$\sum_m \left(\sum_{j\in G_m, m\neq s} \frac{\alpha_j \omega_j^\infty}{\sqrt{\lambda_i}|a_{j_m} - a_{j_s}|} \right)^2 = o \left(\sum \frac{\omega_i^{\infty^2}}{\lambda_j|a_i - a_j|} + \sum \frac{\varepsilon_{ij}^3}{|a_i - a_j|} \right).$$

a_{j_m} and a_{j_s} are the concentration points of the least concentrated mass in G_m and G_s respectively. The above relation is useful for groups reduced to a single mass. Combined with (K''), they provide (K).

We turn now to Proposition 8 and the σ_i-equations which we want to derive when σ_i is a rotation preserving the polar axis for each mass ω_i. The issue is to show that these equations, for each mass ω_i, yield a relation with coefficients equal to $O(\sum_{j\neq i} \varepsilon_{ij}^3)$ and not $O(\sum_{k\neq t} \varepsilon_{kt}^3)$ as Proposition 8 displays (observe that for the rotations preserving the polar axis $\nabla\omega_i^\infty = 0$). Five terms need to be estimated for this purpose:

$$\sum_{k\neq j, k, j\neq i} \int \omega_k^4 |\omega_j| |\frac{\partial\omega_i}{\partial\sigma_i}|, \quad \int |\omega_k|^3 |\omega_i|^2 |\frac{\partial\omega_i}{\partial\sigma_i}|, \quad \int \omega_k^4 \omega_i \frac{\partial\omega_i}{\partial\sigma_i},$$

$$\int \left(\sum_{k\neq i} \alpha_k \omega_k \right)^2 \omega_i^3 \frac{\partial\omega_i}{\partial\sigma_i}, \quad \text{and} \quad \int \nabla v \nabla \frac{\partial\omega_i}{\partial\sigma_i},$$

more generally, $\int \nabla v \nabla \frac{\partial\omega_i}{\partial\sigma_i}$ for $\varphi = \frac{1}{\lambda_i}\frac{\partial\omega_i}{\partial a_i}, \lambda_i\frac{\partial\omega_i}{\partial\lambda_i}, \frac{\partial\omega_i}{\partial\sigma_i^j}$.

Let us start with the last term and observe that, for a rotation preserving the polar axis, $|\frac{\partial\varphi}{\partial\sigma_i}| \leq \frac{C\delta_i^2}{\sqrt{\lambda_i}}$. Thus, $\int \nabla v \nabla \frac{\partial\varphi_i}{\partial\sigma_i} = -\int \Delta v \frac{\partial\varphi_i}{\partial\sigma_i} =$

$O(\int \Delta v O(\frac{\delta_i^2}{\sqrt{\lambda_i}}))$. Using the equation satisfied by v and the L^∞ bounds on v, this is $O(\varepsilon_{ij}^2)$: the only term to fear is $\int \omega_i^4 \omega_j \frac{\partial \varphi}{\partial \sigma_i}$ which is easily seen after rescaling *etc* to be $O(\sum_{j \neq i} \varepsilon_{ij}^3)$. The other terms would contain either a δ_k^2, $k \neq i$, to the least or a v. If an L^∞-estimate on v produces a δ_i, a coefficient ε_{ik} multiplies it and yields our claim. Thus $\int \nabla v \nabla \frac{\partial \varphi_i}{\partial \sigma_i} = O(\sum_{j \neq i} \varepsilon_{ij}^2)$. This term is multiplied by $\partial J(u) \cdot \varphi$ which is $O(\sum_{j \neq i} \varepsilon_{ij})$. We thus find $O(\sum_{j \neq i} \varepsilon_{ij}^3)$.

Considering now $\int \omega_k \omega_\ell \omega_i^3 \frac{\partial \omega_i}{\partial \sigma_i}$, $k, l \neq i$, we rescale first ω_k to $\bar{\omega}_k$ and derive

$$\int \omega_k \omega_l \omega_i^3 \frac{\partial \omega_i}{\partial \sigma_i} = \bar{\omega}_k(\tilde{a}_i) \int \overbrace{\omega_l \omega_i^3 \frac{\partial \omega_i}{\partial \sigma_i}} + O\left(\int |x - \tilde{a}_i| \overbrace{\omega_l \omega_i^3 \frac{\partial \omega_i}{\partial \sigma_i}} \right)$$

$$= \bar{\omega}_k(\tilde{a}_i) \bar{\omega}_k(\tilde{\tilde{a}}_i) \int \tilde{\tilde{\omega}}_i^3 \frac{\partial \tilde{\tilde{\omega}}_i}{\partial \sigma_i} + \bar{\omega}_k(\tilde{a}_i) \int O(|x - \tilde{\tilde{a}}_i|) |\tilde{\tilde{\omega}}_i^3 \frac{\partial \tilde{\tilde{\omega}}_i}{\partial \sigma_i}|$$

$$+ O\left(\int |x - \tilde{a}_i| \tilde{\delta}_l \frac{\tilde{\delta}_i^5}{\sqrt{\tilde{\lambda}_i}} \right) = \bar{\omega}_k(\tilde{a}_i) \int O\left(\int |x - \tilde{\tilde{a}}_i| \frac{\tilde{\tilde{\delta}}_i^5}{\sqrt{\tilde{\tilde{\lambda}}_i}} \right)$$

$$+ O\left(\int |x - \tilde{a}_i| \tilde{\delta}_l \frac{\tilde{\delta}_i^5}{\sqrt{\tilde{\lambda}_i}} \right)$$

$$= \frac{\bar{\omega}_k(\tilde{a}_i)}{\tilde{\lambda}_i^2} + \frac{1}{\sqrt{\tilde{\lambda}_i}} O\left(\left(\int \tilde{\delta}_l^2 \tilde{\delta}_i^4 \right)^{1/2} \left(\int |x - \tilde{a}_i|^2 \tilde{\delta}_i^6 \right)^{1/2} \right)$$

$$= O(\varepsilon_{il}^4 + \varepsilon_{ik}^4).$$

Next, we consider

$$\int \omega_k^4 \omega_j \frac{\partial \omega_i}{\partial \sigma_i} = \bar{\omega}_j(\tilde{a}_i) \int |\tilde{\omega}_k^4| |\frac{\partial \tilde{\omega}_i}{\partial \sigma_i}| + O\left(\int |x - \tilde{a}_i| \tilde{\omega}_k^4 \frac{\tilde{\delta}_i^2}{\sqrt{\tilde{\lambda}_i}} \right)$$

$$= \frac{\bar{\omega}_j(\tilde{a}_i)}{\sqrt{\tilde{\lambda}_i}} \int \tilde{\omega}_k^4 \tilde{\delta}_i^2 + \frac{1}{\tilde{\lambda}_i} O\left(\int \tilde{\omega}_k^4 \tilde{\delta}_i \right)$$

$$= O\left(\varepsilon_{ik}^2 \varepsilon_{ij} + \varepsilon_{ij}^2 \frac{\varepsilon_{ki}}{\sqrt{\tilde{\lambda}_i}} \right) = O\left(\sum \varepsilon_{it}^3 \right).$$

Similarly, we find that

$$\int |\omega_k|^3 |\frac{\partial \omega_i}{\partial \sigma_i}| \omega_i^2 = O\left(\int |\omega_k|^3 \frac{\delta_i^4}{\sqrt{\lambda_i}} \right) = O\left(\int \frac{\tilde{\delta}_i^{\,4}}{\sqrt{\tilde{\lambda}_i}} \right)$$

$$= O\left(\frac{1}{\tilde{\lambda}_i \sqrt{\tilde{\lambda}_i}} \right) = O(\varepsilon_{ik}^3).$$

Finally, we consider

$$\int \omega_k^4 \omega_j \frac{\partial \omega_i}{\partial \sigma_i}$$

$$= \bar{\omega}_j(a_i) \int_B \tilde{\omega}_i \frac{\partial \tilde{\omega}_i}{\partial \sigma_i} \tilde{\omega}_k^3 + O\left(\int_B |x - \tilde{a}_i| \frac{\tilde{\delta}_i^{\,3}}{\sqrt{\tilde{\lambda}_i}} |\bar{\omega}_k|^3 \right) + O\left(\int_{B^C} \bar{\omega}_k^4 \frac{\tilde{\delta}_i^{\,2}}{\sqrt{\tilde{\lambda}_i}} \right).$$

Observe that using the expansion of $\tilde{\omega}_i$ and oddness argument,

$$\int_B \tilde{\omega}_i \frac{\partial \tilde{\omega}_i}{\partial \sigma_i} = O(\frac{1}{\tilde{\lambda}_i^{\,3/2}}) = O(\varepsilon_{ik}^3), \quad \int_B |x - \tilde{a}_i| \frac{\tilde{\delta}_i^{\,3} |\bar{\omega}_k|^3}{\sqrt{\tilde{\lambda}_i}} = o\left(\frac{1}{\tilde{\lambda}_i \sqrt{\tilde{\lambda}_i}} \right),$$

$$\int_{B^C} \bar{\omega}_k^4 \frac{\tilde{\delta}_i^{\,2}}{\sqrt{\tilde{\lambda}_i}} = O(\frac{1}{\tilde{\lambda}_i \sqrt{\tilde{\lambda}_i}}), \text{ so that } \int \omega_k^4 \omega_i \frac{\partial \omega_i}{\partial \sigma_i} = O(\sum \varepsilon_{ik}^3).$$

In all, we see that we find a σ_i-relation for ω_i which has only ε_{ij}^3, $j \neq i$, as coefficients. This relation does not disappear as we divide by $\sum_{j \neq i} \varepsilon_{ij}^3$. A limit relation exists. The conclusion follows. A similar proof can be built under Conjecture 2′, Conjecture 3′.

Bibliography

Bahri, A., Critical Points at infinity in some variational problems, Pitman Research Notes in Mathematics Series No. 182 Longman Scientific and Technical, Longman, London (1989).

Bahri, A.- Chanillo S., The difference of topology at infinity in changing sign Yamabe problems on S^3 (the case of two masses) *Comm. Pure Appl. Math.* **54**:450–478, (2001).

Bahri, A. and Coron, J.M., On a non-linear elliptic equation involving the critical Sobolev exponent: The effect of the topology of the domain, *Comm. Pure Appl. Math.*, **41**:253–284, (1988).

Aubin, T., Equations differentielles non lineaires et probleme de Yamabe concernant la courbure scalaire, *J. Math. Pures appl.*, **55**:269–296, (1976).

Schoën, R., The existence of weak solutions with prescribed singular behavior for a conformally invariant scalar equation, *Comm. Pure and Appl. Math XLI*, 317–392, (1988).

Schoën, R., S. T. Yau, Conformally flat manifolds, Kleinian groups and scalar curvature, *Invent. Math.* **92**:47–71, (1988).

Xu, Y., A note on an Inequality with applications to Sign-Changing-Yamabe type problems, *Annales I.H.P.*, Paris, France, to appear.

Chapter 2

Contact Form Geometry

2.1　General Introduction

The second part of this book is about Contact Form Geometry via Legendrian curves. This is a direction of work which we have started early on [Bahri 1988]. We have published a number of results in this direction [Bahri 1988], [Bahri 1998], [Bahri-1 2003].

The main outcome of this previous work is the definition of a homology in the variational framework which we study. This variational framework is as follows:

Given (M^3, α) a three dimensional compact orientable manifold and α a contact form on M^3, we choose a vector-field v in $\ker \alpha$ which we assume (an assumption which we study in this work) to be non singular.

We introduce the form β dual to α, $\beta = d\alpha(v, \cdot)$ and the space of Legendrian curves of β:

$$\mathcal{L}_\beta = \{x \in H^1(S^1, M) | \dot{x} = a\xi + bv\}.$$

More specifically, we will be considering

$$C_\beta = \{x \in \mathcal{L}_\beta | \alpha_x(\dot{x}) = C > 0\}.$$

Let also

$$C_\beta^+ = \{x \in \mathcal{L}_\beta | \alpha_x(\dot{x}) \geq 0\}.$$

On C_β, we consider the variational problem

$$J(x) = \int_0^1 \alpha_x(\dot{x}) dt.$$

Assuming (an assumption which is also discussed in this work) that β is also a contact form with the same orientation than α, we have established

201

in our earlier work that the critical points of J on C_β were the periodic orbits of the Reeb vector-field of α which we denote ξ.

We have also established that this variational problem has asymptots (critical points at infinity). Asymptots in a variational problem depend on the definition of a pseudo-gradient.

There is a special pseudo-gradient Z_0 for J on C_β which we have built in [Bahri 1998]. The main feature of Z_0 is that the number of zeros of b, the v-component of $\dot{x}$ (x is the curve under deformation) does not increase along a decreasing flow-line.

Z_0 has some flavour of the curve-shortening flow (viewed in three dimensions. Legendrian curves, in a restricted framework, may be viewed as lifts of two dimensional immersed curves). In fact, there is a pseudo-gradient Z_1 for J which behaves just as the curve-shortening flow does. We have to modify it because it has too many asymptots.

Z_0 has the additional property that the L^1-norm of b (the v-component of $\dot{x}$) is bounded on any given flow-line. Furthermore, if blow-up occurs at a time T, then $b(s,t)$ converges to $\sum_{i=1}^m c_i \delta_{t_i}$ as s tends to T i.e. b converges to a finite combination of Dirac masses.

Accordingly, the curve $x(s,t)$ converges (in graph) to a curve made of ξ and $\pm v$-pieces. Among these curves, there are more specific curves which are the **critical points at infinity** of Z_0. There is a vast zoology for them which we have described in [Bahri-1 2003].

We have also explained in [Bahri-1 2003] how the geometric curve associated to one of them could support several critical points at infinity of various indexes. We have defined for these critical points at infinity $\bar{x}^\infty$'s a stable and unstable manifold, a Morse index (made of two parts $i_0 + i_\infty$) and a maximal number of zeros of b, $\theta(\bar{x}^\infty)$, on their unstable manifolds.

We also have explored the flow-lines from the periodic orbits of ξ to these asymptots and the flow-lines from the asymptots to the periodic orbits of ξ.

Accordingly, we have defined a homology using the periodic orbits of ξ and some of the aymptots. We have not computed this homology. Several problems stand in the way of making of this homology a practical tool in the study of contact structures.

We address some of these problems in this work.

The first issue is the generality of this method. It has been suggested that the two basic hypotheses which we introduced for the sake of simplicity, namely that v is non singular and that β is a contact form with the same orientation than α, were severe limitations to our method which would thereby be confined to the framework of unit sphere cotangent bundles.

We address this issue in the first part of this work in a simplified framework.

Namely, we first recognize that the hypothesis that β is a contact form is tied to the amount of rotation of ker α along the flow-lines of v. Typically, around some hyperbolic periodic orbits of v (we provide a normal form for (α, v)), ker α has a finite amount of rotation. Then β cannot be a contact form, at least a contact form with the same orientation than α.

We focus on such a framework, assume that v is a Morse-Smale flow having an attractive and a repulsive orbit and a family of hyperbolic orbits, some of them of the type described above.

We then prove that we can modify the contact form using the large amount of rotation around the attractive and repulsive orbits of v and build "mountains" around the "bad" hyperbolic orbits keeping the flow-lines of our variational problem away from these neighborhoods where β might not exist.

The method is interesting because it relies on a quantitative argument; the introduction of a large amount of rotation is completed via a second order differential equation of the form:

$$[v, [v, \xi]] = -\xi + \gamma(s)[\xi, v] + \theta v$$

s is the time along the flow-line of v.

As a large amount of rotation is introduced, the quantity $\tau = d\alpha([\xi, v], [\xi, [\xi, v]])$ remains bounded above. This in turn keeps $\int_0^1 |b|$ bounded on a given flow-line, even as this large rotation is introduced. This result generalizes to include the case when v might have attractive, repulsive or hyperbolic circles of zeros (the natural framework for vector-fields v in ker α).

We believe that we have in this way partially answered the questions about the limitations of our techniques.

The second part of this work is devoted to a compactness result about the flow-lines of our variational problem. This is the first result of this type for this technique. It is at this stage slightly imperfect: we establish here that there are no flow-lines going from a periodic orbit of Morse index $2k$ (for J on C_β) to a critical point at infinity of index $2k - 1$; there is a technical restriction at this point on this critical point at infinity, namely that it has no characteristic ξ-piece (see [Bahri-1 2003] for the definition of a characteristic piece, also the introduction of Part V of the present book) of strict index less than or equal to 1.

The same result holds for flow-lines from periodic orbits of Morse index $2k + 1$ to critical points at infinity of index $2k$. This result is only a generalization, with some more specific arguments, of the result about the flow-lines from x_{2k} to x^∞_{2k-1}.

The result established here is thus far-reaching because it indicates that there is a homology, **invariant under deformation of contact forms,** which involves only the periodic orbits of ξ.

We provide, at the end of this work—entitled compactness—the classifying map σ for the S^1-action on $C^*_\beta = C_\beta - \{\text{periodic orbits of } \xi\}$. This map should be helpful for the computation of the homology.

The third part of this work is devoted to two basic technical results related to a phenomenon which we have discovered in [Bahri-1 2003] and which we are coming, step by step, to understand better and more.

We use in our technique a procedure which we call **transmutation**. With this procedure, we change, given a critical point at infinity $\bar{x}^\infty$, the maximal number $\theta(\bar{x}^\infty)$ of zeros of the v-component of $\dot{x}, b$, on $W_u(\bar{x}^\infty)$, while keeping $\bar{x}^\infty$ isolated in its "species" i.e. among the critical points at infinity which have the same number of characteristic ξ-pieces (see [Bahri-1 2003]) than $\bar{x}^\infty$.

However, along this process, we create another critical point at infinity $\bar{y}^\infty$ having an additional ξ-piece when compared to $\bar{x}^\infty$.

In order to track down the changes through the collapse of $\bar{x}^\infty$ and $\bar{y}^\infty$, we build a suitable basis for $J''(\bar{x}^\infty)$ made of three parts: one is related to the H^1_0-index i_0 i.e. the index of $\bar{x}^\infty$ along variations where the v-orbits which are part of the graph of $\bar{x}^\infty$ are unchanged while the ξ-pieces are perturbed among the curves x having $\dot{x} = a\xi + bv, a > 0$ (thinking of geodesics, this would the index of a geodesic with fixed end points).

A second part is related to the characteristic pieces of $\bar{x}^\infty$; one vector $\tilde{v}_i$ is defined for each of them. It is built using the degeneracy of a characteristic piece (its end points are in the "cut-locus" one of the other).

A third part is related to the $\pm v$-orbits and the non-degenerate ξ-pieces of $\bar{x}^\infty$.

It turns out that the two last parts of this basis are $J''(\bar{x}^\infty)$-orthogonal and that, after a perturbation of the contact form near $\bar{x}^\infty$, we may assume that the index of $J''(\bar{x}^\infty)$ is distributed as we please among these two parts (given the dimension restrictions on each part). There is therefore a large flexibility in the interpretation of the Morse index at infinity i_∞ which may be read as $i^1_\infty + i^2_\infty$; i^1_∞ is related to the characteristic pieces, i^2_∞ to the $\pm v$-jumps and the non-degenerate ξ-pieces of $\bar{x}^\infty$. The values of i^1_∞ and i^2_∞ can be changed as long as $i^1_\infty + i^2_\infty$ remains unchanged and that each i^j_∞ is less than the dimension of the related space where it lives.

This technical result allows us to track very precisely the changes of indexes between $\bar{x}^\infty$ and $\bar{y}^\infty$ through a collapse and transmutation. We understand thereby what part changed in which part (i.e. among i_0 and i_∞, what piece has increased by 1 and what piece has decreased by 1, for each of $\bar{x}^\infty$ and $\bar{y}^\infty$).

For the full compactness argument, as we are writing it in [Bahri-2 2003], we need to complete such transmutations and understand how a transmutation occurs technically.

To conclude, we simply will say that we have uncovered here new phenomena related to the rotation of ker α along v and we have also started to establish compactness results in order to compute our homology see Proposition 25 Section 2.6 of the second part of this book and its generalizations.

These new results are to us a clear indication that we are progressing in a meaningful direction.

Significant improvements in the understanding of the behavior of contact structures can now be expected via these techniques, after due progress is made and our present results are pushed to their natural ends.

2.2 On the Dynamics of a Contact Structure along a Vector Field of its Kernel

2.2.1 *Introduction*

Let M^3 be a three dimensional manifold and α be a contact form on M. Let v be a vector-field in ker α which we assume throughout this work to have a finite number of non degenerate periodic orbits and also a finite number of circles of zeros.

Let us consider near a point x_0 of M a frame (v, e_1, e_2) transported by v. ker α defines a trace in Span (e_1, e_2) generated by $u = \alpha(e_2)e_1 - \alpha(e_1)e_2$. The fact that α is a contact form translates into a property of monotone rotation of u along v-transport, see [Bahri 1988, Proposition 9 p. 24] for more details. Thus, given a point $y_0 \in M$ and the v-orbit through y_0, there is a definite amount of rotation of ker α on the positive v-orbit and on the negative v-orbit.

It is natural to ask whether these amounts are infinite and the answer to this question is negative since one can produce (see [Bahri 1988, Section 12]) non singular codimension 1 foliation transverse to contact structures. If v generates the intersection of the tangent plane to the foliation with the kernel of the contact structure, the amount of rotation has to be less than π on any positive or negative v-orbit.

On the other hand, having an infinite amount of rotation for all half v-orbits can be quite useful: introducing the dual form $\beta = d\alpha(v, \cdot)$, α and β are transverse, both have v in their respective kernels. If ker α rotates infinitely along v, so does ker β. ker β could have some reverse rotations but it must essentially be a contact form with the same rotation than α.

When β is a contact form with the same rotation than α, a very interesting framework sets in: we introduce the space $\mathcal{L}_\beta = \{x \in H^1(S^1, M)$ s.t. $\beta_x(\dot{x}) \equiv 0\}$ of Legendrian curves of β and also the more constrained space $C_\beta = \{x \in \mathcal{L}_\beta$ s.t. $\alpha(\dot{x}) = $ a positive constant$\}$.

On C_β, the action functional $J(x) = \int_0^1 \alpha_x(\dot{x})dt$ has the periodic orbits of ξ, the Reeb vector-field of α, as critical points (of finite Morse index).

The variational problem is not compact (there are asymptots) but one can nevertheless, after the construction of a special flow [Bahri-1 2003], define a homology related to the periodic orbits of ξ [Bahri-1 2003]. It is therefore interesting to establish, given a contact structure α, that one can find a vector-field v in $\ker\alpha$ such that the amount of rotation on each half-orbit is infinite.

We consider hence vector-fields v in $\ker\alpha$ which have an ω-limit set reduced to their periodic orbits and their circles of zeros i.e. essentially Morse-Smale-type vector-fields (with lines of zeros allowed). Near the attractive orbit of v, after possibly perturbing slightly, v, an infinite amount of rotation is warranted on all half v-orbits attracted by this periodic orbit of v. There is a similar statement for the repulsive periodic orbit.

On the other hand, one can produce models of hyperbolic periodic orbits for v and models of contact forms having v in their kernel such that the amount of rotation of $\ker\alpha$ along these hyperbolic orbits is finite.

This type of orbits is called in this paper "bad hyperbolic orbits". They do not allow to set the variational problem J on C_β properly. Ideally, we would like to get rid of them or to the least to be able to consider the variational problem J on C_β away from these bad regions. In order to achieve this goal, a natural idea which comes to mind is to use the large rotations available (after perturbation possibly) near the attractive or near the repulsive periodic orbits. A diffeomorphism would then redistribute this large rotation over other regions of M, for example around the bad hyperbolic orbit. In this way, the bad hyperbolic orbit could be "surrounded" by a large rotation of $\ker\alpha$ along v, either coming from the attractive orbit or from the repulsive orbit.

This approach has a defect: it does not keep bounds. The price to pay for redistributing the rotation from the attractive or repulsive orbit becomes exponentially high with the amount of rotation.

The bounds carefully built [Bahri 1998] on the L^1-length of b as we deform curves of C_β along a pseudo-gradient for J (one of them is "curve shortening flow" which we do not use because of its "bad" behavior at blow-up) and which rely on a bound from above on $\tau = -d\alpha([\xi, v], [\xi, [\xi, v]])$ collapse.

We need therefore to find another way to introduce a large rotation. We consider two nested tori surrounding the attractive orbit for example. We introduce a second order differential equation which takes the form:

$$[v, [v, \xi]] = -\xi + \gamma(s)[\xi, v] - \gamma'(s)ds(\xi)v.$$

The unknown is ξ and the solution provides us with an extension of α. This differential equation has a unique solution under the condition that we should match α (up to a multiplicative constant) on the boundary of each torus. This differential equation, with v properly rescaled, generates a large amount of rotation. We may introduce this rotation and keep the existing periodic orbits of ξ unperturbed. Some new ones may appear but they are precisely localized and they appear in cancelling pairs. Furthermore, the bounds on all relevant quantities to the variation problem J on C_β (on $|\bar{\mu}|, |d\bar{\mu}|$ see [Bahri 1998] or τ from above) hold, unchanged.

This is the first part of this work.

We then move, in the second part of this work, to set the variational problem J on C_β using this large rotation.

We introduce a "Hamiltonian" $\lambda = e^{\sum \theta_i \delta_i s_i}$. s_i is a measure of the rotation of $\ker \alpha$ completed at a given point on a v-orbit originating at the boundary of one of the tori. λ is localized near the stable and unstable manifold of the bad hyperbolic orbit. Replacing α by $\lambda \alpha$, we build "mountains" around the bad hyperbolic orbit, i.e. regions where the Reeb vector-field of $\lambda \alpha$ is extremely small while the action is large. We prove that the bound from above still holds on τ, independently of λ and that the variational problem J on C_{β_λ} can be defined. Furthermore, we consider compact subsets of C_{β_λ} enjoying bounds independent of λ. Under decreasing deformation along the flow-lines of the pseudo-gradient, these compact sets never enter small pre-assigned neighborhoods of the bad orbits.

This holds in particular for all the flow-lines which start at the (unperturbed) periodic orbits of ξ. The definition of our homology follows and is independent of λ.

These results indicate that the assumption that $\beta = d\alpha(v, \cdot)$ is a contact form with the same orientation than α all over M is not needed in order to define the homology of [Bahri-1 2003].

One may question the generality of this method. We consider in this book only the simpler case of a single bad hyperbolic orbit and we assume that its stable and unstable manifolds are caught by the attractive and repulsive orbits, in short that there is no flow-line connecting hyperbolic orbits.

However, such connecting flow-lines can easily be added, as well as circles of zero as long as they are attractive, repulsive or hyperbolic not of mixed behavior.

The only constraint lies with the hypothesis that the ω-limit set of v is made of periodic orbits and circles of zeros.

Some thought shows that this hypothesis is not needed, but it makes our study much easier. This hypothesis can be weakened and a more general behavior allowed; we expect that there is always, given a contact structure, a vector-field v in its kernel with this behavior.

We proceed now with the statement of our results and their proofs.

Let us consider the differential equation:

$$[v, [v, \xi]] = -\xi + \gamma(s)[\xi, v] - \gamma'(s)ds(\xi)v$$

where s is, in this first step, the time along v.

More generally, for $\varphi = \varphi(s) > 0$, we consider the differential equation:

$$[\varphi v, [\varphi v, \xi]] = -\xi + \gamma(s)[\xi, \varphi v] - \gamma'(s)ds(\xi)\varphi v \qquad (*)$$

$\gamma(s)$ could be replaced by a function $\gamma(x_0, s)$ where x_0 is an initial data for the flow-line of v and s is a monotone increasing function on this flow-line. Observe that $\gamma'(s)ds(\xi) = \xi \cdot \gamma$.

Let us define α by

$$\alpha(v) = 0; \quad \alpha([\xi, \varphi v]) = 0; \quad \alpha(\xi) = 1.$$

This is possible if $v, [\xi, \varphi v]$ and ξ are independent. Writing $(*)$ in a φv-transported frame, with $\varphi v = \frac{\partial}{\partial s_1}, \xi = A\frac{\tilde{\partial}}{\partial x} + B\frac{\tilde{\partial}}{\partial y} + C\frac{\partial}{\partial s_1}$, we derive:

$$\frac{\partial^2 A}{\partial s_1^2} + A + \gamma\frac{\partial A}{\partial s_1} = 0 \qquad \frac{\partial^2 B}{\partial s_1} + B + \gamma\frac{\partial B}{\partial s_1} = 0.$$

Thus,

$$\frac{\partial}{\partial s_1}\left(A\frac{\partial B}{\partial s_1} - B\frac{\partial A}{\partial s_1}\right) = -\gamma\left(A\frac{\partial B}{\partial s_1} - B\frac{\partial A}{\partial s_1}\right).$$

Thus, if $\varphi v, [\xi, \varphi v]$ and ξ are independent at time zero (which we will assume) they are independent thereafter.

We then have:

Lemma 1

(i) $d\alpha(\varphi v, [\xi, \varphi v]) = -1$

(ii) $d\alpha(\varphi v, [\varphi v, [\xi, \varphi v]]) = \gamma$ *(denoted $\bar{\mu}$ usually)*

(iii) $[\varphi v, [\xi, [\xi, \varphi v]]] = -\gamma[\xi, [\xi, \varphi v]] + h\varphi v$.

Corollary 1

(i) *If* $[\xi, [\xi, \varphi v]](0)$ *is collinear to* v, *then so is* $[\xi, [\xi, \varphi v]](s)$ *and* ξ *is the contact vector-field of* α.

(ii) *If* $\gamma = 0$ *on an open set, then* $d\tau(v) = \tau_v$ *is zero on this set* $([\xi, [\xi, \varphi v]] = -\tau\varphi v)$.

Proof. Since $\alpha(v) = \alpha([\xi, \varphi v]) = 0, d\alpha(\varphi v, [\xi, \varphi v]) = -\alpha([\varphi v, [\xi, \varphi v]]) = \alpha(-\xi + \gamma[\xi, \varphi v] - \xi \cdot \gamma\varphi v) = -1$

(i) follows.

Next, we observe that α is a contact form since

$$\alpha \wedge d\alpha(\varphi v, [\xi, \varphi v], \xi) = -1.$$

Let ξ_r be its Reeb vector-field. Since $d\alpha(\varphi v, \xi) = -\alpha([\varphi v, \xi]) = 0$ and $d\alpha(\xi, \xi) = 0$,

$$\xi_r = \xi + \nu\varphi v$$

and using [Bahri 1998],

$$\bar{\mu} = d\alpha(\varphi v, [\varphi v, [\xi_r, \varphi v]]) = d\alpha(\varphi v, [\varphi v, [\xi, \varphi v]])$$

$$= d\alpha(\varphi v, \xi - \gamma[\xi, \varphi v] + z\varphi v) = -d\alpha(\varphi v, [\xi, \varphi v]) \cdot \gamma = \gamma$$

(ii) follows.

Next, we compute:

$$[\xi, [\varphi v, [\varphi v, \xi]]].$$

Using (*), it is equal to:

$$\gamma[\xi, [\xi, \varphi v]] + \xi \cdot \gamma[\xi, \varphi v] - \xi \cdot \gamma[\xi, \varphi v] + h\varphi v$$

$$= \gamma[\xi, [\xi, \varphi v]] + h\varphi v.$$

Using the Jacobi identity, it is equal to:

$$-[[\varphi v, \xi], [\xi, \varphi v]] + [\varphi v, [\xi, [\varphi v, \xi]]]$$

$$= -[\varphi v, [\xi, [\xi, \varphi v]]].$$

Thus,

$$-[\varphi v, [\xi, [\xi, \varphi v]]] = \gamma[\xi, [\xi, \varphi v]] + h\varphi v.$$

(iii) follows. $\qquad\square$

Proof. [Proof of Corollary 1] Set $\varphi v = \frac{\partial}{\partial s_1}$

(iii) reads

$$\frac{\partial U}{\partial s_1} = -\gamma U + h\frac{\partial}{\partial s_1}$$

i.e.

$$\frac{\partial}{\partial s_1}\left(e^{\int_0^{s_1}\gamma}\,U\right) = k\frac{\partial}{\partial s_1}$$

$$\frac{\partial}{\partial s_1}\left(e^{\int_0^{s_1}\gamma}\,U + \int_0^{s_1} k\frac{\partial}{\partial s_1}\right) = 0.$$

The claim follows.
Observe that

$$\alpha([\xi,\xi,\varphi v]]) = -d\alpha(\xi,[\xi,\varphi v]).$$

Set

$$\xi_r = \xi + \nu\varphi v.$$

Then,

$$-d\alpha(\xi,[\xi,\varphi v]) = -d\alpha(\xi_r - \nu\varphi v, [\xi_r - \nu\varphi v, \varphi v])$$

$$= \nu d\alpha(\varphi v, [\xi_r, \varphi v]) = -\nu.$$

Thus, if $[\xi,[\xi,\varphi v]]$ is collinear to v, ν is zero and $\xi = \xi_r$ $\square$

2.2.2   *Introducing a large rotation*

We consider now α_0 and v near the repelling (or attracting) periodic orbit of v. Their normal form, see Appendix 1, is:

$$(\alpha_0)\qquad \alpha_0 = dx + \frac{1}{20}(y + \bar\gamma x)d\theta \quad (\bar\gamma > 0)$$

$$(v)\qquad v = \left(20\frac{\partial}{\partial\theta} - (y + \bar\gamma x)\frac{\partial}{\partial x} + (x - \bar\gamma y)\frac{\partial}{\partial y}\right)\frac{1}{\sqrt{1+\bar\gamma^2}}.$$

Then,

$$\bar\xi = \frac{\partial}{\partial x} - \bar\gamma\frac{\partial}{\partial y}$$

and

$$[\bar\xi, v] = \sqrt{1+\bar\gamma^2}\,\frac{\partial}{\partial y} \text{ so that } d\alpha_0(v,[\bar\xi,v]) = -1.$$

Observe that $[\bar{\xi}, [\bar{\xi}, v]] = 0$ while

$$[v, [v, \bar{\xi}]] = -\left[20\frac{\partial}{\partial\theta} - (y + \bar{\gamma}x)\frac{\partial}{\partial x} + (x - \bar{\gamma}y)\frac{\partial}{\partial y}, \frac{\partial}{\partial y}\right]$$

$$= -\frac{\partial}{\partial x} - \bar{\gamma}\frac{\partial}{\partial y} = -\left(\frac{\partial}{\partial x} - \bar{\gamma}\frac{\partial}{\partial y}\right) - 2\bar{\gamma}\frac{\partial}{\partial y} = -\bar{\xi} - \frac{2\bar{\gamma}}{\sqrt{1+\bar{\gamma}^2}}[\bar{\xi}, v].$$

Thus,

$$[v, [v, \bar{\xi}]] = -\bar{\xi} + \frac{2\bar{\gamma}}{\sqrt{1+\bar{\gamma}^2}}[v, \bar{\xi}] \qquad (**)$$

(**) is the same form than (*).

Indeed, if we set

$$\gamma(s) = \frac{-2\bar{\gamma}}{\sqrt{1+\bar{\gamma}^2}},$$

then $\gamma' = 0$ and (**) is a special case of (*).

We are going to modify (**), keeping the framework of (*) but introducing a function $\gamma(s)$ which has a flat piece where it is equal to zero. We later will use this flat piece in order to introduce a large rotation of γ. However, ξ and $[\xi, v]$ are modified once γ is modified and we need to complete this modification so that the modified data for $\xi, [\xi, v]$ will glue up after a certain time $\bar{s}$ with the former data.

Observe that $\bar{\xi}$ for α_0 is in $\text{Span}\{\frac{\partial}{\partial x}, \frac{\partial}{\partial y}\}$. Observe also that $\text{Span}\{\frac{\partial}{\partial x}, \frac{\partial}{\partial y}\}$ is invariant by the one-parameter group of v. It is easy to construct two vector-fields in $\text{Span}\{\frac{\partial}{\partial x}, \frac{\partial}{\partial y}\}$ which commute to v. They need to satisfy:

$$\left[20\frac{\partial}{\partial\theta} - (y + \bar{\gamma}x)\frac{\partial}{\partial x} + (x - \bar{\gamma}y)\frac{\partial}{\partial y}, A_0\frac{\partial}{\partial x} + B_0\frac{\partial}{\partial y}\right] = 0.$$

This yields

$$\frac{\partial A_0}{\partial s_1} + \bar{\gamma}A_0 + B_0 = 0$$

$$\frac{\partial B_0}{\partial s_1} + A_0 + \bar{\gamma}B_0 = 0.$$

Taking $\begin{pmatrix} A_0 \\ B_0 \end{pmatrix}(0) = \begin{pmatrix} 1 \\ 0 \end{pmatrix}$ or $\begin{pmatrix} 0 \\ 1 \end{pmatrix}$, we derive two vector-fields $\frac{\tilde{\partial}}{\partial x}$ and $\frac{\tilde{\partial}}{\partial y}$ which have components on $\frac{\partial}{\partial x}, \frac{\partial}{\partial y}$ depending only on s_1, not on the initial point (this does not hold if γ changes into $\gamma(s, x_0)$).

Since $\bar{\xi}$ for α_0 is $\frac{\partial}{\partial x} - \gamma \frac{\partial}{\partial y}$, $\bar{\xi}$ reads as

$$a_0 \frac{\tilde{\partial}}{\partial x} + b_0 \frac{\tilde{\partial}}{\partial y} \text{ with } a_0 = a_0(s_1), b_0 = b_0(s_1)$$

while

$$[v, \bar{\xi}] = \frac{\partial \bar{\xi}}{\partial s_1} = a_0' \frac{\tilde{\partial}}{\partial x} + b_0' \frac{\partial}{\partial y} \left(= \sqrt{1 + \bar{\gamma}^2} \frac{\partial}{\partial y} \right)$$

Coming back to ξ and (*), with a general $\gamma = \gamma(s) = \gamma(s(s_1))$, we split ξ on the basis $\frac{\tilde{\partial}}{\partial x}, \frac{\tilde{\partial}}{\partial y}, \frac{\partial}{\partial s_1} = v$:

$$\xi = A \frac{\tilde{\partial}}{\partial x} + B \frac{\tilde{\partial}}{\partial y} + C \frac{\partial}{\partial s_1}.$$

The equations satisfied by A and B are:

$$\frac{\partial^2 A}{\partial s_1^2} + \gamma \frac{\partial A}{\partial s_1} + A = 0$$

$$\frac{\partial^2 B}{\partial s_1^2} + \gamma \frac{\partial B}{\partial s_1} + B = 0.$$

We integrate these equations on an interval $[\bar{s}_1, \bar{s}_2]$, with initial data at $\bar{s}_1$ equal to $(a_0, a_0')(\bar{s}_1)$ for $\left(A, \frac{\partial A}{\partial s_1} \right)(\bar{s}_1)$ and $(b_0, b_0')(\bar{s}_1)$ for $\left(B, \frac{\partial B}{\partial s_1} \right)(\bar{s})$. $\gamma(s)$ near $\bar{s}_1$ is $\frac{-2\bar{\gamma}}{\sqrt{1+\bar{\gamma}^2}} \cdot \left(C, \frac{\partial C}{\partial s_1} \right)(\bar{s}_1) = (0, 0)$ so that ξ near $\bar{s}_1$ is $\bar{\xi}$.

$-\gamma$ will behave as follows:

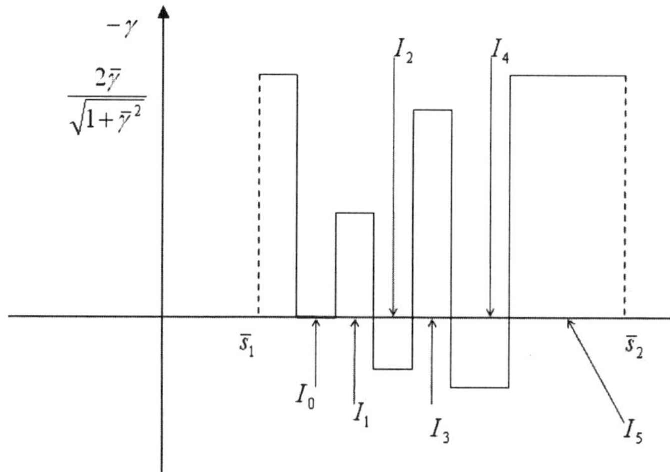

Observe that we need only to worry about the components of $\xi, [\xi, v]$ on $\frac{\tilde{\partial}}{\partial x}$ and $\frac{\tilde{\partial}}{\partial y}$. Indeed, would they match as well as their s_1-derivatives with those of ξ, then ξ and $\bar{\xi}$, $[\xi, v]$ and $[\bar{\xi}, v]$ would only differ by some $\mu v, \mu_1 v$ for $s \geq \bar{s}_2$. They satisfy the same differential equation after $\bar{s}_2$. Thus α and α_0 would match also. Observe that α needs not match with α_0. It suffices that it matches with some $c\alpha_0$ (observe that $d(c\alpha_0)(v, [\frac{\bar{\xi}}{c}, v]) = -1$ and $d(c\alpha_0)(v, [v, [\frac{\bar{\xi}}{c}, v]]) = d\alpha_0(v, [v, [\bar{\xi}, v]])$. We would then extend α near the repelling (respectively attracting) periodic orbit of v with $c\alpha_0$. We need thus only to match $\xi, [\xi, v]$ and $\bar{\xi}, [\bar{\xi}, v]$ in directions with the same ratio of length (not necessarily equal to 1). We prove below that this is possible.

We have:

Proposition 1   *No new periodic orbit of ξ is created in this process.*

The proof of the above proposition requires the three following claims which follow from the construction of γ

1. γ can be constructed so that $\int |\gamma'| \leq 20$.
2. As $\bar{s}_2 - \bar{s}_1$ becomes smaller and smaller, $|\gamma|$ remains bounded by 2.
3. α glues up with $c\alpha_0, c$ tending to 1 as $\bar{s}_2 - \bar{s}_1$ tends to zero.

Proof.   [Proof of Proposition 1] As $\bar{s}_2 - \bar{s}_1$ becomes small, ξ and $\bar{\xi}$ are $o(1)$-close. This is clear from the equations satisfied by the components over $\frac{\tilde{\partial}}{\partial x}$ and $\frac{\tilde{\partial}}{\partial y}$ of ξ and $\bar{\xi}$ and from the third claim stated above.

For the C-component on $\frac{\partial}{\partial s_1}$, it depends on $\gamma'(s)ds(\xi) = \gamma'(s)(Ads(\frac{\tilde{\partial}}{\partial x}) + Bds(\frac{\tilde{\partial}}{\partial y}) + C)$. Observe that, because $\frac{\tilde{\partial}}{\partial x}, \frac{\tilde{\partial}}{\partial y}$ and $\frac{\partial}{\partial s_1}$

commute, $ds(\frac{\tilde{\partial}}{\partial y})$ and $ds(\frac{\tilde{\partial}}{\partial y})$ are independent of s_1. They are bounded uniformly. So are A and B.

Since $\int |\gamma'|$ is bounded by 20, then

$$\int_{\bar{s}_1}^{s_1} \int_{\bar{s}_1}^{\tau} |\gamma'| \leq 20 \left(\sum_{j=0}^{s} |I_j| \right) = o(1).$$

After some work, this implies that the v-components of ξ and $\bar{\xi}$ are close up to $o(1)$. Indeed, C (the v-component of ξ) satisfies:

$$\frac{\partial^2 C}{\partial s_1^2} + C + \gamma \frac{\partial C}{\partial s_1} = \gamma'(s)ds(\xi) = \gamma'(s)\left(Ads\left(\frac{\tilde{\partial}}{\partial x}\right) + Bds\left(\frac{\tilde{\partial}}{\partial y}\right) + C \right)$$

while $\bar{C}$, the v-component of $\bar{\xi}$ satisfies

$$\frac{\partial^2 \bar{C}}{\partial s_1^2} + \bar{C} + \bar{\gamma} \frac{\partial \bar{C}}{\partial s_1} = 0$$

with $C(\bar{s}_1) = \bar{C}(\bar{s}_1)$, $\frac{\partial C}{\partial s_1}(\bar{s}_1) = \frac{\partial \bar{C}}{\partial s_1}(\bar{s}_1)$. $\qquad\qquad \square$

The claim follows.

2.2.3   *How γ is built*

We start with the differential equation with a constant $\bar{\gamma}_0$

$$\frac{\partial^2 u}{\partial s_1^2} + \bar{\gamma}_0 \frac{\partial u}{\partial s_1} + u = 0.$$

We set it in a matricial form with $v = -\frac{\partial u}{\partial s}$.
Then,

$$\frac{\partial}{\partial s} \begin{pmatrix} u \\ v \end{pmatrix} = \begin{pmatrix} 0 & -1 \\ 1 & -\bar{\gamma}_0 \end{pmatrix} \begin{pmatrix} u \\ v \end{pmatrix}.$$

We claim that:

Lemma 2   *Consider with $|\bar{\gamma}_0| < 2$,*

$$e^{t \begin{pmatrix} 0 & -1 \\ 1 & -\bar{\gamma}_0 \end{pmatrix}}.$$

For t small, it reads up to a multiplicative factor as

$$\frac{1}{\cos \varphi} \begin{pmatrix} \cos(\beta t + \varphi) & -\sin(\beta t) \\ \sin(\beta t) & \cos(\beta t - \varphi) \end{pmatrix}$$

with $|\beta| < 1, \beta = \cos\varphi.$

Proof. Since $\bar\gamma_0$ is small ($|\bar\gamma_0| < 2$), $\begin{pmatrix} 0 & -1 \\ 1 & -\bar\gamma_0 \end{pmatrix}$ reduces to a matrix of rotation $\begin{pmatrix} \alpha & -\beta \\ \beta & \alpha \end{pmatrix}$ with $\alpha^2 + \beta^2 = 1.$

$$\begin{pmatrix} 0 & -1 \\ 1 & -\bar\gamma_0 \end{pmatrix} = Q^{-1} \begin{pmatrix} \alpha & -\beta \\ \beta & \alpha \end{pmatrix} Q$$

with

$$Q = \begin{pmatrix} 1 & \alpha \\ 0 & \beta \end{pmatrix}, Q^{-1} = \frac{1}{\beta}\begin{pmatrix} \beta & -\alpha \\ 0 & 1 \end{pmatrix}, 2\alpha = -\bar\gamma_0. \qquad \square$$

Then,

$$e^{t\begin{pmatrix} 0 & -1 \\ 1 & -\bar\gamma_0 \end{pmatrix}} = Q^{-1} e^{t\begin{pmatrix} \alpha & -\beta \\ \beta & \alpha \end{pmatrix}} Q = e^{t\alpha} Q^{-1} e^{t\begin{pmatrix} 0 & -\beta \\ \beta & 0 \end{pmatrix}} Q$$

$$= e^{t\alpha} Q^{-1} \begin{pmatrix} \cos\beta t & -\sin\beta t \\ \sin\beta t & \cos\beta t \end{pmatrix} Q$$

$$= \frac{e^{t\alpha}}{\beta} \begin{pmatrix} \beta & -\alpha \\ 0 & 1 \end{pmatrix} \begin{pmatrix} \cos\beta t & \alpha\cos\beta t - \beta\sin\beta t \\ \sin\beta t & \alpha\sin\beta t + \beta\cos\beta t \end{pmatrix}$$

$$= \frac{e^{t\alpha}}{\beta} \begin{pmatrix} \beta\cos\beta t - \alpha\sin\beta t & -\sin\beta t \\ \sin\beta t & \alpha\sin\beta t + \beta\cos\beta t \end{pmatrix}.$$

Set $\beta = \cos\varphi, \alpha = \sin\varphi.$ We find:

$$e^{t\begin{pmatrix} 0 & -1 \\ 1 & -\bar\gamma_0 \end{pmatrix}} = \frac{e^{t\alpha}}{\beta} \begin{pmatrix} \cos(\beta t + \varphi) & -\sin\beta t \\ \sin\beta t & \cos(\beta t - \varphi) \end{pmatrix}.$$

Observe that the multiplicative factor tends to 1 as $\bar\gamma_0$ tends to zero. $\square$

Let now A be an arbitrary 2 by 2 matrix close to $\begin{pmatrix} 1 & 0 \\ 0 & 1 \end{pmatrix}$ which reads

$$A = \begin{pmatrix} a & -c^2 \\ c^2 & b \end{pmatrix},$$

has complex eigenvalues and determinant equal to 1.

Lemma 3 *There exists then $\beta_1 = \cos\varphi_1 > 0$ such that*

$$A = \frac{1}{\cos\varphi_1} \begin{pmatrix} \cos(\beta_1 t_1 + \varphi_1) & -\sin\beta_1 t_1 \\ \sin\beta_1 t_1 & \cos(\beta_1 t_1 - \varphi_1) \end{pmatrix} = e^{-t_1\sin\varphi_1} e^{t_1 \begin{pmatrix} 0 & -1 \\ 1 & 2\sin\varphi_1 \end{pmatrix}}.$$

Proof. We identify

$$a\cos\varphi_1 = \cos(\beta_1 t_1 + \varphi_1) \quad b\cos\varphi_1 = \cos(\beta_1 t_1 - \varphi_1).$$

$\square$

Thus,

$$\cos(\beta_1 t_1) = \frac{a+b}{2}$$

$$\sin(\beta_1 t_1) = \frac{(b-a)}{2\tan\varphi_1}.$$

The first equation can be solved since A has complex eigenvalues and $|tr\,A| < 2$.

The second equation yields then:

$$\tan\varphi_1 = \frac{b-a}{2\sqrt{1 - \frac{(a+b)^2}{4}}} = \frac{b-a}{\sqrt{4 - (a+b)^2}}.$$

We also have

$$c^2 = \frac{\sin\beta_1 t_1}{\cos\varphi_1}$$

i.e.

$$c^2 \cos\varphi_1 = \frac{(b-a)\cos\varphi_1}{2\sin\varphi_1}$$

$$\frac{b-a}{2} = c^2 \sin\varphi_1.$$

Since $\det A = ab + c^4 = 1$, c^2 is equal to $\sqrt{1 - ab}$.

Thus,

$$\sin\varphi_1 = \frac{b-a}{2\sqrt{1-ab}}, \quad \cos\varphi_1 = \sqrt{1 - \frac{(b-a)^2}{4(1-ab)}}$$

$$\tan\varphi_1 = \frac{b-a}{\sqrt{4(1-ab) - (b-a)^2}} = \frac{b-a}{\sqrt{4 - (a+b)^2}}.$$

The compatibility follows.

$\square$

Observation 1. $1 - ab > \frac{(b+a)^2}{4} - ab = \frac{(b-a)^2}{4}$.

Observation 2. $\beta_1 = \cos\varphi_1$ can tend to zero here. Then since $\beta_1 t_1$ can be chosen close to zero ($\cos\beta_1 t_1 = \frac{a+b}{2}$), $\sin\beta_1 t_1$ is of the order of $\beta_1 t_1 = (\cos\varphi_1)t_1$. Thus,

$$(\cos\varphi_1)t_1 \sim \frac{b-a}{2}\,\frac{\cos\varphi_1}{\sin\varphi_1}.$$

i.e. $t_1 \sim \frac{b-a}{2\sin\varphi_1} = \sqrt{1-ab}$. a and b have both to tend to 1 so that t_1 tends to zero. On the other hand, coming back to $e^{\,t_1\begin{pmatrix} 0 & -1 \\ 1 & -\bar\gamma_0 \end{pmatrix}}$, we have

$$2\alpha_1 = -\bar\gamma_0$$

and $|\alpha_1|$ tends to 1 so that $|\bar\gamma_0|$ is at most 2.

Observation 3. t_1 is positive and tends to zero as c^2 tends to zero. (This follows readily if β_1 does not tend to zero. If β_1 tends to zero, the claim follows from Observation 2).

Next, we prove:

Lemma 4 Let $A = \begin{pmatrix} a & -\gamma \\ \delta & b \end{pmatrix}$, with $a+b < 2, ab < 1, a, b$ close to 1, $0 < \delta < \gamma, ab+\gamma\delta = 1, \delta(\gamma-\delta)$ small enough. Then, there exists $\tilde a, \tilde b, c$ with $\tilde a + \tilde b < 2, \tilde a, \tilde b$ close to $1, \tilde a\tilde b < 1, \tilde a\tilde b + c^4 = 1$ and $\alpha, \beta, \beta_1, \alpha < 1, \alpha$ close to $1, \alpha^2 + \beta^2\beta_1^2 = 1$ such that

$$\begin{pmatrix} a & -\gamma \\ \delta & b \end{pmatrix} = \begin{pmatrix} \tilde a & -c^2 \\ c^2 & \tilde b \end{pmatrix}\begin{pmatrix} \alpha & -\beta_1^2 \\ \beta^2 & \alpha \end{pmatrix}.$$

Proof. We have

$$a = \tilde a\alpha - c^2\beta^2 \quad b = \tilde b\alpha - c^2\beta_1^2$$
$$\alpha^2 + \beta^2\beta_1^2 = 1.$$
$$\delta = c^2\alpha + \tilde b\beta^2 \quad \gamma = \tilde a\beta_1^2 + c^2\alpha$$

$\square$

We may replace the condition $\delta = c^2\alpha + \tilde b\beta^2$ with $\tilde a\tilde b + c^4 = 1$. We then have

$$\tilde b = \frac{b+c^2\beta_1^2}{\alpha}, \qquad \tilde a\beta_1^2 = \gamma - c^2\alpha$$
$$\alpha^2 + \beta^2\beta_1^2 = 1$$
$$a = \tilde a\alpha - c^2\frac{(1-\alpha^2)}{\beta_1^2} \quad \tilde a\tilde b + c^4 = 1$$

which rereads

$$\beta_1^2 = \frac{\alpha(\gamma - c^2\alpha) - c^2(1 - \alpha^2)}{a} = \frac{\alpha\gamma - c^2}{a}$$

$$\tilde{b} = \frac{b + c^2\beta_1^2}{\alpha}$$

$$\tilde{a} = \frac{a(\gamma - c^2\alpha)}{\alpha\gamma - c^2} \qquad \alpha^2 + \beta^2\beta_1^2 = 1$$

$$(b + c^2\beta_1^2)(\gamma - c^2\alpha) = \alpha\beta_1^2(1 - c^4).$$

The last equation yields

$$\beta_1^2(\alpha - \gamma c^2) = b(\gamma - c^2\alpha). \tag{***}$$

Combining this equation with the first equation, we find

$$(\alpha - \gamma c^2)(\alpha\gamma - c^2) = ab(\gamma - c^2\alpha). \tag{****}$$

This equation ties α and c.

If we can solve (****) with c small, α close to $1(\alpha^2 < 1)$, then (***) gives us $\beta_1^2 > 0$ (provided $c^2 < \gamma$). We can then find $\tilde{a}$ and $\tilde{b}$. As α tends to 1, with $\gamma - c^2$ bounded away from zero, $\frac{\tilde{a}}{a}$ tends to 1. If γc^2 is small enough, $\tilde{b}$ will be very close to b. β can be computed from

$$\beta^2 = \frac{1 - \alpha^2}{\beta_1^2}$$

(****) rereads

$$\gamma c^4 - c^2(\alpha\gamma^2 + \alpha - ab\alpha) + \gamma(\alpha^2 - ab) = 0 \quad \text{i.e.}$$
$$c^4 - c^2\alpha(\gamma + \delta) + (\alpha^2 - 1 + \gamma\delta) = 0.$$

The discriminant is

$$(\alpha\gamma^2 + \alpha - ab\alpha)^2 - 4\gamma^2(\alpha^2 - ab) \geq \alpha^2((\gamma^2 + 1 - ab)^2 - 4\gamma^2(1 - ab)).$$

Observe that

$$(\gamma^2 + 1 - ab)^2 - 4\gamma^2(1 - ab) = (\gamma^2 + \gamma\delta)^2 - 4\gamma^2 \cdot \gamma\delta = (\gamma^2 - \gamma\delta)^2 = \gamma^2(\gamma - \delta)^2 > 0.$$

Hence (****) has two positive roots as $|\alpha|$ tends to $1(ab < 1)$. For $\alpha = 1$, the equation becomes

$$c^4 - c^2(\gamma + \delta) + \delta\gamma = 0.$$

The two solutions are

$$c^2 = \gamma \text{ and } c^2 = \delta.$$

Assume $\delta < \gamma$, we choose $c^2 = \delta$ and derive from (****) that

$$\beta_1^2 = \frac{b(\gamma - \delta)}{1 - \gamma\delta} = \frac{\gamma - \delta}{a} > 0.$$

Furthermore, since $\beta^2 = 0$,

$$\tilde{a} = a, \tilde{b} = b + \frac{\delta(\gamma - \delta)}{a}.$$

Thus, if $\delta(\gamma - \delta)$ is small enough, $\tilde{a}$ and $\tilde{b}$ are close to a, b and satisfy our requirements.

If $\alpha < 1$ is very close to 1, all those arguments proceed with a solution c^2 as close as we may wish to δ and $\beta^2 = \frac{1 - \alpha^2}{\beta_1^2}$ close to zero.

We thus need $\gamma > \delta$ in order to solve our equation. $\gamma - \delta$ can be as close as we wish to zero, α will be taken closer to 1.

We consider now the case $\delta > \gamma$. We observe that

$$\begin{pmatrix} a & -\gamma \\ \delta & b \end{pmatrix} = \begin{pmatrix} a & \delta \\ -\gamma & b \end{pmatrix}^t = \begin{pmatrix} a & \bar{\gamma} \\ -\bar{\delta} & b \end{pmatrix}^t \text{ with } \bar{\delta} < \bar{\gamma}.$$

We solve

$$\begin{pmatrix} a & \bar{\gamma} \\ -\bar{\delta} & b \end{pmatrix} = \begin{pmatrix} \tilde{a} & c^2 \\ -c^2 & \tilde{b} \end{pmatrix} \begin{pmatrix} \alpha & \beta_1^2 \\ -\beta^2 & \alpha \end{pmatrix}.$$

This yields

$$a = \tilde{a}\alpha - c^2\beta^2, b = \tilde{b}\alpha - \beta_1^2 c^2, \bar{\gamma} = \tilde{a}\beta_1^2 + c^2\alpha, \bar{\delta} = c^2\alpha + \tilde{b}\beta^2$$

exactly as above, with $\bar{\delta} < \bar{\gamma}$.

Thus, this equation may be solved and consequently, we may write

Lemma 5

$$\begin{pmatrix} a & -\gamma \\ \delta & b \end{pmatrix} = \begin{pmatrix} \alpha & -\beta^2 \\ \beta_1^2 & \alpha \end{pmatrix} \begin{pmatrix} \tilde{a} & -c^2 \\ c^2 & \tilde{b} \end{pmatrix} \text{ for } \delta > \gamma.$$

We want to show how to generate the matrices

$$\begin{pmatrix} \alpha & -\beta_1^2 \\ \beta^2 & \alpha \end{pmatrix} \quad \alpha^2 + \beta^2\beta_1^2 = 1$$

with β_1 and β close to zero, $\beta_1^2 > \beta^2$ or vice-versa.

We compute the product of two matrices A, A_1

$$A = \begin{pmatrix} a & -\sqrt{1-ab} \\ \sqrt{1-ab} & b \end{pmatrix} \text{ and } A_1 = \begin{pmatrix} a_1 & -\sqrt{1-a_1 b_1} \\ \sqrt{1-a_1 b_1} & b_1 \end{pmatrix}.$$

We find

$$AA_1 = \begin{pmatrix} aa_1 - \sqrt{1-ab}\sqrt{1-a_1 b_1} & -a\sqrt{1-a_1 b_1} - b_1\sqrt{1-ab} \\ a_1\sqrt{1-ab} + b\sqrt{1-a_1 b_1} & bb_1 - \sqrt{1-ab}\sqrt{1-a_1 b_1} \end{pmatrix}.$$

If $a = b_1$ and $b = a_1$, we find

$$AA_1 = \begin{pmatrix} 2ab - 1 & -2a\sqrt{1-ab} \\ 2b\sqrt{1-ab} & 2ab - 1 \end{pmatrix}.$$

Clearly if $aa_1 = bb_1$ then

$$AA_1 = \begin{pmatrix} \alpha & -\beta_1^2 \\ \beta^2 & \alpha \end{pmatrix}$$

with $\alpha^2 + \beta^2 \beta_1^2 = 1$.

Furthermore, if a is close to b_1 and b is close to a_1, then α is close to $2ab - 1$, which is close to 1 if ab is close to 1. We thus need to worry about

$$\beta_1^2 > \beta^2 \text{ or } \beta^2 > \beta_1^2.$$

We then observe that, since $aa_1 = bb_1$,

$$\beta_1^2 = a\sqrt{1-a_1 b_1} + b_1\sqrt{1-ab} = a\sqrt{1-a_1 b_1} + \frac{aa_1}{b}\sqrt{1-ab}$$

$$= \frac{a}{b}\left(a_1\sqrt{1-ab} + b\sqrt{1-a_1 b_1}\right) = \frac{a}{b}\beta^2.$$

Taking $a > b$ or $b > a$, we achieve the two occurrences.

Consider now a matrix

$$\bar{A} = \begin{pmatrix} \bar{a} & -\bar{c}^2 \\ \bar{c}^2 & \bar{b} \end{pmatrix} \text{ with } \begin{array}{c} \bar{a}\bar{b} + \bar{c}^4 = 1 \\ 0 < \bar{a} + \bar{b} < 2 \end{array}.$$

$\bar{a}, \bar{b}$ close to 1, fixed.

Consider a small angle $t_2 > 0$ and the product

$$A_{t_2} = \begin{pmatrix} \bar{a} & -\bar{c}^2 \\ \bar{c}^2 & \bar{b} \end{pmatrix}\begin{pmatrix} \cos t_2 & \sin t_2 \\ -\sin t_2 & \cos t_2 \end{pmatrix} = \begin{pmatrix} \bar{a}\cos t_2 + \bar{c}^2 \sin t_2 & \bar{a}\sin t_2 - \bar{c}^2 \cos t_2 \\ \bar{c}^2 \cos t_2 - \bar{b}\sin t_2 & \bar{b}\cos t_2 + \bar{c}^2 \sin t_2 \end{pmatrix}$$

As t_2 tends to zero, this matrix gets closer and closer to $\bar{A}$ and assumes the form

$$\begin{pmatrix} a_0 & -\gamma \\ \delta & b_0 \end{pmatrix}.$$

Assume that

$$\bar{b} > \bar{a} \text{ i.e } \delta < \gamma. \tag{2.1}$$

The other case is similar.

We apply then Lemma 4 and write

$$A_{t_2} = \begin{pmatrix} \tilde{a} & -c^2 \\ c^2 & \tilde{b} \end{pmatrix} \begin{pmatrix} \alpha & -\beta_1^2 \\ \beta^2 & \alpha \end{pmatrix}$$

c solves

$$c^4 - c^2\alpha(\gamma + \delta) + (\alpha^2 - 1 + \gamma\delta) = 0.$$

Assuming that

$$1 - \gamma\delta < \alpha^2 < 1, \gamma + \delta = 2\bar{c}^2 \cos t_2 - (\bar{a} + \bar{b}) \sin t_2 > 0. \tag{2.2}$$

We find

$$c^2 = \frac{\alpha(\gamma + \delta) - \sqrt{\alpha^2(\gamma + \delta)^2 - 4(\alpha^2 - 1 + \gamma\delta)}}{2}$$

$$= \frac{\alpha(\gamma + \delta) - \sqrt{\alpha^2(\gamma - \delta)^2 + 4(1 - \alpha^2)(1 - \gamma\delta)}}{2}.$$

We then have

$$a_0\beta_1^2 = \alpha\gamma - c^2 = \frac{\alpha(\gamma - \delta) + \sqrt{\alpha^2(\gamma - \delta)^2 + 4(1 - \alpha^2)(1 - \gamma\delta)}}{2}. \tag{2.3}$$

The positivity is warranted by $\bar{b} > \bar{a}$.

Observe that

$$|\delta(\gamma - \delta)| \leq |\bar{a} - \bar{b}| |\sin t_2|.$$

For fixed $\bar{a}$ and $\bar{b}$, this can be made as small as we wish by taking t_2 to be small. Observe also that, as we reduce $\begin{pmatrix} a & -\gamma \\ \delta & b \end{pmatrix}$ into

$\begin{pmatrix} \tilde{a} & -c^2 \\ c^2 & \tilde{b} \end{pmatrix} \begin{pmatrix} \alpha & -\beta_1^2 \\ \beta^2 & \alpha \end{pmatrix}, \tilde{a} = a\frac{(\gamma - c^2\alpha)}{\alpha\gamma - c^2} = \frac{a}{\alpha} + \frac{ac^2}{\alpha}\frac{1 - \alpha^2}{\alpha\gamma - c^2} = \frac{a}{\alpha} + \frac{ac^2}{\alpha}(1 - \alpha^2)O\left(\frac{1}{\sqrt{1-\alpha^2}}\right) = \frac{a}{\alpha} + O(\sqrt{1 - \alpha^2})$. Thus, if we can choose α very close to 1, for $\bar{a}$ and $\bar{b}$ fixed with $\bar{a}\bar{b} < 1, \bar{a} + \bar{b} < 2$, then $\tilde{a}\tilde{b} < 1, \tilde{a} + \tilde{b} < 2(\tilde{a}, \tilde{b}$ very close to $\bar{a}, \bar{b})$. Indeed, as α tends to 1, $\tilde{a}$ tends to a, c^2 tends to δ and $c^2\beta_1^2$ is $O(\delta(\gamma - \delta) + \delta\sqrt{1 - \alpha^2})$. This tends to zero and α tends to 1.

We then have

Lemma 6　　*If $\bar{a}, \bar{b}$ are chosen appropriately, there exists $t_2 > 0$ small such that the matrix* $\begin{pmatrix} \alpha & -\beta_1^2 \\ \beta^2 & \alpha \end{pmatrix}$ *can be written as*

$$\begin{pmatrix} \frac{a}{\sqrt{1-ab}} & -\sqrt{1-ab} \\ \sqrt{1-ab} & b \end{pmatrix} \begin{pmatrix} \frac{b}{\sqrt{1-ab}} & -\sqrt{1-ab} \\ \sqrt{1-ab} & a \end{pmatrix} \quad \text{with } 0 < a + b < 2.$$

Observation.　In the reduction of Lemmas 4, 5, α is a free parameter close enough to 1. β_1^2 depends on $\alpha, \bar{a}, \bar{b}, t_2$. Lemma 6 states that we can find $\bar{a}, \bar{b}$ **and** α, also a and b so that the equation $A_{t_2} = \begin{pmatrix} \tilde{a} & -c^2 \\ c^2 & \tilde{b} \end{pmatrix} \begin{pmatrix} \frac{a}{\sqrt{1-ab}} & -\sqrt{1-ab} \\ \sqrt{1-ab} & b \end{pmatrix} \begin{pmatrix} \frac{b}{\sqrt{1-ab}} & -\sqrt{1-ab} \\ \sqrt{1-ab} & a \end{pmatrix}$ is solvable in t_2.

Proof.　We then should have

$$\alpha = 2ab - 1 \tag{2.4}$$

$$\beta_1^2 = 2a\sqrt{1 - ab} \tag{2.5}$$

(2.2) becomes

$$4ab(1 - ab) < \gamma\delta, ab < 1, \gamma + \delta > 0 \tag{2.6}$$

and (2.3) becomes

$$4aa_0\sqrt{1 - ab} = (2ab-1)(\gamma-\delta)+\sqrt{(2ab - 1)^2(\gamma - \delta)^2 + 16(1 - \gamma\delta)ab(1 - ab)}. \tag{2.7}$$

Assume that

$$4aa_0\sqrt{1 - ab} > (2ab - 1)(\gamma - \delta). \tag{2.8}$$

Then, (2.7) yields:

$$16a^2a_0^2\sqrt{1 - ab} + \frac{(2ab - 1)^2(\gamma - \delta)^2}{\sqrt{1 - ab}} - 8aa_0(2ab - 1)(\gamma - \delta) \tag{2.9}$$

$$= \frac{(2ab - 1)^2(\gamma - \delta)^2}{\sqrt{1 - ab}} + 16(1 - \gamma\delta)ab\sqrt{1 - ab}$$

i.e.

$$2aa_0^2\sqrt{1 - ab} - a_0(2ab - 1)(\gamma - \delta) = 2(1 - \gamma\delta)b\sqrt{1 - ab}. \tag{2.10}$$

Thus,

$$2\sqrt{1-ab}(aa_0^2 - b(1-\gamma\delta)) = a_0(2ab-1)(\gamma-\delta). \tag{2.11}$$

Observe that

$$a_0 = \bar{a}\cos t_2 + \bar{c}^2\sin t_2,\ \gamma = \bar{c}^2\cos t_2 - \bar{a}\sin t_2,\ \delta = \bar{c}^2\cos t_2 - \bar{b}\sin t_2 \tag{2.12}$$

t_2 should be positive and small.

Replacing in (2.11) and using $\bar{a}\bar{b} = 1 - \bar{c}^4$, we derive

$$2\sqrt{1-ab}\,\bar{a}(a\bar{a} - b\bar{b}) = (\bar{a}\cos t_2 + \bar{c}^2\sin t_2)(2ab-1)(\bar{b}-\bar{a})\sin t_2 \tag{2.13}$$

$$+ 2\sin t_2\sqrt{1-ab}\left(\sin t_2(\bar{c}^4(b-a) + \bar{a}(a\bar{a} - b\bar{b}))\right)$$

$$+ \cos t_2\bar{c}^2(a(b-a) + b\bar{b} - a\bar{a})\,.$$

Observe that

$$b - a = \frac{a}{\bar{b}}(\bar{a} - \bar{b}) + \frac{b\bar{b} - a\bar{a}}{\bar{b}}. \tag{2.14}$$

Thus,

$$2\sqrt{1-ab}\,\bar{a}(a\bar{a} - b\bar{b}) = (\bar{b} - \bar{a})\sin t_2\left((\bar{a}\cos t_2 + \bar{c}^2\sin t_2)(2ab-1)\right) \tag{2.15}$$

$$-2\frac{a}{\bar{b}}\sqrt{1-ab}\,\bar{c}^4\sin t_2 - 2\frac{a\bar{a}}{\bar{b}}\sqrt{1-ab}\,\bar{c}^2\cos t_2\right) + 2(a\bar{a} - b\bar{b})\sqrt{1-ab}\times\sin t_2$$

$$\times\left(\bar{a}\sin t_2 - \sin t_2\frac{\bar{c}^4}{\bar{b}} - \cos t_2\bar{c}^2\left(\frac{\bar{a}}{\bar{b}} + 1\right)\right).$$

Assume now that

$$\bar{b} > \bar{a},\ a\bar{a} > b\bar{b},\ a\bar{a} - b\bar{b} = O((\bar{b}-a)^2). \tag{2.16}$$

$$ab < 1,\ a + b < 2,\ a, b \text{ close to } 1. \tag{2.17}$$

Since $1 - ab$ is small and $2ab - 1$ is close to 1 while $a\bar{a} - b\bar{b} > 0$, $a\bar{a} - b\bar{b} = O((\bar{b}-\bar{a})^2)$, we can solve (2.15) by implicit function theorem and find $t_2 > 0$ small. Indeed (2.15) rewrites under (2.16):

$$2\sqrt{1-ab}\,\bar{a}\frac{(a\bar{a} - b\bar{b})}{\bar{b} - \bar{a}} = \sin t_2(\bar{a}(2ab-1) + o(1)). \tag{2.18}$$

We need therefore to fulfill (2.16) and (2.17), also (2.8).

Consider

$$1 > \bar{b} > \bar{a}. \tag{2.19}$$

Let $\bar{c}$ be such that $\bar{c}^4 = 1 - \bar{a}\bar{b}$.

Take

$$a = \frac{b\bar{b}}{\bar{a}} + \varepsilon, \varepsilon > 0 \text{ tending to zero}, b < 1. \tag{2.20}$$

For ε small enough, if $\frac{\bar{b}}{\bar{a}}$ is close enough to 1 (in function of b)

$$ab = \frac{b^2\bar{b}}{\bar{a}} + \varepsilon b < 1. \tag{2.21}$$

Also

$$a + b = \frac{b\bar{b}}{\bar{a}} + \varepsilon + b = b\left(\frac{\bar{b}}{\bar{a}} + 1\right) + \varepsilon < 2. \tag{2.22}$$

We then need

$$a\bar{a} - b\bar{b} = O((\bar{b} - \bar{a})^2) \tag{2.23}$$

i.e.

$$\varepsilon = O((\bar{b} - \bar{a})^2) = O((\frac{\bar{b}}{\bar{a}} - 1)^2)$$

and this is easy to satisfy.

The proximity of a and b to 1 depends only on the proximity of b and $\frac{\bar{b}}{\bar{a}}$ to 1.

Finally, we need (2.6) i.e.

$$4ab(1 - ab) < (\bar{c}^2 \cos t_2 - \bar{a} \sin t_2)(\bar{c}^2 \cos t_2 - \bar{b} \sin t_2) \tag{2.24}$$

$$2\bar{c}^2 \cos t_2 - (\bar{b} + \bar{a}) \sin t_2 > 0 \tag{2.25}$$

(2.25) follows from the fact that t_2 is small.

For (2.24), we observe that as $\frac{\bar{b}}{\bar{a}}$ tends to 1 and $1 - ab$ to zero, $\bar{a}\bar{b}$ can be kept away from 1 so that $\bar{c}^2$ is far from zero and t_2 tends to zero. (2.24) follows.

We also assumed (2.8) i.e.

$$(2ab - 1)(\bar{b} - \bar{a}) \sin t_2 < 4a(\bar{a} \cos t_2 + \bar{c}^2 \sin t_2)\sqrt{1 - ab}. \tag{2.25'}$$

Using (2.18), this rereads:

$$2\sqrt{1 - ab}\,(a\bar{a} - b\bar{b})(1 + o(1)) < 4a\bar{a}\sqrt{1 - ab}(1 + o(1)) \tag{2.26}$$

which follows readily. $\square$

We now build γ:

We pick up a small interval J and we consider the v-transport over J which is given in Span $\left(\frac{\tilde{\partial}}{\partial x}, \frac{\tilde{\partial}}{\partial y}\right)$ by

$$e^{\bar{t}\begin{pmatrix} 0 & -1 \\ 1 & \frac{-2\bar{\gamma}}{\sqrt{1+\bar{\gamma}^2}} \end{pmatrix}}.$$

Assume $-\bar{\gamma} > 0$ for example; $\bar{\gamma}$ will be assumed — there is, after a simple argument, no restriction in this assumption — to be as small as we please. Use Lemma 2 to read this matrix up to a multiplicative factor under the form

$$\begin{pmatrix} \bar{a} & -\bar{c}^2 \\ \bar{c}^2 & \bar{b} \end{pmatrix} \quad with \, \bar{a} + \bar{b} < 2, \bar{a}, \bar{b} \text{ close to } 1, \bar{a}\bar{b} + \bar{c}^4 = 1.$$

Choose $\beta = \cos\varphi > 0$. Since $-\bar{\gamma} > 0, \alpha = \sin\varphi$ is positive and since $\bar{t}$ is small,

$$\bar{b} = \frac{\cos(\beta\bar{t} - \varphi)}{\cos\varphi} > \frac{\cos(\beta\bar{t} + \varphi)}{\cos\varphi} = \bar{a}.$$

Also

$$\bar{a}\bar{b} = \frac{1}{\cos^2\varphi}(\cos^2\beta\bar{t}\cos^2\varphi - \sin^2\beta\bar{t}\sin^2\varphi) < 1 \tag{2.27}$$

$$\frac{\bar{b}}{\bar{a}} \text{ tends to 1 as } \bar{\gamma} \text{ tends to zero since } \sin\varphi = \frac{|\bar{\gamma}|}{2}.$$

Pick up $b < 1$. Set $a = \frac{b\bar{b}}{\bar{a}} + \left(\frac{\bar{b}}{\bar{a}} - 1\right)^3$. Adjust $|\bar{\gamma}|$ so small that $\frac{\bar{b}}{\bar{a}} - 1$ is as small as we please with $\bar{b}, \bar{a}$ away from 1 ($\bar{t}$ is given, small, positive). $a, b, \bar{a}\bar{b}$ satisfy (2.16) and (2.17). (2.6) and (2.8) follow if t_2 is small positive, see (2.24), (2.25), (2.25'), (2.26).

Using Lemma 6, we rewrite

$$\begin{pmatrix} \bar{a} & -\bar{c}^2 \\ \bar{c}^2 & \bar{b} \end{pmatrix} =$$

$$\begin{pmatrix} \tilde{a} & -c^2 \\ c^2 & \tilde{b} \end{pmatrix} \begin{pmatrix} a & -\sqrt{1-ab} \\ \sqrt{1-ab} & b \end{pmatrix} \begin{pmatrix} b & -\sqrt{1-ab} \\ \sqrt{1-ab} & a \end{pmatrix} \begin{pmatrix} \cos t_2 & -\sin t_2 \\ \sin t_2 & \cos t_2 \end{pmatrix}$$

with t_2 positive small, $0 < \tilde{a} + \tilde{b} < 2, 0 < a + b < 2, \tilde{a}\tilde{b} < 1, ab < 1$.

Using then Lemma 3, we may write

$$\begin{pmatrix} \bar{a} & -\bar{c}^2 \\ \bar{c}^2 & \bar{b} \end{pmatrix} = \theta e^{t_5 \begin{pmatrix} 0 & -1 \\ 1 & 2\sin\varphi_1 \end{pmatrix}} e^{t_4 \begin{pmatrix} 0 & -1 \\ 1 & 2\sin\varphi_2 \end{pmatrix}} e^{t_3 \begin{pmatrix} 0 & -1 \\ 1 & 2\sin\varphi_3 \end{pmatrix}} e^{t_2 \begin{pmatrix} 0 & -1 \\ 1 & 0 \end{pmatrix}}$$

with $\theta, t_2, t_3, t_4, t_5 > 0$.

All multiplicative factors tend to 1 as t_1 tends to zero. $\int |\gamma'|$ is clearly bounded. $\qquad\square$

2.2.4   *Modification of α into α_N*

We focus on the interval I_0 and we pick up $\varepsilon > 0$ and N large, with $\varepsilon = o\left(\frac{|I_0|}{N}\right)$. We pick up a real

$$\frac{|I_0|}{2\pi(N+1)} < \theta_N < \frac{|I_0|}{2\pi N}$$

in a way which will become clear in a moment.

We build a fucntion ℓ on I_0 as follows:

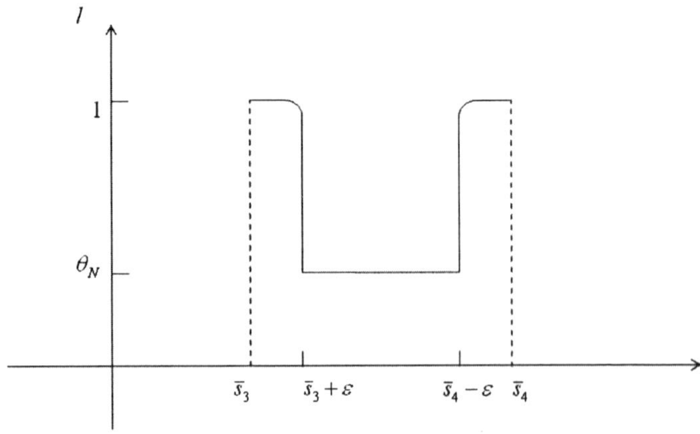

θ_N is chosen so that $\int_{\bar{s}_3}^{\bar{s}_4} \frac{ds_1}{\ell(s_1)} = 2\pi N + |I_0|$.

Setting

$$s_2 = \int_{\bar{s}_3}^{s_1} \frac{d\tau}{\ell(\tau)} \quad \text{i.e.}$$

$$ds_2 = \frac{ds_1}{\ell(s_1)}, \quad \frac{\partial}{\partial s_2} = \ell(s_1)\frac{\partial}{\partial s_1}$$

we consider the differential equation

$$\ell(s_1)\frac{\partial}{\partial s_1}\left(\ell(s_1)\frac{\partial}{\partial s_1}\right)u + u = 0 \quad \text{on } I_0$$

which rereads

$$\frac{\partial^2 u}{\partial s_2^2} + u = 0 \text{ on an interval of length } 2\pi N + |I_0|$$

starting at $\bar{s}_3$ i.e. on $[\bar{s}_3, \bar{s}_4 + 2\pi N]$.

Thus, the solutions of

$$[v, [v, \xi]] = -\xi \text{ on } I_0$$

and of

$$[\ell v, [\ell v, \xi]] = -\xi \text{ on } I_0$$

with the same initial data $\xi, [v, \xi]$ match at $\bar{s}_4$.

2.2.5 *Computation of ξ_N*

We compute ξ_N.

ξ_N satisfies the differential equation:

$$[\varphi v, [\varphi v, \xi_N]] = -\xi_N \quad \varphi = \ell.$$

We know that $\xi_N(\bar{s}_3) = \bar{\xi}$.

We need to compute $[-\varphi v, \xi_N](\bar{s}_3) = [-v, \xi_N](\bar{s}_3)$.

Lemma 7 $-[v, \xi_N](\bar{s}_3) = -2\dfrac{(x - \bar{\gamma}y)(0)}{\sqrt{1 + \bar{\gamma}^2(x^2 + y^2)(0)}}v + \sqrt{1 + \bar{\gamma}^2}\dfrac{\partial}{\partial y}.$

Proof. ξ satisfies

$$[-v, [-v, \xi]] = \xi + \gamma[\xi, v] - (\xi \cdot \gamma)v$$

and γ incurs a jump at $\bar{s}_3$ from $\dfrac{-2\bar{\gamma}}{\sqrt{1 + \bar{\gamma}^2}} = -2\bar{\gamma}_1$ to 0. γ is a function of s, which is a function of τ, the time along $-v$. We are here taking $s = x^2 + y^2$

and thinking of $\gamma' ds(\xi)$ as $d\gamma(\xi)$. We have

$$\frac{d\gamma}{d\tau} = +\frac{d\gamma}{ds} \cdot \frac{ds}{d\tau}$$

so that

$$\gamma' = +\frac{d\gamma}{d\tau} \cdot \frac{d\tau}{ds} = \frac{d\gamma}{d\tau} \times \frac{1}{2\bar{\gamma}(x^2 + y^2)}.$$

Thus,

$$(\bar{s}_3^+) - [-v, \xi](\bar{s}_3^-) = -\frac{2\bar{\gamma}}{\sqrt{1 + \bar{\gamma}^2}} \times \frac{1}{2\bar{\gamma}(x^2 + y^2)} ds(\xi)(\bar{s}_3)v$$

$$= -\frac{2(x - \bar{\gamma}y)(0)}{\sqrt{1 + \bar{\gamma}^2}(x^2 + y^2)(0)} \cdot v.$$

$\square$

On the other hand

$$[-v, \xi](\bar{s}_3^-) = \left[-20\frac{\partial}{\partial\theta} + (y + \bar{\gamma}x)\frac{\partial}{\partial x} - (x - \bar{\gamma}y)\frac{\partial}{\partial y}, \frac{\partial}{\partial x} - \bar{\gamma}\frac{\partial}{\partial y} \right]$$

$$\cdot \frac{1}{\sqrt{1 + \bar{\gamma}^2}} = \sqrt{1 + \bar{\gamma}^2}\frac{\partial}{\partial y}.$$

Thus, ξ_N satisfies

$$\begin{cases} [-\varphi v, [-\varphi v, \xi_N]] = -\xi_N \\ \xi_N(\bar{s}_3) = \bar{\xi} \\ [-\varphi v, \xi_N](\bar{s}_3) = \sqrt{1 + \bar{\gamma}^2}\frac{\partial}{\partial y} - 2\frac{(x - \bar{\gamma}y)(0)}{\sqrt{1 + \bar{\gamma}^2}(x^2 + y^2)(0)}v. \end{cases} \tag{2.28}$$

Let

$$e_1 = 20\frac{\partial}{\partial\theta}, e_2 = y\frac{\partial}{\partial x} - x\frac{\partial}{\partial y}, e_3 = \varphi v$$

φ is a function of $s = x^2 + y^2$. Thus,

$$ds(e_1) = ds(e_2) = 0$$

and

$$[e_1, v] = [e_2, v] = [e_1, \varphi v] = [e_2, \varphi v] = 0$$

since

$$\left[x\frac{\partial}{\partial x} + y\frac{\partial}{\partial y}, y\frac{\partial}{\partial x} - x\frac{\partial}{\partial y} \right] = 0.$$

Observe that

$$x\frac{\partial}{\partial x} + y\frac{\partial}{\partial y} = -\frac{v - 20\frac{\partial}{\partial\theta} + (y\frac{\partial}{\partial x} - x\frac{\partial}{\partial y})}{\bar\gamma}\cdot\sqrt{1+\bar\gamma^2} = \frac{e_1 - e_2 - \varphi v}{\bar\gamma}\sqrt{1+\bar\gamma^2}.$$

Observe also that

$$\frac{\partial}{\partial x} = \frac{x(x\frac{\partial}{\partial x} + y\frac{\partial}{\partial y}) + y(\frac{\partial}{\partial x} - x\frac{\partial}{\partial y})}{x^2 + y^2} = \frac{x}{x^2 + y^2}\sqrt{1+\bar\gamma^2}\frac{e_1 - e_2 - \varphi v}{\bar\gamma}$$

$$+ \frac{y}{x^2 + y^2}e_2\frac{\partial}{\partial y} = \frac{y}{x^2 + y^2}\cdot\frac{e_1 - e_2 - \varphi v}{\bar\gamma}\cdot\sqrt{1+\bar\gamma^2} - \frac{xe_2}{x^2 + y^2}.$$

Thus,

$$\begin{cases} \bar\xi = \frac{\partial}{\partial x} - \bar\gamma\frac{\partial}{\partial y} = \frac{x - \bar\gamma y}{x^2+y^2}\sqrt{1+\bar\gamma^2}\frac{(e_1-e_2-\varphi v)}{\bar\gamma} + \frac{(y+\bar\gamma x)e_2}{x^2+y^2} \\[2mm] [-\varphi v, \xi_N](\bar s_3) = \sqrt{1+\bar\gamma^2}\frac{y}{x^2+y^2}\frac{(e_1-e_2-\varphi v)}{\bar\gamma} - \frac{\sqrt{1+\bar\gamma^2}xe_2}{x^2+y^2} - 2\frac{(x-\bar\gamma y)}{x^2+y^2}\frac{\varphi v}{\sqrt{1+\bar\gamma^2}} \end{cases}.$$

Here, x, y are taken at s_3 and we will label them $x(0), y(0)$.

We solve then (2.28). We derive:

Proposition 2 *Let s_2 be the time parameter along $-\varphi v$*

$$\xi_N = \left(\frac{(x - \bar\gamma y)(0)}{(x^2 + y^2)(0)}\frac{1}{\sqrt{1+\bar\gamma^2}}\left(\frac{e_1 - e_2 - \varphi v}{\bar\gamma} \right) + \frac{(y + \bar\gamma x)(0)}{(x^2 + y^2)(0)}e_2 \right)\cos_2$$

$$+ \left(\sqrt{1+\bar\gamma^2}\frac{y(0)}{(x^2 + y^2)(0)}\frac{e_1 - e_2 - \varphi v}{\bar\gamma} - \frac{\sqrt{1+\bar\gamma^2}x(0)e_2}{(x^2 + y^2)(0)} \right.$$

$$\left. - 2\frac{(x - \bar\gamma y)(0)}{(x^2 + y^2)(0)}\frac{\varphi v}{\sqrt{1+\bar\gamma^2}} \right)\sin s_2.$$

Corollary 2 *Given a fixed connection on M, $\xi_N \cdot \xi_N, \xi_N \cdot [\varphi v, \xi_N]$, $[\varphi v, \xi_N] \cdot \xi_N$ and $[\varphi v, \xi_N] \cdot [\varphi v, \xi_N]$ are bounded independently on M transversally to v.*

Proof. Observe that $ds_2(\varphi v) = -1$ and $ds_2(e_1) = ds_2(e_2) = 0$. The only unbounded terms could come from a derivative taken on φ. But φ is multiplied by v. Hence the claim. $\qquad\square$

Observe that

$$[\varphi v, e_1] = [\varphi v, e_2] = 0.$$

Thus, denoting γ_{s_2} the one-parameter group of $-\varphi v$

$$D\gamma_{s_2}(e_1) = e_1 \quad D\gamma_{s_2}(e_2) = e_2$$

$(x, y, \theta)(s_2)$ is derived from $(x, y, \theta)(0)$ through the use of γ_{s_2}.
We can use this fact and reread the differential equation

$$\overline{(x, y, \dot\theta)} = \xi_N(x, y, \theta).$$

Indeed, we then have

$$D\gamma_{s_2}\left(\overline{(x, y, \dot\theta)(0)}\right) - \dot{s}_2\varphi v = A\varphi v$$

$$+ \cos_2\left(\frac{(x - \bar\gamma y)(0)}{(x^2 + y^2)(0)} \times \frac{1}{\sqrt{1 + \bar\gamma^2}} D\gamma_{s_2}\left(\frac{e_1 - e_2}{\bar\gamma}\right) + \frac{(y + \bar\gamma x)(0)}{(x^2 + y^2)(0)} D\gamma_{s_2}(e_2)\right)$$

$$+ \left(\sqrt{1 + \bar\gamma^2}\frac{y}{x^2 + y^2}(0)D\gamma_{s_2}\left(\frac{e_1 - e_2}{\bar\gamma}\right) - \frac{\sqrt{1 + \bar\gamma^2}x(0)}{(x^2 + y^2)(0)}D\gamma_{s_2}(e_2)\right)\sin s_2.$$

Set

$$x(0) = \rho\cos\psi$$

$$y(0) = \rho\sin\psi.$$

Then,

$$e_2 = y(0)\frac{\partial}{\partial x} - x(0)\frac{\partial}{\partial y} = -\rho\frac{\partial}{\partial\psi}.$$

and

$$\overline{(x, y, \dot\theta)(0)} = \frac{\cos s_2}{\rho}\left(\frac{\cos\psi - \bar\gamma\sin\psi}{\bar\gamma\sqrt{1 + \bar\gamma^2}}\left(20\frac{\partial}{\partial\theta} + \rho\frac{\partial}{\partial\psi}\right) - (\sin\psi + \bar\gamma\cos\psi)\rho\frac{\partial}{\partial\psi}\right)$$

$$+ \frac{\sin s_2}{\rho}\left(\sqrt{1 + \bar\gamma^2}\frac{\sin\psi}{\bar\gamma}\left(20\frac{\partial}{\partial\theta} + \rho\frac{\partial}{\partial\psi}\right) + \sqrt{1 + \bar\gamma^2}\cos\psi\rho\frac{\partial}{\partial\psi}\right).$$

Observe now that

$$\overline{(x, y, \dot\theta)(0)} = \dot\theta\frac{\partial}{\partial\theta} + \rho\dot\psi\frac{\partial}{\partial\psi}$$

since $\rho^2 = (x^2 + y^2)(0)$ is a constant.

We thus derive:

Proposition 3 *The differential equation $\overline{(x, y, \theta)} = \xi_N(x, y, \theta)$ rereads*

$$\rho\dot\psi = \cos s_2 \left(\frac{\cos\psi - \bar\gamma\sin\psi}{\bar\gamma\sqrt{1 + \bar\gamma^2}} - (\sin\psi + \bar\gamma\cos\psi) \right)$$
$$+ \sin s_2 \left(\sqrt{1 + \bar\gamma^2}\frac{\sin\psi}{\bar\gamma} + \sqrt{1 + \bar\gamma^2}\cos\psi \right)$$

$$\rho\dot\theta = 20 \left(\cos s_2 \cdot \frac{\cos\psi - \bar\gamma\sin\psi}{\bar\gamma\sqrt{1 + \bar\gamma^2}} + \frac{\sin s_2\sqrt{1 + \bar\gamma^2}\sin\psi}{\bar\gamma} \right)$$

$$\rho\dot s_2 = \cos s_2 \left(\frac{\cos\psi - \bar\gamma\sin\psi}{\bar\gamma\sqrt{1 + \bar\gamma^2}} \right) + \left(\sqrt{1 + \bar\gamma^2}\frac{\sin\psi}{\bar\gamma} + 2\frac{\cos\psi - \bar\gamma\sin\psi}{\sqrt{1 + \bar\gamma^2}} \right) \sin s_2.$$

If we set $\frac{\partial}{\partial\tau} = \rho\bar\gamma\frac{\partial}{\partial t}$, this becomes

$$\frac{\partial\psi}{\partial\tau} = \cos s_2 \left(\frac{\cos\psi - \bar\gamma\sin\psi}{\sqrt{1 + \bar\gamma^2}} - \bar\gamma(\sin\psi + \bar\gamma\cos\psi) \right)$$
$$+ \sin s_2 \left(\sqrt{1 + \bar\gamma^2}\sin\psi + \bar\gamma\sqrt{1 + \bar\gamma^2}\cos\psi \right)$$

$$\frac{\partial\theta}{\partial\tau} = 20 \left(\cos s_2\frac{\cos\psi - \bar\gamma\sin\psi}{\sqrt{1 + \bar\gamma^2}} + \sqrt{1 + \bar\gamma^2}\sin s_2 \sin\psi \right)$$

$$\frac{\partial s_2}{\partial\tau} = \frac{\cos\psi - \bar\gamma\sin\psi}{\sqrt{1 + \bar\gamma^2}}\cos s_2 + \left(\sqrt{1 + \bar\gamma^2}\sin\psi + 2\bar\gamma\frac{(\cos\psi - \bar\gamma\sin\psi)}{\sqrt{1 + \bar\gamma^2}} \right) \sin s_2.$$

The first and the last equation define an autonomous differential equation. We conjecture that, generically on $\bar\gamma$, this differential equation will have at most a countable number of nondegenerate periodic orbits.

In order to have periodic orbits in (ψ, θ, s_1), we need the additional condition

$$\begin{cases} 2k\pi = 20 \int_0^T \left(\cos s_2 \frac{\cos\psi - \bar\gamma\sin\psi}{\sqrt{1 + \bar\gamma^2}} + \sqrt{1 + \bar\gamma^2}\sin s_2 \sin\psi \right) d\tau \\ k \in \mathbb{Z} \end{cases}.$$

We conjecture that, generically on $\bar\gamma$, this condition is not satisfied, hence that there are no periodic orbits in (ψ, θ, s_2).

We cannot rule out other periodic orbits which would be partly made of orbits of ξ_N continued by orbits of ξ.

Such orbits have $|\Delta\theta| \geq C > 0$ since we can assume that there are no periodic orbits of ξ closing up near a repelling or an attractive orbit of v.

We claim that:

Propositions 3' $|\Delta\theta| \geq 2\pi$ on such periodic orbits.

Proof. We come back to the differential equations corresponding to the flow of ξ_N. We first claim that an orbit of ξ_N, under the energy bound, cannot go from the inner boundary of the torus of modification to the outer boundary of this torus. Indeed, we have

$$\dot{s}_2 = O\left(\frac{1}{\rho\bar{\gamma}}\right).$$

Thus,

$$|\Delta s_2| = \left|O\left(\frac{1}{\rho\bar{\gamma}}\right)\right| = \left|-\int \frac{ds_1}{\varphi}\right| \geq \frac{2\pi N}{|I_0|} \times \frac{|I_0|}{2} = \pi N$$

which is impossible for N large enough.

Thus, a piece of orbit of ξ_N which contributes to a periodic orbit of the contact vector-field goes from the outer boundary to the outer boundary, i.e. from $s_2 = 0$ to $s_2 = 0$ and has

$$|\Delta\theta| \geq C > 0.$$

Coming back to the equations defining the flow of ξ_N, we find

$$\dot{s}_2 = \frac{\dot{\theta}}{20} + O\left(\frac{1}{\rho}\right).$$

Thus,

$$0 = \Delta s_2 = \frac{\Delta\theta}{20} + O\left(\frac{\Delta t}{\rho}\right)$$

and this implies that

$$|\Delta t| \geq C_\rho.$$

Hence, since $\frac{\partial}{\partial\tau} = \rho\bar{\gamma}\frac{\partial}{\partial t}$

$$\Delta\tau \geq \frac{C_\rho}{\rho\bar{\gamma}} = \frac{C}{\bar{\gamma}}.$$

In the variable τ

$$\frac{\partial\theta}{\partial\tau} = 20\cos(s_2 - \psi) + O(\bar{\gamma})$$

$$\frac{\partial}{\partial \tau}(s_2 - \psi) = \bar{\gamma}\sin(\psi + s_2) + O(\bar{\gamma}^2).$$

We start at $s_2 = 0$ and we have an interval of time at least equal to $\frac{C}{\bar{\gamma}}$ ahead of us. Either $|\cos(s_2 - \psi)|$ or $|\sin(\psi + s_2)|$ is therefore larger than $\frac{1}{2}$ as we start.

Assume first that $|\cos(s_2 - \psi)| \geq \frac{1}{2}$ as we start. Since $\frac{\partial}{\partial \tau}(s_2 - \psi) = O(\bar{\gamma})$, $|\cos(s_2 - \psi)|$ will remain larger than $\frac{1}{4}$ for a time interval I of length larger than $\frac{c_1}{\bar{\gamma}}$.

Thus, taking I of length $c_1/\bar{\gamma}$,

$$|\Delta\theta| = 2\theta\left| \int_I \cos(s_2 - \psi) + O(\bar{\gamma})|I| \right| \geq \frac{c_1}{4\bar{\gamma}} - C$$

and the conclusion follows.

The argument extends to the case where, at any time τ,

$$|\cos(s_2 - \psi)| \geq c > 0$$

with c any prescribed positive constant ($\bar{\gamma}$ small in relation to c).

Thus, we may assume that

$$|\cos(s_2 - \psi)| \leq c \quad \text{small}$$

on the entire piece of ξ_N-orbit.

Then, we have

$$\frac{\partial}{\partial \tau}\left(\psi - \frac{\theta}{20} \right) = \bar{\gamma}\sin(s_2 - \psi) + O(\bar{\gamma}^2)$$

and

$$\left(\psi - \frac{\theta}{20} \right)(\tau) = \left(\psi - \frac{\theta}{20} \right)(0) \pm \bar{\gamma}\tau(1 + 0(1)).$$

Also

$$\frac{\partial}{\partial \tau}(s_2 - \psi) = \bar{\gamma}\sin(\psi + s_1) + O(\bar{\gamma}^2)$$

$$= \bar{\gamma}(\sin 2\psi \cos(s_2 - \psi) + \cos 2\psi \sin(s_2 - \psi)) + O(\bar{\gamma}^2)$$

$$= \pm\bar{\gamma}\cos 2\psi + O(\bar{\gamma}).$$

Observe that the constraint $|\cos(s_2 - \psi)| \leq c$ forces

$$\left| s_2 - \psi + \frac{(2k + 1)\pi}{2} \right| \leq c.$$

Thus, at the entry and at the exit point,

$$\psi = \frac{(2k+1)\pi}{2} \pm c$$

and

$$x = \rho \cos\left(\frac{(2k+1)\pi}{2} \pm c\right)$$

$$y = \rho \sin\left(\frac{(2k+1)\pi}{2} \pm c\right).$$

After integration, we derive

$$\int_I \frac{\partial}{\partial \tau}(s_2 - \psi) = \pm\bar{\gamma} \int_I \cos 2\psi + o(\bar{\gamma})|I|.$$

Since $\Delta(s_2 - \psi) = o(1)$, we must have

$$\int_I \cos 2\psi = o(|I|)$$

for any I such that

$$|I| \geq \frac{c}{\bar{\gamma}}.$$

Otherwise, there will be an I, with $|I| \geq \frac{C}{\bar{\gamma}}$, such that

$$\left| \pm\bar{\gamma} \int_I \cos 2\psi + o(\bar{\gamma})|I| \right| \geq c_1\bar{\gamma}|I| \geq c_1 C$$

yielding a contradiction.

Hence, on any such interval I, there exists an integer q and a certain time τ_1 if I such that

$$2\psi + \frac{(2q+1)\pi}{2} = o(1).$$

Thus,

$$\psi + \frac{(2q+1)\pi}{4} = o(1).$$

Comparing, we derive that

$$|\psi(0) - \psi(\tau_1)| \geq \frac{\pi}{4}(1 + o(1)).$$

Thus,

$$\left|\frac{\theta}{20}(\tau_1) - \frac{\theta}{20}(0) \pm \bar{\gamma}\tau_1(1 + o(1))\right| \geq \frac{\pi}{4}(1 + o(1)).$$

If τ is the entire time spent on this ξ_N-piece of orbit, either

$$|\bar{\gamma}\tau| \geq \frac{\pi}{8}(1 + o(1))$$

which forces

$$|\theta(\tau) - \theta(0)| \geq 20 \cdot \frac{\pi}{8}(1 + o(1)) \geq 2\pi$$

since

$$\psi(\tau) - \psi(0) = o(1).$$

Or

$$|\bar{\gamma}\tau| \leq \frac{\pi}{8}(1 + o(1)).$$

Thus,

$$|\bar{\gamma}\tau_1| \leq \frac{\pi}{8}(1 + o(1)).$$

Thus,

$$|\theta(\tau_1) - \theta(0)| \geq 20\frac{\pi}{8}(1 + o(1)) \geq 2\pi \quad \text{again.} \qquad \square$$

2.2.6 *Conformal deformation*

Let λ be a positive function on M. We consider the contact form $\lambda \alpha_N$ where α_N is α modified by the construction of this large rotation.
We assume that

$$d\alpha_N(v_N, [\xi, v_N]) = -1 \quad \text{with } v_N = \varphi_N v$$

in the region of M where we will carry out our constructions and computations. For simplicity and generality, we come back here to the following notations

$$v \text{ instead of } v_N$$

$$\xi_0 \text{ instead of } \xi_N$$

$$\alpha_0 \text{ instead of } \alpha_N.$$

It must though be kept clear to the mind that in the application below —
which is our main purpose — $v = \varphi_N v, \xi_N = \xi_0, \alpha_N = \alpha_0$. Later, we will
have v_N and v, ξ_N and ξ_0, α_N and α_0. This is why we want to avoid any
confusion.

α is $\lambda \alpha_0$ (it will be $\lambda \alpha_N$ thereafter).

We thus assume that

$$d\alpha_0(v, [\xi_0, v]) = -1$$

in the region of M where we will carry out out constructions and computations.

We start with

Lemma 8 $\xi = \frac{\xi_0}{\lambda} + \frac{d\lambda(v)}{\lambda^2}[\xi_0, v] - \frac{d\lambda([\xi_0, v])}{\lambda^2} v.$

Proof.   We compute

$$(d\lambda \wedge \alpha_0 + \lambda d\alpha_0)(\xi, v) = (d\lambda \wedge \alpha_0)(\xi, v) + \lambda d\alpha_0(\xi, v)$$

$$= -\frac{d\lambda(v)}{\lambda} + \frac{d\lambda(v)}{\lambda} d\alpha_0([\xi_0, v], v) = 0$$

$$(d\lambda \wedge \alpha_0 + \lambda d\alpha_0)(\xi, [\xi_0, v]) = (d\lambda \wedge \alpha_0)(\xi, [\xi_0, v]) + \lambda d\alpha_0(\xi, [\xi_0, v])$$

$$= -\frac{d\lambda([\xi_0, v])}{\lambda} - \frac{d\lambda([\xi_0, v])}{\lambda} d\alpha_0(v, [\xi_0, v]) = 0.$$

$\qquad \square$

We now compute

$$d\alpha(v, [\xi, v]) = \lambda d\alpha_0(v, [\xi, v]).$$

Lemma 9

$$-\gamma = d\alpha(v, [\xi, v]) = -\lambda \left(\frac{1}{\lambda} + \left(\frac{1}{\lambda} \right)_{vv} + \left(\frac{1}{\lambda} \right)_v d\alpha_0(v, [[\xi_0, v], v]) \right).$$

Proof.

$$\lambda d\alpha_0(v, [\xi, v]) = \lambda d\alpha_0 \left(v, \left[\frac{\xi_0}{\lambda} + \frac{d\lambda(v)}{\lambda^2}[\xi_0, v] - \frac{d\lambda([\xi_0, v])}{\lambda^2} v, v \right] \right)$$

$$= d\alpha_0(v, [\xi_0, v]) - \lambda \left(\frac{d\lambda(v)}{\lambda^2} \right)_v d\alpha_0(v, [\xi_0, v]) + \frac{d\lambda(v)}{\lambda} d\alpha_0(v, [[\xi_0, v], v])$$

$$= -\lambda \left(\frac{1}{\lambda} + \left(\frac{1}{\lambda} \right)_{vv} + \left(\frac{1}{\lambda} \right)_v d\alpha_0(v, [[\xi_0, v], v]) \right).$$

$\qquad \square$

Corollary 3 *Set* $\lambda_t = \frac{1}{t+\frac{1-t}{\lambda}}$. *If* $d\alpha(v,[\xi,v])(x) < 0$, *then so is* $d\alpha_t(v,[\xi_t,v])(x)$ *for* $\alpha_t = \lambda_t \alpha_0$.

Proof.
$d\alpha_t(v,[\xi_t,v])(x) = -\lambda_t \left(\frac{1-t}{\lambda} + t + \left(\frac{1-t}{\lambda}\right)_{vv} + \left(\frac{1-t}{\lambda}\right)_v d\alpha_0(v,[[\xi_0,v],v]) \right)$ and the result follows. $\qquad\square$

Assume now that

$$\lambda\left(\frac{1}{\lambda}\right)_v , \lambda\left(\frac{1}{\lambda}\right)_{vv} \quad \text{are } o(1).$$

Recall that

$$\gamma(x) = 1 + \lambda\left(\frac{1}{\lambda}\right)_{vv} + \lambda\left(\frac{1}{\lambda}\right)_v d\alpha_0(v,[[\xi_0,v],v])$$

and

$$\tilde{v} = \frac{v}{\sqrt{\gamma(x)}}$$

so that

$$d\alpha(\tilde{v},[\xi,\tilde{v}]) = \frac{1}{\gamma(x)} d\alpha(v,[\xi,v]) = -1.$$

We compute in the sequel

$$\tilde{\mu} = d\alpha(\tilde{v},[\tilde{v},[\xi,\tilde{v}]])$$
$$\tilde{\mu}_\xi = d\tilde{\mu}(\xi)$$
$$\tilde{\mu}_v = d\tilde{\mu}(v)$$

and also $\tilde{\tau}$, where $[\xi,[\xi,\tilde{v}]] = -\tilde{\tau}\tilde{v}$.

Lemma 10 $\tilde{\mu} = \frac{1}{\gamma(x)^{3/2}} \left(d\alpha_0(v,[v,[\xi_0,v]]) \right)$

$\left(1 + 2\lambda\left(\frac{1}{\lambda}\right)_{vv} + \frac{d\lambda(v)}{\lambda}(2+\gamma(x)) - \lambda\left(\frac{1}{\lambda}\right)_{vvv} + \frac{\lambda_v}{\lambda} d\alpha_0(v,[v,[[\xi_0,v],v]]) \right)$.

Proof. Clearly,

$$\tilde{\mu} = \frac{1}{\gamma(x)^{3/2}} d\alpha(v,[v,[\xi,v]]).$$

We have

$$d\alpha(v,[v,[\xi,v]]) = (d\lambda \wedge \alpha_0 + \lambda d\alpha_0)(v,[v,[\xi,v]]) = d\lambda(v)\alpha_0([v,[\xi,v]])$$

$$+ \lambda d\alpha_0(v,[v,[\xi,v]]) = \gamma(x)\frac{d\lambda(v)}{\lambda} + \lambda d\alpha_0\left(v,\left[v,\left[\frac{\xi_0}{\lambda},v\right]\right]\right)$$

$$+ \lambda d\alpha_0 \left(v, \left[v, \left[\frac{d\lambda(v)}{\lambda^2}[\xi_0, v], v \right] \right] \right)$$

$$= \gamma(x)\frac{d\lambda(v)}{\lambda} - \lambda \left(\frac{1}{\lambda} \right)_v d\alpha_0(v, [v, \xi_0]) + \lambda d\alpha_0 \left(v, \left[v, \frac{1}{\lambda}[\xi_0, v] \right] \right)$$

$$+ \lambda d\alpha_0 \left(v, \left[v, \left[\frac{d\lambda(v)}{\lambda^2}[\xi_0, v], v \right] \right] \right) = d\alpha_0(v, [v, [\xi_0, v]]) - 2\lambda \left(\frac{1}{\lambda} \right)_v d\alpha_0(v, [v, \xi_0])$$

$$+ \gamma(x)\frac{d\lambda(v)}{\lambda} + \lambda d\alpha_0 \left(v, \left[v, \left(\frac{1}{\lambda} \right)_{vv}[\xi_0, v] \right] \right) + \frac{\lambda_v}{\lambda}d\alpha_0(v, [v, [[\xi_0, v], v]])$$

$$- \lambda \left(\frac{1}{\lambda} \right)_{vv} d\alpha_0(v, [[\xi_0, v], v]) = d\alpha_0(v, [v, [\xi_0, v]]) \left(1 + 2\lambda \left(\frac{1}{\lambda} \right)_{vv} \right) - \lambda \left(\frac{1}{\lambda} \right)_{vvv}$$

$$- 2\lambda \left(\frac{1}{\lambda} \right)_v + \gamma(x)\frac{d\lambda(v)}{\lambda} + \frac{\lambda_v}{\lambda}d\alpha_0(v, [v, [[\xi_0, v], v]]).$$

$$\square$$

Observe that

$$\lambda \left(\frac{d\lambda(v)}{\lambda^2} \right)_v = \frac{\lambda_{vv}}{\lambda} - 2\frac{\lambda_v^2}{\lambda^2}$$

$$\lambda \left(\frac{d\lambda(v)}{\lambda^2} \right)_{vv} = \frac{\lambda_{vvv}}{\lambda} - 5\frac{\lambda_v \lambda_{vv}}{\lambda^2} + 4\frac{\lambda_v^3}{\lambda^3}.$$

Next, we compute $\tilde{\tau}$. We know that $-\tilde{\tau}$ is the collinearity coefficient of $[\xi, [\xi, \tilde{v}]]$ on $\tilde{v}$.

We will therefore compute $[\xi, [\xi, \tilde{v}]]$ in the $(\xi_0, v, [\xi_0, v])$ basis and we will track down the component on v, throwing away the other components.

We have:

Lemma 11   *Let $\nu(x) = \sqrt{\gamma(x)}$. Then,*

$$\tilde{\tau} = -\frac{1}{\nu(x)} \left[\frac{dA(\xi_0)}{\lambda} - \frac{B\tau}{\lambda} + \frac{d\lambda(v)}{\lambda^2}dA([\xi_0, v]) - A\bar{\mu}_{\xi_0}\frac{d\lambda(v)}{\lambda^2} - \frac{d\lambda([\xi_0, v])}{\lambda^2}dA(v) \right.$$

$$\left. + A \left(\frac{d\lambda([\xi_0, v])}{\lambda^2} \right)_v - B\frac{d\lambda([\xi_0, v])}{\lambda^2}\bar{\mu}_{\xi_0} + B \left(\frac{d\lambda([\xi_0, v])}{\lambda^2} \right)_{[\xi_0, v]} \right]$$

with

$$A = d\nu(\xi) - \bar{\mu}_{\xi_0}\nu\frac{d\lambda(v)}{\lambda^2} + \nu \left(\frac{d\lambda([\xi_0, v])}{\lambda^2} \right)_v, \quad B = \nu \left(\frac{1}{\lambda} + \bar{\mu}\frac{d\lambda(v)}{\lambda^2} + \left(\frac{1}{\lambda^2} \right)_{vv} \right)$$

Proof.

$$[\xi,[\xi,\tilde{v}]] = \left[\frac{\xi_0}{\lambda} + \frac{\lambda_v}{\lambda^2}[\xi_0,v] - \frac{d\lambda([\xi_0,v])}{\lambda^2}v, \left[\frac{\xi_0}{\lambda} + \frac{\lambda_v}{\lambda^2}[\xi_0,v] - \frac{d\lambda([\xi_0,v])}{\lambda^2}v, \tilde{v}\right]\right]$$

with $\tilde{v} = \nu(x)v$. Observe that

$$[v,[\xi_0,v]] = \xi_0 - \bar{\mu}[\xi_0,v] + \bar{\mu}_{\xi_0}v.$$

Indeed,

$$\alpha_0([v,[\xi_0,v]]) = 1$$
$$d\alpha_0([v,[\xi_0,v]],v) = -\bar{\mu}$$
$$d\alpha_0([\xi_0,v],[v,[\xi_0,v]]) = \bar{\mu}_{\xi_0} = d\bar{\mu}(\xi_0).$$

We then compute

$$[\xi,\tilde{v}] = d\nu(\xi)v + \nu\left(-\left(\left(\frac{1}{\lambda}\right)_v + \frac{\lambda_v}{\lambda^2}\right)\xi_0 + [\xi_0,v]\left(\frac{1}{\lambda} + \left(\frac{1}{\lambda}\right)_{vv} + \bar{\mu}\frac{\lambda_v}{\lambda^2}\right)\right.$$

$$\left. + v\left(\bar{\mu}_{\xi_0}\left(\frac{1}{\lambda}\right)_v + \left(\frac{d\lambda([\xi_0,v])}{\lambda^2}\right)_v\right)\right)$$

$$= v\left(d\nu(\xi) + \bar{\mu}_{\xi_0}\left(\frac{1}{\lambda}\right)_v + \nu\left(\frac{d\lambda([\xi_0,v])}{\lambda^2}\right)_v\right)$$

$$+ [\xi_0,v]\nu\left(\frac{1}{\lambda} + \left(\frac{1}{\lambda}\right)_{vv} + \bar{\mu}\frac{d\lambda(v)}{\lambda^2}\right) = Av + B[\xi_0,v].$$

Set $[\xi_0,[\xi_0,v]] = -\tau v$

$$[\xi,[\xi,\tilde{v}]] = \left[\frac{\xi_0}{\lambda} + \frac{\lambda_v}{\lambda^2}[\xi_0,v] - \frac{d\lambda([\xi_0,v])}{\lambda^2}v, Av + B[\xi_0,v]\right]$$

$$= v\left(\frac{dA(\xi_0)}{\lambda} - \frac{B\tau}{\lambda} + \frac{\lambda_v}{\lambda^2}dA([\xi_0,v]) - \frac{d\lambda([\xi_0,v])}{\lambda^2}dA(v) - \bar{\mu}_{\xi_0}A\frac{\lambda_v}{\lambda^2}\right.$$

$$\left. + A\left(\frac{d\lambda([\xi_0,v])}{\lambda^2}\right)_v + +B\left(\frac{d\lambda([\xi_0,v])}{\lambda^2}\right)_{[\xi_0,v]} - B\bar{\mu}_{\xi_0}\frac{d\lambda([\xi_0,v])}{\lambda^2}\right).$$

Thus,

$$\tilde{\tau} = -\frac{1}{\nu}\left(\frac{dA(\xi_0)}{\lambda} - \frac{B\tau}{\lambda} + \frac{d\lambda(v)}{\lambda^2}dA([\xi_0,v])\right.$$

$$-A\bar{\mu}_{\xi_0}\frac{d\lambda(v)}{\lambda^2} - \frac{d\lambda([\xi_0,v])}{\lambda^2}dA(v) + A\left(\frac{d\lambda([\xi_0,v])}{\lambda^2}\right)_v$$

$$+B\left(\frac{d\lambda([\xi_0,v])}{\lambda^2}\right)_{[\xi_0,v]} - B\bar{\mu}_{\xi_0}\frac{d\lambda([\xi_0,v])}{\lambda^2}\right).$$

Coming back to α_N, ξ_N, v_N, we observe that $\bar{\mu}_N$ is bounded as well as its derivative since $\bar{\mu}_N$ is identically zero wherever we introduce the large, $2\pi N$ rotation and $\bar{\mu}_N$ equals $\bar{\mu}_0$ outside of the set where this modification occurs. Similarly, τ_N is bounded as well as its derivatives independently of N by construction. $\qquad\qquad\square$

Proposition 4 $|\bar{\mu}_N|+|d\bar{\mu}_N|+|\tau_N| \leq C$, *where C does not depend on N.*

Next, we show how to build λ so that $\tilde{\tau}$ remains bounded and "mountains" are built around the hyperbolic orbit. These "mountains" keep the variations away from this hyperbolic orbit. Since this is a quite surprising result, we complete our construction carefully.

2.2.7   *Choice of λ*

The construction of $\xi_{N,\lambda}, \alpha_{N,\lambda} = \lambda\alpha_N$ involves the definition of the function λ. We would like to choose this function carefully with respect to $v(v_N)$ so that $\lambda d\alpha_N(v_N, [\xi_{N,\lambda}, v_N]), \tilde{\bar{\mu}}_N, \tilde{\bar{\mu}}_{N,v} = d\tilde{\bar{\mu}}_N(\tilde{v}_N), \tilde{\bar{\mu}}_{N,\xi_{N,\lambda}} = d\tilde{\bar{\mu}}_N(\xi_{N,\lambda}), \tilde{\tau}_N$ enjoy appropriate bounds.

To avoid unnecessary complicated notations, we again use $\xi, \alpha, \xi_0, \alpha_0$ etc. The main issue is that we cannot hope that $\beta = d\alpha(v, \cdot)$ is a contact form (with the same orientation than α).

Assuming that v is nonsingular, it is reasonable to first consider the case when the ω-limit set of v is made of periodic orbits only. Around "elliptic" (attractive or repelling) periodic orbits, $\ker\alpha$ "turns well" (see [Bahri 1988, p. 26]) so that the existence of such a β (with appropriate choices of λ) follows.

Around hyperbolic periodic orbits, $\ker\alpha$ "turns well" in most of the cases, except for one case which yields a precise (local) normal form of α and v (see Appendix 2.3). Then, locally, $\ker\alpha_0$ behaves (nearly) as a foliation. There is no hope for such a λ and such a β to exist near such orbits, with this behavior of α_0 and v.

We thus need to keep our homology away from such neighborhoods or to extend it using the ideas of [Bahri-1 2003, Chapter V.1]. These ideas can be pushed and worked out. They still require a certain amount of work to become practical. We explore here another direction: we aim at keeping the unstable manifolds of the periodic orbits of ξ_0 away from such neighborhoods by creating "mountains" around them. These "mountains" are built by increasing to a high value the Hamiltonian λ around them so that the curves on the unstable manifolds of the periodic orbits of ξ_0 are unable to penetrate them.

We need for this a lot of rotation of $ker\alpha$ around v. This will allow us to keep control of $\tilde{\bar{\mu}}, \tilde{\tau}, d\alpha(v, [\xi, v])$ and derivatives with respect to $v, [\xi_0, v]$... We cannot get such a rotation from the neighborhood of a "bad" hyperbolic orbit since $ker\,\alpha_0$ turns very little around v in such a neighborhood. We have to seek for it in the neighborhood of attractive or repelling orbits and "bring it back" to our neighborhoods.

For the sake of simplicity, we will assume in a first step that the stable and unstable manifolds of our "bad" hyperbolic orbit θ do not intersect the unstable and stable manifolds of another hyberbolic orbit. We will discuss the case later.

Thus, ($\mathcal{H}1$) the stable (respectively unstable) manifold of $\mathcal{O}$ is part of the unstable manifold of the repelling (respectively attracting) periodic orbit.

We will assume — a very natural hypothesis which we will see to hold after a minor modification of $ker\,\alpha_0$ and v if needed — that ($\mathcal{H}2$) α_0, v have the normal forms provided in (α_0), (v) above near the repelling and attracting orbit.

The construction of the function λ is ultimately quite involved. However, in order to describe a basic step in this direction, we first take the following example. The construction is refined later.

2.2.8 *First step in the construction of* λ

Let $\mathcal{W}_u(\mathcal{O})$ be unstable manifold of $\mathcal{O}$ and let ∂V be the boundary of a small basin for the attractive orbit. ∂V is a section to v. We consider $\mathcal{W}_u(\mathcal{O}) \cap \partial V = \Gamma$ and a small neighborhood $\mathcal{L}$ of it in ∂V. Let $\mathcal{L}'$ be an even smaller neighborhood, $\theta = \theta(x_0), x_0 \in \mathcal{L}$ be a C^∞- function valued in $[0,1]$, equal to 1 on $\mathcal{L}'$ and to zero outside of $\mathcal{L}$.

We set:

$$\lambda(x) = e^{\delta\theta(x_0)s(\tau)},$$

for all x of $W_s(\mathcal{L})$ i.e. for all the $x's$ of the flow-lines of v abutting in $\mathcal{L}$.

Such $x's$ are parametrized by a base point x_0 in $\mathcal{L}$ and a time τ on the (reverse) flow-line of v abuting at x_0. δ is a small number which we will choose later.

More generally,

$$\lambda(x) = e^{\delta(\sum \theta_i(x_i) s_i(\tau_i))}$$

where the (x_i, τ_i) are various sets of parameters traking a point x of M through its reference point x_i in a section to v and a time τ_i on the flow-line of v abuting at x_i.

Clearly

$$\frac{\lambda_v}{\lambda} = \delta\left(\sum \theta_i \frac{\partial s_i}{\partial \tau_i}\right), \quad \frac{\lambda_{vv}}{\lambda} = \delta \sum \theta_i \frac{\partial^2 s_i}{\partial \tau_i^2} + \delta^2 \left(\sum \theta_i \frac{\partial s_i}{\partial \tau_i}\right)^2$$

$$\lambda\left(\frac{1}{\lambda}\right)_{vvv} = \left(\lambda\left(\frac{1}{\lambda}\right)_{vv}\right)_v - \lambda_v\left(\frac{1}{\lambda}\right)_{vv} = \left(\lambda\left(\frac{1}{\lambda}\right)_{vv}\right)_v - \frac{\lambda_v}{\lambda}\lambda\left(\frac{1}{\lambda}\right)_{vv}$$

$$= \left(\left(\lambda\left(\frac{1}{\lambda}\right)_v\right)_v - \frac{\lambda_v}{\lambda}\lambda\left(\frac{1}{\lambda}\right)_v\right)_v - \frac{\lambda_v}{\lambda}\lambda\left(\frac{1}{\lambda}\right)_{vv}$$

$$= \delta O\left(\sum \theta_i \left(\left|\frac{\partial^3 s_i}{\partial \tau_i^3}\right| + \left|\frac{\partial s_i}{\partial \tau_i}\right|\left|\frac{\partial^2 s_i}{\partial \tau_i^2}\right|\left|\frac{\partial^2 s_i}{\partial \tau_i^2}\right| + \left|\frac{\partial s_i}{\partial \tau_i}\right|^3\right)\right).$$

$$(2.29)$$

We assume that

$$\left|\frac{\partial s_i}{\partial \tau_i}\right| + \left|\frac{\partial^2 s_i}{\partial \tau_i^2}\right| + \left|\frac{\partial^3 s_i}{\partial \tau_i^3}\right| = O(1). \tag{2.30}$$

Then,

$$\frac{\lambda_v}{\lambda}, \lambda\left(\frac{1}{\lambda}\right)_{vv}, \lambda\left(\frac{1}{\lambda}\right)_{vvv} \quad \text{are } O(\delta).$$

In this way, $\gamma(x)$ and $\tilde{\bar{\mu}}$ are under control. We need to worry about $d\tilde{\bar{\mu}}$ and $\tilde{\tau}$. Coming back to the formula of $\tilde{\tau}$ in Lemma 11, to A and B as well a the formulae for ν and ξ, we see that these formulae involve derivatives of $\frac{1}{\lambda}$. According to our choice of λ above, λ is larger than or equal to 1, might tend to $+\infty$. Because we are only considering negative powers of λ and derivatives of such quantities, we do not fear the increase of λ to infinity.

The derivative of $\delta \sum \theta_i(x_i) s_i(\tau_i)$ yield more problems because s_i may be very large and derivatives of θ may also be very large. Since we want λ to be very large when $\theta_i = 1$ and we are at the "end" of the (reverse) flow-lines, we require

$$\delta s_i(\bar{\tau}_i) = \text{Log } \bar{\lambda}$$

where $\bar{\lambda}$ is some large number. The flow-lines are defined on $[o, \bar{\tau}_i]$. Thus δ cannot tame $s_i(\tau_i)$.

We observe that, as we modify α_0 into α_N and ξ_0 in ξ_N, see Proposition 2, $\bar{\mu}_N = d\alpha_N(v_N, [v_N, [\xi_N, v_N]])$ is zero in the domain where the modification takes place.

Also, since in this domain $[v_N, [v_N, \xi_N]] = -\xi_N$,

$$d\alpha_N(v_N, [v_N, [[\xi_N, v_N], v_N]]) = -d\alpha_N(v_N, [v_N, \xi_N]) = -1.$$

Thus, in the domain where ξ_0 is modified into ξ_N

$$\begin{cases} \bar{\mu}_N = 0 \\ d\alpha_N(v_N, [v_N, [[\xi_N, v_N], v_N]]) = -1 \\ \gamma_N(x) = 1 + \lambda\left(\frac{1}{\lambda}\right)_{v_N v_N} \\ \tilde{\bar{\mu}}_N = \frac{1}{\gamma_N^{3/2}}\left(\frac{d\lambda(v_N)}{\lambda}(2 + \gamma_N(x)) - \lambda\left(\frac{1}{\lambda}\right)_{v_N v_N v_N} - \frac{\lambda_{v_N}}{\lambda}\right). \end{cases} \tag{2.31}$$

We then have:

Proposition 5 *Assume s_i is only a function of τ_i, the time along v_N. Then, there exists C independent of $N, \delta, \bar{\lambda}$ such that, given N and $\bar{\lambda}$,*

$$|\tilde{\bar{\mu}}_N| + |d\tilde{\bar{\mu}}_N(\xi_N)| + |d\tilde{\bar{\mu}}_N(v_N)| + |d\tilde{\bar{\mu}}_N([\xi_N, v_N])| + |d\tilde{\bar{\mu}}_N(\xi_{N,\lambda})|$$

$$+ d\tilde{\bar{\mu}}_N([\xi_{N,\lambda}, v_N]) + |d(d\tilde{\bar{\mu}}_N(\xi_{N,\lambda}))(\xi_{N,\lambda})| \leq C.$$

Proof. Recall that $\lambda = e^{\delta(\sum \theta_i s_i)}$. $\qquad\qquad\qquad\qquad\qquad\square$

Either x is in the domain where ξ_0 has been modified into ξ_N.

Then, $\xi_N \cdot \xi_N, \xi_N \cdot [\xi_N, v_N], [\xi_N, v_N] \cdot \xi_N, [\xi_N, v_N] \cdot [\xi_N, v_N]$ split over e_1, e_2 with bounded coefficients, the bounds being C^1 and independent of N.

We do not claim any control on the v_N-components of these vectors. $\tilde{\bar{\mu}}_N$ is expressed using $\lambda\left(\frac{1}{\lambda}\right)_{v_N}, \lambda\left(\frac{1}{\lambda}\right)_{v_N v_N}$ and $\lambda\left(\frac{1}{\lambda}\right)_{v_N v_N v_N}$. Since the θ_i's have a zero derivative along v_N, all these expressions read as products

$$\delta\left(\sum \theta_i \frac{\partial^m s_i}{\partial \tau_i^m}\right) \quad m = 1, 2, 3.$$

By construction $ds_i(e_1) = ds_i(e_2) = 0$ since $d\tau_i(e_1) = d\tau_i(e_2) = 0$ while $d\tau_i(\varphi_N v) = 1$. $\xi_N, v_N, [\xi_N, v_N]$ split on e_1, e_2 and $\varphi_N v$ with bounded coefficients. Thus $d\theta_i(\xi_N), d\theta_i(v_N), d\theta_i([\xi_N, v_N])$ are clearly bounded and $|\tilde{\bar{\mu}}_N| + |d\tilde{\bar{\mu}}(\xi_N)| + |d\tilde{\bar{\mu}}_N(v_N)| + |d\tilde{\bar{\mu}}_N([\xi_N, v_N])|$ is bounded and even $0(\delta)$ in such a region.

For $d(d\tilde{\bar{\mu}}_N(\xi_{N,\lambda}))(\xi_{N,\lambda})$ we come back to the expression of $\xi_{N,\lambda}$

$$\xi_{N,\lambda} = \frac{\xi_N}{\lambda} + \frac{d\lambda(v_N)}{\lambda^2}[\xi_N, v_N] - \frac{d\lambda([\xi_N, v_N])}{\lambda^2}v_N.$$

We thus have to take derivatives which are typically expressions such as

$$\frac{\delta}{\lambda}\left(\sum \theta_i \frac{\partial^m s_i}{\partial \tau_i^m}\right)_{\xi_N} + \frac{\delta d\lambda(v_N)}{\lambda^2}\left(\sum \theta_i \frac{\partial^m s_i}{\partial \tau_i^m}\right)_{[\xi_N, v_N]}$$
$$- \delta \frac{d\lambda([\xi_N, v_N])}{\lambda^2}\left(\sum \theta_i \frac{\partial^m s_i}{\partial \tau_i^m}\right)_{v_N}$$

and then take again a derivative of such expressions along $\xi_{N,\lambda}$. On $\frac{\partial^m s_i}{\partial \tau_i^m}$, the e_1 and e_2 components of each derivative do not give any contribution. It is only the $\varphi_N v$-components which give a contribution. These are bounded and have a bounded derivative along $\xi_N, [\xi_N, v_N], v_N$. The problems come only after taking a first derivative of θ_i and then going on with a second derivative of this expression. Typically, we need to estimate

$$(d\theta_i(\xi_N))_{\xi_N}, (d\theta_i([\xi_N, v_N]))_{\xi_N}, (d\theta_i([\xi_N, v_N]))_{v_N},$$
$$(d\theta_i([\xi_N, v_N]))_{[\xi_N, v_N]}, (d\theta_i(\xi_N))_{[\xi_N, v_N]} \text{ etc.}$$

We recall now that

$$\xi_N \cdot \xi_N, [\xi_N, v_N] \cdot [\xi_N, v_N], \xi_N \cdot [\xi_N, v_N], [\xi_N, v_N] \cdot \xi_N$$

are bounded transversally to v_N. Furthermore, $d\theta_i(v_N) = 0$. Thus,

$$(d\theta_i(\xi_N))_{\xi_N}, (d\theta_i([\xi_N, v_N]))_{\xi_N} \text{ etc.}$$

are bounded independently of N.

For $d\theta_i([\xi_N, v_N])_{v_N}$ and the like, we observe that

$$d\theta_i([\xi_N, v_N])_{v_N} = d\theta_i([v_N, [\xi_N, v_N]])$$

since $d\theta_i(v_N) = 0$ and the conclusion follows again.

Thus,

$$|d(d\tilde{\bar{\mu}}_N(\xi_{N,\lambda})(\xi_{N,\lambda})| = O(\delta^{m+1} \sum \bar{\tau}_i^m) = \delta O(\operatorname{Log}^m \bar{\lambda}).$$

We can bound $d\tilde{\bar{\mu}}_N(\xi_{N,\lambda})$ in a similar way.

For $d\tilde{\bar{\mu}}_N([\xi_{N,\lambda}, v_N])$, we observe that

$$[\xi_{N,\lambda}, v_N] = \frac{[\xi_N, v_N]}{\lambda} + \frac{d\lambda(v_N)}{\lambda^2}\xi_N + \frac{\lambda_{v_N}}{\lambda}[[\xi_N, v_N], v_N]$$
$$- \left(\frac{\lambda_{v_N}}{\lambda}\right)_{v_N}[\xi_N, v_N] + \left(\frac{d\lambda([\xi_N, v_N])}{\lambda^2}\right)_{v_N} v_N.$$

The condition follows again since we have an additional δ coming from the expression of $\tilde{\bar{\mu}}_N$.

Finally, if x is not in the domain where ξ_0 has been modified into ξ_N, then $d\alpha_0(v, [v, [\xi_0, v]])$, $d\alpha_0(v, [v, [\xi_0, v], v]])$ are C^∞-functions and even though $\tilde{\tilde{\mu}}$ is not expressed only with the use of $\lambda\left(\frac{1}{\lambda}\right)_v$ and other terms of the same type, $\tilde{\tilde{\mu}}$ is a product of these terms with these C^∞-functions, which are independent of N. The above argument extends verbatim.

Before proceeding with the estimate on $\tilde{\tau}$, we make the following four observations:

Observation 1. As we take a derivative of s_i along $\xi_N, [\xi_N, v_N]$ or v_N (which we see as split on the basis (e_1, e_2, v_N)), $\varphi_N v = v_N$ is absorbed in $\frac{\partial s_i}{\partial \tau_i}$ or other derivatives of the same type, but higher order $(d\theta_i(v) = 0)$. v_N gone, we are left with the coefficient of v_N which is C^1-bounded independently of N. We can take safely another derivative along $\xi_N, [\xi_N, v_N]$ or v_N. We will not hit φ_N with a derivative.

Observation 2. Since $d\theta_i(v) = 0, v_N \cdot d\theta_i(w) = d\theta_i([v_N, w])$. We may then take one more derivative along a direction such as $\xi_N, [\xi_N, v_N]$. We know that $\xi_N \cdot \xi_N, \xi_N \cdot [\xi_N, v_N], [\xi_N, v_N] \cdot \xi_N, [\xi_N, v_N] \cdot [\xi_N, v_N]$ are bounded transversally to v and that $d\theta_i(v) = 0$. We get then bounds on such expressions which depend on $|\theta_i|_{C^2}$ and are independent of N.

Observation 3. If we take a derivative of s_i along v_N, ξ_N or $[\xi_N, v_N]$, we free a δ. Furthermore, in all our computations, we never take a v_N-deriative after taking two derivatives along ξ_N or $[\xi_N, v_N]$. Otherwise, we might end up with terms such as $d\theta_i(d\varphi_N(v_N) \cdot (\xi_N \cdot v))$. This never happens.

Observation 4. Thus, if a derivative is taken along v_N, either it goes onto s_i and frees a δ, or it goes onto $d\theta_i(\xi_N)$ or $d\theta_i([\xi_N, v_N])$. Since $d\theta_i(v_N) = 0$, we end up with $d\theta_i$ of a Lie bracket $([v_N, \xi_N]$ or $[v_N, [\xi_N, v_N]] = \xi_N)$, hence with an expression of the same type.

Taking more derivatives along v_N will not change this pattern. We can then always take one more derivative along ξ_N or $[\xi_N, v_N]$ and use Observation 1 if the expression which we have contains a $d\theta_i(\xi_N)$ or $d\theta_i([\xi_N, v_N])$; or this expression contains only θ_i and we can then take two derivatives along ξ_N and/or $[\xi_N, v_N]$. The result is bounded independently of N.

Using the four observations above, we turn to $\tilde{\tau}$ and estimate it, firstly in the domain where ξ_0 has been replaced by ξ_N.

Then, in Lemma 11, A and B reduce to

$$A = d\nu(\xi) + \nu\left(\frac{d\lambda([\xi_0, v])}{\lambda^2}\right)_v$$

$$B = \nu\left(\frac{1}{\lambda} + \left(\frac{1}{\lambda}\right)_{vv}\right)$$

and $\tilde{\tau}$ reads

$$\tilde{\tau} = -\frac{1}{\nu(x)}\left[\frac{dA(\xi_0)}{\lambda} - \frac{B\tau}{\lambda} + \frac{d\lambda(v)}{\lambda}dA([\xi_0,v]) - \frac{d\lambda([\xi_0,v])}{\lambda^2}dA(v)\right.$$
$$\left. + A\left(\frac{d\lambda([\xi_0,v])}{\lambda^2}\right)_v + B\left(\frac{d\lambda([\xi_0,v])}{\lambda^2}\right)_{[\xi_0,v]}\right]$$

ξ_0 is in fact ξ_N and $v = v_N$. Recall that $\bar{\mu}_N$ is zero. Thus, since $\bar{\mu}_{N,\xi_N\xi_N} + \tau_N\bar{\mu}_N = -d\tau_N(v_N)$ [Bahri 1998], $d\tau_N(v) = 0$ and τ_N is constant on a flow-line of v. They all abut to a point where $\varphi_N = 1$ and τ_N is the original τ of α_0. Thus, τ in the expression of $\tilde{\tau}$ above is τ_N and is bounded.

$d\nu(\xi)$ is equal to

$$\frac{1}{\lambda}d\nu(\xi_0) + \frac{d\lambda(v)}{\lambda^2}d\nu([\xi_0,v]) - \frac{d\lambda([\xi_0,v])}{\lambda^2}d\nu(v).$$

All of this involves derivatives of $\lambda\left(\frac{1}{\lambda}\right)_{vv}$ along $\xi_0, [\xi_0, v], v$. In computing $\tilde{\tau}$, we take one further derivative along ξ_0 of $d\nu(\xi)$.

Taking into account our four observations, we derive that

$$(d\nu(\xi))_{\xi_0} = O(\delta \, \mathrm{Log}^m \bar{\lambda})$$

since the initial derivative $\lambda\left(\frac{1}{\lambda}\right)_{vv}$ frees a δ. This O depends on $d\theta_i, d^2\theta_i$.

The same estimate holds for the contribution of the $\nu\left(\frac{1}{\lambda}\right)_{vv}$ part of B. Also, taking ν-derivatives in $\nu\left(\frac{d\lambda([\xi_0,v])}{\lambda^2}\right)_v$ (the second part of A) or in $\frac{\nu}{\lambda}$ (this first part of B) yields the same estimate since $\nu = 1 + 0(\delta)$. The same holds true of $\frac{d\lambda(v)}{\lambda}dA([\xi_0,v])$.

Thus,

$$\tilde{\tau} = O(\delta \, \mathrm{Log}^m \bar{\lambda} + 1)$$
$$-\left(\frac{1}{\lambda}\left(\left(\frac{d\lambda([\xi_N,v_N])}{\lambda^2}\right)_{v_N}\right)_{\xi_N} + \frac{d\lambda([\xi_N,v_N])}{\lambda^2}\left(\frac{d\lambda([\xi_N,v_N])}{\lambda^2}\right)_{v_Nv_N}\right.$$
$$\left. -\left(\frac{d\lambda([\xi_N,v_N])}{\lambda^2}\right)^2_{v_N} - \frac{1}{\lambda}\left(\frac{d\lambda([\xi_N,v_N])}{\lambda^2}\right)_{[\xi_N,v_N]}\right).$$

As we compute $\left(\frac{d\lambda([\xi_N,v_N])}{\lambda^2}\right)_{v_N}$ or $\left(\frac{d\lambda([\xi_N,v_N])}{\lambda^2}\right)_{v_Nv_N}$, all v_N- derivatives have to be taken on $d\theta_i([\xi_N, v_N])$. Otherwise, a δ is freed either because v_N has been absorbed in s_i, yielding $\frac{\partial s_i}{\partial \tau_i} = O(1)$, or because $[\xi_N, v_N]$ has been applied to s_i in the first place, with the same conclusion. Such contributions can be included into $O(\delta \, \mathrm{Log}^m \bar{\lambda})$. Observe also that v_N cannot be applied to a simple θ_i since $d\theta_i(v_N) = 0$.

Thus, since $d\theta_i(v_N) = 0$

$$\left(\frac{d\lambda([\xi_N, v_N])}{\lambda^2}\right)_{v_N} = \frac{(d\lambda([\xi_N, v_N]))_{v_N}}{\lambda^2} + O(\delta \ \mathrm{Log}^m \bar{\lambda})$$

and since $v_N \cdot d\theta([\xi_N, v_N]) = d\theta_i([v_N, [\xi_N, v_N]]) = d\theta_i(\xi_N)$,

$$\left(\frac{d\lambda([\xi_N, v_N])}{\lambda^2}\right)_{v_N} = \frac{d\lambda(\xi_N)}{\lambda^2} + O(\delta \ \mathrm{Log}^m \bar{\lambda}).$$

(In $d\lambda(\xi_N)$, either the ξ_N-derivative is taken on θ_i or, if not, a δ is freed. The additional contribution is thrown into $O(\delta \ \mathrm{Log}^m \bar{\lambda})$).
   Similarly,

$$\left(\frac{d\lambda([\xi_N, v_N])}{\lambda^2}\right)_{v_N v_N} = \frac{d\lambda([v_N, \xi_N])}{\lambda^2} + O(\delta \ \mathrm{Log}^m \bar{\lambda}).$$

Thus,

$$\tilde{\tau} = -\left(\frac{1}{\lambda}\left(\frac{d\lambda(\xi_N)}{\lambda^2}\right)_{\xi_N} + \frac{1}{\lambda}\left(\frac{d\lambda([\xi_N, v_N])}{\lambda^2}\right)_{[\xi_N, v_N]}\right.$$
$$\left. - \left(\frac{d\lambda(\xi_N)}{\lambda^2}\right)^2 - \left(\frac{d\lambda([\xi_N, v_N])}{\lambda^2}\right)^2\right) + O(\delta \ \mathrm{Log}^m \bar{\lambda} + 1).$$

   Observe now that

$$-\frac{1}{\lambda}\left(\frac{d\lambda(\xi_N)}{\lambda^2}\right)_{\xi_N} = -\frac{1}{\lambda^2}\left(\frac{d\lambda(\xi_N)}{\lambda}\right)_{\xi_N} + \left(\frac{d\lambda(\xi_N)}{\lambda^2}\right)^2$$

$$-\frac{1}{\lambda}\left(\frac{d\lambda([\xi_N, v_N])}{\lambda^2}\right)_{[\xi_N, v_N]} = -\frac{1}{\lambda^2}\left(\frac{d\lambda([\xi_N, v_N])}{\lambda}\right)_{[\xi_N, v_N]} + \left(\frac{d\lambda([\xi_N, v_N])}{\lambda^2}\right)^2.$$

Thus, using the identities above and the form observations stated earlier

$$\tilde{\tau} = O(\delta \ \mathrm{Log}^m \bar{\lambda} + 1) - \frac{1}{\lambda^2}\left(\left(\frac{d\lambda(\xi_N)}{\lambda}\right)_{\xi_N} + \left(\frac{d\lambda([\xi_N, v_N])}{\lambda}\right)_{[\xi_N, v_N]}\right)$$
$$= O(\delta \ \mathrm{Log}^m \bar{\lambda} + 1) - e^{-2\delta \sum \theta_i s_i} \sum_i \delta s_i \left(\sum (d\theta_i([\xi_N, v_N]))_{[\xi_N, v_N]}\right.$$
$$\left. + (d\theta_i(\xi_N))_{\xi_N}\right).$$

Let π be the pull back map from x towards the torus $T_2 - T_1$ (outside T_1, inside of T_2) where the modification takes place (the introduction of the large rotation) onto $\partial T_2 \cdot \theta_i$ is in fact defined on ∂T_2 and should be thought of as $\theta_i \circ \pi$. Thus,

$$\tilde{\tau} = O(\delta \, \mathrm{Log}^m \bar{\lambda} + 1) - e^{-2\delta \sum \theta_i s_i} \sum_i \delta s_i \left([\xi_N, \upsilon_N] \cdot (d\theta_i \circ d\pi)([\xi_N, \upsilon_N]) \right.$$

$$+ \, \xi_N \cdot (d\theta_i \circ d\pi)(\xi_N) \big) .$$

We then split

$$[\xi_N, \upsilon_N] = d\pi([\xi_N, \upsilon_N]) + \alpha_N \upsilon_N$$
$$\xi_N = d\pi(\xi_N) + \beta_N \upsilon_N$$

α_N and β_N are easily seen to be bounded and independent of N. Thus, since $d\theta_i \circ d\pi(\upsilon_N) = 0$,

$$\alpha_N \upsilon_N \cdot d\theta_i \circ d\pi([\xi_N, \upsilon_N]) = \alpha_N d\theta_i \circ d\pi([\upsilon_N, [\xi_N, \upsilon_N]]) = \alpha_N d\theta_i \circ d\pi(\xi_N)$$
$$\beta_N \upsilon_N \cdot d\theta_i \circ d\pi(\xi_N) = \beta_N d\theta_i \circ d\pi([\upsilon_N, \xi_N])$$

and

$$\tilde{\tau} = O(\delta \, \mathrm{Log}^m \bar{\lambda} + 1) - e^{2\delta \sum \theta_i s_i} \sum_i \delta s_i \left(d\pi([\xi_N, \upsilon_N]) \cdot d\theta_i \circ d\pi([\xi_N, \upsilon_N]) \right.$$

$$+ \, d\pi(\xi_N) \cdot d\theta_i \circ d\pi(\xi_N) + O(|d\theta_i \circ d\pi|) \,)$$

$$= O(\delta \, \mathrm{Log}^m \bar{\lambda} + 1) - e^{-2\delta \sum \theta_i s_i} \sum_i \delta s_i \left(d^2\theta_i \left(d\pi([\xi_N, \upsilon_N]), d\pi([\xi_N, \upsilon_N]) \right) \right.$$

$$+ \, d^2\theta \left(d\pi(\xi_N), d\pi(\xi_N) \right) + d\theta_i \left(d\pi([\xi_N, \upsilon_N]) \cdot d\pi([\xi_N, \upsilon_N]) \right)$$

$$+ \, d\pi(\xi_N) \cdot d\pi(\xi_N)) + O(|d\theta_i \circ d\pi|) \,) .$$

It is easy to see that $d\pi([\xi_N, \upsilon_N]) \cdot d\pi([\xi_N, \upsilon_N])$ and $d\pi(\xi_N) \cdot d\pi(\xi_N)$ are bounded independently of N so that

$$\tilde{\tau} = O(\delta \, \mathrm{Log}^m \bar{\lambda} + 1) - e^{-2\delta \sum \theta_i s_i} \sum_i \delta s_i \left(d^2\theta_i \left(d\pi([\xi_N, \upsilon_N]), d\pi([\xi_N, \upsilon_N]) \right) \right.$$

$$+ \, d^2\theta_i(d\pi(\xi_N, d\pi(\xi_N))) + O(|d\theta_i \circ d\pi|) \,) .$$

From the formula for ξ_N, see Proposition 2, it is clear that $d\pi([\xi_N, \upsilon_N])$ and $d\pi(\xi_N)$ have bounded lengths and lengths bounded away from zero and their determinant is bounded away from zero ($\bar{\gamma}$ is small). Indeed

$$(x^2 + y^2)(0)\bar{\gamma}d\pi(\xi_N) =$$

$$\left(\frac{(x - \bar{\gamma}y)(0)}{\sqrt{1 + \bar{\gamma}^2}}\cos s_2 + \sqrt{1 + \bar{\gamma}^2}y(0)\sin s_2\right)(e_1 - e_2)$$

$$+ \bar{\gamma}\left((y + \bar{\gamma}x)(0)\cos s_2 - \sqrt{1 + \bar{\gamma}^2}x(0)\sin s_2\right)e_2$$

$$(x^2 + y^2)(0)\bar{\gamma}d\pi([v_N, \xi_N]) =$$

$$\left(-\frac{(x - \bar{\gamma}y)(0)}{\sqrt{1 + \bar{\gamma}^2}}\sin s_2 + \sqrt{1 + \bar{\gamma}^2}y(0)\cos s_2\right)(e_1 - e_2)$$

$$- \bar{\gamma}\left((y + \bar{\gamma}x)(0)\sin s_2 + \sqrt{1 + \bar{\gamma}^2}x(0)\cos s_2\right)e_2.$$

On the basis $((e_1 - e_2), e_2)$, the determinant is:

$$- \bar{\gamma}\left(\left((y + \bar{\gamma}x)(0)\sin s_2 + \sqrt{1 + \bar{\gamma}^2}x(0)\cos s_2\right)\right.$$

$$\left(\frac{(x - \bar{\gamma}y)(0)}{\sqrt{1 + \bar{\gamma}^2}}\cos s_2 + \sqrt{1 + \bar{\gamma}^2}y(0)\sin s_2\right)$$

$$+ \left((y + \bar{\gamma}x)(0)\cos s_2 - \sqrt{1 + \bar{\gamma}^2}x(0)\sin s_2\right)$$

$$\left.\left(-\frac{(x - \bar{\gamma}y)(0)}{\sqrt{1 + \bar{\gamma}^2}}\sin s_2 + \sqrt{1 + \bar{\gamma}^2}y(0)\cos s_2\right)\right)$$

$$= -\bar{\gamma}\left((y(0)\sin s_2 + x(0)\cos s_2)^2 + (y(0)\cos s_2 - x(0)\sin s_2)^2\right.$$

$$\left. + \bar{\gamma}O\left((x^2 + y^2)(0)\right)\right) = -\bar{\gamma}(x^2 + y^2)(0)\left(1 + O(\bar{\gamma})\right).$$

On the other hand, setting $(X = A(e_1 - e_2) + Be_2)\|X\|^2 = A^2 + \frac{1}{\bar{\gamma}^2}B^2$, we have:

$$\left(\frac{(x - \bar{\gamma}y)(0)\cos s_2}{\sqrt{1 + \bar{\gamma}^2}} + \sqrt{1 + \bar{\gamma}^2}y(0)\sin_2\right)$$

$$+ \left((y + \bar{\gamma}x)(0)\cos s_2 - \sqrt{1 + \bar{\gamma}^2}x(0)\sin s_2\right)^2$$

$$= (x(0)^2 + y(0)^2)(1 + O(\bar{\gamma})) = (x^2 + y^2)^2(0)\bar{\gamma}^2\|d\pi(\xi_N)\|^2.$$

The claim follows.

We are ready to prove:

Proposition 6   *There exists a constant C independent of $N, \bar{\lambda}$ such that*

$$\tilde{\tau} \leq C.$$

Proof.   It suffices to build $\theta_i \circ \pi$ (independent of $N, \bar{\lambda}$ etc.) so that

$$d^2\theta_i\left(d\pi(\xi_N), d\pi(\xi_N)\right) + d^2\theta_i(d\pi([v_N, \xi_N]), d\pi([v_N, \xi_N])) + O(|d\theta_i \circ d\pi|) \geq 1$$

if $0 \leq \theta_i \leq \frac{1}{2}$. This is possible in view of our claim above.

Furthermore, there exists a constant C_1 independent of N such that

$$d^2\theta_i\left(d\pi(\xi_N), d\pi(\xi_N)\right) + d^2\theta_i(d\pi([v_N, \xi_N]), d\pi([v_N, \xi_N])) + O(|d\theta_i \circ d\pi|) \leq C_1.$$

Finally, we choose δ and $\bar{\lambda}$ so that

$$O(\delta \, \text{Log}^m \bar{\lambda} + 1) \leq C_1.$$

The estimate on $\tilde{\tau}$ follows.

Assume now, in a first step, that no periodic orbit generating our homology intersects the stable or the unstable manifold of $\mathcal{O}$. We first complete a diffeomorphism of M and spread the rotation which we have introduced near the attracting and repelling orbit along the stable and unstable manifold of $\mathcal{O}$. We are pointing out, on the drawing below the zones where φ_N is non constant, dropping from 1 to a value $O\left(\frac{1}{N}\right)$ or climbing back to 1.

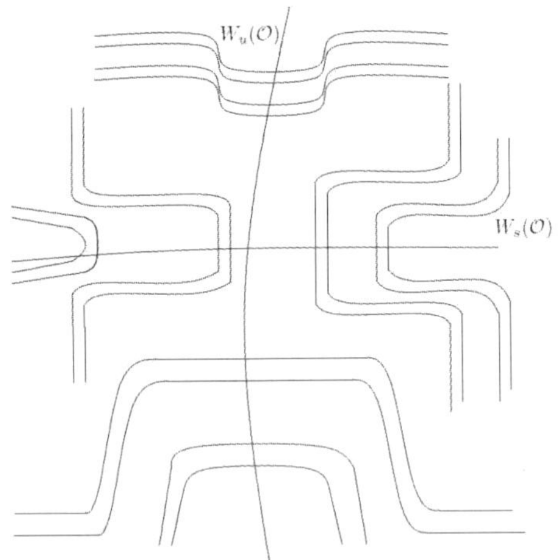

We create, half-way between each pair of strips, a surface S. We cut then in this picture a thin hyperbolic neighborhood of $W_u(\mathcal{O}) \cup W_s(\mathcal{O})$ and a thinner one.

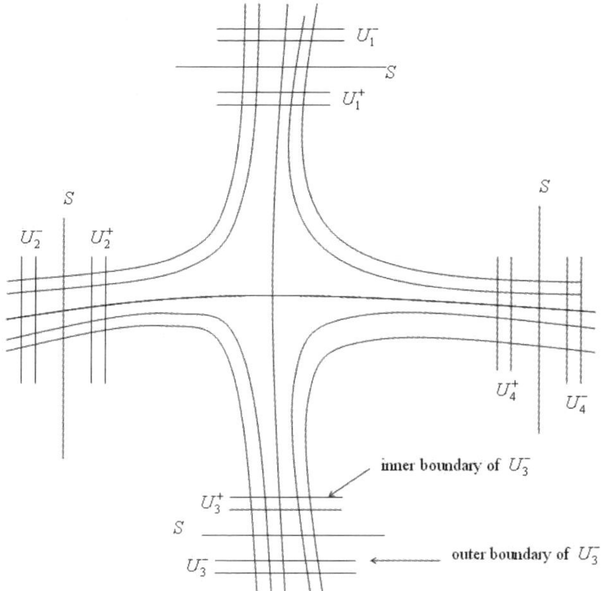

Between U_i^- and S, $\ker \alpha_N$ turns considerably along $v_N = \varphi_N v$. We can build, with all required bounds on $\frac{\partial^m s_i}{\partial \tau_i^m}$, a function s_i equal to zero on the outer boundary of U_i^- and equal to a large value ℓ_N as we reach S. We can also build $\theta_i = \theta$, a function equal to zero ouside of the larger neighborhood of $W_u(\mathcal{O}) \cup W_s(\mathcal{O})$ and equal to 1 on the smaller one. We need here only two functions s_i, s_1 and s_2, s_1 for the repelling orbit and s_2 for the attracting one, with $\theta_1 = \theta_2 = \theta$.

As we reach S, s_1 and s_2 are equal to ℓ_N, $\alpha_{N,\lambda} = e^{\delta\theta\ell_N}\alpha$. For $\theta = 1, \alpha = e^{\delta\ell_N}\alpha = \lambda_N\alpha$, λ_N tending to $+\infty$ with N. Thus, our form extends to all of M. $\qquad\square$

We claim now:

Proposition 7 *Let us consider the periodic orbits of ξ_0 which define the homology at some fixed index k_0 and their unstable manifolds in C_β. The curves on these unstable manifolds do not enter a fixed and small neighborhood of $\mathcal{O}$.*

Proof. All curves x of this type have a tangent vector

$$\dot{x} = a\xi_{N,\lambda} + bv_{N,\lambda}$$

with

$$a \le a_0; \quad \int_0^1 |b| \le C.$$

a_0 and C are independent of N, a_0 for energy reasons, C because of the bound on $\tilde{\tau}$. Suppose x enters the inner chore. Then, $\dot{x} = O\left(\frac{1}{\lambda_N}\right) + bv_{N,\lambda}$. For N very large, this is basically a piece of orbit of v. If x enters the inner chore from the side i.e. from the boundary of the hyperbolic neighborhood, it stays away from $\mathcal{O}$ since similar orbits of v do not approach $\mathcal{O}$.

On the other hand, if these curves enter the inner chore through the interior boundary of U_i^-, then $v_{N,\lambda} = \frac{\varphi_{N}v}{\sqrt{\gamma(x)}}$ between this interior boundary and S, i.e. $v_{N,\lambda} = O\left(\frac{1}{N}\right)v$. Since $\int_0^1 |b|$ is bounded and $\xi_N = O\left(\frac{1}{\lambda_N}\right)$, such a curve can hardly move. It cannot enter, assuming it starts in U_i^- or between U_i^- and S, a smaller neighborhood of $\mathcal{O}$ since the piece of orbit of v it spans is so small.

Next, the curve x has no point between U_i^- and S, i.e. lies entirely between S and $\mathcal{O}$. By continuity, we may assume that x starts near S and is therefore entirely contained between S and the outer-boundary of U_i^+, away from the side-boundaries.

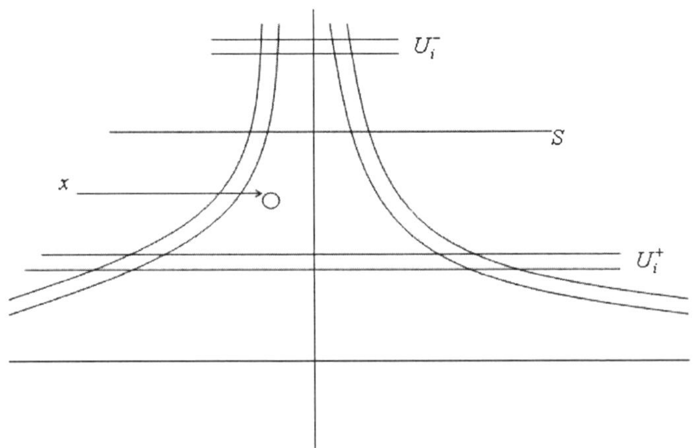

Then,

$$\dot{x} = \frac{\xi_0}{\lambda_N} + bvO\left(\frac{1}{N}\right), \quad \text{with} \quad \int_0^1 |b| \le C.$$

In this region, $\alpha_N = \lambda_N \alpha_0$, λ_N a constant and C_β for (α_N, v_N) and (α_0, v) coincide. The curve is tiny and a pseudo-gradient for $\int_0^1 \alpha_0(\dot{x})dt$ on C_{β_0} is a pseudo-gradient for $\int_0^1 \alpha_N(\dot{x})dt$ on C_{β_N}. It is easy to see that such a pseudo-gradient (for $\int_0^1 \alpha_0(\dot{x})dt$ on C_{β_0}) will drive such tiny curves to points locally i.e. keeping away from $\mathcal{O}$. $\qquad\square$

We now have to face the possibility that the periodic orbits of ξ_0 might intersect $W_u(\mathcal{O})$ and $W_s(\mathcal{O})$.

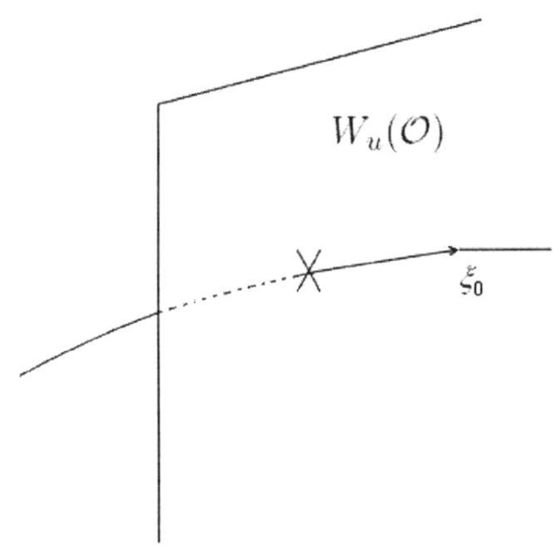

If we try then to carry the rotation from the attractive or repulsive periodic orbit of v to the hyperbolic one, we perturb ξ_0, push away the periodic orbit. If we change the Hamiltonian, we change completely the periodic orbit since we go beyond the effect of a diffeomorphism (carrying rotation is completed through a diffeomorphism once the modification of α_0 into α_N is achieved). If we remove a flow-line neighborhood, as small as we may wish, of the flow-lines of v originating in such periodic orbits (which intersect $W_u(\mathcal{O})$ or $W_s(\mathcal{O})$), we can carry out the rotation of α_N on the complement. How large a neighborhood of the hyperbolic orbit are we carrying then? How much are we missing?

Suppose for example that no periodic of ξ_0 intersects, $W_u(\mathcal{O})$, but that several periodic orbits of ξ_0 intersect $W_u(\mathcal{O})$, typically one for simplicity. Then, the rotation from the repelling orbit can be carried out beyond the hyperbolic orbit. These flow-lines (which carry a lot of rotation) fill in a neighborhood of $W_u(\mathcal{O})$ after removing $W_u(\mathcal{O})$.

Using a view from top, we have:

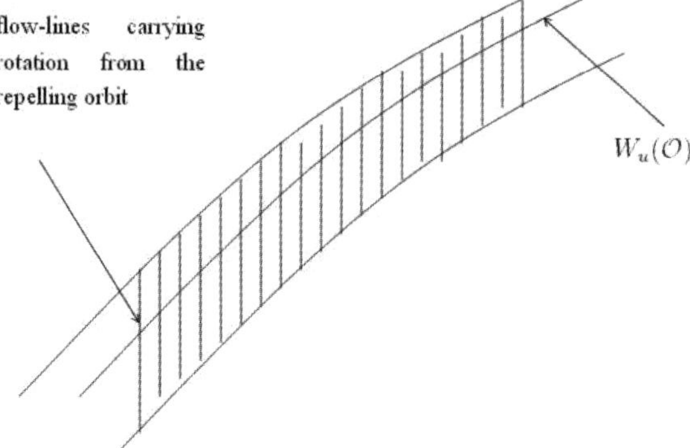

On the other hand, our periodic orbit intersects $W_u(\mathcal{O})$:

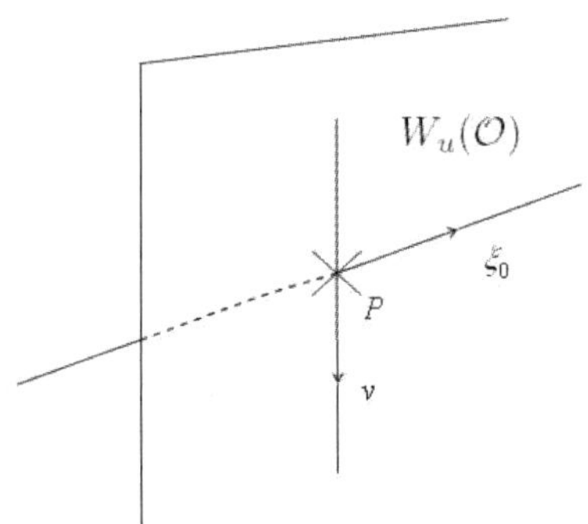

Thus, if we remove a neighborhood of the flow-line of v through P, we can safely bring rotation from the attracting orbit as well.

Using a view from top

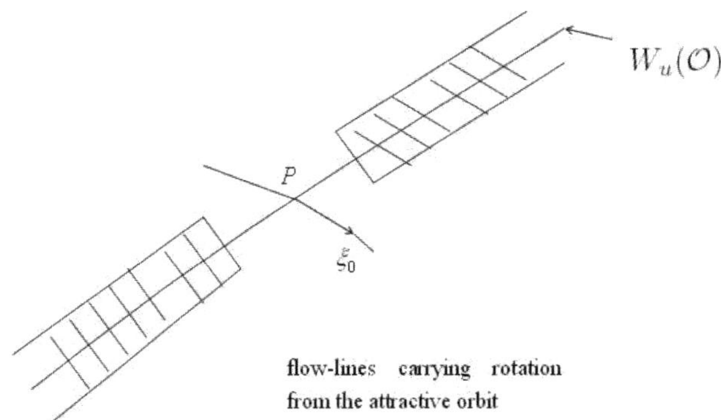

Thus, below P, we can build a lot of rotation after combining the rotation which we can safely bring from the attractive orbit with the rotation from the repelling orbit:

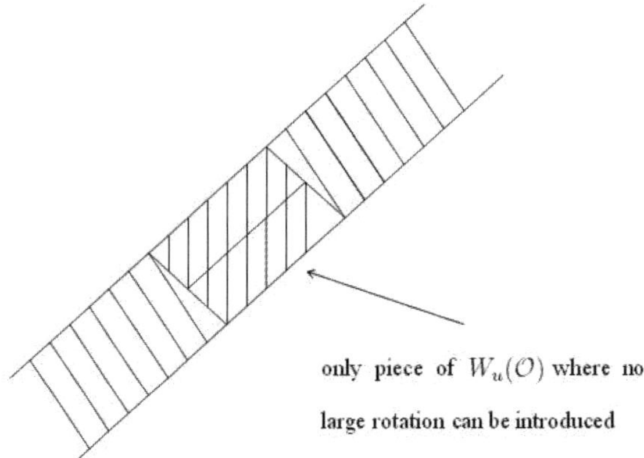

The piece which is left is as small as we wish, we can think of it as a tiny neighborhood of the (downwards) flow-line of v through P. We thus only need to fill this hole. On the boundaries of this hole, we have a lot of rotation distributed as follows

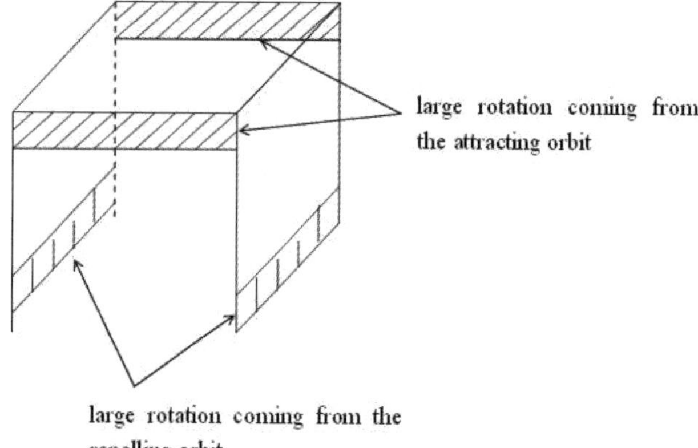

If several periodic orbits intersect $W_u(\mathcal{O})$ and none $W_s(\mathcal{O})$, this hole becomes several holes, but the basic process does not change.

If one (or several) periodic orbit intersects $W_u(\mathcal{O})$ and one (or several) periodic orbit intersect $W_s(\mathcal{O})$, the situation changes since orbits very close to $W_u(\mathcal{O}) \cup W_s(\mathcal{O})$ connect these orbits then.

Any such v-orbit cannot be filled with rotation:

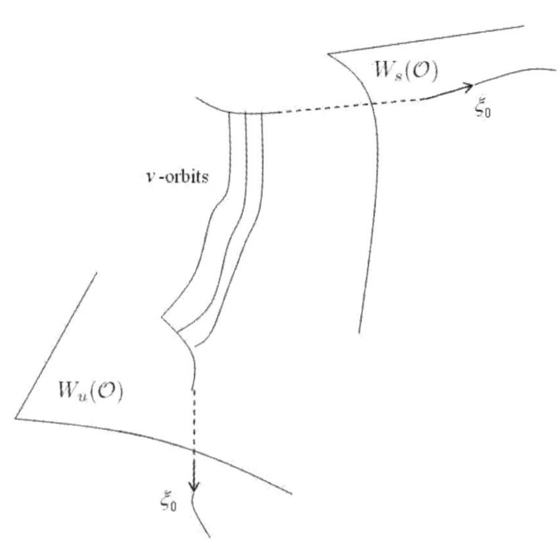

If we remove these flow-lines, we can fill in every remaining flow-line with rotation.

Combining, we can certainly fill with rotation near the intersection point of a periodic orbit with $W_s(\mathcal{O})$ or $W_u(\mathcal{O})$ the following (shaded) set of flow-lines:

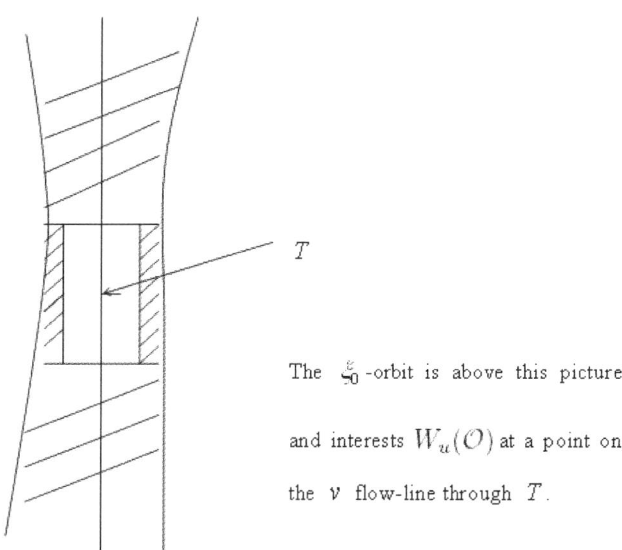

The ξ_0-orbit is above this picture

and interests $W_u(\mathcal{O})$ at a point on

the v flow-line through T.

The ξ_0-orbit is above this picture and intersects $W_u(\mathcal{O})$ at a point on the v-flow-line through T.

The periodic orbit lies above this picture. We cannot fill the hole more because some v-flow-lines of the hole connect this periodic orbit (which intersects $W_u(\mathcal{O})$ here) with another periodic orbit (intersecting $W_s(\mathcal{O})$ then). We can take this hole to be as small as we wish though after taking thinner neighborhood of the allowed set of flow-lines.

Indeed, any space between the v-orbits connecting the two periodic orbits can be filled in, see the drawing below.

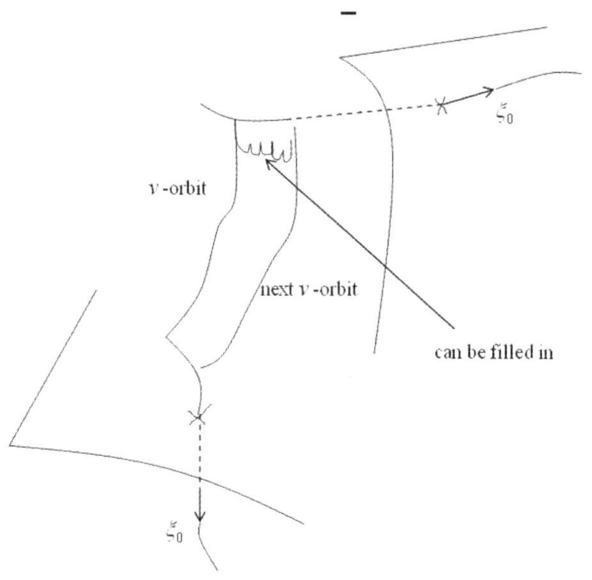

Such spaces are as close as we wish from P. Combining with the rotation brought from above P (repelling or attracting orbit) — carefully removing first the v-flow-lines of the periodic orbit — we derive the "hole-neighborhoods".

Thus, in all the cases studied above, we have derived sets of flow-lines surrounding the hyperbolic periodic orbit and carrying as much rotation as we please but for a finite number of hole-neighborhoods of the following type:

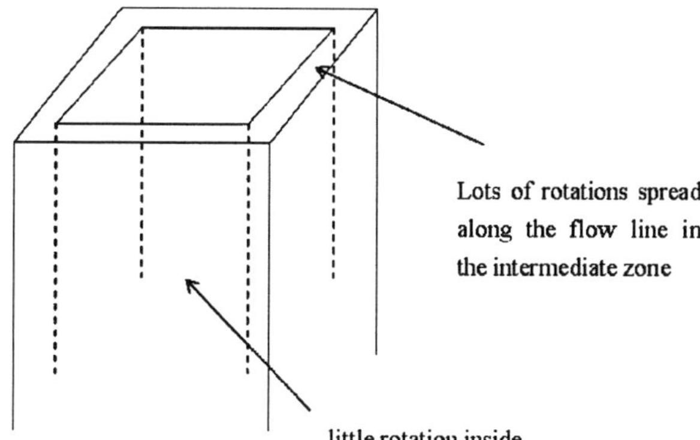

We remove the content of the hole i.e. $\alpha = \alpha_N$ is now defined only at the top of the hole and in the outer neighborhood of flow-lines where there is a lot of rotation.

In the empty hole, we now build a new α_N which rotates considerably before reaching the hyperbolic periodic orbit:

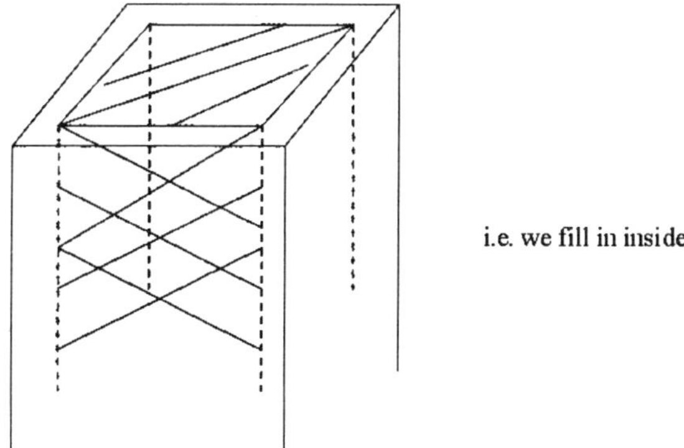

For this purpose, we use (*). We only need to use the appropriate γ and to glue the new rotation so that we have a globally defined α_N, with all required bounds etc.

The first step is to get $\bar{\mu} = 0$ on the boundaries of smaller holes, including the top boundary. Next, we need to rescale the large rotation that we have on each lateral wall so that it stays large but becomes the same all around instead of being split between the top part and the bottom part (see Figure (A)) according to the wall which we are considering.

These two steps are completed in the space between an inner hole and an outer hole (which is smaller than the initial hole, (see Figure (B) below). Then, we can fill the inner hole with a uniform, large rotation.

We first observe that the boundary ∂S, in the flow-box, of the set $S = \{x \in \text{flow-box}, \varphi(x) \neq 1\}$ is independent of N. It depends only on v, which remains unchanged with N, on the intervals I_0 and on how we carry the rotation below the periodic orbit (coming from the attractive as well as repulsive orbits).

Second, v is transverse to ∂S since v is transverse to the boundaries of the tori where the insertion of a large rotation has taken place.

Third, on each flow-line of v in the box, there are at most three intervals; one where φ is not 1, then one where φ is 1 and a last one where φ is not 1 again.

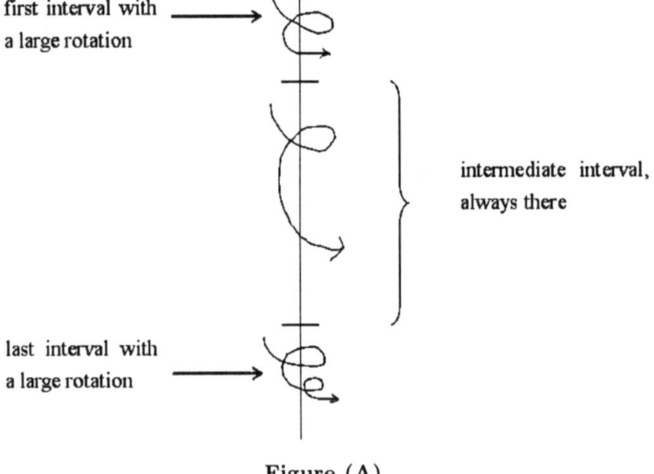

Figure (A)

The only interval which is always present is the one where φ is 1, the "intermediate" one since the large rotations take place in the vicinity of the top and of the bottom of the box. The flow-lines run from top to bottom near (only near) the lateral sides of the box. Each of these flow-lines carries a large rotation which is borrowed either from its top portion or from its bottom one, or from both. Using the large rotations coming from the attractive v-orbit and the repelling v-orbit, we can construct a box around the hole where a lot of rotation is carried around its boundary (all of it):

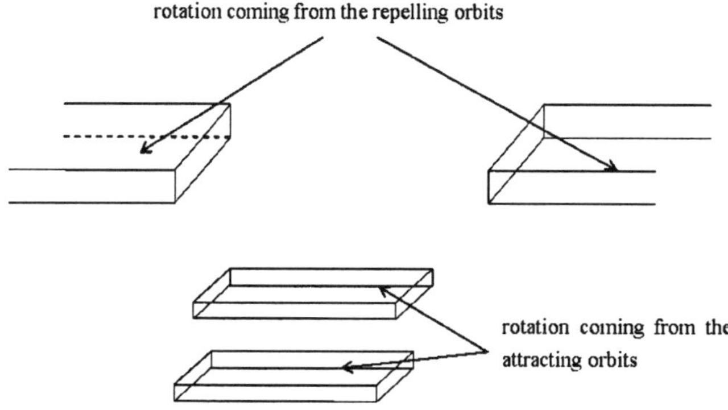

We draw then the following two layers:

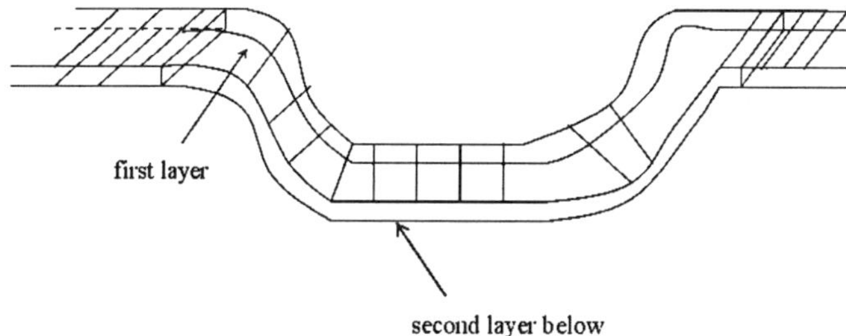

which cut into the upwards rotation, go down to the downwards one and then come back to the upwards one. The top one stays some more upstairs as the lower one speeds up to the lower level to collect rotation from there. If we cut then the central piece and flatten it, we find a thin box:

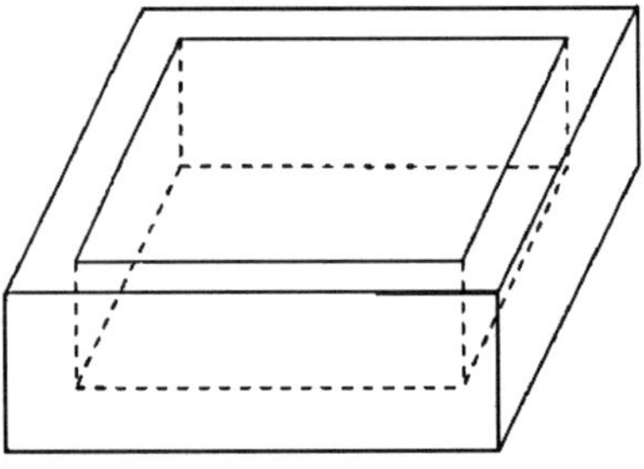

Figure (B)

This carries rotation all around its boundary.

Our arguments apply to this box.

If there is a flow-line (and then several) running from a periodic orbit of $\bar{\xi}$ cutting $W_u(\mathcal{O})$ to a peridic orbit of $\bar{\xi}$ cutting $W_s(\mathcal{O})$, there are two constructions as the one carried out above, one for the top with the periodic orbit cutting $W_u(\mathcal{O})$, the other one for the bottom with the periodic orbit cutting $W_s(\mathcal{O})$. We can match the parts containing a lot of rotation from the top and from the bottom (the boundaries parts). We then fill in partially inside (without matching, leaving a hole which is a neighborhood of the flow-line which connects the two periodic orbits) as if we had only a top or only a bottom.

The basic double picture for top and bottom together (not thickened, just flat) is:

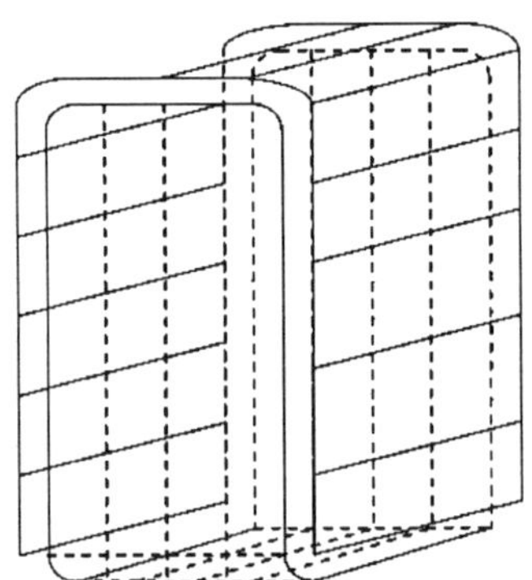

i.e. there is a top:

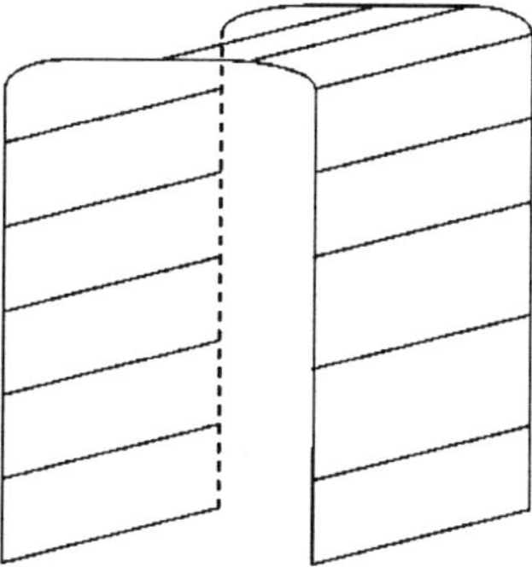

and a bottom:

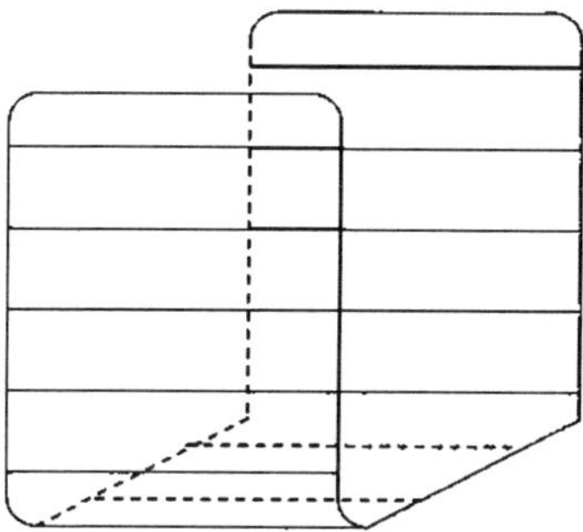

which basically bound the same boundary. Near the boundary, for both of them, there is a lot of rotation. Top and bottom fit together to define a flow-box. The v-flow-line is jailed in the box.

In order to see better how to build our boxes, we draw two ends together and mark with bold lines the two caps of the box from inside which we add rotation, sealing the whole box with rotation all around except for a hole inside it.

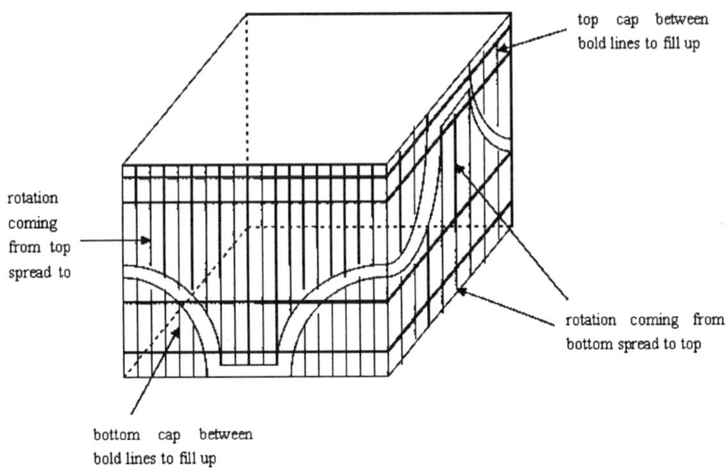

Observation. Between the two bold lines defining either of the top or the bottom caps, the v-flow-lines carry a lot of rotation near the boundary. This fact is used to extend the rotation inside the box, sealing it off; a hole is left inside.

We now have our initial flow-box and an inner, smaller one without bottom. α is defined in the space between the two boxes and or the top side.

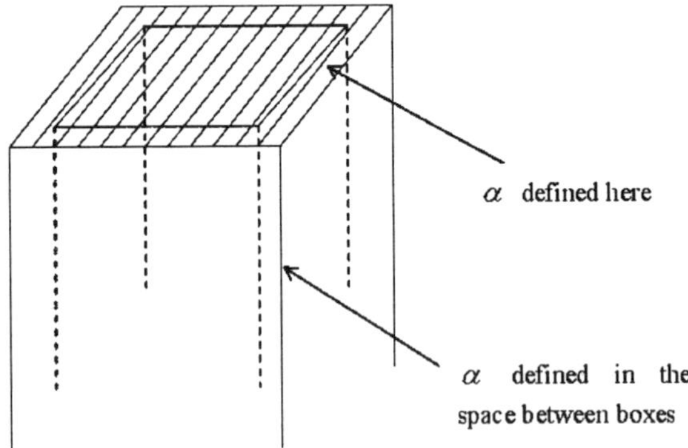

The only space where ξ and α are not defined is a parallelepiped with a top and with a bottom. We build another yet smaller parallelepiped:

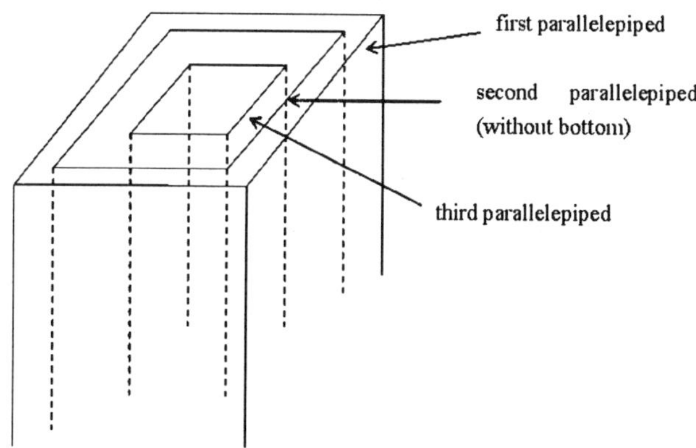

Since ∂S is independent of N, we can easily extend the function φ (the parametrization along v) between parallelepiped 2 and 3 and extend as well $\bar{\mu}, \xi$ etc. The uniformity of ∂S allows us to keep all bounds. $\bar{\mu}$ (extended using the function γ and (*), ξ, α are extended using (*)) is kept equal to zero on the extension of S (which we may complete as we please between box 2 and box 3 as long as it matches with the boundary data — of ∂S — on the boundaries — top and lateral — of box 2).

Such an extension of ξ, α etc. enjoys the same bounds. Indeed, outside of S, ξ was $\bar{\xi}$. Thus, on $\partial S, \xi = \bar{\xi}, [\varphi v, \xi] = [v, \bar{\xi}]$, the v- interval outside of S is "large" as pointed out (independent of N) so that (*) provides C^{∞}- bounds depending only on γ. τ is therefore bounded outside of S. Inside S, it is bounded because $\bar{\mu} = 0$, thus $\tau_v (= -\bar{\mu}_{\xi\xi} - \tau\bar{\mu})$ is zero and τ equals the value it has on the boundaries of the "intermediate, large" interval.

One issue to worry about is the glueing of the data of $\xi, [\varphi v, \xi]$ derived from the initial conditions near the top of the box after the use of (*) once we reach the bottom part of the box.

Since box 2 has no bottom side, we can sidestep this problem here, but one can easily overcome it manipulating (*) above the boxes.

We may assume now that (the extension of $\bar{\mu}$ into γ is as we please, subject to $\gamma = 0$ on the extension of S) γ is zero identically on the boundaries of box 3, including the top side.

One can get $\bar{\mu} = 0$ on the top side after a modification of $\bar{\mu}$ into zero, using (*) and γ, along small flow-lines originating in this top part — curving then the top part, we can build a flow-box 1 which has the same lateral boundaries than the former flow-box 1 and still carries a large rotation on all flow-lines between box 1 and box 2, while $\bar{\mu} = 0$ on the top portion of box 3.

We then observe that the rotation of γ on the lateral boundaries of box 3 is large, either on the top portion or on the bottom one. This is embedded in the construction and is due to the fact that $\int \frac{dx}{\varphi(x)}$ (φ is the parametrization along v) is large on each of these flow-lines. We then extend φ so that it becomes constant (small obviously) on all the lateral sides of a yet smaller parallelepiped box 4. From there, the extension inside box 4 is immediate.

We need to check that this last modification, the spreading of the rotation so that it becomes uniform, keeps all bounds holding true. This rescaling is typically derived through the diffeomorphism of $[0,1]$.

$$[0, 1] \longrightarrow [0, 1]$$

$$x \longrightarrow \frac{t \int_0^x \frac{ds}{\varphi(s)} + (1 - t)x}{\int_0^1 \frac{ds}{\varphi(s)}}$$

$[0,1]$ is the time along the v-flow-line from bottom to top, φ is the function built with the rotations, as such it depends on the base point z of the flow-line. $t = t(z)$ depends also on the base point of the flow-line z, which is on the top of the box. t is zero on the lateral boundary of box 2 and 1 on the lateral boundary of box 3.

This gives rise to a diffeomorphism $\gamma_{s(y)}(y)$ of the space between box 2 and box 3. γ_s is the one-parameter of v.

$D\gamma_s$ is of course bounded. ds is the differential of

$$\frac{t(z) \int_0^x \frac{ds}{\varphi(s)} + (1 - t(z))x}{\int_0^1 \frac{ds}{\varphi(s)}} - x.$$

Observe that $\frac{1}{\varphi}$ is at most CN, hence is upperbounded by $C_1 \int_0^1 \frac{ds}{\varphi(s)}$, since we may assume that the total rotation of these flow-lines, between box 2 and box 3, is at least $c_0 N$.

We also claim that

$$\left| \int_0^x \frac{\partial \varphi}{\partial z} \frac{ds}{\varphi^2} \right| \leq C_1 \int_0^1 \frac{ds}{\varphi(s)}.$$

Indeed, the top of the flow-box is transverse to v. φ is a function of s and as such, between the two tori, does not depend on the flow-lines. The dependency on the flow-lines is due to the transformation (given, independent of N) which brings the rotation to the flow-box. Thus,

$$\left| \frac{\partial \varphi}{\partial z} \right| \leq C |\varphi'(s)|$$

and

$$\left| \int_0^x \frac{\partial \varphi}{\partial z} \frac{ds}{\varphi^2} \right| \leq C \int_0^x \frac{|\varphi'|}{\varphi^2} ds \leq C' \, \mathrm{Max} \, \frac{1}{\varphi}.$$

Thus, again,

$$\frac{\left| \int_0^x \frac{\partial \varphi}{\partial z} \frac{ds}{\varphi^2} \right|}{\int_0^1 \frac{ds}{\varphi(s)}} \leq C_1$$

dt is also bounded independently of N. Thus, ds is bounded independently of N.

We thus have ($\tilde{\varphi}$ is the parametrization which we built).

Proposition 8 $\xi, [\tilde{\varphi}v, \xi], \bar{\mu}, \tau$ *are bounded. Furthermore,*

$$\xi \cdot \xi, [\tilde{\varphi}v, \xi] \cdot [\tilde{\varphi}v, \xi], \xi \cdot [\tilde{\varphi}v, \xi], [\tilde{\varphi}v, \xi] \cdot \xi$$

are bounded independently of N transversally to v.

Proof. Since $D\gamma_{s(y)} + ds(\cdot)v$ is bounded, $\xi, [\tilde{\varphi}v, \xi]$ are bounded. $\bar{\mu}$ and τ are bounded by construction.

$\xi \cdot \xi$ etc are initially bounded but $\gamma_{s(y)}(y)$ could have an effect. However, denoting $\tilde{\xi}$ the initial ξ,

$$\xi = D\gamma_{s(y)}(\tilde{\xi}) + ds(\tilde{\xi})v = D(\gamma_{s(y)})(\tilde{\xi}).$$

Any further derivative taken on ξ through a vector-field which reads

$$D\gamma_{s(y)}(X) + ds(X)v = D(\gamma_{s(y)})(X)$$

would yield derivatives of $\tilde{\xi}$ along X which are bounded transversally to v (and v is mapped onto θv by $D(\gamma_{s(y)})$) if X splits on $\tilde{\xi}, [\varphi v, \tilde{\xi}]$, i.e. if $D\gamma_{s(y)}(X) + ds(X)v$ splits on $\xi, [\tilde{\varphi}v, \xi]$, derivatives of v, which are bounded and would yield derivatives of $D(\gamma_{s(y)})$ which might be unbounded.

But

$$D(\gamma_{s(y)}) = D\gamma_{s(y)} + dsv.$$

Derivatives fo $D\gamma_{s(y)}$ are bounded as well as derivatives of v. Derivatives of ds might be unbounded; but these are multiplied by v and our estimate is transversal to v.

Our construction of the functions s_1, s_2 etc proceeds then as in the case when no periodic orbit of the contact vector-field intersected $W_u(\mathcal{O})$ and $W_s(\mathcal{O})$. Proposition 8 holds. $\square$

Lemma 12   *The function x^2+y^2 has no local maximum near the repelling or attracting orbits along the trajectories of $\bar{\xi}$.*

Proof.   We recall that $\bar{\xi} = \frac{\partial}{\partial x} = \bar{\gamma}\frac{\partial}{\partial y}$ so that

$$\bar{\xi} \cdot \bar{\xi} \cdot (x^2 + y^2) = 2\bar{\xi} \cdot (x - \bar{\gamma}y) = 2(1 + \bar{\gamma}^2) > 0.$$

$\qquad\square$

Corollary 4   *There is no "small" $\bar{\xi}$-trajectory exiting from T_2 and coming back after a short time.*

Next, we establish the following qualitive result:

Lemma 13   *For $\bar{\gamma}$ small enough, τ_N is negative in $T_2 - T_1$.*

Proof.   We write

$$\xi_N = (A \cos s_1 + B \sin s_1)e_2 + (A_1 \cos s_1 + B_1 \sin s_1)\varphi v + De_1$$

φv is $\frac{\partial}{\partial s_1}, e_2 = y\frac{\partial}{\partial x} - x\frac{\partial}{\partial y}, e_1 = 20\frac{\partial}{\partial \theta}$.
We know that

$$[\xi_N, [\varphi v, \xi_N]] = \tau_N \varphi v$$

i.e.

$$\left[\xi_N, \frac{\partial \xi_N}{\partial s_1}\right] = \tau_N \frac{\partial}{\partial s_1}.$$

Also

$$\frac{\partial e_1}{\partial s_1} = \frac{\partial e_2}{\partial s_1} = 0.$$

Since $ds_1(e_1) = ds_1(e_2) = 0$ and since $\frac{\partial}{\partial \theta}A_1 = \frac{\partial}{\partial \theta}B_1 = 0, e_1$ has no contribution in $[\xi_N, \frac{\partial \xi_N}{\partial s_1}]$ and

$$\left[\xi_N, \frac{\partial \xi_N}{\partial s_1}\right] = \left[(A \cos s_1 + B \sin s_1)e_2 + (A_1 \cos s_1 + B_1 \sin s_1)\frac{\partial}{\partial s_1},\right.$$

$$\left.(-A \sin s_1 + B \cos s_1)e_2 + (-A_1 \sin s_1 + B_1 \cos s_1)\frac{\partial}{\partial s_1}\right]$$

$$= \frac{\partial}{\partial s_1}\left(-(A_1 \cos s_1 + B_1 \sin s_1)^2 - (B_1 \cos s_1 - A_1 \sin s_1)^2\right.$$

$$+(A \cos s_1 + B \sin_1)(-e_2 \cdot A_1 \sin s_1 + e_2 \cdot B_1 \cos s_1)$$

$$-(-A \sin s_1 + B \cos s_1)(e_2 \cdot A_1 \cos_1 + e_2 \cdot B_1 \sin s_1))$$

$$= \left(-(A_1^2 + B_1^2) + Ae_2 \cdot B_1 - Be_2 \cdot A_1\right) \frac{\partial}{\partial s_1}.$$

Thus,

$$\tau_N = -(A_1^2 + B_1^2) + Ae_2 \cdot B_1 - Be_2 \cdot A_1.$$

From Proposition 2, we derive

$$A = \frac{-(x - \bar{\gamma}y)}{\bar{\gamma}(x^2 + y^2)\sqrt{1 + \bar{\gamma}^2}}(0) + \frac{(y + \bar{\gamma}x)(0)}{(x^2 + y^2)(0)}$$

$$B = \frac{-y(0)\sqrt{1 + \bar{\gamma}^2}}{\bar{\gamma}(x^2 + y^2)(0)} - \frac{\sqrt{1 + \bar{\gamma}^2}x(0)}{(x^2 + y^2)(0)}.$$

$$A_1 = -\frac{(x - \bar{\gamma}y)(0)}{\bar{\gamma}(x^2 + y^2)(0)\sqrt{1 + \bar{\gamma}^2}}$$

$$B_1 = -\frac{y(0)\sqrt{1 + \bar{\gamma}^2}}{\bar{\gamma}(x^2 + y^2)(0)} - 2\frac{(x - \bar{\gamma}y)(0)}{(x^2 + y^2)(0)\sqrt{1 + \bar{\gamma}^2}}.$$

Observe that

$$e_2 \cdot (x^2 + y^2)(0) = 0$$

$$\bar{\gamma}(x^2 + y^2)(0)A = -x(0) + O(\bar{\gamma})(|x| + |y|)$$

$$\bar{\gamma}(x^2 + y^2)(0)B = -y(0) + O(\bar{\gamma})(|x| + |y|)$$

$$\bar{\gamma}(x^2 + y^2)(0)A_1 = -x(0) + O(\bar{\gamma})(|x| + |y|)$$

$$\bar{\gamma}(x^2 + y^2)(0)B_1 = -y(0) + O(\bar{\gamma})(|x| + |y|)$$

$$e_2 \cdot (\bar{\gamma}(x^2 + y^2)(0)A_1) = -y(0) + O(\bar{\gamma})(|x| + |y|)$$

$$e_2 \cdot (\bar{\gamma}(x^2 + y^2)(0)B_1) = x(0) + O(\bar{\gamma})(|x| + |y|).$$

Thus,

$$\tau_N = \frac{1}{\bar{\gamma}(x^2 + y^2)(0)^2}\left(-2(x(0)^2 + y(0)^2)(1 + O(\bar{\gamma}^2))\right) = -2\frac{(1 + O(\bar{\gamma}^2))}{\bar{\gamma}^2(x^2 + y^2)(0)} < 0. \qquad \square$$

Lemma 14 *If all the θ_i's involved in the construction of $\tilde{\tau}$ are small at a point x of M ($\theta_i \leq c, c$ independent of N), then*

$$\tilde{\tau} \leq \frac{-2}{\bar{\gamma}^2(x^2+y^2)(0)}(1+O(\bar{\gamma}^2))e^{-\delta\sum\theta_i s_i} + O(\delta \, Log^m \bar{\lambda}).$$

Proof. Coming back to $\tilde{\tau}$, we recognize that the term $O(1)$ in its expression comes from $\frac{B\tau}{\nu(x)\lambda}$ and that, for θ_i small,

$$[\xi_N, v_N] \cdot (d\theta_i \circ d\pi)([\xi_N, v_N]) + \xi_N \cdot (d\theta_i \circ d\pi)(\xi_N)$$

is positive. The claim follows.

$W_s(\mathcal{O})$ and $W_u(\mathcal{O})$ are tangent to v. Let us for example focus here on $W_u(\mathcal{O})$. Along v, as we move away from $\mathcal{O}$ to go to the attracting orbit of $v, \bar{\xi}, \xi_N$ rotate as well as $[\bar{\xi}, v]$ and $[\xi_N, v_N]$. This builds a sequence of lines (closed lines) of tangency of $\bar{\xi}(\xi_N), [\bar{\xi}, v]([\xi_N, v_N])$ to $W_u(\mathcal{O})$:

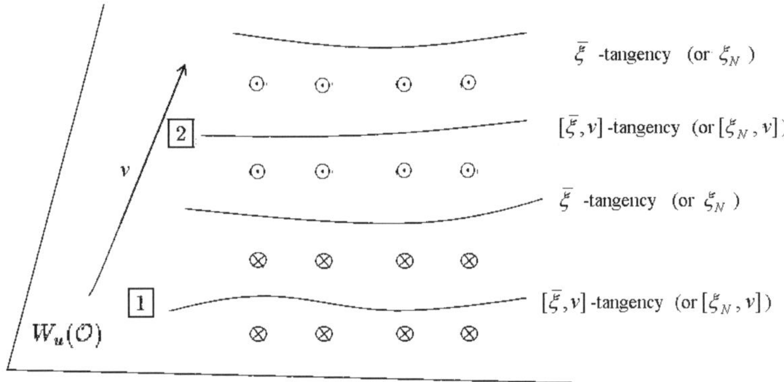

These lines are sizably spaced along v for $\bar{\xi}, [\bar{\xi}, v]$, along v_N for $\xi_N, [\xi_N, v_N]$.

$\otimes$ designates the region where $\bar{\xi}$ (or ξ_N) points into the paper across $W_u(\mathcal{O})$ while $\odot$ designates the region where $\bar{\xi}$ (or ξ_N) points towards us.

Let us consider the vector-field

$$-s\delta d\theta([\xi, v])v$$

where $\xi = \xi_{N,\lambda}$ and $v = v_{N,\lambda}$. θ here is zero above the piece of paper and builds up to 1 as we approach it:

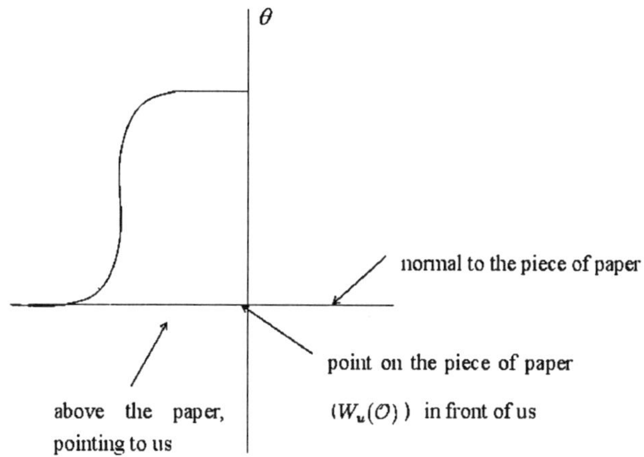

Lemma 15 *Between* $\boxed{1}$ *and* $\boxed{2}$, $-s\delta d\theta([\xi, v])v$ *points downwards from* $\boxed{2}$ *and* $\boxed{1}$. *Below* $\boxed{1}$ *until the next $\bar{\xi}$-tangency line,* $-s\delta d\theta([\xi, v])v$ *points upwards towards* $\boxed{1}$.

Proof. Observe that, since $d\theta(v) = 0$,

$$v \cdot d\theta(\xi) = d\theta([v, \xi]).$$

From $\boxed{1}$ to $\boxed{2}$ along a v-flow-line, $d\theta(\xi)$ decreases (it is positive near $\boxed{1}$, negative near $\boxed{2}$) so that

$$d\theta([v, \xi]) < 0, d\theta([\xi, v]) > 0 \text{ between } \boxed{1} \text{ and } \boxed{2}.$$

The claim follows.
We now have:

Lemma 16 *Assume that $\delta s \geq M$. There exists $c(M) > 0, c(M)$ tending to zero with $\frac{1}{M}$ and δ, such that any piece of $\xi_{N,\lambda}$-orbit entering and exiting the region of modification between $\boxed{1}$ and $\boxed{2}$ stays in a $c(M)$-neighborhood of the $\bar{\xi}(\xi_N)$ line of tangency.*

Proof. We take local coordinates between $\boxed{1}$ and $\boxed{2}$ where $W_u(\mathcal{O})$ is $y = 0, v$ is $\frac{\partial}{\partial z}$ $\left(v_N = \frac{\partial}{\partial z}\right)$ and $\frac{\partial}{\partial x}$ is tangent to the $\bar{\xi}$ or ξ_N tangency line. $\xi_{N,\lambda}$ reads up to the factor $\frac{1}{\lambda}$ as

$$\bar{\xi} - s\delta d\theta([\xi, v])v + O(\delta\theta)[\bar{\xi}, v]$$

$\bar{\xi}$ has near the tangency line a non-zero component on $\frac{\partial}{\partial x}$. It thus reads as

$$\begin{cases} \theta_0 + a_1(x - x_0) + b_1 y + c_1 z + \text{ higher order} \\ -z\gamma(x, y, z) \qquad , \text{ with } \theta_0 \neq 0, \gamma(x_0, 0, 0) \neq 0, \text{ positive} \\ \mu_0 + a_2(x - x_0) + b_2 y + c_2 z + \text{ higher order.} \end{cases}$$

near $(x_0, 0, 0)$. The y-axis goes from left to right through $W_u(\mathcal{O})$. Thus $\xi_{N,\lambda}$ reads after setting $\theta = \bar{M}(y + \eta)^{+4}$ for example

$$O(\bar{M}(y+\eta)^{+4}\delta) + \begin{cases} \theta_0 + a_1(x - x_0) + b_1 y + c_1 z + \text{ higher order} \\ -z\gamma(x, y, z) \\ \mu_0 + a_2(x - x_0) + b_2 y + c_2 z - \delta s 4\bar{M}(y + \eta)^{+3}\bar{c}(x, y, z) \\ \qquad + \text{ higher order.} \end{cases}$$

with $\bar{c}(x_0, 0, 0) > 0$ and $\bar{M} = \bar{M}(x, y, z)$ very-large.

This provides the general form (up to meaningless details) of $\xi_{N,\lambda}$ near the line of tangency of $\bar{\xi}$. The size of the neighborhood where this form holds does not depend on $\bar{M}, M$. The higher terms are independent on $M, \bar{M}$.

Observe that if $(\bar{c}_0 \leq \bar{c})$

$$4M\bar{M}(y + \eta)^3 \bar{c}_0 \geq C_1,$$

C_1 a fixed constant, then

$$\dot{z} < 0.$$

Thus, if at such point $z < O(\bar{M}(y+\eta)^{+4}\delta)$, $\dot{y} = -z\gamma + O(\bar{M}(y+\eta)^{+4}\delta) > 0$. $4M\bar{M}(y + \eta)^3 \bar{c}_0$ is larger than C_1 thereafter, z remains less than $O(\bar{M}(y + \eta)^{+4}\delta)$, the $\xi_{N,\lambda}$ piece of orbit cannot exit without crossing the chore.

Thus, we need

$$4M\bar{M}(y + \eta)^3 \bar{c}_0 \leq C_1 \text{ as long as } z < -c_2\bar{M}\delta(y + \eta)^{+4}.$$

Assume now that

$$z(0) \leq -c(M), y + \eta(0) \geq 0.$$

Then,

$$z(t) \leq -c(M) + \bar{c}\Delta t \quad , \bar{c} \text{ independent of } M, \bar{M}.$$

Thus,

$$z\left(\frac{c(M)}{2\bar{c}}\right) \leq -\frac{c(M)}{2}$$

and for $0 \leq t \leq \frac{c(M)}{2\bar{c}}$, taking $\bar{M}\delta < 1$:

$$y + \eta(t) \geq \frac{c(M)}{4} \underline{\gamma} t.$$

Thus,

$$\begin{cases} (y + \eta)\left(\frac{c(M)}{2\bar{c}}\right) \geq \frac{c(M)^2 \underline{\gamma}}{8\bar{c}} \\ z\left(\frac{c(M)}{2\bar{c}}\right) \leq -\frac{c(M)}{2}. \end{cases}$$

It suffices then to take

$$\begin{cases} c(M) > 2c_2 \bar{M} \delta \\ 4M\bar{M}\bar{c}_0 \frac{c(M)^6 \underline{\gamma}^3}{\underline{\gamma}^3 \bar{c}^3} = 2C_1 \end{cases}$$

and we have a contradiction.

We thus need to have

$$0 \geq z(0) \geq -c(M)$$

at the time of entry.

We now follow the piece of orbit of $\xi_{N,\lambda}$. As z reaches the value $\frac{\bar{c}(M)}{2} -$ if it does; if it does not, we are done — either $\dot{z} < 0$ and z becomes less than $\frac{\bar{c}(M)}{2}$ or $\dot{z}$ is nonnegative. This forces $(\overline{s\delta} = \delta$ so that $s\delta = s\bar{\delta} + 0(\delta) = s\bar{\delta}(1 + o(1)))$

$$(y + \eta)^+ \leq \left(\frac{C_1}{4s\bar{\delta}\bar{M}\bar{c}_0}\right)^{1/3}.$$

As long as z remains larger than $\frac{\bar{c}(M)}{4}$,

$$\overline{(y + \eta)^+} \leq -\frac{\bar{c}(M)}{8}$$

and

$$0 \leq (y + \eta)^+ \leq -\frac{\bar{c}(M)}{8} \Delta t + \left(\frac{C_1}{4s\bar{\delta}\bar{M}\bar{c}_0}\right)^{1/3}. \tag{2.32}$$

This forces

$$\Delta t < \frac{8}{\bar{c}(M)}\left(\frac{C_1}{4s\bar{\delta}\bar{M}\bar{c}_0}\right)^{1/3} = \Delta t_{\text{max}}.$$

Assume that $z(t_0) = \frac{\bar{c}(M)}{2}$ and for $t \in [t_0, t_1]$,

$$z(t) \geq \frac{\bar{c}(M)}{4}.$$

Then,

$$(y + \eta)^+ \leq \left(\frac{C_1}{4s\bar{\delta}\bar{M}\bar{c}_0}\right)^{1/3} \quad \text{for } t \in [t_0, t_1]$$

and

$$\dot{z}(t) \geq -C_2$$

so that

$$z(t) \geq z(t_0) - C_2(t - t_0) = \frac{c(M)}{2} - c_2(t - t_0).$$

$\frac{\bar{c}(M)}{2} - c_2(t - t_0)$ is larger than $\frac{\bar{c}(M)}{4}$ if

$$t - t_0 \leq \frac{\bar{c}(M)}{2c_2}.$$

We thus know that until $\frac{\bar{c}(M)}{2c_2}$, $z(t)$ is larger than $\frac{\bar{c}(M)}{4}$. We thus ask that

$$\Delta t_{\max} = \frac{8}{\bar{c}(M)} \left(\frac{C_1}{4s\bar{\delta}\bar{M}\bar{c}_0}\right)^{1/3} \leq \frac{\bar{c}(M)}{2c_2}$$

e.g.

$$\bar{c}(M) \sim c_3 \left(\frac{1}{M\bar{M}}\right)^{1/6}.$$

Then (2.32) holds until the time of exit i.e.

$$t_1 - t_0 \leq \frac{8}{\bar{c}(M)} \left(\frac{C_1}{4s\bar{\delta}\bar{M}\bar{c}_0}\right)^{1/3}$$

and

$$z(t_1) \leq z(t_0) + C(t_1 - t_0) = \frac{\bar{c}(M)}{2} + \left(\frac{\tilde{C}}{s\bar{\delta}\bar{M}}\right)^{1/6} \leq K\bar{c}(M).$$

$\square$

2.3 Appendix 1

2.3.1 *The normal form for (α, v) when α does not turn well*

We consider a hyperbolic orbit $\mathcal{O}$ of v. We establish:

Proposition 9 *There is, up to diffeomorphism, a unique local model for (α, v) around $\mathcal{O}$ such that α does not turn well.*

Proof. Let σ be a section to v at $x_0 \in \mathcal{O}$. Since $\mathcal{O}$ is hyperbolic, it has a stable and an unstable manifold; they can be seen as two foliations $\mathcal{F}_u, \mathcal{F}_s$ with traces A_u, A_s in σ.

Since α does not turn well along v, $\ker \alpha$ along $\mathcal{O}$ is never tangent to $\mathcal{F}_u$, neither is it tangent to $\mathcal{F}_s$. Otherwise the mononicity of the rotation of $\ker \alpha$ would imply an infinite amount of rotation. Thus, $\ker \alpha$ is contained between $T\mathcal{F}_u$ and $T\mathcal{F}_s$, it lies in exactly one of the sectors defined by the tangent spaces along $\mathcal{O}$ to $\mathcal{F}_u$ and $\mathcal{F}_s$.

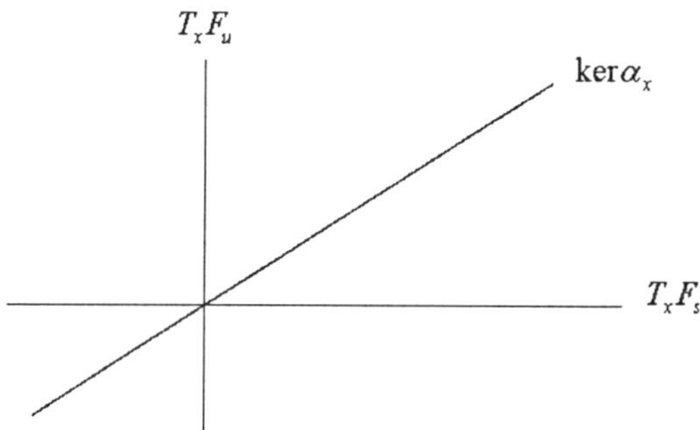

This property extends, by continuity, to a small neighborhod of $\mathcal{O}$. Let σ be a small section to v at x_0 and let ℓ be the Poincaré return map. Along the v-orbit from $x \in \sigma$ to $\ell(x)$, $\ker \alpha_y$ rotates monotonically with respect to the tangent spaces to the foliations $T_y \mathcal{F}_u$ and $T_y \mathcal{F}_s$, since these tangent spaces are transported by v. If we declare $T_x \mathcal{F}_u$ and $T_x \mathcal{F}_s$ to be orthogonal to each other, we have

$$0 < \theta(x) < \frac{\pi}{2}$$

$\theta(x)$ designates the amount of rotation of $\ker\alpha_y$ from x to $\ell(x)$. Given two distinct contact structures, both having v in their kernel and both not turning well along v, we have two functions $\theta_1(x), \theta_2(x)$, both between 0 and $\frac{\pi}{2}$.

We complete around $\mathcal{O}$ a rotation which maps $\ker\alpha_1$ to $\ker\alpha_2$ at x_0. Along the rotation, $\ker\alpha_1$ remains in the same quadrant for all x in a small neighborhood of $\mathcal{O}$. It is then easy to scale the speed of the rotation of $\ker\alpha_1$ so that it coincides along $\mathcal{O}$ with $\ker\alpha_2$. This rescaling is a diffeomorphism of a neighborhood of $\mathcal{O}$ which may be achieved through a reparametrization of the v-orbits.

$\ker\alpha_1$ and $\ker\alpha_2$ are now close in a whole neighborhood of $\mathcal{O}$ and they both have v in their kernel.

Since both planes rotate monotonicaly, are very close and have the same limits at infinity (due to the hyperbolic behavior of v), it is possible to bring one onto the other through a reparametrization of each v-orbit. $\qquad\square$

2.4 The Normal Form of (α, v) Near an Attractive Periodic Orbit of v

Let $\mathcal{O}$ be a periodic orbit of v. We establish:

Proposition 10 *There are suitable coordinates (θ_1, x, y), θ_1 being an angular coordinate along $\mathcal{O}$ such that α reads $\lambda(\theta_1, x, y)(xd\theta_1 + dy)$.*

Proof. $\ker\alpha$ is tangent to $\mathcal{O}$ and there is an additional direction along $\mathcal{O}$ defining $\ker\alpha$. v and this additional direction define an orientable frame in $\ker\alpha$. If we add to these two vectors the contact vector-field ξ of α, we build a frame for M^3 along $\mathcal{O}$. We thus can take, along $\mathcal{O}$, this additional direction to be $\frac{\partial}{\partial x}$ and ξ to be $\frac{\partial}{\partial y}$, $v = \frac{\partial}{\partial\theta}$.

α then takes the form:

$$\big(a(\theta)x + b(\theta)y + O(x^2 + y^2)\big)d\theta + \big(1 + a_2(\theta)x + b_2(\theta)y + O(x^2 + y^2)\big)dy$$

$$+ \big(a_1(\theta)x + b_1(\theta)y + O(x^2 + y^2)\big)dx.$$

Using Gray's theorem, its use leaves $\mathcal{O}$ unchanged as can be checked, we can get rid of all second order terms after the introduction of a related diffeomorphism.

Rescaling, α reads (up to a multificative factor):

$$\big(a(\theta)x + b(\theta)y\big)d\theta + \big(a_1(\theta)x + b_1(\theta)y\big)dx + dy.$$

Since $\alpha \wedge d\alpha$ is a volume form, $a(\theta)$ is non-zero for every $\theta \in [0, 2\pi]$.

We rewrite α (rescaled) as:

$$\big(a(\theta)x + b(\theta)y\big)d\theta + \big(1 - b_1(\theta)x\big)dy + d\left(a_1(\theta)\frac{x^2}{2} + b_1(\theta)xy\right)$$

$$- \frac{x^2}{2}a_1'(\theta)d\theta - b_1'(\theta)xy\,d\theta.$$

We remove as above $\frac{x^2}{2}\,a_1'(\theta)d\theta - b_1'(\theta)xy\;d\theta$ since it is second order. $d(a_1(\theta)\frac{x^2}{2} + b_1(\theta)xy)$ is closed and $o(1)$. We may also remove it using again Gray's theorem.

We are left, after rescaling with:

$$\big(a(\theta)x + b(\theta)y\big)d\theta + dy \quad \text{with} \quad a(\theta) \neq 0.$$

Setting

$$d\theta_1 = a(\theta)d\theta,$$

we find

$$\big(x + \tilde{b}_1(\theta)y\big)d\theta_1 + dy.$$

The family

$$\big(x + t\tilde{b}_1(\vartheta)y\big)d\vartheta_1\,dy$$

is a one-parameter family of contact forms. For all of them, there is a vector-field v_t in their kernel having $\mathcal{O}$ as a periodic orbit,. The family is constant equal to dy on $\mathcal{O}$.

We thus can use Gray's theorem and reduce up to rescaling, α to

$$xd\theta_1 + dy$$

as claimed.

Let us consider two vector-fields v_1, v_2 having $\mathcal{O}$ as a periodic orbit. Up to reparameterization, they read:

$$\frac{\partial}{\partial\theta_1} - x\frac{\partial}{\partial y} + \delta x_i\frac{\partial}{\partial x}.$$

Assuming that $\mathcal{O}$ is attractive for both of them, we can find functions $a_1(\theta), a_2(\theta), b_1(\theta), b_2(\theta), c_1(\theta), c_2(\theta)$ such that $a_i > 0, b_i^2 - 4a_ic_i < 0$ and $d(a_i(\theta)x^2 + b_i(\theta)xy + c_1(\theta)^2)(v_i) < 0$ for $x^2 + y^2 > 0$, small.

This reads:

$$(a_i' - b_i)x^2 + (b_i' - 2c_i)xy + c_i'y^2 + 2a_i(\theta)\left(x + \frac{b_i}{2a_i}\right)\delta x_i < 0.$$

Set

$$X = x + \frac{b_i}{2a_i}y \qquad Y = y$$

so that

$$x = X - \frac{b_i}{2a_i}Y.$$

The above equation rereads:

$$(a'_i - b_i)(X - \frac{b_i}{2a_i}Y)^2 + (b'_i - 2c_i)(X - \frac{b_i}{2a_i}Y)Y + c'_iY^2 + 2a_i(\theta)X\delta x_i < 0.$$

Setting

$$\delta x_i = A_i(\theta)X + B_i(\theta)Y + \quad \text{higher order,}$$

we find

$$(a'_i - b_i + 2a_iA_i)X^2 + (b'_i - 2c_i + 2a_iB_i - \frac{b_i}{a_i}(a'_i - b_i))XY < 0$$

$$(\frac{b_i^2}{4a_i^2}(a'_i - b_i) + c'_i - \frac{b_i}{2a_i}(b'_i - 2c_i))Y^2 < 0,$$

i.e.

$$\begin{cases} (b'_i - 2c_i + 2a_iB_i - \frac{b_i}{a_i}(a'_i - b_i))^2 - 4(c'_i - \frac{b_i}{2a_i}(b'_i - 2c_i) + (\frac{\ell_i}{2a_i})^2(a'_i - b_i)) \\ (a'_i - b_i + 2a_iA_i) < 0 \\ 2a_iA_i + a'_i - b_i < 0. \end{cases}$$

This implies:

$$\left(\frac{b_i}{2a_i}\right)^2(a'_i - b_i) + c'_i < \frac{b_i}{2a_i}(b'_i - 2c_i).$$

Furthermore,

$$2a_iA_i + a'_i - b_i \qquad \text{must be sufficiently negative,}$$

once B_i is given.

This last condition is easy to satisfy as we build a convex-combination of v_1 and v_2 in $\ker\alpha$. $\mathcal{O}$ should be an attractive orbit for the convex-combination. We thus consider the condition:

$$\left(\frac{b_i}{2a_i}\right)^2(a'_i - b_i) + c'_i < \frac{b_i}{2a_i}(b'_i - 2c_i)$$

i.e.

$$c_i' + \frac{b_i}{a_i}\left(c_i - \frac{b_i^2}{4a_i}\right) < \left(\frac{b_i b_i'}{2a_i}\right) - \left(\frac{b_i}{2a_i}\right)^2 a_i' = \left(\frac{b_i^2}{4a_i}\right)'.$$

Setting

$$c_i - \frac{b_i^2}{4a_i} = \psi,$$

we derive:

$$\psi' < -\frac{b_i}{a_i}\psi.$$

Thus we need to find, given a_i, b_i, a positive periodic function ψ such that

$$\frac{\psi'}{\psi} < -\frac{b_i}{a_i}.$$

The only conditiion is therefore to have:

$$\int_0^1 \frac{b_i}{a_i} < 0.$$

Assuming such a condition is fulfilled (it has to be for $i = 1, 2$, but we are considering more general a_i, b_i), ψ, i.e., c_i is easy to build.

There, v_1 and v_2 can be deformed one onto the other among vector-fields of $\ker \alpha$ for which $\mathcal{O}$ is attractive. $\qquad\square$

2.5 Compactness

Let (M^3, α) be a three-dimensional compact orientable manifold without boundary and let α be a contact form on M^3. We consider the Legendrian framework developed in [Bahri 1988], [Bahri 1998], [Bahri-1 2003]: we assume that there is a non-singular vector-field v in $\ker \alpha$ such that $\beta = d\alpha(v, \cdot)$ is a contact form with the same orientation than α. We introduce the action functional $J(x) = \int_0^1 \alpha_x(\dot{x})dt$ on the space of Legendrian curves $C_\beta = \{x \in H^1(S^1, M) \, s.t. \, \beta_x(\dot{x}) \equiv 0, \alpha(\dot{x}) = \text{positive constant}\}$.

The various hypotheses involved in this construction are discussed in other papers [Bahri-1 2003, Section IV].

The critical points of J are the periodic orbits of ξ which is the Reeb vector-field of α. We refer to [Bahri 1998] for the construction of a special decreasing pseudo-gradient which we built for this variational problem.

We prove in this work that, under two suitable hypotheses, denoted Hypothesis (A) and Hypothesis (B) described below — which we conjecture to be always satisfied — compactness holds for some flow-lines (all in fact, see [Bahri]) on the unstable manifold of a periodic orbit.

This result is far reaching because it indicates the existence of an invariant sub-Morse complex made of flow-lines connecting periodic orbits, with no additional asymptots involved.

We have defined in [Bahri-1 2003] a homology related to the periodic orbits using J or C_β. The existence of this invariant sub-Morse complex can be equivalently rephrased into the more technical statement that this homology has only periodic orbits as intermediate critical points (at infinity).

We believe that this homology is equal to the homology of $P\mathbb{C}^\infty$ for the standard contact form on S^3. Some of the tools required in order to compute this homology in this particular case extend to the general case. We provide a brief account of one of these tools in Section 2.5.7.

This work is organized as follows:

In Section 2.5.1, we recall (without proof) some basic facts about α, v, J and C_β. Our proof uses little of the analytical framework. In Section 2.5.2 we consider a periodic orbit of index m in C_β and we build a model for its unstable manifold. In Section 2.5.3, we introduce Hypothesis (A), Hypothesis (B), and we state our results. In Sections 2.5.4 and 2.5.5, we proceed with the proofs. In Section 2.5.6, we extend the result to flow-lines from x_{2k+1} to x_{2k}^∞ (In Sections 2.5.4, 2.5.5 we had only considered the case $x_{2k} - x_{2k-1}^\infty$). In Section 2.5.7, we provide the classifying map for the S^1-action on the space C_β. This should be an essential tool in the computation of this homology. In Section 2.5.8, we discuss oscillations and how to enlarge them or tame them whenever possible, if they are "high". In Section 2.5.9, we discuss iterates (of critical points at infinity, not iterates of periodic orbits). In Sections 2.5.10 and 2.5.11 we overcome the Fredholm and transversality difficulties which we had set aside in the previous sections.

2.5.1 *Some basic facts*

α is a contact form, v is a vector-field in ker α. We are assuming that $\beta = d\alpha(v, \cdot)$ is a contact form with the same orientation than α. We rescale v so that

$$\beta \wedge d\beta = \alpha \wedge d\alpha.$$

The following results are established in [Bahri 1998], [Bahri-1 2003].

Proposition 11

i) $d\alpha(v, [\xi, v]) = -1$,

ii) $[\xi, [\xi, v]] = -\tau v$,

iii) The contact vector-field of β is $w = -[\xi, v] + \bar{\mu}\xi$ where $\bar{\mu} = d\alpha(v, [v, [\xi, [\xi, v]]])$.

Let

$$C_\beta = \{x \in H^1(S^1, M) \text{ s.t. } \beta_x(\dot{x}) \equiv 0, \alpha_x(\dot{x}) = \text{a positive constant.}\}$$

A tangent vector z to M reads

$$z = \lambda\xi + \mu v + \eta w.$$

If x belongs to C_β, then $\dot{x} = a\xi + bv$, with a being a positive constant. A tangent vector z at x to C_β reads $z = \lambda\xi + \mu v + \eta w$ with:

$$\begin{cases} \overline{\dot{\lambda} + \bar{\mu}\eta} = b\eta - \int_0^1 b\eta \\ \\ \dot{\eta} = \mu a - \lambda b. \end{cases} \qquad \lambda, \mu, \eta \quad 1 - \text{periodic}$$

Let $J(x) = \int_0^1 \alpha_x(\dot{x})dt$.

We have established in [Bahri 1998], [Bahri-1 2003]:

Proposition 12   *There exists a decreasing pseudo-gradient Z for J on C_β such that*

i) The number of zeros of b does not increase on the decreasing flow-lines of Z.

ii) On each flow-line, $\int_0^1 |b|(s) \leq C + \int_0^1 |b|(0)$.

iii) At the blow-up time, $b(s, t)$ converges weakly to $\sum_{i=1}^m c_i\delta_{t_i}$, with $|c_i| \geq c_0 > 0$.

We also recall, see [Bahri 1988], [Bahri 1998], [Bahri-1 2003], that if a vector z is transported by v, then its components verify (derivatives are taken with respect to the time along v):

$$\begin{cases} \overline{\dot{\lambda} + \bar{\mu}\eta} = \eta \\ \dot{\eta} = -\lambda. \end{cases}$$

On the other hand, if a vector v is transported by ξ, then its components satisfy (derviatives are taken along ξ):

$$\begin{cases} \dot{\mu} + \eta\,\tau = 0 \\ \dot{\eta} = \mu. \end{cases}$$

2.5.2   *A model for $W_u(x_m)$, the unstable manifold in C_β of a periodic orbit of index m*

Let x_m be a periodic orbit of index m. We provide in what follows a model for its unstable manifold in C_β.

Proposition 13   *The unstable manifold of x_m can be achieved in Γ_{2m} for $m \geq m_0 (m_0 \geq 4)$.*

Proof.   Let us assume in a first step that x_m is a simple periodic orbit. We indicate at the end of the proof how to extend the result to iterates. We start with the case where $m = 2k+1$. x_m can be elliptic or hyperbolic. Assume that it is hyperbolic to start with. Then, if ℓ is the Poincaré-return map of the periodic orbit and u is a real eigenvector of $d\ell$ in $\ker \alpha$, u is transported by ξ while v rotates (considerably: $m \geq m_0$).

Setting $x_m(0)$ at a point where v and u coincide, setting η to be the $-[\xi, v]$-component of u, we have

$$\eta(0) = \eta(1) = 0$$

η can easily be seen to have at least $2k$ genuine zeros, also $\dot{\eta}(1)\dot{\eta}(0) < 0$ because the periodic orbit is hyperbolic of odd Morse index. The zeros of η are $t_1 = 0, t_2, \ldots, t_{2k+1}, t_{2k+2} = 1$.

Setting $\eta = \eta_i|_{[t_i, t_{i+1}]}, i = 1, \ldots, 2k+1$ we derive $2k+1$-functions which are pairwise orthogonal and are orthogonal to themselves. Each η_i defines a tangent vector z_i with

$$\ddot{\eta}_i + a^2 \eta_i \tau = c_i \delta_{t_i} + c_{i+1} \delta_{t_{i+1}}.$$

Thus, Span $\{z_1, \ldots, z_{2k+1}\}$ is achieved in Γ_{2m}.

If x_{2k+1} is of odd Morse index $2k+1$ but elliptic, the $[\xi, v]$- component η of any transported vector u has always $2k, 2k+1$ or $2k+2$ zeros exactly. Indeed, it needs to have at least $2k$, at most $2k+2$, zeros and if the number of zeros were to change, there would be a point where $u(v)$ would be mapped onto $\gamma u(\gamma v)$ with $\gamma < 0$, yielding $2k+1$ zeros, $2k$ of which are genuine i.e. a hyperbolic orbit.

Taking the base point at one of the nodes of the most oscillating eigenfunction, η must have $2k+1$ zeros to the least. This yields $2k$ function

$\eta_1, \ldots, \eta_{2k}$ which are zero at

$$t_1 = 0, t_2, \ldots, t_{2k+1}$$

and yield $2k$ vectors $z_1, \ldots, z_{2k}$ which are pairwise orthogonal and orthogonal to themselves. For each of these vectors, we have

$$\ddot{\eta}_i + a^2 \eta_i \tau = c_i \delta_{t_i} + c_{i+1} \delta_{t_{i+1}}$$

so that $\mathrm{Span}\{z_1, \ldots, z_{2k}\}$ is achieved in Γ_{4k+2}.

We still need an additional vector z_{2k+1}. At $x_m(0)$, we introduce the solution of

$$d\ell(u) - u = v.$$

This yields a new vector $\tilde{z}_{2k+1}$. η of $\tilde{z}_{2k+1}$, which we denote $\tilde{\eta}_{2k+1}$, is not zero at $x(0)$ or at $t_1 = 0$ since the orbit is not hyperbolic. If we compute the second derivative $J''(x_m) \cdot \tilde{z}_{2k+1} \cdot z$ (the $-[\xi, v]$- component of z is η), we find

$$J''(x_m) \cdot \tilde{z}_{2k+1} \cdot z = -\int_0^1 (\ddot{\tilde{\eta}}_{2k+1} + a^2 \tilde{\eta}_{2k+1} \tau) \eta = \eta(0).$$

Thus, $\tilde{z}_{2k+1}$ is $J''(x_m)$-orthogonal to $z_1, \ldots, z_{2k}$.

We claim that we can take, after deformation of the contact form, $\tilde{\eta}_{2k+1}(0)$ to be negative and $\tilde{z}_{2k+1}$ in the negative eigenspace of $J''(x_m)$. Then, $J''(x_m)$ is non positive on $\mathrm{Span}\{z_1, \ldots, z_{2k}, \tilde{z}_{2k+1}\}$ which is again achieved in Γ_{4k+2}.

In order to prove our claim, we complete a deformation of the contact form so that x_m, without degenerating, changes from elliptic to hyperbolic with eigenvalues equal to -1 and back to elliptic i.e. x_m iterated twice degenerates but not x_m which stays of odd Morse index. At the switch, $d\ell|_{\ker \alpha}$ is -Id and we can set $x(t_1) = x(0)$ at any point. We may solve continuously, through the degeneracy, the equation $d\ell(u) - u = v$.

$\tilde{\eta}_{2k+1}(0)$ is zero at the switch and changes sign through it. We consider the side of the switch where it is negative. $z_1, \ldots, z_{2k}$ still exist but z_{2k+1} might have disappeared (when the Poincaré-return map has an eigenvalue equal to -1 and 0 is at the node, z_{2k+1} exists). The negativity of $\tilde{\eta}_{2k+1}(0)$ means that $J''(x_m) \cdot \tilde{z}_{2k+1} \cdot z_{2k+1} < 0$. We thus will have produced a space of dimension $2k+1$, $\mathrm{Span}\,\{z_1, \ldots, z_{2k}, \tilde{z}_{2k+1}\}$ where $J''(x_m)$ is non positive.

If $x_m = x_{2k}$ is of Morse index $2k$, the most negative eigenfunction of $-(\ddot{\eta} + a^2 \eta \tau)$ has $2k$ zeros. An elliptic orbit is, as we will see, of odd Morse index. Thus x_{2k} is hyberbolic. We pick an eigenvector u. Since $m \geq 4, u$ coincides with v at some points on the periodic orbit. We set $x(0)$ at such a time. We transport v along the periodic orbit. This yields an η-component of the transported vector. Using Sturm-Liouville arguments, η vanishes at $2k + 1$ times, $t_1 = 0, t_1, \ldots, t_{2k+1} = 0$. This yields $2k$ functions $\eta_1, \ldots, \eta_{2k}$ with

$$\ddot{\eta}_i + a^2 \eta_i \tau = c_i \delta_{t_i} + c_{i+1} \delta_{t_{i+1}}$$

and $2k$ vectors $z_1, \ldots, z_{2k}$ which are pairwise $J''(x_m)$-orthogonal and orthogonal to themselves.

Again,

$$\text{Span } \{z_1, \ldots, z_{2k}\} \text{ is achieved in } \Gamma_{2m}.$$

In each occurence, $J''(x_m)$ is non positive on the vector-space which we build and we therefore can take these directions to achieve the unstable manifold of x_m.

Let us enter into some more details and check that our representation of $W_u(x_m)$ in Γ_{2m} works. We will consider for simplicity first the case of a hyperbolic orbit of index $2k$.

We then have $2k$ nodes $\bar{t}_1, \bar{t}_2, \cdots, \bar{t}_{2k}$ where we locate the v-jumps. The basic equation is

$$\ddot{\eta} + a^2 \eta \tau = \sum_{i=1}^{2k} c_i \delta_{\bar{t}_i},$$

η reads as $\sum_{i=1}^{2k} c_i \bar{\eta}_i$ where each $\bar{\eta}_i$ solves

$$-(\ddot{\bar{\eta}}_i + a^2 \bar{\eta}_i \tau) = \delta_{\bar{t}_i}$$

$$\bar{\eta}_i 1 - \text{periodic}.$$

We may consider the above equation when the t_i's are in the vicinity of the $\bar{t}_i$. We find in this way $2k$ functions $\eta_1, \eta_2, \ldots, \eta_{2k}$. The second derivative $\int_0^1 \dot{\eta}^2 - a^2 \eta^2 \tau$ in $\text{Span}\{\eta_1, \ldots, \eta_{2k}\}$ reads as $c^t A c = \sum_{i \neq j} c_i c_j \left(\eta_i(t_j) + \eta_j(t_i) \right) + \sum c_i^2 \eta_i(t_i)$. Clearly after integration by parts, $\eta_i(t_j) = \eta_j(t_i)$.

When all the t_i's are located at the $\bar{t}_i$'s, this quadratic form is identically zero. We claim that

$$\sum_i \left| c_i \left(2 \sum_{j \neq i} c_j \frac{\partial}{\partial t_i} \eta_j(t_i) + c_i \frac{\partial}{\partial t_i} \eta_i(t_i) \right) \right|_{(\bar{t}_1, \ldots, \bar{t}_{2k})} \right| \geq \bar{c} \sum c_i^2$$

when $\bar{c}$ is a fixed positive constant. Thus, by moving slightly for each direction $(c_1, \ldots, c_{2k})$ the location of the Dirac masses, we may achieve that $J''(x_{2k})$ is negative in this space.

Assume that the above inequality does not hold. Then, for each i such that $|c_i| \geq \bar{c}_1 \sqrt{\sum c_k^2}$, where $\bar{c}_1$ is a small constant($\bar{c}_1 = \sqrt{\bar{c}}$ for example),

$$c_i \frac{\partial}{\partial t_i} \eta_i(t_i) + 2 \sum_{j \neq i} c_j \frac{\partial}{\partial t_i} \eta_j(t_i) \mid_{(\bar{t}_1,\ldots,\bar{t}_{2k})} = o\left(\left(\sum c_k^2\right)^{1/2}\right).$$

In the remainder getting rid of the other c_i's ($\frac{\partial}{\partial t_i} \eta_j(t_i)$ is bounded as we will see) we derive that the matrix (off diagonal terms should be multiplied by 2)

$$\left(\frac{\partial}{\partial t_i} \eta_j(t_i)\right) \mid_{(\bar{t}_1,\ldots,\bar{t}_{2k})}$$

or some nontrivial sub-matrix of the above one should have a zero eigen-value. For the sake of simplicity, we will assume that the number of points $(\bar{t}_1,\ldots,\bar{t}_{2k})$ has not been reduced in the above process. We can compute this matrix as follows: we set a section to ξ at t_1, tangent to $ker\alpha$ at $x_{2k}(t_1)$. Let ψ be the Poincaré-return map. η_1 is obtained by solving the equation

$$D\psi(z) - z = v,$$

z is in $ker\alpha_{x_{2k}(t_1)}$ and reads $z = \mu_1 v - \eta_1[\xi, v]$.

Let ψ_0 be the Poincaré-return map at $\bar{t}_1$. x_{2k} is hyperbolic so that $D\psi_0(v) = \gamma v$.

$D\psi$ read as $\sigma \circ D\psi_0 \circ \sigma^{-1}$ and our equation above becomes

$$D\psi_0 \circ \sigma^{-1}(z) - \sigma^{-1}(z) = \sigma^{-1}(v).$$

Let $\Delta t = -t_1 + \bar{t}_1$ which we can assume to be small,. We then have, using the transport equations ($\dot{\mu} = -\eta\tau, \dot{\eta} = \mu$) : $\sigma^{-1}(z) = (\mu_1 + O(\eta_1 \Delta t))v - (\eta_1 + \mu_1 \Delta t)[\xi, v] + o(\Delta t)$. Observe that $\bar{\eta}_1(\bar{t}_1) = 0$ so that $\eta_1(t_1) = o(1)$ and $\sigma^{-1}(z) = \mu_1 v - (\eta_1 + \mu_1 \Delta t)[\xi, v] + o(\Delta t)$.

The matrix of $D\psi_0$ on $ker\alpha$ reads $\begin{pmatrix} \gamma & \bar{\alpha} \\ 0 & \frac{1}{\gamma} \end{pmatrix}$ in the $(v, -[\xi, v])$ basis so that

$$D\psi_0 - Id = \begin{pmatrix} \gamma - 1 & \bar{\alpha} \\ 0 & \frac{1}{\gamma} - 1 \end{pmatrix}.$$

Our equations on z then re-read:

$$(\gamma - 1)\mu_1 + \bar{\alpha}\eta_1 = 1 + o(\Delta t)$$

$$\frac{1 - \gamma}{\gamma}(\eta_1 + \mu_1 \Delta t) = \Delta t(1 + o(1)).$$

Thus, since $\Delta t = -t_1 + \bar{t}_1$ and $\bar{\eta}_1(\bar{t}_1) = 0$, we find:

$$\frac{\partial}{\partial t_1} \eta_1(t_1) \mid_{\bar{t}_1} = -\frac{\gamma+1}{1-\gamma} = \frac{\gamma+1}{\gamma-1},$$

and this extends to give

$$\frac{\partial}{\partial t_i} \eta_i(t_i) \mid_{\bar{t}_i} = \frac{\gamma+1}{\gamma-1}.$$

If we want to compute now $\frac{\partial}{\partial t_i} \eta_j(t_i) \mid_{\bar{t}_i}$ for $i \neq j$, we use that fact that $\dot{\eta}_j = \mu_j$ so that $\frac{\partial}{\partial t_i} \eta_j(t_i) \mid_{\bar{t}_i} = \mu_j(\bar{t}_i)$. Between $\bar{t}_j$ and $\bar{t}_i$, along ξ, v is mapped onto $\theta_j^i v$ so that

$$\mu_j(\bar{t}_i) = \theta_j^i \mu_j(\bar{t}_j) = \frac{\theta_j^i}{\gamma-1}.$$

Our matrix thus reads

$$\frac{1}{\gamma-1} \begin{pmatrix} \gamma+1 & 2\theta_2^1 & \cdots & \cdots & 2\theta_{2k}^1 \\ 2\theta_1^2 & \gamma+1 & 2\theta_3^2 & \cdots & 2\theta_{2k}^2 \\ \cdots & \cdots & \cdots & \cdots & \cdots \\ \cdots & \cdots & \cdots & \cdots & \cdots \\ 2\theta_1^{2k} & 2\theta_2^{2k} & \cdots & \cdots & \gamma+1 \end{pmatrix}.$$

The θ_i^j satisfy the relations (we go from $x(\bar{t}_j)$ to $x(\bar{t}_i)$ along $+\xi$): $\theta_i^j \theta_j^i = \gamma$; $\theta_i^j \theta_j^\ell = \theta_i^\ell$ if $x_{2k}(t_j)$ is in between $x_{2k}(t_i)$ and $x_{2k}(t_\ell)$, $\theta_i^j \theta_j^\ell = \gamma \theta_i^\ell$ otherwise.

We have to compute the determinant of this matrix. We multiply the matrix by $\gamma - 1$ and each line i by θ_i^1 ($\theta_1^1 = 1$). We find then the following determinant (after a manipulation and a transposition with the above notations):

$$\begin{vmatrix} \gamma+1 & 2\theta_2^1 & \cdots & \cdots & 2\theta_{2k}^1 \\ 2\gamma & (\gamma+1)\theta_2^1 & 2\theta_3^1 & \cdots & 2\theta_{2k}^1 \\ 2\gamma & 2\theta_2^1\gamma & (\gamma+1)\theta_3^1 & \cdots & 2\theta_{2k}^1 \\ \cdots & \cdots & \cdots & \cdots & \cdots \\ 2\gamma & 2\theta_2^1\gamma & 2\theta_3^1\gamma & \cdots & (\gamma+1)\theta_{2k}^1 \end{vmatrix}$$

$$= \theta_2^1 \cdots \theta_{2k}^1 \begin{vmatrix} \gamma+1 & 2 & 2 & 2 & 2 \\ 2\gamma & (\gamma+1) & 2 & \cdots & 2 \\ 2\gamma & 2\gamma & \gamma+1 & \cdots & 2 \\ \cdots & \cdots & \cdots & \cdots & \cdots \\ 2\gamma & 2\gamma & 2\gamma & \cdots & \gamma+1 \end{vmatrix}.$$

which is not zero for $\gamma \neq 1$ and $\gamma \neq -1$.

The proof extends verbatim for x_{2k+1} hyperbolic.

We turn now to the case of elliptic orbits to prove our claims and prove also that we can produce for $W_u(x_m)$ an m-dimensional manifold — with $\pm v$-jumps which can be tracked down — in Γ_{2m} where $J''(x_m)$ is negative. Using a scheme similar to the one used in Lemma 11 of [Bahri-1 2003], we can see that an elliptic orbit must be of odd Morse index. Indeed, deforming the contact form as in Lemma 11 of [Bahri-1 2003], we can change this elliptic orbit into a hyperbolic orbit: we rotate $Cv(0)$ on $\lambda v(1), \lambda < 0$; C is the Poincaré-return map. We cannot complete the other move allowed by Lemma 11 of [Bahri-1 2003] with a critical point of infinity i.e. bring $Cv(0)$ onto $\lambda v(1), \lambda > 0$ since the Poincaré-return map C would then degenerate for a periodic orbit, while it does not for a critical point at infinity. We thus have changed our elliptic orbit into a hyperbolic orbit with eigenvalue $\gamma = -1$. Let us follow the proof of Lemma 11 of [Bahri-1 2003]: -1 is of course a double eigenvalue at the switch; $C =$-Id in fact. Beyond the switch, see the proof of Lemma 11 of [Bahri-1 2003], x_m is still an elliptic orbit of index m ($|trC|$ remains less than 2 since C remains conjugate to a rotation). But $d\alpha(Cv(0), v(0))$ has changed sign.

The analysis which we completed above for hyperbolic orbits applies; but it does not lead to the conclusion which we desire since $\gamma = -1$ (observe that $\bar{\alpha} = 0$).

We consider a base point 0. Its location is not important since $D\psi_0 = -Id$. Taking the solution of $\ddot{\eta} + a^2\eta\tau = 0, \eta(0) = 0, \dot{\eta}(0) = 1$, we claim that we can produce $2k$ oscillations, thus produce $2k$ functions η_i in the null eigenvalue of the quadratic form $\int_0^1 \dot{\eta}^2 - a^2\eta^2\tau$ using $2k+1 \pm v$-jumps. We still need an additional direction $\tilde{z}_{2k+1}$ which we create by solving $D\psi(z) = z + v$ at 0. Since $\eta_i(0) = 0$ for $i = 1,\ldots,2k, \tilde{z}_{2k+1}$ and the z_i's associated to the η_i's are orthogonal for $J''(x_m)$.

We claim that

$$J''(x_m)\tilde{z}_{2k+1} \cdot \tilde{z}_{2k+1} < 0$$

on one side of the switch: indeed, $\tilde{\eta}_{2k+1}(0)$ is zero at the switch. Solving the equation for $\tilde{\eta}_{2k+1}$ continuously through $C = -Id, \tilde{\eta}_{2k+1}(0)$ becomes negative on one side or the other since $d\alpha(Cv(0), v(0))$ changes sign (in the argument of Lemma 11 of [Bahri-1 2003], $Cv(0)$ crosses $\mu v(1), \mu \in \mathbb{R}$ — here $\mu = -1$ — through the deformation).

Since

$$J''(x_m) \cdot \tilde{z}_{2k+1} \cdot \tilde{z}_{2k+1} = -\int_{S^1} \left(\ddot{\tilde{\eta}}_{2k+1} + a^2\tilde{\eta}_{2k+1}\tau\right)\tilde{\eta}_{2k+1}$$

$$= \tilde{\eta}_{2k+1}(0) < 0,$$

our claim is established.

We now claim — this will conclude the proof of Proposition 1 after a perturbation argument bringing us back to the elliptic case — that, at the switch,

$$\sum_i \left| 2 \frac{\partial}{\partial t_i} \left(\sum_{\ell \neq j} c_\ell c_j \tilde{\eta}_\ell(t_j) \right) + \frac{\partial}{\partial t_i} c_i^2 \tilde{\eta}_i(t_i) \right| \geq \bar{c} \sum c_j^2$$

for $z \in \text{Span } \{z_1, \ldots, z_{2k}\}$. z_i is built using η_i. The $\tilde{\eta}_j$'s are not here the η_j's of the z_j's. They are the $\tilde{\eta}_j$'s corresponding to the various solutions of $D\psi(z) - z = v$ at the various nodes.

Computing as in the case when x_m was hyperbolic (here x_m is elliptic turned to hyperbolic with $\gamma = -1$), we find that such an inequality holds for all vectors $(c_1, \ldots, c_{2k})$ which are not in the vicinity of unit vectors such that an odd number of c_i's are non zero. In fact, the zero eigenvectors of the matrices A — these matrices may be $p \times p$, for any $p \leq m$ if some c_i's are $o(\sqrt{\sum c_\ell^2})$ — are equal to $(1, -1, 1, -1, \ldots)$ with an odd number of components (some intermediate c_i's might be zero).

The η-component of all these eigenvectors read $\sum c_j \tilde{\eta}_j$. Since the $\tilde{\eta}_j$ are equal to $\sum_j^{j+2k} (-1)^{i-j} \eta_i$ with $\eta_i > 0$, such a combination $\sum c_j \tilde{\eta}_j$ with the c_j's as above (an odd number of them is non zero and equal to 1 in absolute value) is far from Span $\{z_1, \ldots, z_{2k}\}$: a vector in Span$\{z_1, \ldots, z_{2k}\}$ has no component on η_{2k+1}.

The conclusion follows.

Another line of proof uses Proposition 29, page 198 of [Bahri 1998]: once $C = -Id$, we can perturb slightly C and bring C to have real eigenvalues very close to -1 but different from -1. x_m has become hyperbolic and the framework developed for hyperbolic orbits applies. This method has the definite advantage that the representation of $W_u(x_m)$ extend to iterates since the Morse index of iterates grows then linearly. Turning all simple periodic orbits into hyperbolic orbits might be the best choice for this homology. $\square$

Observe that the unstable manifold of x_m, as we start near x_m, is provided by $m \pm v$-jumps which have a location than can be tracked down.

$*$ will designate below the location of such a $\pm v$-jump

2.5.3 *Hypothesis (A), Hypothesis (B), Statement of the result*

J on C_β has critical points at infinity which are curves of $U_k \Gamma_{2k}$. Each Γ_{2k} is the space of curves made of k pieces of ξ-orbits alternated with k pieces of $\pm v$-orbits. Denoting a_i the time spent along ξ on the i-th ξ-piece of a curve of Γ_{2k}, we introduce the functional J_∞ equal to $\sum a_i$.

The critical points at infinity of J on C_β are the critical points of J_∞ on $U_k\Gamma_{2k}$. They are described in [Bahri-1 2003], there is a vast zoology among them.

Let us recall the two following definitions:

Definition 1 A v-jump between two points x_0 and $x_1 = x_{s_1}$ is a v-jump between conjugate points if, denoting φ_s, the one-parameter group of v, we have:

$$\left(\varphi_{s_1}^* \alpha\right)_{x_1} = \alpha_{x_1}.$$

Such points live generically on a hypersurface $\sum$ of M.

Definition 2 A ξ-piece $[y_0, y_1]$ of orbit is characteristic if v has completed exactly k $(k \in \mathbb{Z})$ half revolutions from y_0 to y_1.

The description of the critical points at infinity x^∞ of J goes as follows [Bahri-1 2003]:

• If x_∞ has no characteristic ξ-piece, all its v-jumps are v-jumps between conjugate points.

• If x^∞ has some characteristic ξ-pieces, some additional conditions have to be satisfied, see [Bahri-1 2003]. Any v-jump between non characteristic pieces is a v-jump between conjugate points.

To each x^∞ is associated a suitable Poincaré-return map C (see [Bahri-1 2003]) which preserves area.

We have shown in [Bahri-1 2003], Proposition 28 and Lemma 11, how, **without modifying** C, we can redistribute the v-rotation from a non-characteristic ξ-piece of orbit to another one at the expense of creating additional critical points at infinity with more characteristic ξ-pieces than x^∞. We may thus redistribute all the v-rotation from all the characteristic ξ-pieces on a single one (up to $\varepsilon > 0$; a small amount of rotation is to be left on each ξ-piece).

We then introduce:

Hypothesis (A)

Assume that x^∞ has at most one characteristic ξ-piece. As the number of ξ-pieces of x^∞ tends to infinity, the amount of rotation of v derived after rescaling the rotation from all the non-characteristic ξ-pieces onto a single one of them is at least 2π (the ξ-piece where the rotation is relocated is of our choice).

Hypothesis (A) is not needed in the study of the flow-lines from a periodic orbit x_m to a critical point at infinity having at least one characteristic piece x^∞_{m-1} with m odd, i.e. in the Sections 2.5.4 and 2.5.5 of the present work. It is needed later as we study flow-lines from x_m to x^∞_{m-1}, with m even or with x^∞_{m-1} having only non degenerate ξ-pieces, see Section 2.5.6.

We also need to state our results another hypothesis, Hypothesis (B) as "companions" develop. The concept of "companions" is explained below in 2.5.5.2, it is related to the existence of "false critical points at infinity" which we bypass by introducing small $\pm v$-jumps near one of the edges of one of their characteristic pieces, see [Bahri 1998], [Bahri-1 2003] for more details. "Companions" have all the same orientation and build "families". The $\pm v$-jumps of two distinct families do not overlap, see 2.5.5.2. Some thought shows that it is reasonable to expect that, as we decrease J and we deform our curves in the $\Gamma'_{2s}s$ and if the curves under deformation are not in the vicinity of a critical point at infinity, the relative sizes of the various $\pm v$-jumps should be bounded above and below by constants C and $\frac{1}{C}$, unless the ξ-distance between two of them becomes very small (then, one of them might disappear if it has been created as a companion of the other one). As we approach a critical point at infinity, we cannot expect such estimates to hold uniformly since some of the $\pm v$-jumps become large and contribute to the edges while other ones live on the ξ-pieces and are small. However, we may assume that,

Hypothesis (B)

On a given characteristic piece, all the $\pm v$-jumps belonging to the same family and which are a little bit inside(counting using v-rotation) the characteristic piece are of comparable relative sizes

The statement of Hypothesis (B) is very natural as long as some of the $\pm v$-jumps of a family do not concur to build an edge as the associated configurations approach the critical point at infinity. However, if a given companion is part of one of the edges, therefore might be large and enters a characteristic ξ-piece, we may "tame" it as it enters so that it becomes comparable to the other companions or members of the other families in the second case which are already inside. This can be done with the use of a decreasing normal see Section 2.5.4.2 below or see [Bahri-1 2003], if the critical point at infinity is true. If it is false, this issue does not arise since a decrease can then be engineered with the introduction of a companion.

Under Hypothesis (A) and Hypothesis (B), we prove the following result:

Theorem 1	*Let y^∞_{m-1} be a critical point at infinity of index $m-1$ having at least one characteristic piece. Assume that the maximal number of zeros of b on its unstable manifold is $2[\frac{m}{2}]$. Assume that, if m is odd, m is large and that all ξ-pieces are of (strict if degenerate)H^1_0-index at least 1. Let y_m be a periodic orbit of ξ of index m. Then, the intersection number $i(y_m, y^\infty_{m-1})$ for the flow of [Bahri-1 2003] is zero.*

Observation: The proof of Theorem 1 which we present here does not require any restriction on the H_0^1-index of the ξ-pieces if the flow-lines from y_m to a neighborhood of y_{m-1}^∞ do not involve companions, see 2.5.5.2 for the definition of this notion. The case of zero H_0^1-index is discussed in Appendix 1. Hypothesis (A) and Hypothesis (B) are discussed in [6]. Flow-lines from y_m to y_{m-1}^∞ with y_{m-1}^∞ having only non characteristic ξ-pieces have already been studied and ruled out for m large under Hypothesis (A), in [Bahri-1 2003].

2.5.4 *The hole flow*

2.5.4.1 *Combinatorics*

We start with an abstract result which may be viewed as an (elementary) observation in combinatorics.

We consider a sequence of $2k$ points each bearing a sign, $+$ or $-$. Such a sequence, together with the assigned distribution of signs, is called in the sequel a **configuration**.

A configuration contains at most $2k$ sign changes. A configuration with $2k$ sign changes is called **maximal**.

Given a configuration, we consider two consecutive points. We assume $k \geq 2$. We are given a sign of rotation (the same for all configurations, which are placed on a curve). Then, one of these points is the first one (using the positive rotation) and the other one is the second one. Let us assume that the sign $+$ is assigned to the first one:

$$+$$
$$\cdot \ \ \cdot \ \ \cdot$$
$$1 \ \ 2 \ \ 3$$

The **hole flow** assigns signs to the intervals between the points as follows:

Starting from the $+$ assigned to 1 and **independently** of the configuration which we are facing, we assign signs to the remaining $2k - 1$ points so that there is an alternance each shift between the $+$ and $-$ signs. Thus, we have

$$+ \ + \ - \ +$$
$$\cdot \ \ \cdot \ \ \cdot \ \ \cdot$$
$$1 \ \ 2 \ \ 3 \ \ 4$$

signs of the configuration

$$1\ \ 2\ \ 3\ \ 4$$
$$\cdot\ \ \cdot\ \ \cdot\ \ \cdot$$
$$+\ -\ +\ -$$

alternating distribution

We then introduce on each interval $[i, i+1]$ the sign of i in the alternating distribution.

We claim:

Proposition 14 *The original configuration of signs with the additional intermediate signs has at most 2k sign changes.*

Proof. We reverse the process and start with the alternating configuration. We add in between its jump the signs of the original configuration, inserting between 1 and 2, the sign of 2 for the original configuration and so forth. The result is the same. Viewed in this way, the claim of the Proposition 14 is obvious. $\square$

Next, we discuss the choice of the starting point 1. We assume that we have some freedom of choice on 1 but the sign assigned to the choice should be + (i.e. the same, it could also be -) for all possible choices in the configuration. We then claim:

Proposition 15 *Let us consider two distinct choices 1 and 1′ for 1 extracted from the same configuration. Then either the configuration has a sign repetition between 1 and 1′ or the two hole flows, the one corresponding to 1 and the one corresponding to 1′, coincide.*

Proof. The alternating distributions for 1 and 1′ coincide if and only if there is an odd number of points between 1 and 1′. Then (and only then), the hole flows coincide.

If there is an even number of points between 1 and 1′, then there is a forced sign repetition in the configuration since 1 and 1′ bear the same sign. $\square$

An obvious observation which we will be using below states:

Observation 1. Given a configuration, assume that between two identical signs points there is an even number of points or that between two reverse signs points there is an odd number of points, then the configuration is forced to contain a sign repetition.

Next, we define:

Definition 3   Given a configuration, an **elementary** operation on this configuration is the reversal of the sign of one and only one point in the configuration.

Proposition 16   *Given a configuration with a choice of 1 and the addition of the intermediate signs of the hole flow, any elementary operation completed on a point of the configuration distinct from 1 does not increase the number of sign changes of the configuration beyond 2k.*

Proof.   The hole flow is unperturbed. We are simply modifying the original configuration outside of 1.   □

Let us assume that we are considering several configurations σ which all have the sign $+$ on point j and have a forced repetition between positive signs in $[i,j]$, $i < j$. Assume that two distinct alternating distributions starting from a $+$ can be defined on σ. We then have

Proposition 17   *Choose a hole flow using one of the alternating distributions. Introduce the corresponding intermediate signs in all intervals $[k, k + 1]$ except those contained in $[i, j]$. Introduce furthermore a sign $-$ between j and $j + 1$. The distribution of signs derived in this way has at most 2k sign changes.*

Another more obvious Proposition states:

Proposition 18   *Given all configurations σ which have a sign repetition, we can introduce once an arbitrary additional sign on the interval of our choice without increasing the number of sign changes beyond 2k.*

Observation 2.   For Proposition 18, since we are not assuming that we proceed as for Proposition 14, we need to be careful and not use the hole flow as we are introducing the additional sign. We may use, however, elementary operations on the configurations σ as long as these elementary operations do not destroy one sign repetition at least.

Proof.   [Proof of Proposition 17]   □

In the case where the hole flow used is not the one provided by the use of j as initial point, then this hole flow gives a $-$ between j and $j + 1$ (since the one of j gives a $+$ in this interval). Thus this hole flow is compatible with the additional introduction of a $-$ between j and $j + 1$ and the number of sign changes does not increase beyond 2k.

We consider now the hole flow provided by the use of j as initial point. We introduce a $-$ between j and $j + 1$. This might increase the number of zeros beyond 2k.

From i to j, we then have one sign repetition to the least. The full use of the hole flow associated to j (before the introduction of the $-$ between

j and $j+1$) will result in an additional sign change between i and j with respect to σ. This is obvious if this sign repetition occurs between j and $j-1$. Then the hole flow would introduce a $-$ between $j-1$ and j which are both positive.

Otherwise, starting from the $+$ sign introduced by the hole flow between j and $j+1$, we evolve left towards i, alternating the signs. Before hitting a repetition of sign, the sign introduced by the hole flow agrees with the sign of the left edge of the interval $[k, k+1]$ considered; otherwise an additional sign change occurs and the claim follows:

$$+ \quad + \quad - \quad - + +$$
$$\cdot \qquad\quad \cdot \qquad \cdot \qquad \cdot$$
$$j-2 \quad j-1 \quad j \quad j+1$$
$$\longleftarrow$$

etc.

Accordingly, the sign introduced has to disagree with the sign of the right edge of the interval. But then, we arrive at the repetition, a sign change has to occur.

We thus see that if we had used the hole flow between i and j, we would have introduced a sign change, i.e. two additional zeros. The total number of sign changes would be $2k$. However, we have not used this hole flow between i and j. The introduction of the sign $-$ between j and $j+1$ would only provide for one sign change. Proposition 17 follows.

Observation 3. We could have introduced the $+$ of the hole flow of j or $[j, j+1]$, only we would need to complete it close to j and introduce the $-$ after it is the interval. With this provision, the statement of Proposition 5 can be modified into "Introduce the corresponding intermediate signs — except those contained in $[i, j]$."

2.5.4.2   *Normals*

Given a critical point as infinity x^∞, with a ξ-piece $[x_0^-, x_0^+]$ which is characteristic, i.e. along which v has completed exactly a certain number m of half-revolutions in the ξ-transport, we may define $m+1$ nodes,

$$y_0 = x_0^- < y_1 < \cdots < y_m = x_0^+.$$

These nodes are the points y_i along the characteristic ξ-piece at which v, starting from $y_0 = x_0^+$, has completed exactly i half-revolutions. To each interval (y_i, y_{i+1}), are associated **a decreasing normal to the right** and a **decreasing normal to the left**, in fact continuously varying families of these as follows.

We choose a point z in (y_i, y_{i+1}). We consider the vector v at z and we transport it to $y_0 = x_0^-$ for the normal to the right and to $y_m = x_0^+$ for the normal to the left. We derive vectors which have components on $[\xi, v]$ since z is not a node. On the other hand, this characteristic ξ-piece is preceded and is followed by ξ-pieces:

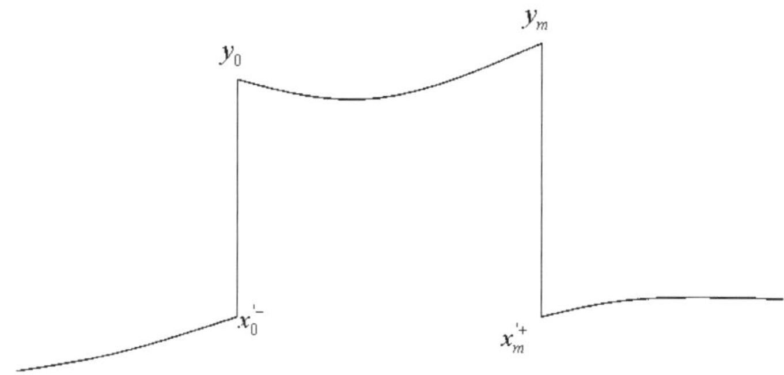

We take ξ at $x_0^{'-}, x_0^{'+}$ and we transport along v to y_0, y_m. We derive vectors at y_0, y_m, equal respectively to $(1 + A_1^-)\xi + B_1^-[\xi, v] + \mu^- v, (1 + \tilde{A}_1^+)\xi + \tilde{B}_1^+[\xi, v] + \mu^+ v$. We scale these vectors so that they compensate exactly the $[\xi, v]$-component of the transport of $v(z)$ at y_0, y_m:

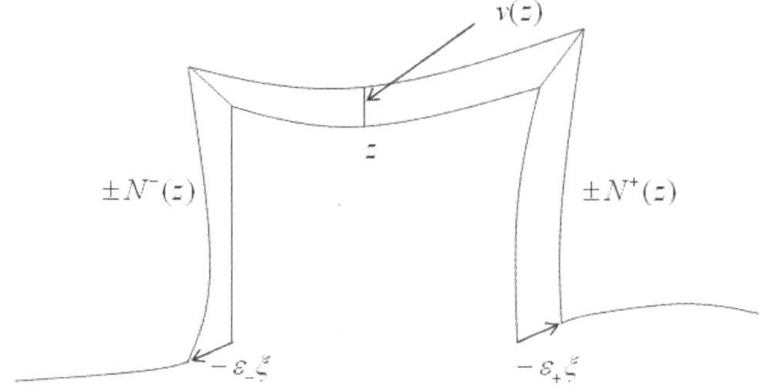

We adjust the ξ and v-components by modifying the lengths of the ξ and v-pieces. We derive in this way a normal to the right and a normal to the left, $\pm N_i^-(z), \pm N_i^+(z)$.

We assign to each of them the orientation which corresponds to decrease J. Indeed, $A_1^-, \tilde{A}_1^+$ are not zero and

$$\left| J'(x^\infty) N^\pm(z) \right| = \varepsilon_\pm A_1^\pm \quad (\tilde{A}_1^+ \text{ for } A_1^+).$$

We derive our normals $N_i^\pm(z)$.

To each of them is associated on orientation of the $\pm$ v-jump which is introduced at z. Observe that there is a way to combine $\pm N_i^-(z)$ with $\mp N_i^+(z)$ in order to build a tangent vector into x^∞; we simply assign the same orientation to the v-jump at z. Because x^∞ is critical in its stratification,

$$J'(x^\infty) \cdot u = 0.$$

Thus, if u is chosen with N_i^- on the left side, it was $-N_i^+$ on the right side. N_i^- and N_i^+ corresponding to $\pm$ v-jumps having reverse orientations. Accordingly, if we think of the curves $x^\infty + \varepsilon N_i^-, x^\infty + \varepsilon N_i^+, \varepsilon > 0$, these two curves will contain at z a $\pm$ v-jump with the **same orientation**. This orientation is the same throughout (y_i, u_{i+1}). We will refer to it in the remainder of this work as the **preferred orientation** of the normals in (y_i, y_{i+1}).

Definition 4 x^∞ will be labeled **false** if the preferred orientation in (y_0, y_1) (or (y_{m-1}, y_m)) corresponds to the orientation of the left edge of the ξ-piece (respectively the orientation of the right edge of this ξ-piece). Otherwise, x^∞ is sign-true.

We then have:

Proposition 19   *If x^∞ is sign-true, then the edge orientations are reversed; otherwise, if m is odd, they agree.*

Proof.   Assume the left edge is positively oriented. Since x^∞ is sign true, $N_0^-(z)$ has $-$ as preferred orientation. The preferred orientation switches from a nodal zone to the next nodal zone. If m is even, $N_{m-1}^-(z)$ has $+$ as preferred orientation. Since x^∞ is sign-true, the right edge has the negative orientation. Proposition 19 follows. $\square$

2.5.4.3   *Hole flow and Normal (II)-flow on curves of Γ_{4k} near x^∞*

Γ_{4k} is the space of curves made of $2k$ ξ-pieces alternating with $2k \pm v$-pieces. The $2k \pm v$-pieces build a configuration. Some of these $\pm v$-jumps might be zero. We are assuming throughout section 2.5.4 that **we can keep track of all the $2k\pm$ v-pieces on our deformation classes.** Let us consider such curves near x^∞. Since they are close in graph to x^∞, they must have a nearly ξ-piece (a "ξ-piece" broken maybe with $\pm v$-jumps) close to the characteristic piece of x^∞ and they must have nearly $\pm v$-jumps corresponding to the two edges of this characteristic piece.

Let us assume, in a first step, that we can associate without ambiguity a $\pm v$-jump on our curves which corresponds to the left edge. We will discuss this point later. We have "nodes" also, $x_0, \ldots, x_m$ which correspond to $y_0, \ldots, y_m$ and can be roughly defined using the v-rotation in the transport along the nearly ξ-piece, up to $o(1)$. We thus have preferred orientations on each (x_i, x_{i+1}) and also normals $N_i^\pm$ defined on our curves (the definition of preferred orientation, normals on x^∞ extends easily).

Let us consider such a curve and the sequence of its $\pm v$-jumps on the nearby ξ-piece. We denote the nodes (up to $o(1)$) by vertical lines, the sequences of $\pm v$-jumps by $*$'s. We will also have a line indicating the **preferred orientation** of each nodal zone, which will be labelled "Area of $+$" or "Area of $-$". We will assume for the sake of simplicity that the left edge is positively oriented and we will introduce the hole flow associated to this left edge. Its distribution of signs will be indicated at the bottom of our drawings:

Preferred orientation$\searrow$

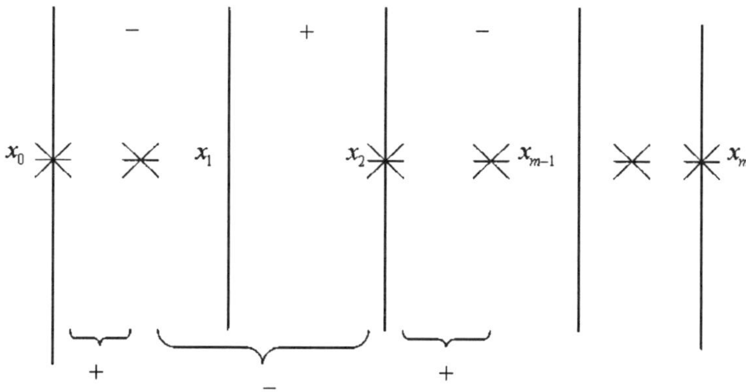

(Area of) Hole flow $\nearrow$

Observe that there is a $*$ with the positive orientation at x_0 or nearby, corresponding to the left edge. These should be also another one at x_m or nearby (up to ambiguity) corresponding to the right edge.

We then have:

Proposition 20 *If at any point between two consecutive $*$'s, not close $o(1)$ to the nodes, there is agreement between the sign of the hole flow and the preferred orientation, a decreasing normal can be defined which will bring the curve below the level of $J(x^\infty)$ without increasing the number of sign changes beyond $2k$.*

If there are several points of this type, these decreasing normals can be convex-combined and the claims remain unchanged.

Proof. The point lies between two nodes (x_i, x_{i+1}). It is not close $o(1)$ to the edges of this interval. We may use $N_i^{\pm}(\xi)$. This use decreases J at a negative rate bounded away from zero. Along this displacement, the nodes may move. We need to decrease J only by an amount equal to $o(1)$ so that the $\pm v$-jump introduced by the normal is small and the nodes move little. The points when N_i^- was introduced was not $o(1)$ close to the nodes so that the argument proceeds above $J(x^\infty) - \varepsilon$. $\qquad\square$

Since the sign(s) introduced by this (these) normals agree with the sign(s) of the hole flow associated to the left edge, the number of sign changes does not increase beyond $2k$.

Corollary 5 *If there is a node with no $*$ close to it, such a normal can be introduced.*

Proof. Indeed, there are then two consecutive $*$'s with points inbetween them not close $o(1)$ to nodes and having reversed preferred orientations. One of them agrees with the sign of the hole flow on the interval. $\qquad\square$

Proposition 21 *Assume that a $*$ is in between nodes, not close $o(1)$ to any of them. Then, a normal can be introduced at the location of this $*$ with the associated preferred orientation. This normal will decrease the curve below $J(x^\infty) - \varepsilon$ without increasing the number of sign changes beyond $2k$.*

The use of several such normals together, also combined with the hole flow, has the same properties.

Proof. This corresponds to elementary operations on a configuration.

Since the $*$ corresponding to the the left edge is not in between nodes, we are not perturbing it. The conclusion then follows from Proposition 16 above. $\qquad\square$

Definition of the $*$ of an edge

We consider the case of the left edge and an approaching configuration σ. We can define an "average left edge" θ which varies continuously with σ. We consider a fixed small number $\delta > 0$ and all the $\pm v$-jumps of σ which are contained in a δ-neighborhood of θ, $\mathcal{V}_\delta$. δ is fixed as σ approaches x^∞.

δ is as small as we may wish. These $\pm v$-jumps are ordered according to the time-parameter on the corresponding curve. c_0 is a fixed small parameter. We consider the last of the $\pm v$-jumps in $\mathcal{V}_\delta$ of size at least $2c_0$ having the orientation of the left edge. If this last $\pm v$-jump is defined unambiguously, the corresponding $*$ is the $*$ of the left edge. As this $\pm v$-jump becomes of size $2c_0$ or starts to move out of $\mathcal{V}_\delta$, we may have an overlap of various choices for this $\pm v$-jump and therefore of various choices for the $*$ defining the left edge. A problem of definition for our deformation arises when these various $*$'s define different hole flows; we will address this issue in our arguments. As we point out below, in the proof of Proposition 24, when the last of these large $\pm v$-jumps becomes small (of size $\frac{c_0}{2} < c < c_0$), we can move it away from this $\mathcal{V}_\delta$-neighborhood while decreasing J. This uses Proposition 30 (below).

2.5.4.4 *Forced repetition*

The construction completed in 2.5.4.3 rests upon the definition of the hole flow which, in turn, rests upon the choice of a representative for the left edge If there is a clear representative for this left edge—as when the left edge of the curve neighboring x^∞ is large and isolated from the other $\pm v$-jumps— this hole flow is well defined. But if there is an ambiguity—as when the left "edge" of the neighboring curve is made of several $\pm v$-jumps having the same orientation—then the hole flow might not be well defined. In view of the proof of Proposition 17, the configuration has then to contain a forced repetition corresponding to the orientation of the left edge (positive). Thus,

Proposition 22   *If the construction of 2.5.4.3 cannot be carried out at a curve x near x^∞, then the configuration σ of x contains a forced repetition (with the positive orientation) near the left edge of x^∞. Outside of the curves x having such configurations, the constructions of 2.5.4.3 can be carried out continuously. Furthermore,*

Proposition 23   *Let x be a curve neighboring x^∞ with a configuration σ for which the hole flow of the left edge is defined without ambiguity. If the construction of 2.5.4.3 cannot be carried on a whole neighborhood of x, then σ contains a forced repetition between the representatives of the left edge and the right edge of x^∞.*

Proof.   In view of 2.5.4.3, we must then have at least a $*$ at each node $y_1, \ldots, y_{m-1}$. Two very close $*$'s can be separated, see Proposition 30, below. Thus, there is exactly one $*$ at each node (close $o(1)$). In view of 2.5.4.3 again, we cannot have then a $*$ between nodes. Finally, before the family of $*$'s at $x_1, \ldots, x_{m-1}$, there must be a $*$ at x_0 (close $o(1)$)) with the orientation of the left edge. A similar statement can be made for the right edge.

We are assuming that x^∞ is sign true. Thus, if m is odd, the edge orientations agree and we have a sign repetition in between. The argument repeats for m even since the edge orientations are then reversed. $\square$

We continue our study of such configurations and observe:

Proposition 24   *A neighborhood U of such configurations may be chosen so that the construction of 2.5.4.3 on ∂U reduces to the use of the normal (II) flow on $*$'s which do not represent the left edge, while all configurations in U contain a forced repetition.*

Proof.   If U is taken small enough, all the configurations will have exactly one $*$ in a neighborhood of a node. On ∂U, one of these $*$'s has to be "away" from its corresponding node and the normal (II) flow can then be applied to this $*$. Since this $*$ might represent the left edge, we have to modify U. We observe that if any of the $*$'s which are located near $x_1, \ldots, x_{m-1}$ moves more than $o(1)$ from the related node, the configuration has crossed in a region where the use of the normal (II) flow on a $*$ which does not represent the left edge is available. We then consider the first $*(*_o)$ before the $*$ at x_1 having the same orientation than the left edge and the first $*(*_m)$ after the $*$ at x_{m-1} having the same orientation than the right edge. If one of these $*$'s advances (for x_1) or recedes (for x_m) towards the neighboring $*$ on the characteristic piece, then this $*$ has to move out of the node and the configuration crosses into a region with a good normal (II) flow.

One of these $*$'s could also exit the characteristic piece and reverse sign or become small after leaving the edge (forcing thereby all preceding $*$'s on the left, all following $*$'s for the right to leave also). But since we need a representative for the edge, a $*$ at the node x_1 of x_{m-1} has to move out of the node, with the same conclusion.

Finally, one of these $*$'s, $*_o$ or x_m, could become smaller and smaller, reaching a size c, $\frac{c_0}{2} < c < c_0$, c_0 small and fixed. Since all our configurations are near x^∞, one or several large $\pm v$-jumps build the edges and none of them is then $*_0$ or $*_m$, whichever has become small. We are assuming, for example, that $*_0$ is becoming small. There is a $*$, denoted $*_1$, within $o(1)$ of x_1 and no $*$ between $*_0$ and $*_1$.

Using Proposition 30 (see below) on such configurations, we can move $*_0$ away from the large $\pm v$-jumps of the edge towards x_0, while decreasing J. Once $*_0$ is a little bit away from the edge, since it does not represent it anymore (it is small and slightly away from the "average edge"), the normal (II) flow can be used on it in order to decrease below x^∞. This covers in particular the situation, near ∂U, when the jump $*_0$ defining the left edge (or $*_m$ for the right edge) is becoming small and is being replaced by another jump, another $*$ for the left edge. In the transition, the initial $*_0$ can be moved inside the characteristic piece. As the definition of the $*$ associated to the left edge changes, the normal (II) flow can be used on this initial $*_0$ to decrease below x^∞. We thus see that we can track down all these configurations and that they contain, throughout, a forced repetition inside the characteristic piece, i.e. between $*_0$ and $*_m$ and that the exit set is through the frontier ∂U of a domain U where a normal (II) flow is available. This includes the case when the jump at $*_0$ or $*_m$ becomes of a size c, $\frac{c_0}{2} < c < c_0$. Then we are changing the definition of $*_0$ or $*_m$ for the left or right edge (respectively). In the transition region, the normal (II) flow (without any use of another hole flow on the introduction of an additional$\pm v$-jump) can be used on the jump $*_0$ or $*_m$ which has become small. $\qquad\qquad\square$

Observation 4. Some further thinking shows that the above reasoning on $*_0$ and $*_m$ is not needed. The basic argument runs as follows: U is defined to be the set on $W_u(x_{2k})$ near x^∞_{2k-1} where there is a $*$ near each node $x_1, \ldots, x_m$. Would there be other $*$'s in between nodes, the normal II flow can be used on them. This is in particular the case on ∂U. Assuming that there is a $*$ near each node and none in between, in $U_1 \subset \overline{U}_1 \subset U$, we can introduce in each zone a decreasing normal corresponding to the preferred orientation. This corresponds to a distribution of signs which behaves exactly as a hole flow, only that the initial 1 is not defined for this distribution of signs. However, since we are introducing this alternating distribution **once**, we do not increase the number of signs changes beyond $2k$. The definition of this alternating distribution is clear and unique. The use of this distribution is limited to U_1, shielded from U^c. It convex - combines naturally with the normal II flow-. The advantage of this argument is that it bypasses the use of Proposition 23 and the forced repetition in U, thus the definition of $*_0$ and $*_m$ as well as of the $*$'s of the edges as this point.

2.5.4.5 *The Global picture, the degree is zero*

Two or more $*$'s can be assumed not to be too close. Otherwise, using Proposition 30 below we can bring them apart (in a global deformation) while decreasing J.

We have now four regions and the transition between them; there is first U and $U_1 \subset U$, a smaller version of U. On U_1 we use the normal $N_0^-(z)$ and extensions of it (alternating sign distribution starting with $N_0^-(z)$ between nodes) related to (x_0, x_1) and we convex-combine it with the normal (II) flow as above on $U - U_1$. Using Proposition 18 and Observation 2, we have not raised the number of zeros beyond $2k$. Outside of U, we use the construction 3. when available. It provides with combinations of hole flow and normal (II) flows in the region where the representative of the left edge is defined properly. In the remainder, we have several possible representatives. We may assume that the hole flows corresponding to these representatives do not coincide. Otherwise, we find no problem in order to define our decreasing deformation. We then must have a forced repetition. In order to decrease J in this region, we introduce again a normal $N_0^-(z)$ related to (x_0, x_1). This is completed as follows: we may assume that there is no * "between nodes" (i.e. not in the immediate vicinity of nodes). Otherwise, we use the normal (II) flow maybe on several nodes at the same time. This flow can easily be seen, after the construction of $N_0^-(z)$, convex-combined through "sliding" with $N_0^-(z)$ (we can use several consecutive copies of $N_0^-(z)$ in the same interval between two *'s to complete the convex-combination). Under this assumption, there is a last * "before the node at x_1"(i.e. before x_1 and not in the vicinity of x_1). This last *($*_0$) is close to the left edge.

$N_0^-(z)$ is introduced betweeen this * and x_1. It is introduced in the "middle" of the first nodal zone. The representatives for the left edge are chosen among the v-jumps of size $\geq c_0$ having the orientation of the left edge and close to the average left edge. Thus $N_0^-(z)$ is introduced **after** any of these representatives. It is also introduced before (on the characteristic piece) the $\pm v$-jump corresponding to the hole flow (whatever it is) in the interval starting with $*_0$. In fact, since the only $\pm v$-jumps needed for the hole flow can be located in the vicinity of the nodes and the other ones taken to be zero, we may assume that $N_0^-(z)$ is introduced **before** any $\pm v$-jumps used by any of the hole flows on the configuration. Thus we do not introduce v-jumps "in the edge" i.e. before $*_0$. Since the competing *'s for the definition of our hole flows are before $*_0$ ($*_0$ possibly included), we do not introduce $\pm v$-jumps between the starting 1's of our hole flows.

We have to worry about the convex-combination of $N_0^-(z)$ and one of the competing hole-flows, namely the one such that the $\pm v$-jump introduced between $*_0$ and $*_1$ (the next *, to the right of $*_0$) is not negative. If $*_1$ is not beyond x_1, no positive v-jump is introduced before $*_1$ since this v-jump would not decrease J. We need to worry only if $*_0$ is a positive v-jump. Otherwise if $*_0$ is negative or zero, $N_0^-(z)$ and $*_0$ are compatible (including if $*_0$ is zero).

If $*_0$ is positive, it generates a hole flow which may be viewed as one of the competing hole flows.

We thus may view $*_0$ as a positive j generating one hole flow and we have another i, $i < j$, with a * at i, positive again, in the left edge, defining the competing hole flow. We thus have replaced our two competing *'s with the $*$ at i and $*_0$. We then invoke Proposition 17: the use of $N_0^-(z)$ is allowed without increase of the number of sign changes beyond $2k$. Along the decreasing deformation, $*_0$ is untouched and therefore its sign unchanged, unless it moves inside the characteristic piece. But, then, $N_0^-(z)$ locates precisely at $*_0$ and becomes the normal II flow which is compatible with all hole flows. At the boundary of the region where there is a hole but not a well-defined representative for the left edge, $N_0^-(z)$ will convex-combine with each of the hole flows (used on disjoint closed regions) and by Proposition 17, this combination will not raise the number of zeros beyond $2k$.

With the support of a drawing:

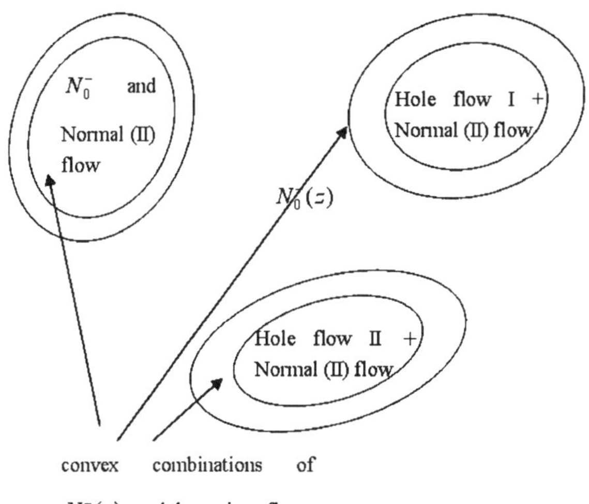

Using now a standard deformation lemma, we may assert that a given compact K in $J_{c_\infty + \varepsilon}$, with $c_\infty = J(x^\infty)$, can be deformed into $J_{c_\infty - \varepsilon'} \cup V$, where V is contained in a neighborhood, as small as we please, of x^∞. In this neighborhood, the nodes are defined up to $o(1)$. If K is in Γ_{4k} and if its $\pm v$-jumps can be tracked down, as is the case fo K's contained in $W_u(x_{2k}^\infty)$, x_{2k} a periodic orbit of index $2k$, then after the use of the deformation lemma, we may apply our construction to the part of the deformed set in V. The result is that all of K is moved below c_∞. Thus—and this is the form under which we will use this conclusion:

Proposition 25 *Let x_{2k-1}^∞ be a true critical point at infinity of index $2k - 1$ (x_{2k-1}^∞ could be a cycle at infinity) with at least one characteristic* *piece. Assume that all the oscillations on $W_u(x_{2k})$ near x_{2k-1} are small (i.e. no Fredholm issue). Then, the intersection number of $W_u(x_{2k})$ with $W_s(x_{2k-1}^\infty)$ is zero if there is no intermediate false critical point at infinity between x_{2k} and x_{2k-1}^∞.*

2.5.5　*Companions*

2.5.5.1　*Their definition, births and deaths*

When we defined normals, we defined also what a "preferred orientation" was. If the preferred orientation on (y_0, y_1) or (y_{m-1}, y_m) coincides with the orientation of the closest edge, x^∞ is labelled false.

False x^∞'s can be bypassed downwards, without increase in the number of sign changes but at the expense of introducing a normal, i.e. a new $\pm v$-jump which is the immediate neighbor of an edge $\pm v$-jump bearing the same orientation. Bypassing such x^∞'s modifies our configurations in one regard: individual jumps are replaced by families of companions around a single original jump. If we wish to extend Proposition 25 to this new situation, we find out very soon that the reasoning has to be different since a family can control several nodes through its various companions.

Let us observe though that as long as there are two companions or more in a family, the original companion is among them and it survives their death, with the same orientation, i.e. as an added companion goes away, its original companion survives its death, with the same orientation. Elsewhere, this original jump might switch signs, but it then has no companion.

This observation is important as can be seen when we try to understand how a true critical point at infinity y^∞ dominates another true critical point at infinity x^∞. If x^∞ has several characteristic pieces, then x^∞ is in fact a cluster of several critical points $x_s^\infty, x_{s-1}^\infty, \ldots, x_{s-\ell}^\infty$. $x_{s-\ell}^\infty$ is the critical point at infinity which has the same graph as x^∞ but its associated cycle does not use any full (half)-unstable manifold associated to a characteristic ξ-piece. $x_{s-\ell+1}^\infty$ builds a cycle through a combination of various characteristic ξ-pieces, one at a time, as pieces of a puzzle. $x_{s-\ell+2}^\infty$ using two of them at a time, etc. As soon as y^∞ dominates x_s^∞, it has to dominate $x_{s-1}^\infty, \ldots, x_{s-\ell}^\infty$ and if we try to cancel $W_u(y^\infty) \cap W_s(x_s^\infty)$, it is in fact $\overline{W_u(y^\infty) \cap W_s(x_{s-\ell}^\infty)}$, a stratified space of top dimension ℓ which we need to move down past the level of x^∞. Along this stratified space, configurations vary, companions are given birth to, then die, etc. It is a full story which we are facing.

The following Proposition which we will prove later reduces considerably this otherwise quite complicated scenery:

Proposition 26 *When a companion in a family which is not a family of an edge, and such that the v-rotation between its extreme companions, on the left and on the right, is less than $\pi - \delta, \delta > c$, becomes extremely small, it can be brought back close to its closest surviving companion (towards the original companion).*

We will prove this Proposition later. Its proof involves slight modifications of the flow which we will indicate later. It is of relevant importance in that no companion can appear suddenly or disappear somewhere. All these births and deaths are within the family.

2.5.5.2 *Families and nodes*

a. *Critical Configurations of families*

The basic observation which we make here is that a family of two companions or more cannot overlap a node which is not x_0 or x_m, i.e. an edge node; or if it does, the associated configuration, in fact all the configurations containing such an overlap, can be moved continuously down, past $J(x^\infty)$.

Indeed, the family covers then at least two distinct preferred orientations; one of them coincides with the orientation of the $\pm v$-jumps of the family or by creating a new companion (obviously with the same orientation) in between existing companions, we can move all these configurations down.

Each time a deformation (obviously a J-downwards deformation) is defined, we study the configurations which are "at the boundary" of such a deformation, i.e. the transitions to other configurations where such a deformation cannot be defined.

For the deformations defined above, their definition is directly related to the overlap. As the overlap fades away and the family recedes (up to $o(1)$) on one side of the node, this deformation cannot be defined anymore. Thus, we have to find another way of decreasing these configurations once the overlap on a node is one-sided up to a small constant c. We will then scale the growth of the companion which we grew or introduced, putting it to a smaller and smaller growth. It does not suffice anymore to move the configuration down. We have to find other, compatible ways to decrease J over such configurations (by deforming them).

The configurations which we consider below are therefore configurations such that any family is within $o(1)$ of at most two consecutive nodes. Indeed, otherwise, there is a definite overlap. On the other hand, we will assume that the hole flow of the left edge is well defined, which amounts to say that the family associated to the left edge is defined without ambiguity, then its hole flow can be used in between families, which implies that continuously defined and decreasing deformations can be defined on configurations such that one node among $x_1, \ldots, x_{m-1}$ is not within $o(1)$ reach of a family. The hole flow convex-combines in a natural way with the previous flow related to the overlaps. They coexist without increasing the number of sign changes in the configuration beyond $2k$.

Lastly, we observe that if a family lies between two consecutive nodes, within $c > 0$, c small and fixed, of each node, then we can think of the whole family as reduced to its original companion.

Such a family is obviously not associated to an edge of the characteristic piece and the use of the normal (II) flow, increasing or decreasing all members of the family at once, whatever suitable, even reversing, all together, their orientation, is available for it. Again this portion of the flow convex-combines with the other portions defined above and there is no familly which cannot have at least a node with $o(1)$ reach.

We thus define:

Definition 5   A critical configuration on a characteristic ξ-piece of x^∞ is a configuration such that every node $x_1, \ldots, x_{m-1}$ is within $o(1)$-reach of one family, every family with support on the characteristic piece is within $o(1)$-reach of a node at least and every such family is within $o(1)$-reach of at most two consecutive nodes.

We then have:

Proposition 27   *Consider all critical configurations such that one family is within $o(1)$-reach of two consecutive non-edge nodes. Assume that the hole flow of the left edge is well defined. Then all these configurations can be deformed below $J(x^\infty)$ using the hole flow of the left edge.*

Proof.   On each side of the area where the family sits (up to $o(1)$), the preferred orientations agree. There is one side where the preferred orientation agrees with the sign of the $\pm v$-jump introduced by the hole flow as we use it. $\square$

We extend Proposition 27 as follows:

Proposition 28   *Assume that a critical configuration σ does not contain a forced sign repetition, with the sign of the left edge, between the family defining this left edge and $x_0 + c$, $c > 0$ small. Then, Proposition 27 extends to all such σ's which contain a family within $o(1)$ of x_0 and x_1.*

Remark 1 $x_0 + c$ *stands for a point close to x_0 inside the characteristic piece.*

Proof.   This family has to have the orientation of the edge, otherwise we could separate it away from the edge (a family with the wrong orientation cannot leave the characteristic piece through the left edge unless it is followed by a family having the orientation of the edge which will take the place of the family of the edge. Such a family would be after x_1 and we would have ample room to separate the edge and the family with the wrong orientation.)

The hole flow starting at this family has to agree, under our assumption (no sign repetition near the left edge, with the sign of the left edge), with the hole flow defined by the family of the edge. This hole flow provides a sign $+$ after x_1 and this agrees with the preferred orientation of (x_1, x_2).

$\square$

Corollary 6   *If a critical configuration σ contains two or more families within $o(1)$-reach of two consecutive nodes, then either σ can be decreased below $J(x^\infty)$ in a continuously defined deformation or σ contains a forced sign repetition, with the sign of the left edge, between the family defining this left edge and $x_0 + c$, $c > 0$ small.*

Proof.   At least one of the families does not control the nodes x_{m-1}, x_m but controls another pair of nodes.

$\square$

b.   Critical configurations with one family within $o(1)$-reach of (x_{m-1}, x_m)

We thus see that the only critical configurations which we have not been able to decrease below $J(x^\infty)$ are those such a family is within $o(1)$-reach of (x_{m-1}, x_m). If an additional family is within $o(1)$-reach of two other nodes, we have defined such a deformation. However, once we define a decreasing deformation for the configurations with only one family within $o(1)$-reach of (x_{m-1}, x_m), we need to check that the flows convex-combine and generate a global decrease below $J(x^\infty)$, without increase of the number of zeros beyond $2k$.

Let us consider such a configuration:

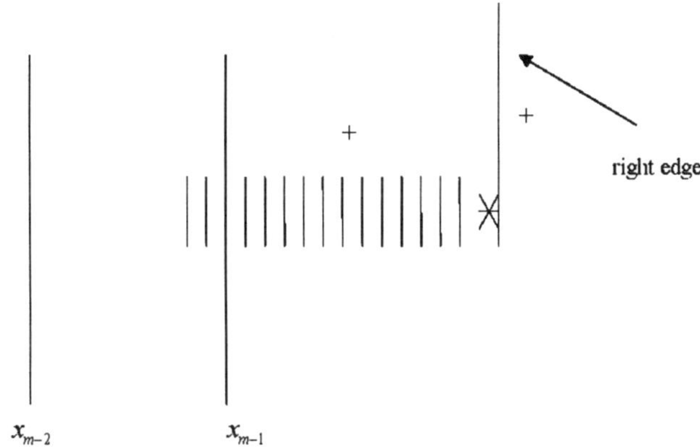

We use in the sequel the hole flow of the right edge, between x_{m-3} and x_{m-1}. It is well defined if there is no repetition in the configuration related to this edge. We make this assumption in a first step. Since we are using the hole flow of the left edge in other parts, we need to convex-combine in some regions and this needs careful checking. This is the reason why we are not starting from a critical configuration and we are simply assuming that a family controls x_{m-1} and x_m.

The normal (II) flow on intermediate families is compatible with both flows. Thus, we assume that every family has to be within $o(1)$-reach of a node and that it does not overlap over a node. We cannot yet assert that every node is controlled by a family.

We focus on x_{m-2}.

Proposition 29   *If no family is within $o(1)$-reach of x_{m-2}, then the two hole flows can be convex-combined near the configuration on the interval preceding (x_{m-1}, x_m).*

Proof.   Since x^∞ is sign-true, the preferred orientation of (x_{m-2}, x_{m-1}) is $+$ and the preferred orientation of (x_{m-1}, x_{m-2}) is $-$. If the two hole flows disagree, then one requires the use of $+$ between the family controlling x_{m-1}, x_m and the previous family, while the other one requires the use of $-$. The introduction of $+$ follows the introduction of $-$ and is compatible with the orientation of the family controlling (x_{m-1}, x_m). The convex-combination can proceed. It makes use of a single interval in the transition but, on each side, the use of the full hole flow is warranted. $\square$

Thus a family must be within $o(1)$-reach of x_{m-2}, on one side or on the other side, with no overlap.

Let U_0 be a small neighborhood of the set of all configurations containing such a behavior, with no repetitions related to the right as well as to the left edges.

Let F_1 be the family within $o(1)$-reach of (x_{m-1}, x_m) and F_2 be the family within $o(1)$-reach of x_{m-2}

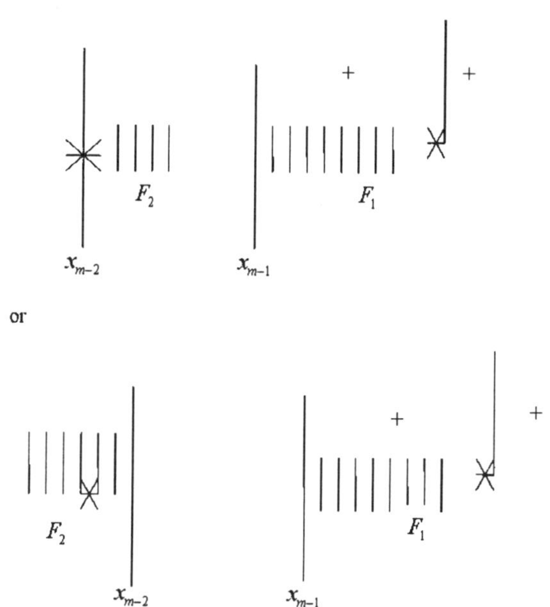

or

We claim:

We consider in what follows three distinct consecutive families of companions F_1, F_2, F_3. F_2 follows F_3 and F_1 follows F_2. The support of F_2 is inside a characteristic piece. We define the **thickness** T of F_2 to be the ξ-length between the first and the last companion of F_2. We assume throughout this part that the maximal size of a companion of F_2 is $o(T)$ and that $T \leq \frac{\pi}{2}$.

We then claim:

Proposition 30 *We can rearrange F_1, F_2, F_3 along a J-decreasing deformatiion so that*

i) The v-rotation between the right edge of F_2 and the left edge of F_1 is at least $\pi - \delta$, where δ is a fixed positive number.

*ii) If $T > 0$, the v-rotation between the left edge of F_2 and the right edge of F_3 is also at least $\pi - \delta$. It is also at least $\pi - \delta$ if the v-rotation between F_1 and F_2 is more than $\pi + \delta$. The first part of ii) has to be understood as a statement involving T **after** the deformation.*

iii) Assume that F_2 had initially a thickness $T \geq \bar{c}$, a fixed positive number $\bar{c} \leq \frac{\pi}{2}$. Assume that F_2 can be separated in two distinct families of consecutive $\pm v$-jumps F_2^-, F_2^+ and that the maximum size of the jumps of F_2^+, s_+ is small o of the maximum size of the jump of F_2^-, s_-. Then, after decreasing deformation, all of F_2^+ is at a ξ-distance equal to $\frac{\bar{c}}{100}$ at most of F_2^-. Furthermore, s_+ changes into s_+^1, with $\frac{1}{c} s_+ \leq s_+^1 \leq c s_+$ and s_- into $s_-(1 + o(1))$.

Proof.

Proof of i) See Section 2.5.8, Proposition 32. Given two consecutive $\pm v$-jumps, the ξ-piece between them has H_0^1-index zero (is minimal) if the v-rotation along this ξ-piece is at most $\pi - \delta$. If it is less than $\pi - \delta$, then we can "widen" the intermediate ξ-piece between them by inserting v-verticals to the left of left jump or to the right of the right jump. We take the side which corresponds to the smallest jump. In this way, no v-jump disappears. We need to take this cautionary step only if the jumps have opposite orientations. This widening process never stops as long as the v-rotation is less than π. There are indeed no "small" rectangles made of two small $\pm v$-jumps and two large ξ-pieces as long as the v-rotation along these ξ-pieces is less than $\pi - \delta$. See Section 2.5.8 for further details.

Proof of ii) Once the distance between F_1 and F_2 is $\pi - \delta$ or more, we turn to F_2 and F_3 and complete the same process. If the thickness of F_2 does not fade away in this process, then it goes to the end and the distance between F_2 and F_3 is at least $\pi - \delta$. It can also happen that the thickness of F_2 is zero or becomes very close to zero while the distance between F_1 and F_2 is more than π.

At the end of the process, if either $T > 0$ or the v-rotation between F_1 and F_2 is more than π, the v-rotation between F_2 and F_3 is at least $\pi - \delta$.

Proof of iii) The claim can be derived from the case when F_2^- is reduced to a single $\pm v$-jump of size c_- and F_2^+ is reduced to a single $\pm v$-jump of size c_+, with $c_+ = o(c_-)$ assume that both $\pm v$-jumps have the positive orientation for example.

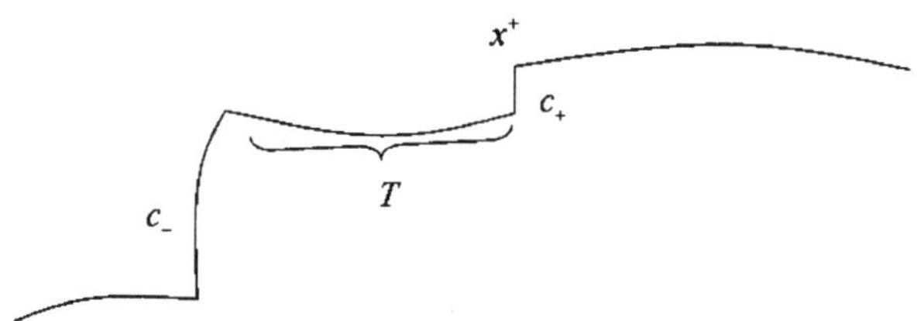

Our deformation is the composition of two deformations. The first one involves a recession of the right v-jump towards the left. This one might not be (in fact is not) J-decreasing. We thus have to compose this deformation with a second one which decreases J. To define the first deformation, we consider the ξ-orbit through x^+ and we consider a distance T^+ along this ξ-orbit before x^+; we find a point x^-:

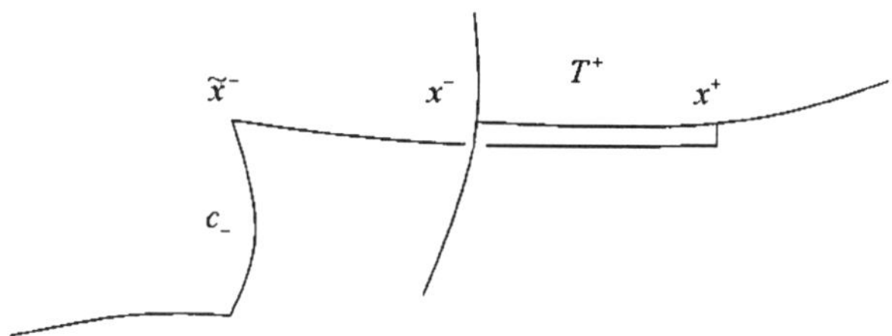

We introduce the v-vertical at x^-. There is a unique piece of ξ-orbit between the v-vertical corresponding to the v-jump of size c_- and the v-vertical at x^-:

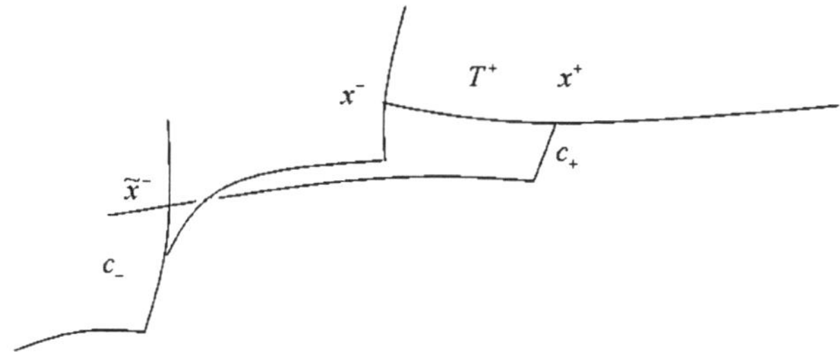

When $T_+ = T$, the distance between the v-verticals through $\tilde{x}^-$ and x^- is $O(c_+)$ and so is the length of the ξ-piece between them. We need, however, to check that the length c_- has not been consumed along this process.

The condition is easy to see:

Let δ be the size of the v-jump at x^-. This v-jump transported $T - T^+$ backwards along ξ and the v-jump transported backwards T along ξ should have the same $[\xi, v]$-component. This gives us the equation:

$$\delta(T - T^+) = c_+ T(1 + o(1)).$$

Then,

$$\delta = \frac{c_+ T}{T - T^+}$$

and the size of the v-jump **removed** from $\tilde{x}^-$ is $O(\delta + c_+)$, i.e.

$$O\left(c_+ + \frac{c_+ T}{T - T^+}\right).$$

We thus want to have:

$$c_+\left(1 + \frac{T}{T - T^+}\right) = o(c_-)$$

i.e.

$$\frac{T}{T - T^+} = o\left(\frac{c_-}{c_+}\right).$$

For example, this works if

$$\frac{T - T^+}{T} = \sqrt{\frac{c_+}{c_-}}.$$

As pointed out above, this deformation is not J-decreasing, it is in fact J-increasing. To compute the increase, we compare J at two nearby curves along the deformation:

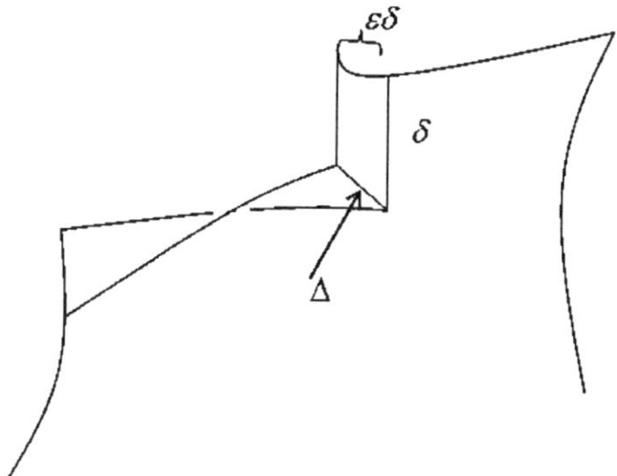

The change in J is $\varepsilon - \int_\Delta \alpha$.

Δ is obtained from $\varepsilon\xi$ by v-transport during a time c. Thus, since

$$\begin{cases} \overline{\dot{\lambda} + \mu\eta} = \eta \\ \dot{\eta} = -\lambda \end{cases}$$

$\varepsilon - \int_\Delta \alpha = O(\delta^2)$.

The size δ of the v-jump has been computed above. It is $\frac{c_+ T}{T - T^+}(1 + o(1))$. Thus,

$$\frac{dJ}{dT^+} = \frac{O(c_+^2 T^2)}{(T - T^+)^2}$$

and

$$\Delta J = O(c_+^2 T)\sqrt{\frac{c_-}{c^+}} = O(c_+^{3/2}\sqrt{c_- T}).$$

In order to define a decreasing deformation, we need to find another deformation of the same curves which would be decreasing and which would convex-combine with the previous one.

We recall for this that F_2 has support in a characteristic piece and that its thickness T is less than $\frac{\pi}{2}$ and larger than $\bar{c} > 0$. Thus, one of these v-jumps (we are assuming for the sake of simplicity that F_2 is made of two jumps. There is no loss of generality in this assumption) is $\bar{c}/2$ to the least away from a node. A decreasing normal can be defined at such a jump and it will warrant a decrease of J at a rate $\geq A\bar{c}$, $A > 0$ fixed. Consuming a size $c^+/100$ of the related v-jump (this is possible since $c_+ = o(c_-)$) which we set aside and do not touch during the first deformation if the normal has to be taken at x^+, we warrant a constant decrease of J. This works if the normal at x^+ or $\tilde{x}^-$ decreases the size of the v-jump.

It also works in the other case at $\tilde{x}^-$. If the normal at x^+ increases the size of the v-jump, then we engineer the decrease of J by increasing the size of the v-jump **as it travels** along the first deformation. We then stop — to avoid entering into technical difficulties — when the right v-jump is at $\frac{\bar{c}}{100}$ of the left v-jump since then the distance to a node might have shrinked to $\frac{\bar{c}}{200}$. We could of course have used other smaller fractions of $\bar{c}$.

The statement about s_+ and s_- follows from the estimate on ΔJ and on δ with $T - T^+ \geq \frac{\bar{c}}{100}$. This proof contains the proof of Proposition 26 as well. $\qquad\qquad\qquad\square$

Proposition 31   *If x_{m-2} is not x_0, all such configurations in U_0 can be deformed below $J(x_\infty)$ and the deformation can be convex-combined, with global decrease and no increase in the number of sign changes beyond $2k$, with the hole flow of the left edge.*

Proof.   The proof of Proposition 31 uses Hypothesis (B). On all the configurations which we are considering, there are three consecutive families. F_1 and F_2 are as above. F_3 is the first family before F_2. We pick up a small constant $\delta = \frac{\bar{c}^3}{200}$ and we reorder these families, along a J-decreasing deformation as described in Proposition 32, Section 2.5.8 below.

F_1 and F_2 are now away by $\pi - \frac{\bar{c}^3}{200}$ to the least. If they are away by more than $\pi + \bar{c}$ or if F_2 has some thickness, then F_2 and F_3 are also away by $\pi - \frac{\bar{c}^3}{200}$ to the least. Finally, the two-edge jumps of F_2 are of comparable size. Otherwise, the thickness of F_2 is less than $\frac{\bar{c}}{100}$. We assume in a first step that x_{m-3} is not x_0. We distinguish five distinct situations according to the behaviour of the right and left edge of F_2, denoted $L.E.F_2$ and $R.E.F_2$ with respect to x_{m-2}. $\qquad\qquad\qquad\square$

1st Case: The distance (all distances are along ξ) of $R.E.F_2$ to x_{m-2} is more than $\bar{c}$. Since the v-rotation between $R.E.F_2$ and $L.E.F_1$ is at least $\pi - \frac{\bar{c}}{2}$, there is a hole at x_{m-1}. We use the hole flow of the left edge in this **hole** to decrease J. Observe that we need to prove Proposition 31 if the hole flows of the right edge and hole flow of the left edge do not coincide. This implies at once that the hole flow of the left edge should yield a $-$ between F_1 and F_2. If it yields a $+$, Proposition 31 is a straightforward statement since this $+$ coincides with the orientation of F_1, and is provided by the hole flow which we are using outside of U_0.

2nd Case: The distance of $R.E.F_2$ to x_{m-2} is less than $\bar{c}$ but more than $\frac{\bar{c}}{2}$ and the distance of $L.E.F_2$ to x_{m-2} is less than $\frac{\bar{c}^3}{100}$.

F_2 has then a thickness larger than $\frac{\bar{c}}{100}$. This implies that its right edge jump and its left edge jump (we may assume that there are only two of them) are of comparable size. If we come back to our computation, we find:

$$\frac{c^+}{c^-} \geq C\bar{c}$$

where C is a constant related to Hypothesis (B). The rate of decrease of a normal at the location of the right edge jump is lowerbound by $C_1\bar{c}$ since the distance of $RE.F_2$ to x_{m-2} is more than $\frac{\bar{c}}{2}$. The rate of variation of a normal at the location of the left edge jump is upperbounded by $C\frac{\bar{c}^3}{100}$ since the distance of $L.E.F_2$ to x_{m-2} is at most $\frac{\bar{c}^3}{100}$.

We can therefore use the two normals together to decrease J and since the size of the jump is comparable, we can get all the $\pm v$-jumps of F_2 to switch signs together along this unhindered J-decreasing deformation. This acts as an normal (II) flow. This flow can be convex-combined with the flow defined in the first case without increase of the number of zeros beyond $2k$ since this would yield a combination of normal (II) flow (generalized) with a hole flow (used partially near x_{m-1}).

3rd Case: The $R.E.F_2$ is within $\frac{\bar{c}}{2}$ of x_{m-2} while $L.E.F_2$ is within $\frac{\bar{c}^3}{100}$ of x_{m-2}. We then introduce a normal between $R.E.F_2$ and x_{m-1} which yields, since this is the preferred orientation of this area, a positive v-jump. (If $+$ was the preferred orientation of (x_{m-1}, x_m), we would grow F_1. Proposition 31 would be a straightforward statement.) This can be considered to be a local use of the other hole flow. While it provides a decrease of J, it is not compatible with the hole flow of the left edge. However, this flow which is used locally and does not require the hole flow of the right edge to be defined, is compatible with the normal (II) flow of the second step. It is incompatible with the local use of the hole flow of the left edge as in the 1st step and the uses of these two flows are shielded one from the other by the flow of the second step. This flow, because it introduces a $+$ after F_2 and before F_1 and because F_1 is positively oriented, is also compatible with the use of the hole flow **before** F_2. It is also compatible with any normal (II) flow (under the form of overlap as well).

4th Case: $L.E.F_2$ is more than $\frac{\bar{c}^3}{100}$ to the left of x_{m-2}. $R.E.F_2$ is less than $\frac{\bar{c}^3}{100}$ to the left of x_{m-2} and less than $\bar{c}$ to its right. It lies in between. We use i) of Proposition 30 ($\delta = \frac{\bar{c}^3}{200}$). Either F_2 overlap x_{m-3} or there is a hole at x_{m-3}. We can use the overlap (a normal (II)-type) flow or the hole flow of the left edge. They are compatible, thus can be convex-combined for transition, with all the flows we defined above. We made a special point about this in the third case.

5th Case: $R.E.F_2$ is more than $\frac{\bar{c}^3}{100}$ to the left of x_{m-2}. We use the hole flow of the left edge near x_{m-2}. It introduces a negative v-jump before

x_{m-2} and is compatible with all previous flows.

The above argument requires some additional work in the second case. Indeed, in order to create a hole at x_{m-2} or at x_{m-3}, that is in order to set up the framework of the discussion above, we need to use the widening process developed above.This process has a disadvantage: if two families such as F_2 and F_3 have the same orientation, then as the v-rotation between them approaches π, the size of the members of these families facing each other increases considerably. This can destroy Hypothesis(B). If they have an opposite orientation , we push away F_3 from F_2; then the v-jumps of F_2 are decreased, we can scatter the decrease over the family and keep Hypothesis (B) satisfied. The argument used above thus works perfectly if F_3 has the positive orientation since when we use Hypothesis (B), see above in the second case, F_2 is negatively oriented. The orientation of the families is not relevant in the other cases.

We thus can use this widening process and the above discussion only if F_3 is positively oriented or if it is tiny, reduced to a single negative v-jump. Again in this latter case, the widening process will not increase the size of the left edge of F_2 considerably.

We need to adjust our contact form now so that this argument will also proceed in the case when two families F_3 and F_2 having the same negative orientation face each other.

In order to solve this problem, we provide the following construction which allows the use of the Normal (II)-flow while decreasing J on receding families.

Let us consider a piece of ξ-orbit, of length $\frac{\pi}{4}$. We can modify ξ into $\tilde{\xi}$, in Darboux coordinates, α into $\tilde{\alpha} = (1 + o(1))\alpha$ near this piece of ξ-orbit so that, in these coordinates , the transport map of $\tilde{\xi}$ from one edge to the other edge of the ξ-orbit becomes

$$\begin{pmatrix} \frac{1}{C} & 0 \\ 0 & C \end{pmatrix}$$

C is a positive constant. The transport matrix is provided in a basis of $ker\,\alpha = ker\,\tilde{\alpha}$ transported by ξ. Details for this construction are provided in Appendix 1.

We use the above construction twice on the characteristic piece: we use it once just after the right edge, on a v-rotation interval of length $\frac{\pi}{2}$. The modification takes place a bit away from the boundaries of this interval. The first vector of the basis is parallel to v at the right edge and is ξ-transported, the second vector of the basis is parallel to $[\xi, v] + \gamma v$ at the right edge and is ξ-transported also. C is a large constant. In this way, the transport matrix of $\tilde{\xi}$ to the left(in the direction of $-\tilde{\xi}$)will yield a large contraction along the $[\xi, v]$.

We "undo" this modification but on another piece of ξ-orbit carrying also a v-rotation equal to $\frac{\pi}{2}$. This piece of ξ-orbit is the first half of the nodal zone (x_{m-2}, x_{m-1}). On this latter piece of ξ-orbit, $C$is replaced by $\frac{1}{C}$. The two vectors of the basis are equal to $v, [\xi, v]$ at x_{m-2}. In this way the total transport matrix along the characteristic piece is unchanged. It is quite obvious from the details of the construction which are carried out in Appendix 2 that we do not create in this way new critical points at infinity. However, we modify greatly the rates of decrease or increase of J along normals between the middle of the first nodal zone, starting from left, until x_{m-2}, which are tamed in this way: the $[\xi, v]$-component of a transported vector to the left edge is divided by $\frac{1}{C}$, while the rate of decrease of a normal after the first half of the nodal zone (x_{m-2}, x_{m-1}) is kept unchanged. The use of the Normal(II)-flow on F_2 as it recedes follows.

These modifications affect v, see Appendix 2 and also the transport matrices of ξ. We prove in Appendix 2 that we can keep the same v in the nodal zone (x_{m-2}, x_{m-1}). v is modified in the first half of the first nodal zone, but this has no impact on our deformation.

We discuss this deformation in details:

In the interval which immediately follows the left edge, we observe that the modification affects the flow very little. Assuming that left edge is positively oriented, small positive v-jumps can be destroyed in this region, this decreases J before and after the modification. Small negative v-jumps can be slid to the right in this first half of the first nodal zone without change in their sizes, J decreases in this way. This follows from the construction of a new type of "Normals" of which ample use is made in the sequel:

Given any$\pm v$-jump inside the characteristic piece, of a given size c, we can transport $\pm \xi$ from the base of this $\pm v$-jump to its top, creating therefore a small vector at its top with an η-component equal to $\pm c(1 + o(1))$. The ξ-component has changed by $O(c^2)$. The v-component is also $O(c^2)$ and therefore, the size of our $\pm v$-jump is not really affected. We then transport this vector back to the left edge. Assume for simplicity that this $\pm v$-jump was the first one after the left edge. The vector at the left edge has then a component equal to $\tilde{C}c$. $\tilde{C}$ is far from zero if the $\pm v$-jump is far from special positions where $[\xi, v]$ is transported from the v-jump to the left edge onto $\pm v$ (up to a multiplicative constant). Transporting $\pm \xi$ from the base of the left edge to its top as if we were building our left normals, we find a variation of our curve which decreases it by $C_1 c$ where C_1 is non zero unless $\tilde{C}$ is zero. But then, the v-jump is in between nodes and using our usual normals, we can decrease and slide over these special positions so that we do not encounter them until the next critical one or until the process is hindered by another $\pm v$-jump, of this family or of another one. Direct computations show that in the first half of the first nodal zone, negative v-jumps must slide to the right, positive v-jumps must slide to the left. This alternates as we move from each half nodal zone to the next half. We refer in the sequel to this deformation as the **sliding deformation**.

Our deformation proceeds then, with the help of the construction and the estimates of Appendix 2, as follows: Let c_∞ be the critical level at infinity. Consider a tiny parameter ϵ positive. We may assert that (assume that x_∞ is the only critical "point" in its level surface) $J_{c+\epsilon}$ retracts by deformation onto $U \cup J_{c-\epsilon}$. U is a neighborhood of x_∞ made of curves y of C_β such that $\partial J(y)$ is appropriately small in function of ϵ.

Two cases are possible: either the number of distinct families on the characteristic piece as F_2 recedes is strictly less than $m - 2$, there is then a node, to the left of x_{m-2}, which is not controlled by a family or there is a family controlling at least two nodes to the left of x_{m-2}. We can build a suitable decreasing deformation. Or there are $m - 2$ distinct families on the characteristic piece; then there are as many families to the left of x_{m-2} as there are interior nodes and none of these families can control two nodes. If y is then a curve under deformation and y is in U, these families can be reduced to tiny $\pm v$-jumps located at these nodes. This reduction follows from the sliding deformation combined with the use of the Normal (II)-flow. In addition we claim that the following estimate holds on the size of these v-jumps(to the left of $x_{m-2}(x_{m-2}$ included):

$$\Sigma |c_i| = O(\epsilon)$$

Indeed, if this estimate did not hold, we could use the sliding deformation(we do not use on the receding side of F_2, since would only "shift" the difficulty without solving it) which has a rate of decrease bounded away from zero when the $\pm v$-jumps are near the interior nodes and move away these curves inside $J_{c-\epsilon}$. Those remaining in U satisfy then this estimate.

We use then the results of Appendix 2 and introduce our modifications. Accordingly, the curves y of U are changed into curves $\tilde{y}$, J is changed into $\tilde{J}$; using the estimate above, $U \cap J_{c-\epsilon}$ is changed into $\tilde{U} \cap \tilde{J}_{c-\epsilon+o(\epsilon)}$, now the transport matrix of $\tilde{\xi}$ from x_{m-2} or points close to this point, the transport matrix to the left edge introduces a contraction equal to $\frac{1}{C}$ along $[\xi, v]$ and the Normal (II)-flow can be used on the receding F_2. Observe that after the use of the Normal (II)-flow on F_2, a $\pm v$-jump near a node remains located near a node. $(\tilde{U}, \tilde{U} \cap \tilde{J}_{c-\epsilon+o(\epsilon)})$ is thus deformed, relative to the second set B of the pair, onto (B, B). From there, we can , using our observation about the location of the $\pm v$-jumps, move back to the curves to the curves y and to J. This defines globally a deformation of $(U, U \cap J_{c-\epsilon})$ onto $(U \cap J_{c-\epsilon}, U \cap J_{c\epsilon})$ and a deformation (almost decreasing) of $J_{c+\epsilon}$ onto

$J_{c-\epsilon}$. The result follows.

When $x_{m-3} = x_0$, the argument adapts: instead of pushing away F_3 which is now the family of the left edge, we can use the above observations and footnote and slide the negative v-jump past the first half of the first nodal zone, after the first modification. The deformation argument extends then, unchanged. The case of degenerate ξ-pieces of H_0^1-index strict 1 and the case of non-degenerate ξ-pieces is discussed later.

We summarized these steps in a chart with $L.E.F_2$ as ordinate y, $R.E.F_2$ as abcissa x. We have $y \le x$. We have introduced the various regions with the essential flows used in each of them. At the frontiers, convex-combinations are used. We have discussed above their compatibility.

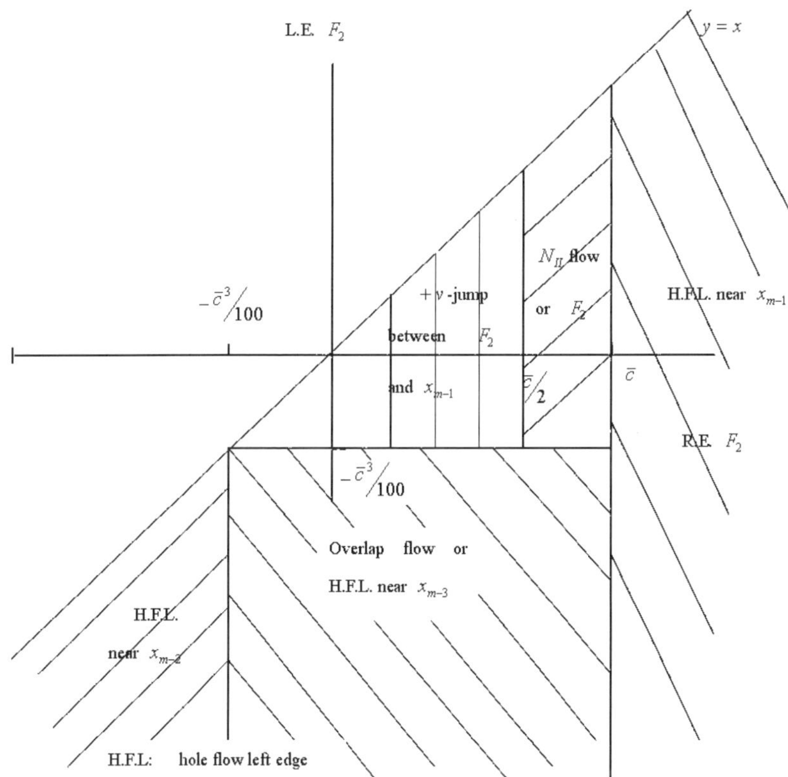

c. The case of H_0^1-index 1 and the non-degenerate case

We now address the cases when the H_0^1-strict index of the characteristic ξ-piece is 1 and we also provide the observations used to extend these arguments to non-characteristic pieces. This is needed in order to establish that transversality holds.

When a characteristic piece is of H_0^1-index 1, the two edges facing each other may each have one or more companions living on this ξ-piece. If there is no additional family is in between, there is a problem of transversality. If one of the families is confined to the interior of the half of the ξ-piece defined by the edge within a family, it is easy to define a decreasing deformation via the introduction of an additional companion for the other family in between the middle of the ξ-piece and the confined family. However, if this does not happen, the downwards deformation becomes very difficult to define. We have been able, along the lines of the argument developed in the previous section, to find such a deformation. The idea of this deformation is the following: if we could, using one of the families, engineer a decrease of J which would not change the orientation of the companions of this family but would cancel the companions of the other edge which are living on the characteristic ξ-piece, we could then complete the deformation downwards since we would then have after only one family on the ξ-piece. This has the flavor of the argument used for the Normal(II)-flow on F_2 in the previous section: we use the strength of a $\pm v$-jump to cancel another one.

The modifications of the last section, those of Appendix 2 become useful, but only after we can identify the $\pm v$-jumps and where they are localized.

The $\pm v$-jump used to cancel the other one must survive the other one all over the configurations under deformation. This is the result of the following construction:

We can always, using the "widening" process between the right edge and the family of the left edge-this is a technical improvement of the classical widening process defined above, which bypasses intermediate small companions of the right edge-assume that the companions of the left edge enter little in the second half of the characteristic piece. We could then use again the widening process, now between the family of the left edge and the companions of the right edge and bring these companions to be at least $\frac{\pi}{4}$ away from the family of the left edge.

As long as a small companion for the left edge survives now, we can engineer a decrease of J combining the sliding deformation with the use of decreasing normals on this companion. We can use this decrease to build up a sizable companion for the right edge on the ξ-piece, evolve therefore to configurations where a(n) "artificial" companion of the right edge is present as long as a companion for the left edge is present. This deformation can be assumed to be J-decreasing. The sliding process might have brought this companion of the left edge a little too much inside the second half of the ξ-piece, we can make use again of the widening processes and bring this companion only a little bit inside this second half, with the companion of the right edge at least $\frac{\pi}{4}$ away. The relative sizes can be assumed to be comparable as the size of the companion of the left edge tends to zero. This companion might recede towards the left edge, the decrease which we can then create is small, but we do not need it then: the middle nodal position and a neighborhood of it are free.

Let us rethink about our deformation process. Again we can deform $J_{c+\epsilon}$ onto $U \cup J_{c-\epsilon}$. The $y's$ of U have their small $\pm v$-jumps on the characteristic piece close to the middle node. Therefore, U can be naturally divided in two distinct connected components or more: those such that this $\pm v$-jump belongs to the family of the left edge, those such that it belongs to the right edge, and in fact also those such there is no such v-jump. These connected components are connected one to the other through $J_{c-\epsilon}$ and we need to define a coherent decreasing deformation on all of them.

For the two last ones, we can choose to introduce a negative v-jump(with the orientation of the right edge) in the first half of the characteristic ξ-piece. For the first, we want to do the same thing, but we need first to get rid of the positive companion of the left edge near the middle nodal point. This is completed using our modifications, as in the previous section and in Appendix 2. The case of characteristic ξ-pieces of non-zero index is thereby completely solved.

In the case of non-characteristic pieces of non-zero H_0^1-index, the deformation uses the representation of Proposition 13 of the $\pm v$-jumps as Dirac masses that give rise(here)toH_0^1 (there periodic)functions η between the edges of the non-characteristic piece. These functions η are used in the second derivative of J. This representation has already been introduced and used in [Bahri-1 2003], the H_0^1-problem has been studied in details.

Once a sequence of small $\pm v$-jumps has been represented in this way, we can use the H_0^1-diffusion flow of the second derivative $(\ddot{\eta} + \eta\tau)$, projected on the L^2-spheres (L^2 in η) to engineer a decrease of J and once this is accomplished, we can turn back to the representation involving $\pm v$-jumps. The number of sign changes never increases along this process. Furthermore if the initial datum is non zero, it remains so.

The number of sign changes, counted locally, between the v-edges of the non-characteristic piece indicates the(minimal) number of families needed to describe this configuration. This number does not increase. Thus the minimal number of families (which might differ by 1 from the number of sign-changes) needed to describe the configurations which we consider and deform never increases. We might have a gap with respect to the actual number of families because there might be sign repetitions between these families. But coming from a number which is less than or equal than this actual number, we can certainly relabel, reorder, create twins and rebuild our original local set of families.

Assuming that the H_0^1-index of our non degenerate ξ-piece is initially non zero, we can use the results of [Bahri-1 2003], manipulate the rotation of v on this ξ-piece so that the maximal number of sign changes on the H_0^1-unstable manifold of this ξ-piece is $i_0^j + 1$. i_0^j is the H_0^1-index of the ξ-piece (after manipulation of the rotation).

Then, if the number of sign changes between the families in this ξ-piece, with the edges added, is less than $i_0^j + 1$, the flow will deform our configurations below the critical level corresponding to this ξ-piece for index reasons. From there, we end up in $J_{c-\epsilon}$. This implies transversality as far as non-characteristic pieces are involved.

If the maximal number of sign changes on the H_0^1-unstable manifold of a non degenerate ξ-piece is i_0^j, then we can assert that the flow deforms our configurations below the critical level only if the number of sign changes between the families is less than i_0^j. We can generate i_0^j sign changes in this framework with $i_0^j - 1$ distinct families. Thus transversality does not follow outright from this argument in such a case. We have to use the Appendix 2 (adapted) again, as in Section 2.5.5.2c: we need to show that we can adjust the value(s) of the Green's function of the operator $\ddot{\eta} + \eta\tau$ under Dirichlet boundary conditions on the non degenerate ξ-piece. We elaborate some more about this at the end of Section 2.5.9, but the main arguments are essentially the same than in 2.5.5.2c and Appendix 2.

d. An additional rule

After this full discussion, there is still an additional **rule** to introduce in order to build a global flow as some $*'s$ turn to families or vice-versa. This rule reads as follows:

If there is a hole at the last node x_{m-1} and no family is "advancing" towards this node from the left, we introduce, in order to engineer a decrease of J, before this last node a $\pm v$-jump with the orientation of the right edge rather than using the hole flow near this node

This is of course impossible if another family advances on this last node, but we can then use the construction of Section 2.5.5.2b and switch to the other flow if the strict H_0^1-index of the characteristic is not 1 and the construction of 2.5.5.2c otherwise, while if we used the hole flow and this hole flow would have directed us to introduce a $\pm v$-jump with the orientation opposite to the one of the right edge in the last nodal zone, then **unless** a family advances from the left, this switch is impossible. Thus, as a family for the right edge advances towards this node, we are less and less able to engineer a decrease and a switch is not available. This issue does not arise with the left edge, because the hole flow provides always for this edge a $\pm v$-jump with the orientation that this edge has.

Using this rule, Proposition 31 follows under the assumption that the hole flow of the left edge is well defined.

If there is a repetition barring the use of the left edge hole flow, we need to introduce a decreasing normal and we are allowed to introduce it with an additional sign change if the repetition is not already exhausted by the use of one of the non-compatible hole flows We have gone over this argument before:

The repetition occurs "in the left edge" in that, in our construction, we may require that the family of the left edge should be at a distance $o(1)$ from the average left edge (which can be defined easily). It follows that no normal is used in between the repetition. In the interval where it is used, this normal identifies then with the orientation of one of the hole flows. With the other one, it might induce an additional sign change but this would correspond exactly to the single additional sign change allowed.

$\square$

2.5.6 *Flow-lines for x_{2k+1} to x_{2k}^∞*

The extension of the results to the odd case is long and delicate. We provide it here.

We are assuming that we have at least two characteristic pieces. Let us assume for simplicity-the arguments are unchanged in the other cases-that the two left edges of the two characteristic pieces are positively oriented. We observe first that the hole flow extended into an alternating sequence does not provide more than $2k$ sign changes even in the case of $2k+1\pm v$-jumps. Indeed ,starting from the first v-jump which we assume for simplicity to be positive ,the alternating sequence ends with the positive sign on the other side of this initial v-jump, we then have a positive v-jump on both sides of the first positive v-jump and we may as well think that this v-jump is not there. The conclusion follows.

Next we have two characteristic pieces, we assume in the sequel that, after the use of the Normal (II)-flow, there is a hole on each of them. We build a decreasing deformation on such configurations that do not raise the number of zeros beyond $2k$. We pick a hole flow on each characteristic piece. We overextend this hole flow as an alternating sequence all around. Either it coincides with the other hole flow and their combination does not provide more than $2k$ sign changes. Or it does not. Then the alternating sequence starts with a $-$ to the right of the positive (for simplicity) starting v-jump of the other hole flow (which starts with a $+$). Considering these two sequences between the edges of the second characteristic piece-starting from the v-jump of this other hole flow-we see that the sequence of sign changes of this other hole flow gives at most as many changes of sign as the alternating sequence. Thus, the two hole flows (restricted to their respective characteristic pieces) can be convex-combined with a number of sign changes at most equal to $2k$.

Third, we consider one of the characteristic pieces and $*_0$, the first $*$ before (not in the vicinity) of the first node. We know how to build a decreasing normal $N_0^-(z)$: we introduce a negative v-jump after $*_0$, we have discussed this matter above; since we are discussing hole flows, we may assume that the $*$'s are not in between nodes. $N_0^-(z)$ is in a natural way part of an alternating sequence on this characteristic piece. Let us assume that there are competing hole flows on this characteristic piece - because there are several possible representatives for the left edge - and let us assume that $N_0^-(z)$ is **not** part of the hole flow which introduces a-sign after the representative most to the right of the left edge (baptized "first" representative). Thus, there is an even number of $*$'s between $*_0$ and this first representative of the left edge. We assume that no sign changes are used for decreasing normals to the left of $*_0$. We then claim that $N_0^-(z)$ is compatible with both hole flows. This is a local argument which can be used as we please. Indeed, it is compatible with the hole flow which includes it. The other hole flow can be identified as the hole flow starting with a - sign at the first representative and has a + sign immediately after $*_0$. There might be a problem with the simultaneous use of $N_0^-(z)$ and this other hole flow if $N_0^-(z)$ and the orientation of $*_0$ do not agree i.e if $*_0$ is a positive v-jump. Then this other hole flow, the distribution of signs of this other hole flow introduces a sign change between this first representative and $*_0$ since there is an even number of $*$'s in between. This sign change is not actually being used according to our rules and we thus have the freedom to use $N_0^-(z)$ and induce thereby a decreasing deformation. This works also if there is more than one repetition in the left edge. Again this argument is of local type, $N_0^-(z)$ is free, available under such circumstances and can be inserted in any other flow.

Thus so far, we have seen that, if there are holes on one of the two characteristic pieces, we can define a decreasing deformation if the hole flow is well defined on at least one of the characteristic pieces and also if this hole flow is not well defined on both, but there is an even number of $*$'s in between $*_0$ and the first representative of the left edge of any of them. This works in case of multiple repetitions in the edge, the local argument also works with the right edge.

Let us now consider the case when there are competing hole flows on each characteristic piece and there each $N_i^-(z)$ is part of the alternating sequence starting with a $-$ at the first representative of each characteristic piece. We may also assume for simplicity that the competing representatives for the left edges on each characteristic piece are consecutive $*$'s. We thus have two pairs of consecutive representatives for the left edges of these two characteristic pieces. These two pairs define on the circle two time intervals. On one of them, there is an even number of $*$'s. This interval is unique since the total number of $*$'s is odd. Let us assume that it starts for example with the left edges of the first characteristic piece. We then claim that the hole flow starting with a $+$ at the first representative of the first characteristic piece, when combined with any other hole flow on the second characteristic piece yields at most $2k - 2$ zeros. Indeed, since the number of intermediate $*$'s is even, this hole flow ends also with a $+$ at the second representative of the left edge of the second characteristic piece. Let us extend, as an alternating sequence all around, the sequence starting with a $-$ at the first representative of the first characteristic piece after identifying the two representatives of the **first** characteristic piece. We then have $2k$ sign changes. We compare: the hole flow starting with a $+$ ends also with a $+$ before the second representative of the second characteristic piece, while the alternating sequence starts and ends with a $-$ at both (positive) ends. We thus have two less zeros for the hole flow starting with a $+$. On the second characteristic piece, the alternating sequence starts with a $-$ sign, opposite to the sign $+$ of the first representative of the second characteristic piece (there are two representatives of the second characteristic piece to bypass, starting with a $-$ sign on one side. We reach the other side with a $-$ sign also. The two representatives of the first characteristic piece have been identified). Hence this alternating sequence introduces at least as many changes on this second interval (running from the first representative of the second characteristic piece to the second representative of the first characteristic piece) as any other alternating sequence restricted to this interval with the given boundary (positive) v-jumps. The claim follows. We thus have a freedom of two zeros with the use of the hole flow starting with a $+$ on the first characteristic piece when combined with any hole flow on the second one. $N_1^-(z)$ is included in all combinations containing the hole flow starting with a sign $-$ on the first characteristic piece. The addition of $N_1^-(z)$ to the hole flow starting with a $+$, when combined with any other hole flow on the second piece does not raise the number of zeros beyond $2k$. Thus, as competing hole flows- with the presence of holes on both characteristic pieces-develop, we reduce our flow to the use of one of the two $N_i^-(z)$'s depending on the configuration. We may then use both of them since, once the other v-jumps of the hole flows are removed, their joint use becomes possible. Along the convex-combination, the number of sign changes does not raise above $2k$. We may assume in the sequel, working on pairs of characteristic pieces one after the other, that there are holes on one characteristic piece at most.

Assume now that, on this characteristic piece, there are two holes or a hole covering two nodes. We then claim that if a repetition develops in the left edge of this characteristic piece and we are forced to switch hole flows, this switch can be completed without increase in the number of sign changes beyond $2k$. Indeed, assume for example that we have two holes, each jumping over exactly one node, each with a starting $*$. We can attach a sign to each $*$, starting from the representative most to the right of the left edge which gets assigned the sign defined by its own orientation, and alternating to the right. Then for a given $*$, either the orientation of this $*$ and the attached orientation agree, or there is an additional repetition; we can use it to introduce two zeros. We have two hole flows. One agrees on the interval immediately after any $*$ with the attached sign, the other one disagrees (signs for hole flows also alternate). Let us consider the hole flow which agrees with the attached orientation of the starting $*$ for each given interval. On the other hand, we have defined a preferred orientation for each nodal region and there is one which is associated to the nodal region immediately following a given $*$. If the attached orientation and this preferred orientation disagree at a hole, they will agree at the (next or previous) hole since there is one more change at a node for a preferred orientation and not for the attached sequence of $*$'s. At the hole where the preassigned orientation of the nodal zone to the right of the $*$ agrees with the attached sign on this $*$, we know that this preferred orientation corresponds to a hole flow- the one of the attached sign- while the preferred orientation of the next nodal zone corresponds to the other hole flow. Typically, we find for example:

$$\text{Attached sign } +$$

$$\begin{array}{ccc} * & + & - \\ HF1 & HF2 & \end{array} .$$

Now eliminate the normals and hole flows in all other intervals and observe the following: If the actual orientation of the $*$ agrees with the attached orientation, HF1 does not introduce any sign change. We can use jointly HF2 on this interval. Otherwise, they disagree and there is an additional repetition and we have a freedom of two zeros which we use to introduce HF1 together with HF2 (HF2 by itself did not then add zeros). We thus can use such a node to switch from a hole flow to another one. Summarizing, if there are two holes on two distinct characteristic pieces, we have built a decreasing deformation. Any change in a configuration is monitored by the Normal(II)-flow. Also if all the characteristic pieces are filled but one and if on the remaining one there are two holes, or a hole over more than one nodal zone, we also have a decreasing deformation since we can switch hole flows. This also holds if there is no repetition in the left edge of the piece with a hole. Finally, if on one of the characteristic besides the one bearing a hole there is a repetition-as when the number of $*$'s in the characteristic piece from $*_0$ till $*_m$ equals, mod2,the total number of $*$'s carried by this characteristic piece- then we can also switch at the hole the two hole flows. Thus a decreasing deformation is impeded (when there are holes) if all characteristic pieces but one are filled up and do not jail a forced repetition ($*_0$ and $*_m$ are different from their associated edges, on one side the number of $*$'s in between is even, on the other side it is odd), there is exactly one hole at exactly one node on this last characteristic piece, there is a repetition in its left edge. It is also impeded if all characteristic pieces have been filled up and, singling out a characteristic piece, characteristic piece number 1, there is no admissible extension (i.e. not raising the number of zeros beyond $2k$) for its basic set of decreasing normals (such as its $N_0^-(z)$). We are going to discuss this below. In all the other cases, after numbering the characteristic pieces, we can start to reduce holes for pieces 1 and 2. Once only one with a hole is left, we move to reduce holes for 1, 2, 3 (only a pair is involved after the first stage), then 1, 2, 3, 4 etc. At the end of this continuous process, at most one with holes is left. If the holes are more than one or contain more than one node, we can build a decreasing deformation , this is in fact part of a global deformation.

The transition between all these situations involving a change in the holes or a change in the number of holes is monitored by the Normal (II)-flow which is compatible with all hole flows. The only issue related to the use of the Normal (II) flow is that it might change the orientation of $*_0$, $*_m$. This becomes important now because we are tracking sign-repetitions. We address this issue: In order to stabilize the repetitions associated to a characteristic piece which has been "filled up" or is being "filled up", we introduce representatives L for its left edge and R for its right edge. Let assume that there is dissymmetry i.e that the number of $*$'s in between $*_0$ and L, for example is odd, while the number of $*$'s in between $*_m$ and R is even. $*_m$ and R have the same orientation before any use of the Normal (II)-flow since the characteristic piece is almost filled up, if not filled up and therefore there must be a repetition between $*_m$ and R. We thus may define an alternating sequence which will start with the sign opposite to the orientation of R immediately to the left of $*_m$ and move backwards along the characteristic piece, alternating at each crossing of a $*$. Since the characteristic piece is sign-true, we derive for this sequence the sign opposite to the orientation of $*_0$ or L immediately to the right of $*_0$. We stop here the alternating sequence and we use it in order to define decreasing normals. Since we have not used one repetition to the least, it is extendable into an admissible alternating sequence .The argument and the repetitions being local, we can use these flows on these dissymmetric pieces together. Furthermore, immediately to the right of $*_0$ and to the left of $*_m$, we have decreasing Normals which have the orientations of the Normal (II)-flow. Thus, we may extend the use of this flow after sliding the position of its action on the v-jumps over $*_0, *_m$ to the position of the decreasing normals. The sliding starts on both sides simultaneously and preserves therefore the repetition between $*_0$ and $*_m$ (no v-jump is introduced before $*_m$, the repetition is to the right of $*_m$). We could also, since we have two repetitions on this characteristic piece, reverse the position of the normals and exhaust the repetition in the middle, leaving the one to the side unperturbed. Again, we can convex-combine all the flows on these dissymmetric pieces. The Normal (II)-flow finds a natural extension on these pieces.

Along the other characteristic pieces, we can either continue the full use of the Normal (II)-flow, a repetition is anyway jailed in there or $*_0$ and $*_m$ confound with the edges; there is no use of the Normal (II)-flow on them and nothing to extend.

In the sequel, [L, R] designates the characteristic piece with a single hole at a single node in it and the left edge repetition; or [L, R] designates, when all characteristic pieces have been filled up the first one in an *a priori* chosen order.

In the second case, on [L, R] we pick up the flow defined by the basic set of normals and we need to extend it only on dissymmetric critical configurations, which is easy to accomplish as explained above. We could even switch, if needed, the use of - between the two repetitions and incorporate the flow inside such pieces into the alternating sequence generated by this basic set of normals. We do not need to extend it over symmetric configurations. This works as long as there is one symmetric configuration on a characteristic piece distinct from [L, R] which we can use in order to turn our sequence into an admissible sequence. A similar argument can be made in the first case, with the same basic set of normals, as we already have a repetition in the left edge of [L, R] and we need another one, distinct from [L, R], in order to switch hole flows. We develop a counting argument which is the same in both cases: Let n be the number of characteristic pieces, let i_0^c be the total strict index on the characteristic pieces, i_0^f be the index on the non-degenerate ξ-pieces. Let $\bar{\gamma}_1$ be the number of free ξ-pieces having either even H_0^1-index and reverse edge orientations or odd H_0^1-index and identical edge orientations, see Definition 4, page 78 of [Bahri-1 2003] for the use of this notion. Let γ be the number of non degenerate ξ-pieces bearing a sign change between their edge orientations. Let ℓ be the number of full (with their full half-unstable manifold) characteristic pieces used once at a time in the definition of x_m^∞. We then have, see [Bahri-1 2003, pp. 139] also page 78 to gather the full information:

$$i_0^c + i_0^f + \bar{\gamma}_1 + 2\ell + 1 = 2k + 1.$$

Using the results of [Bahri-1 2003], [Bahri-2 2003], we can modify α in the vicinity of x^∞ and set $i_0^f = 0$. Thus,

$$i_0^c + 2\ell + \gamma + 1 = 2k + 1.$$

On the other hand, assuming that all characteristic pieces have been filled up, if as [L, R] is becoming symmetric and we need to extend its basic set of normals into an admissible sequence, there is no other symmetric configuration on any other characteristic piece, then on each characteristic piece, including [L, R], there are $i_0^j + 2*$'s. i_0^j is the strict H_0^1 index of the jth characteristic piece, a 1 is added for the dissymmetry ($*_0$ or $*_m$) and another 1 for the incoming edge. Summing up, we find

$$\sum_j (i_0^j + 2) = i_0^c + 2n$$

v-jumps to which we have to add 1 for each non degenerate ξ-piece with a change in the edge orientations. Since the edges could be part of the same family, we do not consider the other non-degenerate ξ-pieces. In all, we find $i_0^c + 2n + \gamma \pm v$-jumps. The count has to be slightly modified when there is a single hole and a repetition in [L, R]. The formula is decreased by one. Thus, in all cases,

$$i_0^c + 2n + \gamma - 1 \leq i_0^c + 2\ell + \gamma + 1$$

and

$$\ell \geq n - 1.$$

If ℓ is strictly less then $n-1$, we find in the second case that, even when [L, R] has turned symmetric, there is an extra symmetric configuration, the basic set of normals can be extended. The same conclusion holds in the first case. This leaves with $\ell = n - 1$ as the only possible case. In fact, we must have:

$$i_0^c + 2n - 2 + \gamma = 2k. \tag{**}$$

otherwise, this is not possible.

Continuing our analysis, we look into more details at what happens as we try to extend our alternating sequence outside of [L, R] assuming that there is no outside symmetric configuration. Since there are more than one characteristic piece, there are outside dissymmetric configurations. Let us assume for simplicity that the edges of one such dissymmetric configuration are positively oriented. We extend using the ξ-orientation i.e. we go from characteristic piece to characteristic piece along ξ. If we reach the left edge of our dissymmetric configuration, after extending the basic set of normals of [L, R], with a +, then we will be able in a natural way to jump over an inside repetition in this configuration, and still include a decreasing normal, in our alternating sequence. The alternating sequence would become admissible. Otherwise, we have to alternate signs inside the full dissymmetric configuration, thereby exhausting the two inside repetitions in order to build an extension to the Normal (II)-flow inside (which we need to do). We then exit on the other side with the + sign, i.e with the orientation of the edge. This repeats starting from R until the next characteristic piece etc. Each occurrence, we reach , if the alternating sequence is not admissible, the left edge with an orientation opposite to this edge and leave the right edge with the orientation of the edge, the only exception being L itself. Thus, on each such interval between two characteristic pieces, there must be a free ξ-piece with a switch in the orientation of the edges:

$$\gamma \geq n - 1.$$

The argument goes now as follows. Assume that n is large. Choose a characteristic piece with $\gamma_j \neq 0, i_0^j = 0$ (H_0^1-index 0, reverse edge orientation). We had previously turned i_0^j to be 0 on this critical point at infinity, see [Bahri-1 2003], [Bahri-2 2003] for the argument used to derive this result. In doing this, we could have kept, since we have a large amount of non-degenerate ξ-pieces, the amount of v-rotation on several of these pieces to be less than π, but only by a tiny bit; so that we can easily rebuild a large rotation on a given ξ-piece. This last part of the argument uses Hypothesis (A). We pile up v-rotation on this ξ-piece which has $i_0^j = 0, \gamma_j \neq 0$. Piling up, we cross $i_0^j = 1$, creating only false critical points at infinity since they have a characteristic piece of strict H_0^1-index 0 and with the orientation of its edges reversed. There is still our critical point at infinity which has now a non degenerate ξ-piece with $i_0^j = 1$. Counting the maximal number of zeros of b on the unstable manifold of this critical point at infinity, we find that we should have if it interacts in our homology (page 78, pp 137-139 of [Bahri-1 2003]:

$$i_0^c + 1 + \gamma - 12\ell' = i_0^c + \gamma 2\ell.$$

Thus,

$$\ell' = \ell.$$

We continue our filling on the same characteristic piece. We cross the H_0^1-index 2 and the maximal number of zeros of b on the unstable manifold of our critical point at infinity (with the same number of characteristic pieces) changes by 2. This critical point at infinity do not interfere with our homology anymore. We still have the new critical points at infinity which we have created. They have $i_0 = 0$ again because the only new phenomenon is taking place on the new characteristic piece. Again, counting see [Bahri-1 2003, pp. 139], also page 78, we should have

$$i_0^c + 1 + \gamma - 1 + 2\ell' = i_0^c + 2\ell + \gamma.$$

Thus, $\ell' = \ell$, which is $n - 1$ while the number of characteristic pieces is now $n + 1$. A contradiction.

n is bounded and so is the number of non degenerate ξ-pieces of x^∞. Otherwise, using Hypothesis (A), we would reach the same conclusion. As k tends to infinity, this forces the H_0^1-strict index of some characteristic piece to tend to infinity. Considering $W_u(x_{2k+1}) \cap W_u(x_{2k}^\infty)$, we observe that ℓ is bounded above independently of k. Let n_0 be this upperbound. Then $W_u(x_{2k+1}) \cap W_u(x_{2k}^\infty)$ is of dimension n_0 at most. Tracking this intersection near x_{2k+1}, using the construction of Proposition 1, we may assume after perturbing the location of the v-jumps in Proposition 1 if needed, that each given configuration has at most n_0 v-jumps which are zero. This could change after the use of the Normal (II)- flow . However, this does not change under the use of the flow at infinity i.e the flow on the Γ_{2s}'s. This flow is the one driving our curves to x^∞. The other flows are used to bypass x^∞ downwards. In this circumstance, we are not going to use the Normal(II)- flow because we want to keep the orientation of the non-zero v-jumps unchanged. We go to the characteristic piece with a larger H_0^1-index. There are several triplets of v-jumps on this characteristic piece having non-zero alternating orientations. Otherwise there are too many repetitions and we may use a basic normal. We pick such a triplet; in fact we may choose a large number of triplets which are "far away" one from the other. On a given configuration, there will be among those triplets, under our assumption that there are few repetitions, a non-zero alternating triplet. Our construction below can then be convex-combined; as the configurations change, we change our triplet and the construction of the normals which we complete below can be in the transition zones carried out on several triplets, the fading and the emerging ones at the same time. We can "widen" the first oscillation of these triplets, using the procedure of Proposition 30 and if along this procedure we find some other v-jumps, we continue the process with the agglomerate v-jump but keeping track of the sign changes allowed. In the end, going left and going right around such a triplet, we can assume that we cover more than one nodal zone. We choose to construct only positive Normals in between. Since we are covering more than one nodal zone, we are always sure that one of them will be decreasing. That is the one that we choose to expand. Our proof is complete in this case.

We are left with the case when x^∞ has only one characteristic piece. Then $\ell = 0$ and $W_u(x_{2k+1}) \cap W_s(x_{2k}^\infty)$ is made of isolated flow-lines. We may assume, perturbing the locations of the v-jumps near x_{2k+1}, that none of them is zero on these flow-lines at the beginning. The flow on the Γ_{2s}'s will preserve this property. Thus no $\pm v$-jump is zero on our single characteristic piece, which is of strict index at least 2. The flow-line is isolated, we are facing a single configuration and we can grow a companion to the appropriate $*$, closest to the first node. If there are two such $*$'s,one to the right, one to the left, we choose the one to the left.

In order to complete our proof, we have to remove the assumption that the characteristic pieces are isolated one from the other. Assuming now that we have sequences of characteristic pieces, we use Proposition 33 below which involves a special construction of a decreasing deformation on consecutive ξ-pieces as an additional $*$ travels along a common edge. After completing this construction, we resume the proof of Theorem 1 in our present framework and finish it, see Section 2.5.8, after the proof of Proposition 33 below.

2.5.7   *The S^1-classifying map*

Consider $C_\beta^* = C_\beta$ - {periodic orbits of ξ}. Assume that the curves x which we consider are H^1 (this follows, continuously, from a simple regularization procedure).

We then have a map

$$C_\beta^* \xrightarrow{\sigma} H^1(S^1, \mathbb{R}) - \{o\}$$
$$x \longrightarrow b$$

where b is the v-component of $\dot{x}$.

Let

$$C_\beta^{**} = \{x \in C_\beta^* \text{such that } b \neq \int_0^1 b\}.$$

We compose σ with the orthogonal projection q on $E^+ \oplus E^-$ where $E^+ \oplus E^-$ is the L^2-orthogonal of the constants in $H^1(S^1, \mathbb{R})$. $q \circ \sigma$ maps C_β^{**} into $E^+ \oplus E^- - \{0\}$. S^1 acts effectively on $E^+ \oplus E^- - \{0\}$. Therefore $q \circ \sigma$ is the classifying map for the S^1-action on C_β^{**}. The use of this observation is deferred to a forthcoming work. We now study the Faddell-Rabinowitz index γ_{FR} of some subsets of C_β.

We have:

Lemma 17   *Let $A = \{x$ s.t b has at most $2r$ zeros and at least two genuine zeros$\}$.*

Then $\gamma_{FR}(A) \leq r - 1$; the classifying map for A is provided by $p_r(-b^- \int_0^1 b^+ + b^+ \int_0^1 b^-)$ where p_r is the L^2-orthogonal projector on $\mathrm{Span}\{\cos 2\pi it, \sin 2\pi it, i = 1, \ldots, r\}$.

Proof. Let $\tilde{b} = b^+ \int_0^1 b^- - b^- \int_0^1 b^+$. $\tilde{b}$ has also at most $2r$ zeros and $\int_0^1 \tilde{b} = 0$. Assume that $p_r(\tilde{b}) = 0$. Then, $\tilde{b} = \sum_{l \geq 2r+1} a_k \sin 2\pi kx + b_k \cos 2\pi kx$.

Rescaling, we may assume that $\int_0^1 \tilde{b}^2 = 1$. We then solve

$$\frac{\partial u}{\partial s} = \ddot{u} + \left(\int_0^1 \dot{u}^2\right) u / \int_0^1 u^2 \qquad u(0,t) = \tilde{b}(t).$$

Observe that the number of zeros of u does not increase with s, that $\int_0^1 u^2 = \int_0^1 \tilde{b}^2$ and that

$$\frac{1}{2}\frac{\partial}{\partial s} \int_0^1 \dot{u}^2 = - \int_0^1 \ddot{u}^2 + \frac{(\int_0^1 \dot{u}^2)^2}{\int_0^1 u^2} \leq 0.$$

Thus $u(s, \cdot)$ converges strongly in L^2 to a rest-point ($\int_0^1 \dot{u}^2 / \int_0^1 u^2$ is a Lyapunov function) which is a solution of $\ddot{\varphi} = \lambda\varphi$.

Observe that

$$\frac{\partial}{\partial s} \int_0^1 u = c(s) \int_0^1 u.$$

Since $\int_0^1 \tilde{b} = 0$, $\int_0^1 u = 0$ and thus this solution verifies $\int_0^1 \varphi = 0$. Thus,

$$\varphi = A \cos 2\pi kt + B \sin 2\pi kt \qquad \text{with } k \geq 2r + 1$$

φ has more than $2r$ zeros, a contradiction. $\qquad\qquad\square$

Lemma 18 Let $B_\varepsilon = \{x \text{ s.t } \dot{x} = a\xi + bv \text{ with } \int_0^1 b^+ \int_0^1 b^- = 0; a \geq \varepsilon\}$. Let $B_1 = B_\varepsilon \cap \left(\overline{U_{z\ periodic\ orbit}\ W_u(z)}\right)$. Then B_1 cannot dominate any periodic orbit of ξ of non zero index and B_1 cannot dominate any critical point at infinity of J of non zero H_0^1 -index. Furthermore, $\gamma_{FR}(B_1) = 0$.

Proof. The first claim follows from the fact that the curves of B_1, have $b \geq 0$ or $b \leq 0$. This extends to $\bar{B}_1$. For the second claim, we argue as follows:

The curves x of B_ε have either $b \geq 0, b \neq 0$ or $b \leq 0, b \neq 0$ or $b \equiv 0$. Near each z, the curves of B_ε can be equivariantly (S^1-equivariantly) pushed, with a unchanged, into a neighborhood as small as we may wish of z and its first unstable direction. Thus, B_1 retracts by deformation, S^1-equivariantly, into an S^1-invariant set, which is stratified and which is of dimension 1 transversally to the S^1-action.

There are no fixed point for the S^1-action on B_1, in fact the S^1-action on B_1 is free since the curves are immersed.

Thus, $\gamma_{FR}(B_1)$ can be directly computed on B_1/S^1. Since this set is stratified of top dimension 1, it cannot carry any Chern class of positive even degree and $\gamma_{FR}(B_1) = 0$. $\qquad\square$

These are the initial steps towards computing on homology in a more general framework.

2.5.8 Small and high oscillation, consecutive characteristic pieces

We study in this section curves $\bar{x}$ of C_β or $C_\beta^+ = \{x \in H^1(S^1, M)$ such that $\dot{x} = a\xi + bv, a \geq 0\}$ supporting "oscillations" i.e. sequences of consecutive $\pm v$-jumps which are either small and close — we then "widen" them i.e. bring the two consecutive $\pm v$-jump further apart — or are large and close (high oscillations).

For high oscillations, we prove that these — which are due to the non-Fredholm character of the variational problem — either can be enlarged while J is decreasing somewhere above x^∞, or they can all be thinned down and brought back to be small oscillations.

In the next section, we will use this technical result in order to include high oscillations in our compactness result.

We also study in this section critical points at infinity containing sequences of consecutive characteristic ξ-pieces.

We start with:

Proposition 32 *There exists $c_0 > 0$ small such that for every $c > 0$ and $0 < \delta < c$, considering all curves x of C_β^+ bearing two consecutive $\pm v$-jumps of size less than c_0 separated by a ξ-piece supporting a v-rotation less than $\pi - c$, a J-decreasing deformation (generated by a vector-field) can be defined in their neighborhood. This deformation sepatates further these $\pm v$-jumps, without cancelling them or reversing their orientation, until the v-rotation on the ξ-piece between them is less than $\pi - \delta$.*

Proposition 33 *Let $\bar{x}_\infty$ or $\bar{x}$ have two consecutive characteristic pieces c_1, c_2 with strict H_0^1-index zero. Let $\tilde{c}_1$ be the full unstable manifold associated to c_1 and $\tilde{c}_2$ be the full unstable manifold associated to c_2. Any cycle*

$\sigma \times \tilde{c}_1 \cup \sigma \times \tilde{c}_2$ *in* $J_{\tilde{c}}, \bar{c} = J_\infty(\bar{x}_\infty)$ *or* $J_\infty(\bar{x})$, *bounds with no increase in the maximal number of zeros of* b.

Observation 5. Proposition 33 extends to $\bar{x}_\infty$ having more than two consecutive characteristic pieces, see the end of the proof of Proposition 33. The construction of Proposition 33 allows to cover the case of sequences of consecutive characteristic pieces left open in Section 2.5.6. The case of characteristic pieces of H_0^1-index zero is discussed in Appendix 1.

Proof. [Proof of Propsition 32]

We have to take care of curves bearing "small and rapid oscillations".

Let us draw such an oscillation with its incoming ξ-piece and let us introduce a v-vertical through A, A close to A_o on the incoming ξ-piece:

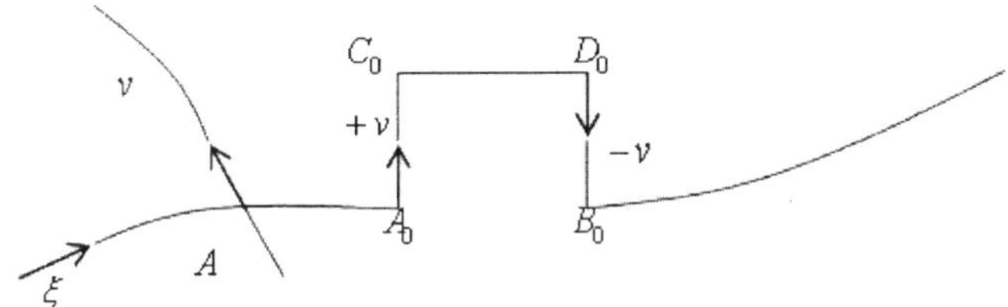

The curve $A - A_o - C_o - D_o$ is a curve whose tangent vector $\dot{x} = a\xi + bv$ splits on ξ and $\pm v$; it goes from the v-vertical through A to the v-vertical through B_o

Since these two verticals are close, since $C_o D_o$ is a piece of ξ-orbit (a "geodesic") from the v-vertical through A_o to the v-vertical through D_o, there is also a ξ-orbit joining these two verticals and its length is less than the addition of the ξ-length of AA_o with that of $C_o D_o$:

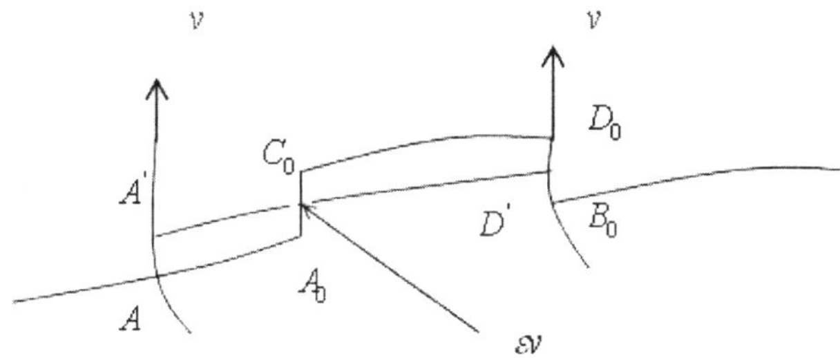

Thus, replacing $A - A_o - C_o - D_o - B_o$ with $A - A' - D' - B_o$ shortens the ξ-length of the curve, keeps the sign distribution for A close to A_o and increases the size of $C_o D_o$ into $A'D'$. The fact that the ξ-piece $A'D'$ is larger than $C_o D_o$ can be seen as follows: we consider the case where A is very close to A_o. $A_o A$ is then a small ξ-piece which we think of as a tangent vector $\tilde{z}$ at A_o. v-transporting it during the time $-\varepsilon$ from A_o to C_o, we get a vector $\tilde{z}_\varepsilon$ going from C_o to the v-orbit through A. We write at first order that

$$\overrightarrow{C_o A'} = \tilde{z}_\varepsilon + \delta v.$$

On the other hand, the variation of the ξ-length is at first order equal to

$$\alpha(\overrightarrow{D_o D}') - \alpha(\overrightarrow{C_o A}') = -\alpha(\tilde{z}_\varepsilon), \text{ which is positive.}$$

This enlargement can be completed without cancelling the base $\pm v$-jumps of the oscillation.

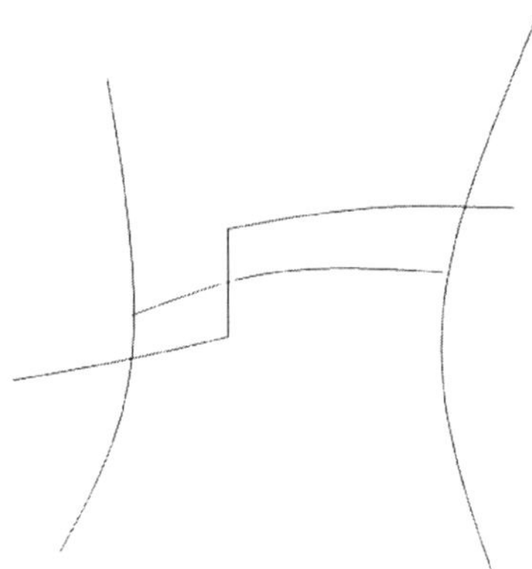

The v-jump which is on the side of the oscillation which we extend does not cancel; this is easy to see.

Thus if we alternate, extending always on the side which corresponds to the smaller v-jump, both v-jumps would cancel together.

This cannot happen because if we then superpose the initial and the final pictures, we get a small rectangle as follows:

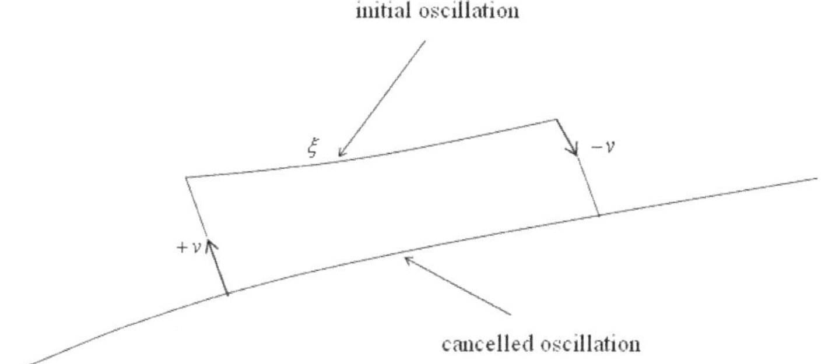

Such rectangles do no exist if the ξ-piece are small, less than character-istic of H_0^1-index zero.

Indeed, we choose Darboux coordinates at x_o (which depend on x_o in a differentiable way) so that $v \cdot v(x_o) = 0$. v then reads up to a multiplicative factor as $\frac{\partial}{\partial y} - x\frac{\partial}{\partial z} + (az+bx)\frac{\partial}{\partial x} + 0(x^2+z^2+xy+yz+g(y))\frac{\partial}{\partial x}$. Completing the Darboux change of coordinates:

$$\begin{cases} z' = z + h(y) \\ x' = x - h'(y) \\ y' = y \end{cases}$$

this vector-field becomes:

$$\frac{\partial}{\partial y'} - x'\frac{\partial}{\partial z'} + (az' + bx' + g(y') - ah(y') + bh'(y') - h''(y')$$

$$+ 0(x'^2 + z'^2 + x'y' + y'z')$$

$$+ 0(y'h(y') + y'h'(y') + 0(h^2(y') + h'^2(y'))\frac{\partial}{\partial x'}.$$

We set

$$\begin{cases} h''(y') - bh'(y') + ah(y') + 0(y'h(y')) \\ \quad + 0(y'h'(y') + 0(h^2(y') + h'^2(y')) = g(y') \\ h'(0) = h(0) = 0. \end{cases}$$

We derive a function h in this way, which depends differentially on $x_0 = (0,0,0)$ in our coordinates.

Then v, up to a multiplicative factor reads:

$$\frac{\partial}{\partial y'} - x'\frac{\partial}{\partial z'} + (az' + bx' + 0(x'^2 + z'^2) + 0(x'y') + 0(y'z'))\frac{\partial}{\partial x'}.$$

Starting from $x_0 = (0,0,0)$, a small rectangle of sides $\delta, \varepsilon, \delta_1, \varepsilon_1$:

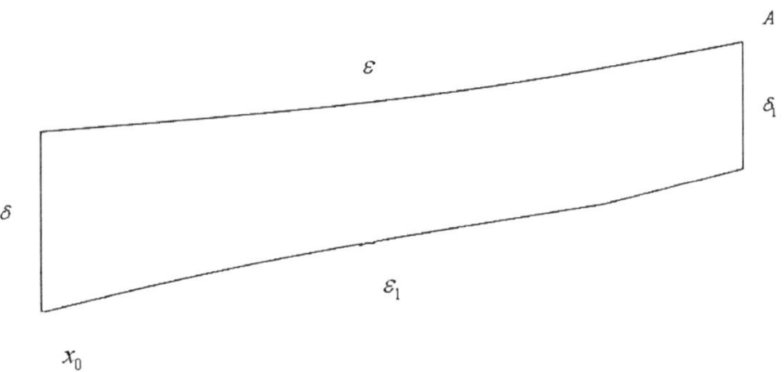

yields coordinates at A equal to $(\varepsilon, \delta, 0)$ on one side, $(\varepsilon_1(1+o(1)), \delta_1, \varepsilon_1\delta_1 + o(\varepsilon_1\delta_1))$ on the other side; this is a contradiction.

These extensions of the oscillations on $W_u(\bar{x}_\infty)$, near $\bar{x}$, are completed in a separate step, without moving otherwise the curves, so that the above argument about rectangles can be applied. A strip in the decreasing flow-lines is defined and in this strip (near $\bar{x}$), the oscillations are enlarged.

If the edge of an oscillation becomes close to a large, basic $\pm v$-jump we obviously cannot extend as we please on the side of the large $\pm v$-jump; but we can use it as if a small $\pm v$-jump were occuring in it. It becomes a side for a larger oscillation which ends at the outer $\pm v$-jump of the inital oscillation.

With more oscillations, some coalescing, the above argument extends.

Taking care of high oscillations

The arguments displayed above as well as the arguments which we will present for the proof of Proposition 22 assume that the oscillations i.e. the intermediate v-jumps are "small", of a size less than a fixed constant θ_0.

If a "large" oscillation develops, we need either to tame it or to use it and move all curves away.

Let us consider such a large and thin oscillation.

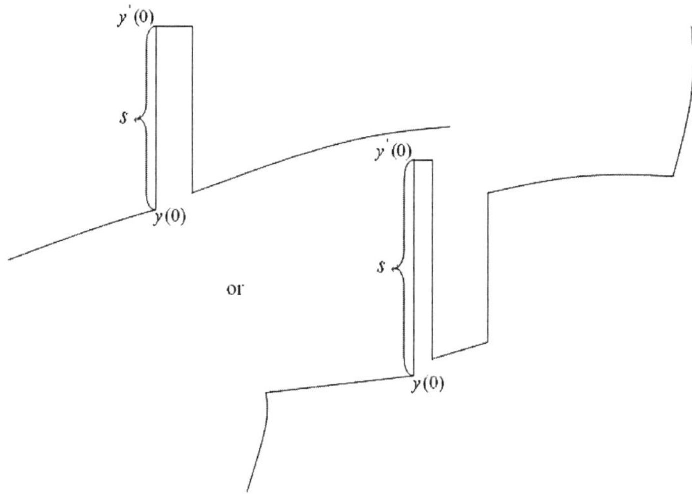

On such curves which support an oscillation of a height larger than θ_0, we can define a decreasing pseudo-gradient z_1 as follows:

We pick up a small ξ-piece of size ε at the edge of the ξ-piece preceeding the almost characteristic piece where this high oscillation takes place and we transport it to $y(0)$:

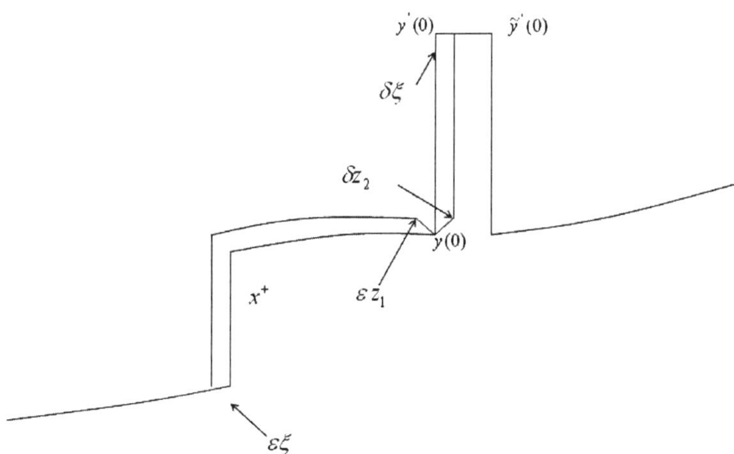

On the other hand, we pick up $\delta\xi$ at $y'(0)$ and we pull it down to $y(0)$. We obtain two small displacements at $y(0), \varepsilon z_1$ and δz_2. If the $[\xi, v]$-component of εz_1 is not zero, we can use it to compensate the $[\xi, v]$-component of δz_2. This will adjust, for a given δ, the value of ε. Adjusting the v and ξ displacement at $y(0)$, we define a tangent vector z. If $J' \cdot z$ is non zero, we have a decreasing direction using z or $-z$.

It might happen though that εz_1 has no $[\xi, v]$-component for special positions of $y(0)$; or that for special values of s (the height of the oscillation), $J' \cdot z$ is zero even though εz_1 has a $[\xi, v]$-component.

The first problem is easy to take care of: instead of transporting $\varepsilon\xi$, we can transport εv from x^+ to $y(0)$, yielding an $\varepsilon z_1'$ at $y(0)$. It cannot happen that both of εz_1 and $\varepsilon z_1'$ have no $[\xi, v]$-component.

If s is special and the z-directions defined above both satisfy $J' \cdot z = 0$ or if one of them cannot be defined while the other one satisfies $J' \cdot z = 0$, we build yet another direction by pulling back $\varepsilon\xi$ from $y(0)$ to $y'(0)$ and we use the small top ξ-piece, transport δv from $\tilde{y}'(0)$ to $y(0)$. We can then glue the $[\xi, v]$-components of the two-displacements after adjusting δ to ε. This new z' satisfies

$$J' \cdot z' \neq 0.$$

If θ_0 is not too large and if $|s| \leq 2\theta_0$, the first z will always work and the decreasing direction will always remove a portion of the small ξ-piece $[y'(0), \tilde{y}'(0)]$ and decrease its height, thereby decreasing and thinning out the oscillation.

The process might however be different for higher oscillations and although there is always a decreasing direction, we might end up with very thin back and forth v-runs without any ξ-piece between them.

We also have to glue our deformation, which we built for "small" oscillations (of size θ_0 at most) with these new deformations (not yet completely specified). For this latter purpose, we slightly modify z and z' to incorporate in them a small "widening" of the base of the oscillation so that it becomes, after the use of z or z' even for a small time, into:

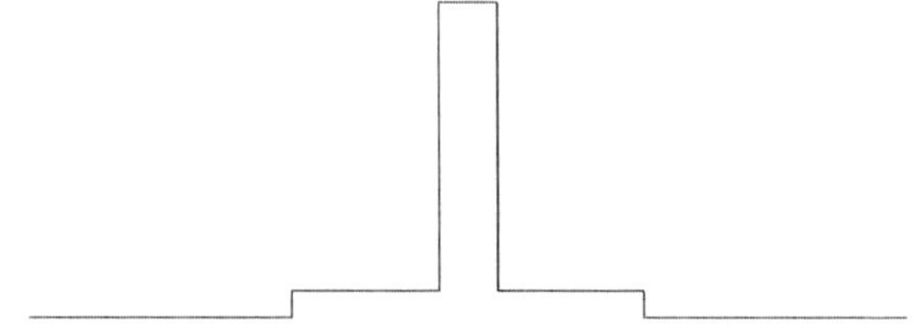

This "base" which we create for our high oscillation can be quite tiny. It is easy to incorporate it into z, z' and even into the flow defined for "small" oscillations. z, z' continue to act, with marginally modified definitions, on these high oscillations even with this additional base.

We will need z here for special values of $y(0)$, which we describe below. z' will be used for the proof of Proposition 22.

After use of these flows, our curves will either move away from $\bar{x}$ as one "high" oscillation widens because it borrows some ξ-displacement "from below". Or the oscillation will become of size less than $2\theta_0$ and will be destroyed, the curve losing its high oscillation. If this oscillation maintains, but becomes very thin when compared to its base, we can cancel it while decreasing J_∞ after expanding the base or one side of it i.e.:

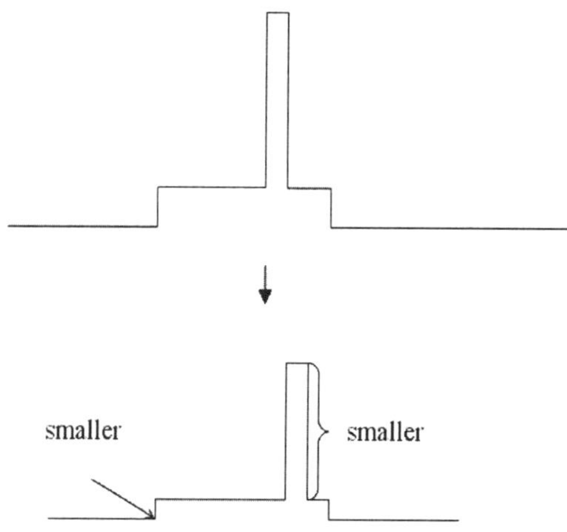

Because the oscillation is so thin when compared to its base, we can build a pseudo-gradient which will cancel the oscillation and only slightly expand its base (on one side, if needed): it will only depend on how thin the oscillation is when compared to its base.

When several oscillations pile up, we might find some difficulty keeping large bases. This difficulty can be overcome if, for each large $\pm v$-jump, we create on each of the two adjacent ξ-pieces (the incoming and the outgoing one) a bump, small in size, becoming smaller as the ξ-piece becomes smaller and larger if it becomes larger (as well as taller, all remaining small):

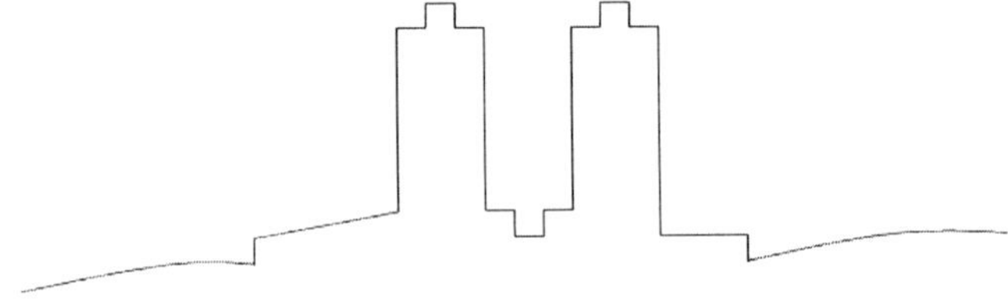

As several high and thin, oscillations pile up (this number is *a priori* bounded), the first and the last one have "sizable" small bumps, on the left for the first, on the right for the last, and we can use them to cancel all the thin oscillations in between.

We now show how to use z_1 or z (the first decreasing direction which we have built) in order to get rid of high oscillations along a characteristic piece with H_0^1-index non zero. It is easy to convex-combine the flow which is defined for small oscillations with the flow on high oscillations; we need to take care of these only when they develop in the vicinity of a subcharacteristic piece.

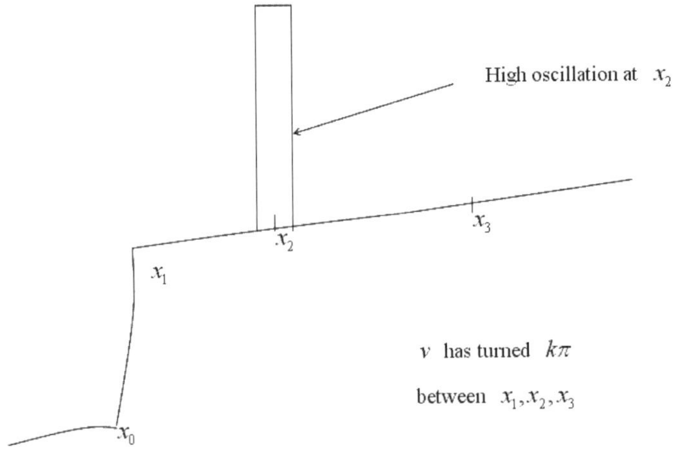

since then the v-jumps of such an oscillation can be used to define the decreasing normal associated to this characteristic piece when one of them is far from x_1, x_2, x_3 etc.

We revisit now the definition of z:

We transport ξ from x_0 to x_1 along v, then from x_1 to x_2 along ξ. We derive at x_2 a vector $\theta(\xi) = a_1\xi + b_1[\xi, v] + c_1 v$.

We pull back ξ from a height s along v above x_2. We derive a vector at x_2: $D\phi_{-s}(\xi) = A_1\xi + B_1[\xi, v] + C_1 v$. $D\phi_{-s}(\xi) - \frac{B_1}{b_1}\theta(\xi)$ has no component along $[\xi, v]$.

Using the ξ-piece abuting at x_2 and the v-jump starting at x_2, we can adjust these other components.

If the high oscillation at x_2 is not high enough, we can increase it (if the deformation which we are going to define is decreasing) and if it is too high, we can introduce a tiny, flat ξ-piece at the right level.

We compute the variation of J along this direction. We find

$$\varphi(s) = 1 - A_1 + \frac{B_1}{b_1}(a_1 - 1) = 1 - \alpha_{x_2}(D\phi_{-s}(\xi)) - \frac{\beta_{x_2}(D\phi_{-s}(\xi))}{b_1}(a_1 - 1).$$

If this expression is negative at any $\bar{s}$, we can introduce a decreasing deformation which expands the tiny high ξ-piece and moves these curves away from the critical one. This deformation convex-combines with the other ones which take place when the base of the high oscillation or the oscillation is far from x_2, x_3 etc.

If it is positive for every s, we can thin down any high oscillation and tame it down to be a small one. The argument preceeds.

If $\varphi(s_0) = 0$ above $x_2, x_3, \ldots$, which are extremities of a subcharacteristic piece, $\varphi'(s_0)$ is non zero generically and the argument proceeds. $\quad\square$

Proof. [Proof of Propsition 33] Assume that the first characteristic piece c_1, lies between two positive edge v-pieces. Assume that its full unstable manifold is used in order to define the dominated cycle. Since its unstable manifolds bounds, there must be another characteristic piece involved, with its full unstable manifold, in the definition of this cycle.

A dominating flow-line must contain, as it spans the full unstable manifold of c_1, a negative small v-jump in addition to all the v-jumps of the base curve; otherwise the full unstable manifold of c_1 cannot be covered. The same argument holds for c_2, with a positive or negative v-jump depending on the signs of the edge v-jumps of c_2, and in fact holds for every c_i as its full unstable manifold is used.

If these additional v-jumps do not travel, they would allow to move down, past $\bar{x}$; they can be used either with the decreasing normals associated to each characteristic piece, or in a direct construction of a decreasing deformation.

Thus, they must travel from a characteristic piece c_1 to another characteristic piece, for example c_2. Assume that c_1 and c_2 follow each other, so that both of them have positive v-edges in particular.

This happens in the following process.

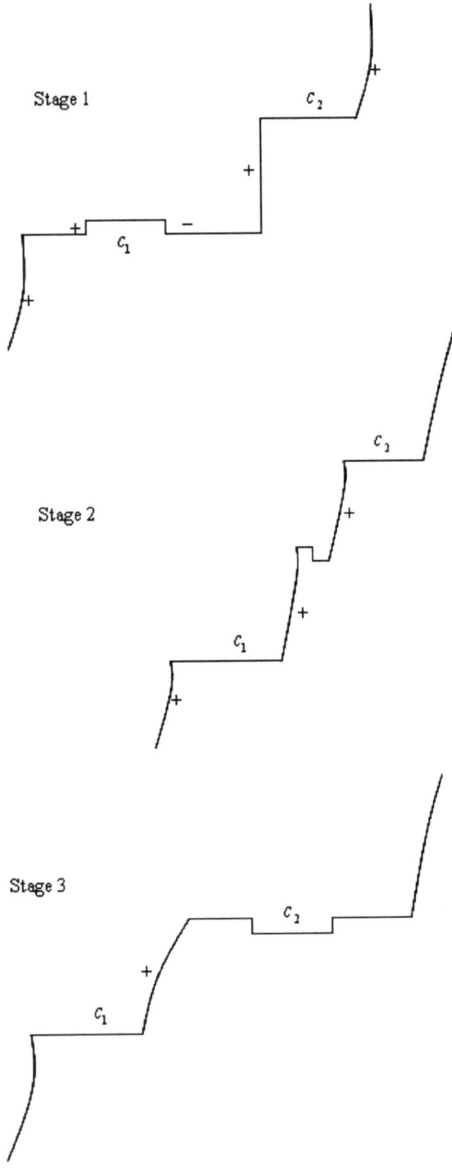

As in stage 2 the small oscillations start their journey from c_1 to c_2, we expand them with tiny sharp oscillations while decreasing the functional so that they all look in stage 2 as:

(*)

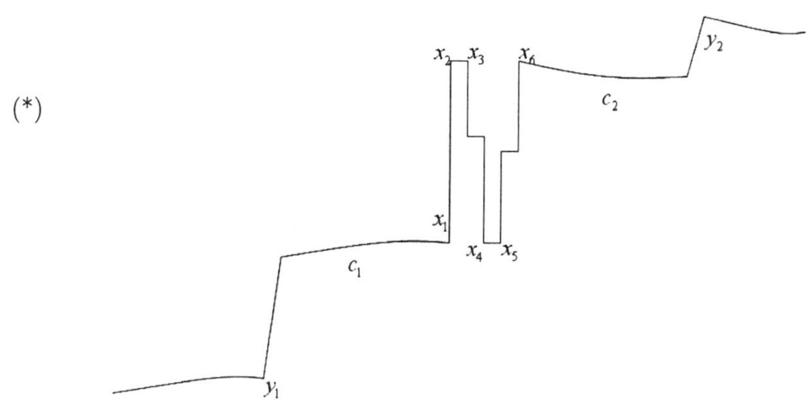

Transporting ξ from x_1 to x_2 and using the tiny ξ-piece $[x_2, x_3]$, we can build a decreasing direction z_2 (reversing this direction if needed).

This ability to decrease might fade away very fast as either the ξ-piece $[x_2, x_3]$ disappears or the sharp oscillation $x_1 - x_2 - x_3 - x_4$ changes sharply its height.

It is easy to stabilize and even to increase the length of the ξ-piece $[x_2, x_3]$. Indeed, there is a tangent direction $\bar{z}$ on the base curve

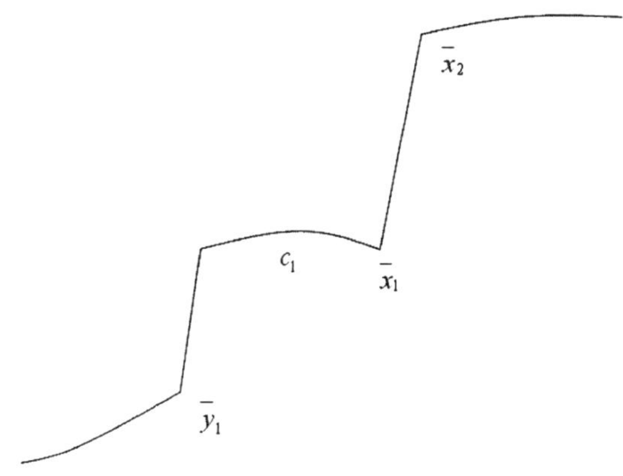

formed by transporting ξ from $\bar{y}_1$ to $\bar{x}_1$, pulling down ξ from $\bar{x}_2$ to $\bar{x}_1$ and matching, after scaling and adding v and ξ-components at $\bar{x}_1$ the two vectors thus derived at $\overline{x}_1$. Since the base curve is critical,

$$J' \cdot \overline{z} = 0$$

$\overline{z}$ or $-\overline{z}$ will increase the length of c_2.

We can use $\overline{z}$ on (*) so that the ξ-length of $[x_2, x_3]$ increases. Since $J' \cdot \tilde{\overline{z}} = o(1)$ while our other direction z_2 has $J' \cdot z_2 \leq -\gamma^2 < 0$, the combination still decreases J and increases $[x_2, x_3]$.

The v-jumps around $[x_2, x_3]$ are considerably perturbed in this way. We need to make sure that as, for example, the oscillation corresponding to $[x_2, x_3]$ (upwards) increases somewhat, we can tame it and bring it down.

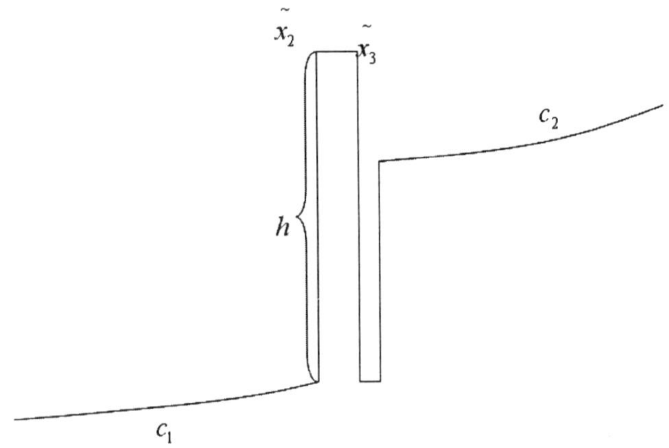

If at any height h above c_1 we can decrease J by expanding $[\tilde{x}_2, x_3]$ after transporting $\pm\xi$ from $[\tilde{x}_2, \tilde{x}_3]$ down to c_1 or to any other part of the base curve, we can infer that we could have scaled (expanded) our oscillations or have introduced, if they went above h a tiny flat ξ-piece at the level h, and from there on use this decreasing direction which expands the ξ-piece at the level h.

This deformation, unlike z_2, does not hinder itself since it expands $[\tilde{x}_2, \tilde{x}_3]$ and does not involve great perturbations of the v-size of the oscillation.

If on the contrary no such h exists, we are able to thin down any such oscillation while decreasing J. Hence, would our oscillation greatly perturb after the use of $z_2, \tilde{\overline{z}}$, we would be able to tame it, reintroduce an $[x_2, x_3]$-piece of a similar size or larger at the level of c_2 and proceed with this downwards deformation.

Let us quantify the above argument:

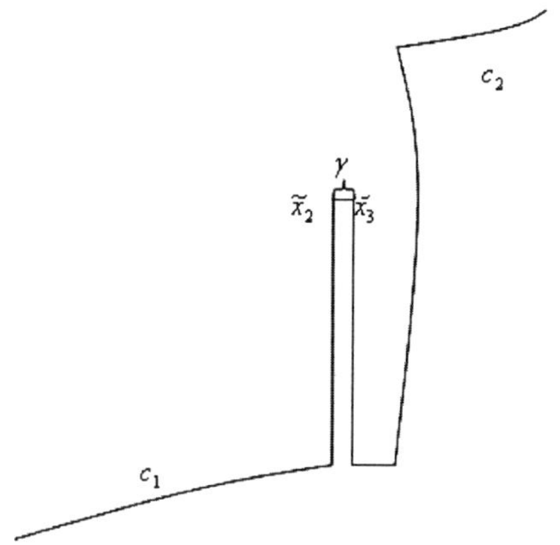

Let us assume that $[\tilde{x}_2, \tilde{x}_3]$ has length γ. We break it into two equal parts

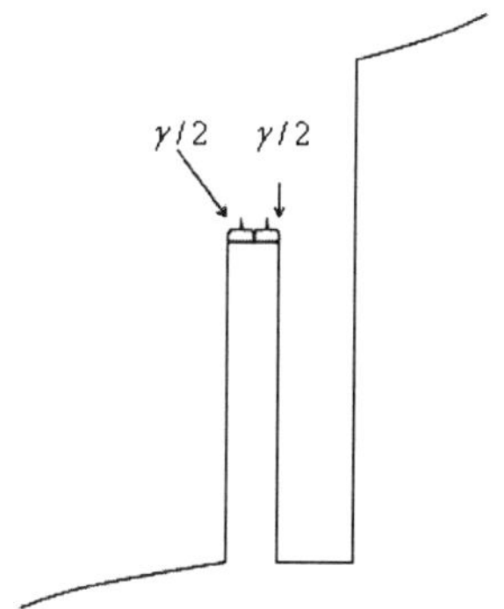

We do not touch the right half-part except to extend it using $\bar{z}$ or $-\bar{z}$. We use the left half-part for z_2, thereby creating an extra $\pm v$-piece which might increase greatly.

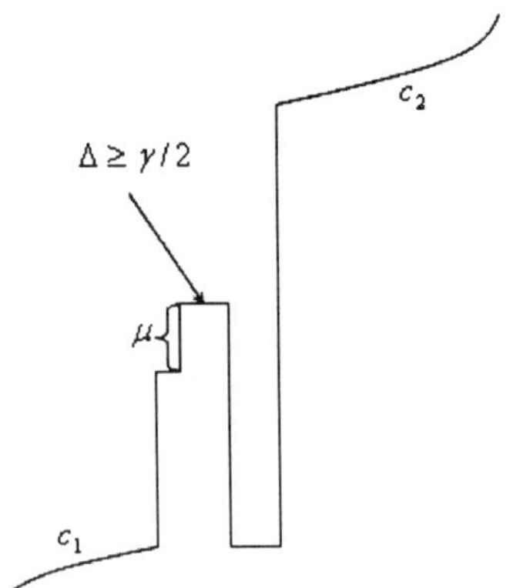

Let us estimate the size μ of the v-jump created versus the size Δ of the right half-piece.

After time t if we use $\pm M\bar{z}$, Δ is at least

$$\frac{\gamma}{2} + Mt.$$

During the same time, the left half-piece length has become $\frac{\gamma}{2} - O(ct)$, due to the use of z_2, c *a priori*-bounded and the size μ of the v-piece created is $O\left(\frac{t}{\gamma}\right)$ as long as $ct \leq \frac{\gamma}{4}$.

We thus need

$$\begin{cases} ct \leq \frac{\gamma}{4} \\ \frac{t}{\gamma} \leq c_0. \end{cases}$$

Thus, $t \leq C\gamma$. After the time $C\gamma$, the right half-piece has length

$$\frac{\gamma}{2} + MC\gamma.$$

If $MC > \frac{1}{2}$, which we may assume since M can be chosen large, we have rebuilt a top ξ-piece of size γ. The direction of the new $\pm v$-jump which we will create as we iterate the construction described above is the same then the one found in the first step because z_2 in each stage is built after transporting ξ along nearly the same large $\pm v$-jump, with maybe small (in total) ξ-pieces inserted in between and back and forth runs along v.

The decreasing deformation described above is thus self sustained and can be used to glue the decreasing deformation along $\tilde{c}_1$ with the decreasing deformation along $\tilde{c}_2$.

Another issue has to be addressed though in the above process: coming back to stage 1 of this process, we observe the need along c_1 to have **two** v-jumps, one positive, the other one negative in order to complete the travel process in stage 2 and after. While the negative v-jump among these two is naturally provided by Span of $\tilde{c}_1$, the positive one may be thought as a companion of the positive left edge of c_1 and might have to be created. This can be achieved using the "small room" for decrease provided by the existence of the negative v-jump , exhausting half of it if needed in order to create the positive one close to the left edge as it is created, moving to the negative one as it becomes larger. The whole process of creation of this v-jump can be embedded in a decreasing flow-line along which on one hand Hypothesis (B) is locally satisfied, on the other hand the flow-line creates the initial conditions of stage 1 and stage 2, i.e a small positive v-jump close to a small negative v-jump, see the figures above.

The first statement of Proposition 33 follows for two characteristic pieces. Observe that the above deformation requires only the existence of an intermediate positive v-jump followed by a negative v-jump, thus the existence on this pair of characteristic pieces of three distinct families, one of each extreme edge and one for the intermediate negative v-jump. This observation will be generalized (with a slight variation) below for more characteristic pieces.

We now establish that the topological statement of Proposition 33 extends for more than two characteristic pieces of strict H_0^1-index zero:

Given two consecutive characteristic pieces and their associated full (half) unstable manifolds $\tilde{c}_1$ and $\tilde{c}_2$, we just proved that

$$\partial(\tilde{c}_1 \otimes \tilde{c}_2) = \partial\sigma$$

where σ carries 2 zeros at most for b. We also have established that $\tilde{c}_1 \otimes \tilde{c}_2$ and σ are homologous relative to their common boundary, without increase in the number of zeros of b relative to $\tilde{c}_1 \otimes \tilde{c}_2$. Assume that we have $k+1$ consecutive characteristic ξ-pieces. Let $\tilde{c}_i$ be their full (half) - unstable manifold. We write:

$$\partial(\tilde{c}_1 \otimes \cdots \otimes \tilde{c}_{k+1}) = \partial\sigma \otimes \tilde{c}_3 \otimes \cdots \otimes \tilde{c}_{k+1} + \tilde{c}_1 \otimes \tilde{c}_2 \otimes \partial(\tilde{c}_3 \otimes \cdots \otimes \tilde{c}_{k+1})$$

$$= \partial(\sigma \otimes \tilde{c}_3 \cdots \otimes \tilde{c}_{k+1}) + (\tilde{c}_1 \otimes \tilde{c}_2 - \sigma) \otimes \partial(\tilde{c}_3 \otimes \cdots \otimes \tilde{c}_{k+1})$$

$$= \partial(\sigma \otimes \tilde{c}_3 \cdots \otimes \tilde{c}_{k+1} + u \otimes \partial(\tilde{c}_3 \otimes \cdots \otimes \tilde{c}_{k+1})).$$

Observe that

$$\sigma \otimes \tilde{c}_3 \cdots \otimes \tilde{c}_{k+1} + u \otimes \partial(\tilde{c}_3 \otimes \cdots \otimes \tilde{c}_{k+1})$$

carries two less zero than $\tilde{c}_1 \otimes \cdots \otimes \tilde{c}_{k+1}$.

The conclusion follows for more than two characteristic consecutive ξ-pieces. We could even multiply by other homology classes (due to the index at infinity or other characteristic ξ-pieces of non zero index) the argument would go through for all classes containing $\partial(\tilde{c}_1 \otimes \cdots \otimes \tilde{c}_{k+1})$ as a factor. The argument goes also through for characteristic pieces of non zero H_0^1 strict index without substantial modifications. The idea is the following: each $c_j, j = 1, 2$ may be represented using $i_0^j + 1 \pm v$-jumps which have sizes varying independently and an additional large $\pm v$-jump in order to represent the left edge. i_0^j of them are located at interior nodes, they provide a representation of the strict H_0^1-unstable manifold, the additional $\pm v$-jump is used to represent the full unstable manifold of the characteristic piece. We can insert this additional $\pm v$-jump in the last nodal zone for c_1 and in the first nodal zone for c_2. Assuming that the common edge is positive, this additional v-jump is negative. Again, we may view $\partial(c_1 \otimes c_2)$ as the boundary of another chain where this additional negative v-jump will travel along the common edge. As in the case of the zero H_0^1-index, this travel requires the use of a positive v-jump which would progressively, as this additional v-jump moves to the next characteristic piece, build up the common edge. If the $\pm v$jump at the last interior node of the first characteristic piece is positive, we can build a companion of it to play this role. Otherwise, we have to introduce it after the i_0^j internal $*'s$. Because the critical point at infinity is not false, i_0^j is even (the two edges are positively oriented) and the introduction of this positive v-jump does not increase the number of sign changes between the left edge and the right edge beyond $i_0^j + 2$ which is the maximal number of sign changes allowed by c_1. The additional negative v-jump can travel now and once it has reached the second characteristic piece, then it can be used to generate c_2. $\partial(c_1 \otimes c_2)$ is generated in this way as the boundary of a chain supporting two zeros less than $c_1 \otimes c_2$. $\qquad\qquad\qquad\qquad\square$

Proof of the results of Section 2.5.6 and Theorem 1 completed We consider now a maximal sequence of consecutive characteristic pieces which we assume to be derived after the constructions of Section 2.5.6. Then, all interior nodes of all characteristic pieces but one node possibly, on a single characteristic piece, are controlled by a $*$. This feature of the configurations which we study is stable. This means that as this feature is about to change, the Normal (II)-flow can be used on one of the $*'s$ which are not *'s of an edge, in fact is an interior $*$, in the transition zone. On these configurations, each edge must have a $*$ associated to it. Otherwise, the edge would be a companion of a neighboring interior $*$ and we could engineer a decrease through a use of a decreasing normal to the immediate right of the $*$ for a right edge, to the immediate left of the $*$ for the left edge. This normal would be an additional companion for this $*$. This holds under the assumption that the H_0^1-strict index of these ξ-pieces is non zero. This argument is adjusted for characteristic pieces of H_0^1-index zero in Appendix 1.

Furthermore, we can state that on these configurations, all $*'s$ are either in the vicinity of the interior nodes or in the vicinity of the edges.

Let us assume in a first step that there is no hole on the sequence of characteristic pieces. The count of Section 2.5.6 is tight and implies then that on each characteristic piece but one, there are $i_0^j + 1*$'s, besides the $*$'s of the edges. On the remaining characteristic piece, there are i_0^j*'s and a $*$ for each edge. We refer to this characteristic piece in the sequel as the "(I)ξ-piece". Let us assume that (I) is in between characteristic pieces. The additional $*$ on each other characteristic piece is in the vicinity of an edge. Thus, on the characteristic piece with only i_0^j*'s, there is an edge which does not have an additional $*$ in its vicinity, neither on this nor on the next characteristic piece. No additional $*$ will ever come close to this edge, unless we can use the Normal (II) flow on it prior to its motion close to this edge. Let us consider the other edge of the same characteristic piece. This other edge has an additional $*$ in its neighborhood, on the next characteristic ξ-piece. There are then two possibilities: either this other edge also does not have an additional $*$ in its vicinity. Then, no additional $*$ will come to its vicinity unless we can use the Normal (II) flow on an interior node to decrease. The (I)ξ-piece jails a repetition which is stable. We can introduce on the next characteristic piece to the right a fixed decreasing Normal N^-, it will be extendable into an admissible alternating sequence because of the stable repetition. These configurations can be decreasing. Or this other edge does have an additional $*$ in its vicinity, not on the (I)ξ-piece, but this could change after a travel of this $*$ along this edge. We then choose to introduce a decreasing Normal N^- in the nodal zone of the (I)ξ-piece neighboring this edge. This requires some justification:

The additional $*$ which we are considering remains in the vicinity of this edge, otherwise we could use the Normal (II) flow on it and decrease the configuration. It can however travel along the edge, from the (I) ξ-piece to the next characteristic piece. When it is in the (I) ξ-piece, the use of N^- is justified, we have used this argument several times before. When the $*$ travels along the edge, it has the orientation of the edge because there are only then two neighboring $*$'s. There is then a repetition and this N^-

can be made part of an admissible alternating sequence. However when this $*$ has traveled to the next characteristic piece, it is disconnected from the use of N^-. This $*$ could change sign and the repetition with the $*$ of the edge would disappear and N^- would not be justified anymore. So we need then to make N^- travel to the next characteristic piece to become next to this additional $*$ and to the other $*$ of the edge, along the process described in the proof of Proposition 33. This requires the introduction of a $\pm v$-jump with the orientation of the edge along which the travel takes place on the other side of N^- than the edge. This $\pm v$-jump can be introduced as a simple counting argument on the $(I)\xi$-piece after the introduction of N^- shows: the various $*$'s, including those of the edges, build with N^- a sequence which admits an alternating representative. One can then check that the introduction of such a $\pm v$-jump in the second nodal zone closest to the edge never increases the total number of families after reordering, relabeling etc. Furthermore, it decreases J. Using the widening process, we can then assume that the node closest to the edge is occupied by this $\pm v$-jump. The next $\pm v$-jump in the direction of the edge either has the orientation of N^- and can be confounded with it(this includes the case when it is zero) . A $\pm v$-jump can now be introduced prior to N^-(the edge being after) with the orientation of the edge.

Or this next $\pm v$-jump has already the orientation of the edge. There is no need to introduce an additional $\pm v$-jump.

The process of Proposition 33 can now begin. It will decrease J and bring N^- to the next characteristic piece as the repetition between the $*$ of the edge and the additional $*$ still holds. Once on the other side, our large $\pm v$-jumps can be assumed to be small, but the new edge. The two $*$'s are now on the other side, follow each other and have the same orientation, we confound them, bring them into the new N^- after the travel which they cross to reappear on the other side close to the edge. This crossing can be accomplished as J is decreasing using the "negative" decreasing Normal. We now relabel $*$ and $*$ the two $*$'s which follow each other and have the orientation of the edge. The new N^- is next to it. The decreasing process goes on, continuity has been achieved.

A similar argument, with some modifications, can be made in the case of a single hole at a single node. There is a repetition in the left edge between exactly two $*$'s, of course this can evolve; but each other characteristic piece has an additional $*$, as above only that it can travel along the edge it neighbors. Let us assume that the ξ-piece next to the characteristic piece with a hole(after it)is also characteristic. An additional $*$ neighbors it. There are exactly i_0^j $*$'s strictly between the representative most to the left of the left edge and the right edge when the additional $*$ is either traveling along the edge or on the next characteristic piece to the right. We are in the same situation than above. We can make an N^- decreasing Normal travel and follow the additional $*$.

The N^- decreasing Normal is part of the Hole flow corresponding to the representative most to the right of the left edge because of the single hole. Thus, there is no need to make N^- travel as this representative is "large". We can declare it then to be left edge and use its Hole flow. N^- can be used then, without travel. As the other representative develops, a repetition occurs as long as the additional $*$ is either traveling or on the next characteristic piece as discussed above and the travel becomes possible. In all, the location of N^- can be adjusted in a continuous way.

If the next ξ-piece to the right is not characteristic, then the right edge of (I) is isolated and can be used to define a Hole flow. This construction works as long as another $*$ does not reach this edge obviously through its non degenerate end, a case which is discussed below.

The conclusion follows in such cases.

The argument works for a (I)ξ-piece which is at the end of a sequence if the extreme edge of the sequence does not "receive" other $*$'s from its non degenerate end. In case of "stable" repetition, we should choose to introduce N^- on the nearest characteristic ξ-piece, to the right or to the left, depending on the configuration. It also works for an isolated ξ-piece, under the condition that both edges do not receive additional $*$'s. For a "stable" repetition and an isolated ξ-piece, we insert N^- on a characteristic piece which is interior to a sequence ξ-piece and does not neighbor (I) (which is here an extreme or isolated characteristic piece). We are assuming that there are sequences with more than three characteristic pieces.

Such a (I)ξ-piece may receive other $*$'s from its non degenerate end(s). Either they come from a non degenerate ξ-piece which is then, at an intermediate time, pierced with a hole; J decreases below the critical level at that intermediate time, the cycle splits. Or they come from other characteristic sequences which are neighbors to (I), from their edge characteristic pieces through the removal of the additional $*$. After this removal, either we have a "stable" repetition on another characteristic ξ-piece or the construction of Proposition 33 can be carried out on this new characteristic ξ-piece which we label (I)'. There might two possible choices for (I)', but (I) was unique and we declare (I)' to be the first one to the right of (I).

Observe that in the case of the "stable" repetition, the traveling $*$ will not reach the interior of (I)if a second repetition does not develop in an edge. This allows to switch between the decreasing flow(use of one decreasing Normal N^- for each of them, or of a Hole flow **outside** of the support of the two repetitions) due to (I) and the decreasing flow due to (I)': each one can be thought as due to alternating sequences with disjoint supports(observe that N^- was always introduced either in the same sequence or interior to a sequence, see the construction above, which we slightly modify for (I)' so that the supports of the constructions for (I) and (I)' are disjoint); none of them includes a Normal in the repetitions which we just pointed out. We can extend one of them, not crossing the support of the other one until it reaches one of the repetition. Extending it inside the repetition if needed, we can make its orientation and the orientation of one of the $\pm v$-jumps of the repetition coincide. It behaves then as a Hole flow. We then overextend the other partial alternating sequence jumping over the **other** repetition. We derive an admissible alternating sequence. However since the other alternating sequence behaves as a Hole flow, we can remove the corresponding part of the first sequence and insert the second one in it. The switch is possible.

On the other hand, the construction of Proposition 33 finds its justification locally and therefore can be used in (I)' in conjunction with the deformation induced on (I). Indeed, typically when there is an additional $*$ on a characteristic ξ-piece, either this $*$ and the related edge do not have the same orientation and the use of N^- is allowed as well as the switch; or they have the same orientation, then the ξ-piece (I)' jails exactly two repetitions. On the side of (I), either we have a stable repetition and we then introduced somewhere else, not on (I)', an N^-. We are also introducing an N^- on (I)'. Two repetitions are spared sometimes over the transition between (I) and (I)', we can switch. Or we are using a Hole flow on (I), it can be inserted in an alternating sequence defined by the N^- on (I)' and jumping over one of the repetitions of (I)', again the switch is possible. The transition, in case of a hole, from this Hole flow to the construction of Proposition 33 on (I) is a reduction to the use of N^- on (I), no justification for the switch is required. If we are using directly this construction on (I) (typically when there is no hole on (I)), either the additional $*$ has the orientation opposite to the edge on (I) or the next piece(after the travel is completed), there is nothing to prove. Or there are two repetitions, in addition to the two repetitions related to (I)'. Two of them will be used by the two N^-'s and two are left, they warrant the compatibility and they make the switch possible. Our proof is complete. $\qquad\square$

We study in this section the behavior of the Morse index at infinity i_∞ of $\bar{x}_\infty$ along iterates.

2.5.9 *Iterates of critical points at infinity*

Lemma 19 *Let x_∞ or $\bar{x}$ be a hyperbolic critical point at infinity with at least one non characteristic ξ-piece, with H_0^1-index i_0 and with index at infinity i_∞. Let x_∞^k be the kth iterate of x_∞ (or $\bar{x}$).*

The H_0^1-index of x_∞^k is ki_0 and its index at infinity is ki_∞.

Corollary 7 *Assume that x_∞ or $\bar{x}$ be a hyperbolic critical point at infinity with H_0^1-index zero. Assume that the maximal number of zeros of b on $W_u(x_\infty^p)$ is $2k$ or $2k - 2$ and that the index of x_∞^p is $2k - 1$. Then, $p = 1$.*

Proof. [Proof of Corollary 7] Using Lemma 3, we have:

$$pi_\infty = 2k - 1, p\gamma(x_\infty) = 2k \text{ or } 2k - 2.$$

Thus, $\gamma(x_0)$ and i_∞ are not equal and this forces p to be 1. $\square$

Proof. [Proof of Lemma 19] x_∞ (or $\bar{x}$) has a nondegenerate ξ-piece and has therefore a Poincaré-return map preserving area. In order to see this claim clearly, we consider the case when x_∞ has exactly one characteristic and one noncharacteristic ξ-piece:

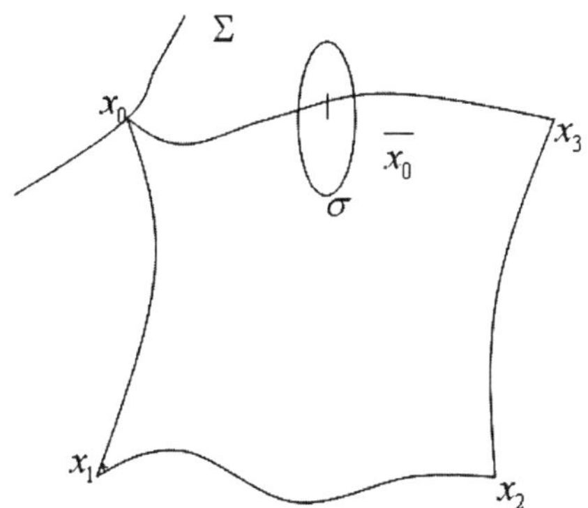

A hypersurface Σ is built at x_0 as follows:
Σ has the following property: let γ be the map which assigns to $y_0 \in V(x_0)$ a neighborhood of x_0 in M, a point y_3 near x_3 as follows. From y_0, we follow a v-orbit till y_1 so that, pulling back ξ from y_1 to y_0, we get:

$$\alpha(D\phi_{-s}(\xi)) = 1.$$

ϕ_s is the one-parameter group of v. We then go from y_1 to y_2 along a characteristic ξ-piece and then from y_2 to y_3 along a v-orbit so that y_3 is to y_2 what y_0 is to y_1.

γ being defined , we pull back ξ from y_0 to y_1 and from y_3 to y_2. We derive two vectors z_1 and z_2.

z_1 splits into:

$$z_1 = (1 + A_1)\xi + B_1[\xi, v] + C_1 v$$

z_2 splits into

$$z_2 = (1 + \tilde{A}_1)\xi + \tilde{B}_1[\xi, v] + \tilde{C}_1 v.$$

Along the characteristic ξ-piece from y_1 to y_2, v is mapped onto θv and $[\xi, v]$ maps onto $\frac{1}{\theta}[\xi, v] + \nu v$.

So that z_1 and $\frac{B_1}{\theta \tilde{B}_1} z_2$ have, after ξ-transport from y_1 to y_2, the same $[\xi, v]$-component.

$\sum$ is then defined to be $\{y_0$ such that $A_1 \tilde{B}_1 \theta = B_1 \tilde{A}_1\}$.

If $\bar{y}_0 \in \sigma$, we then define a Poincaré-return map C as follows: we follow ξ backwards starting from $\bar{y}_0$ until we hit $\sum$, then proceed with γ until y_3 and from y_3 to σ following ξ. It is easy to verify that

$$C^* \alpha = \alpha + df,$$
$$C^* d\alpha|_\sigma = d\alpha|_\sigma$$

from which our claim is derived.

It therefore makes sense to speak about hyperbolic or elliptic critical point (at infinity). We are assuming here that x_∞ is hyperbolic. Then, $DC_{\bar{x}_0}$ has two eigenvalues, λ and $\frac{1}{\lambda}$, both real and different from 1 (and -1, generically).

After multiplication by $\mathbb{C}$, the tangent space has a natural complexification and the quadratic form $d^2 J_\infty(x_\infty)$ becomes a Hermitian form of index $2i_\infty$. Considering now the iterate x_∞^p of order p of x_∞, we introduce the pth-roots of 1:

$$1, \omega_1, \ldots, \omega_{p-1}.$$

The tangent space to x_∞^p splits in a natural way into a direct summand:

$$\oplus_{i=1}^p T_i$$

x_∞^p is parametrized over $[0, p]$. If $z \in T_i$, then $z|_{[j,j+1]} = \omega_i z|_{[j-1,j]}$, with $j = 1, \ldots, p$.

It is easy to see that, for $i \neq j$, T_i and T_j are $d^2 J_\infty(x_\infty^p)$-orthogonal.

The equation of T_i can be written using C, the initial variation in σ, $z(0)$ and then various increments in ξ and v (when compared to C) on the various branches transported to σ.

$$dC_{\bar{x}_0}(z(0)) + X = \omega_i z(0).$$

Then, $z(1) = \omega_i z(0)$.

If z_1 and $z_2 \in T_i$, then

$$\bar{z}_1(0) \cdot \alpha(z_2(0) + \bar{z}_1(0) \cdot \alpha(z_1(0)) - \bar{z}_1(1) \cdot \alpha(z_1(1)) - \bar{z}_2(1) \cdot \alpha(z_2(1)) = 0$$

$$\text{since } \omega_i \bar{\omega}_i = 1.$$

This computation extends to vectors in $T(\mu)$ where $\mu \in \mathbb{C}, |\mu| = 1$ defined by the equation: $dC_{\bar{x}_0}(z(0)) + X = \mu z(0)$.

$d^2 J_\infty$ is defined on T_i and this definition extends to $T(\mu)$ as we will see later.

We consider a path $\mu(t)$ on the unit circle, from 1 to ω_i. $T(\mu)$ evolves then from T_{x_∞}, the tangent space to x_∞, where $d^2 J_\infty(x_\infty)$ has index $2i_\infty$, to T_i. If we can prove that the quadratic form does not degenerate, we are done. Assume, arguing by contradiction, that the quadratic form $d^2 J_\infty$ degenerates at $T(\mu_0)$, with a simple degeneracy. Perturbing a little bit C (as in Proposition 15 of [Bahri-1 2003], for example) and using a genericity argument, we can move μ_o continuously, filling a tiny open interval, where a root of unity ω, with $\omega^q = 1$, can be found.

Thus, we may assume that $\mu_0 = \omega$. Then, $d^2 J_\infty$ degnerates on $T(\omega)$ if and only if $d^2 J_\infty(x_\infty^q)$, on $T_{x_\infty^q}$, degenerates with the direction of degeneracy satisfying:

$$z|_{[j+1,j]} = \omega z|_{[j,j-1]}$$

x_∞^q is a closed curve, with Poincaré-return map $C^q \cdot d^2 J_\infty(x_\infty^q)$ degenerates if and only if C^q degenerates and this is impossible because the eigenvalues of C are real and different from 1 in absolute value. Thus, no such degeneracy may occur and $d^2 J_\infty(x_\infty)|_{T_i}$ has index $2i_\infty$. Lemma 3 follows.

The extension of $d^2 J_\infty$ to $T(\mu)$ is quite simple: there is a formula for $d^2 J_\infty \cdot z \cdot \bar{z}$ involving only the values of a tangent vector z at the edge points $x_j (j = 0, 1, 2, 3$ in the example shown above). This formula extends naturally on $T(\mu)$. When $\mu = \omega$, with $\omega^q = 1$, we find (up to multiplication by q) $d^2 J_\infty(x_\infty^q) \cdot z \cdot \bar{z}$ on the tangent space to x_∞^q. $\qquad\square$

We now have:

Lemma 20 *Let x_∞ (or $\bar{x}$) be of H_0^1-index zero and have a non-degenerate ξ-piece. Assume x_∞ (or $\bar{x}$) is elliptic. Then, all iterates of x_∞ (or $\bar{x}$) of order $p \geq 2$ can be discarded from our homology for $k \geq 4$.*

Proof. If x_∞ or $\bar{x}$ is elliptic, we can complete deformations of α near x_∞ so that x_∞^p degenerates and x_∞^r does not for $1 \leq r < p$. Then the index of x_∞^p changes (x_∞^p maintains since x_∞ maintains) and x_∞^p is replaced by a very close y_∞ which is not an iterate but has as many ξ and $\pm v$-pieces as x_∞^p. Indeed, then C is conjugate to a rotation $\begin{pmatrix} \cos\theta_0 & -\sin\theta_0 \\ \sin\theta_0 & \cos\theta_0 \end{pmatrix}$ and C^p degenerates after the introduction (as in Propostion 15 of [Bahri-1 2003]) of a perturbation of α into $\lambda\alpha$ which changes C into $R_{-\theta_0/p}C$ or into $R_{2\pi-\theta_0/p}C$. We would like β to remain a contact from so that we require that $\theta_0/p < \gamma$, where γ is the v-rotation on this non-degenerate ξ-piece. In the second case, we would like i_0 to remain zero, so that we require that

$$2\pi - \theta_0/p < \pi - \gamma.$$

Clearly, since $0 < \theta_0 < 2\pi$, one of θ_0/γ or $\frac{2\pi-\theta_0}{\pi-\gamma}$ is less than 2 so that this yields $p \geq 2$.

If x_∞^p hinders our deformation downwards, then assuming tht x_∞ has H_0^1-index zero, we have:

$$\begin{cases} p\gamma(x_\infty) = 2k \text{ or } 2k - 2 \\ \text{index } (x_\infty^p) = 2k - 1 \end{cases} \tag{*}$$

This equation might be satisfied for more than one index of iteration p; but since $\gamma(x_\infty)$ is at least 2 ($k \geq 2$), if this occurs for more than one index of iteration, then these indexes will be consecutive indexes p and $p - 1$. Working as above, with $R_{-\theta_0/p-1}$ or $R_{2\pi-\theta_0/p-1}$, C^p and C^{p-1} degenerate (p will be ≥ 3 if $k \geq 4$) and $x_\infty^p, x_\infty^{p-1}$ do not have the Morse index $2k - 1$ anymore. $\qquad\square$

Lemma 21   *Assume that x_∞ (or $\bar{x}$) is a critical point (at infinity) having a non degenerate ξ-piece with H_0^1-index i_0^j.*

If using the procedure of Proposition 15 of [Bahri-1 2003], we can increase i_0^j to $i_0^j + 1$, then x_∞ (or $\bar{x}$), prior to any perturbation, has an index at infinity $i_\infty \geq 1$ and the index at infinity of its iterates x_∞^p is at least p.

Proof.   If i_0^j can be increased to $i_0^j + 1$, then $i_\infty \geq 1$ since the total index is unchanged. Thus, when the H_0^1-index on the non-degenerate ξ-piece is i_0^j, x_∞ has a negative direction at infinity, which is derived after bifurcation from the $\tilde{v}$ associated to the non-degenerate ξ-piece as it becomes degenerate at the transition between i_0^j and i_0^{j+1}.

Very near to this degeneracy position, when the H_0^1-index on this branch is i_0^j, we may view the new negative direction at infinity as obtained from v-tranport from one edge to the next one of this ξ-piece, the $[\xi, v]$-component being compensated by v-transport of ξ from the next ξ-piece back to the ξ-piece which we are considering:

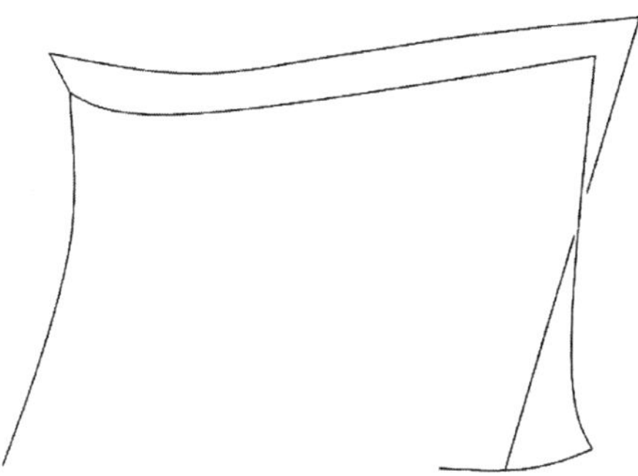

This negative direction at infinity $\bar{\bar{v}}$ has compact support in $[0,1]$. Lemma 5 follows. $\qquad\square$

2.5.10 *The Fredholm aspect*

We know that our variational problem is not Fredholm [Bahri 1998], [Bahri-1 2003]. This translates into the fact that the "oscillations" i.e. a sequence of two consecutive v-jumps of opposite orientation can be large :

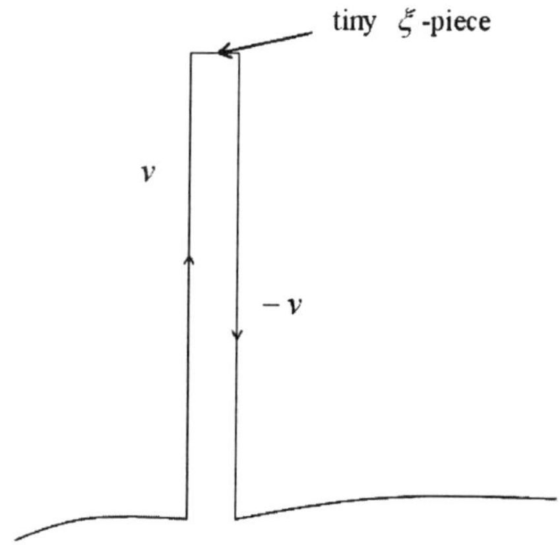

Indeed, as explained in [Bahri 1998], [Bahri-1 2003], some variations which are not L^∞-small can decrease J.

We have described above how such oscillations can be enlarged in order to decrease J; or, if this is not possible, they can be thinned down and brought to be small oscillations to which our compactness arguments apply.

When they cannot be uniformly thinned down, we find ourselves with two distinct methods in order to decrease J: when the oscillations are small and x^∞ is a true critical point at infinity, we would like to use our compactness argument and state that the degree/intersection number is zero. When they are large and cannot be thinned down, we want to proceed as in Section 2.5.8, enlarge them in order to decrease them. Both arguments have to be combined in a single one. It is also important that the part of the compact set K which we are deforming and which will move down, past x^∞, be built only with the original *'s and companions of them or new well defined *'s and their companions, defining $2k$ definite families. This allows to repeat the argument at lower energy levels as we encounter x^∞'s.

Under deformation, we can think of K as made of two pieces: a piece where the oscillations are small and to which the first argument applies and a second one where they are large and J-decreases by "opening" the oscillations (see Section 2.5.8, above) and increasing the size of the ξ-pieces which they jail. We have to glue them and the glueing should be such that, in the parts of K which decrease past x^∞, $2k$ families and associated *'s can be recognized.

Let us discuss what happens as an oscillation, thin and large builds up om a characteristic piece of x^∞.

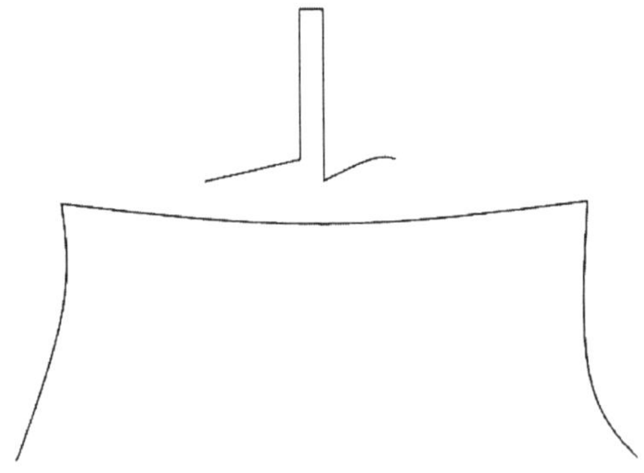

If one of the legs of this oscillation is far from a node, we can increase or decrease it freely, choosing whatever decreases J. This is an extension of the normal (II) flow, compatible with our wishes since no additional v-jump is introduced through this procedure.

If, anywhere along the characteristic piece, there is an interval along which all large oscillations can be thinned down while decreasing (a region where the Fredholm hypothesis is satisfied, this can be quantified geometrically see [Bahri 1998], [Bahri-1 2003]), we can redistribute the rotation along this ξ-piece and assume that all the nodes are in this region. Thus, all large oscillations near the nodes can be brought down to be small.

Our compactness arguments proceed.

We consider now the case when every thin oscillation of the right size (with the orientation of the oscillations which we have) can be enlarged along a decreasing deformation of J in the vicinity of a given node.

Thus, if an oscillation is of the right size or more, we can enlarge it:

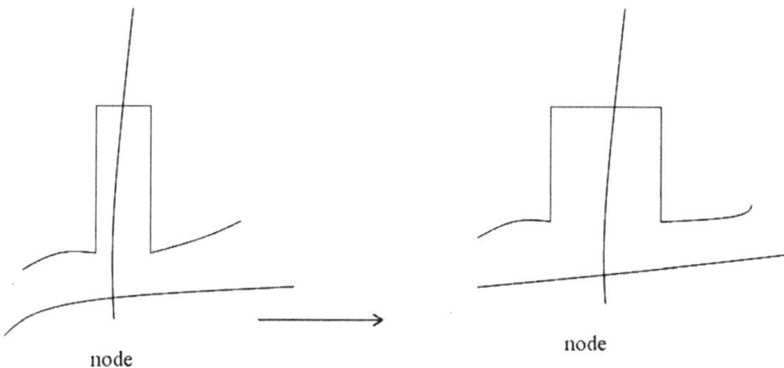

and decrease J.

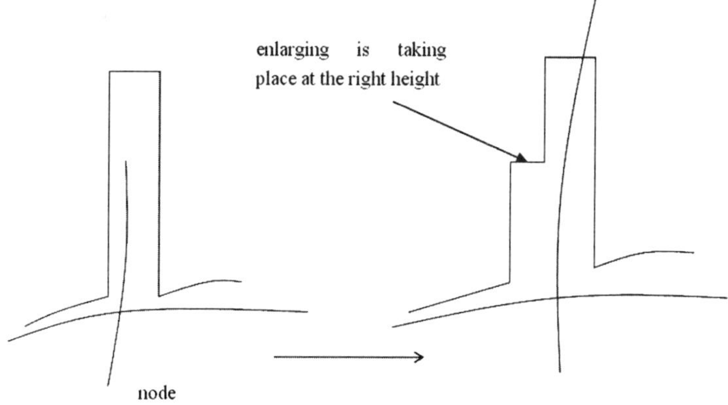

If the oscillation is small but sizable, we can build up on top of it an (very thin) oscillation of the right size or more while decreasing J (we only need a tiny amount which we would borrow from the legs of the oscillation) and enlarge this thin oscillation at the first right height.

However, when the oscillation is quite small, it is the other precedure which we should use, Normal (II) or hole flow etc.

Observe that an oscillation involves two different families since the orientation of the v-jumps of the legs are opposite. There is therefore a hole flow defined between them. As long as such an oscillation is small (the $\pm$ v-jumps of the legs are of size $< c_0$, see Propostion 32), we may assume that it is "wide" (Proposition 32) i.e. containing a v-rotation $\geq c_1 > 0$ (this v-rotation can be assumed to be $< \frac{\pi}{2}$ since a large thin oscillation is building up). Thus, we may decrease J either using the hole flow or the normal II-flow, hole flow on the interval of the oscillation, normal II-flow on one of its legs. As the oscillation builds up, this becomes increasingly impossible if the oscillation surrounds the node because it becomes thin at the same time.

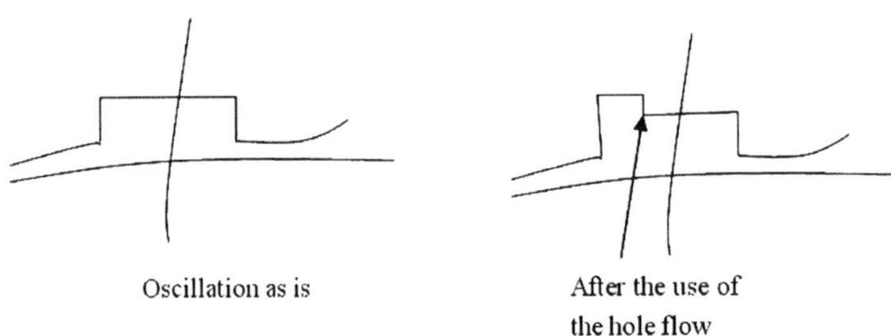

Oscillation as is   After the use of
                           the hole flow

As it thins and builds up:

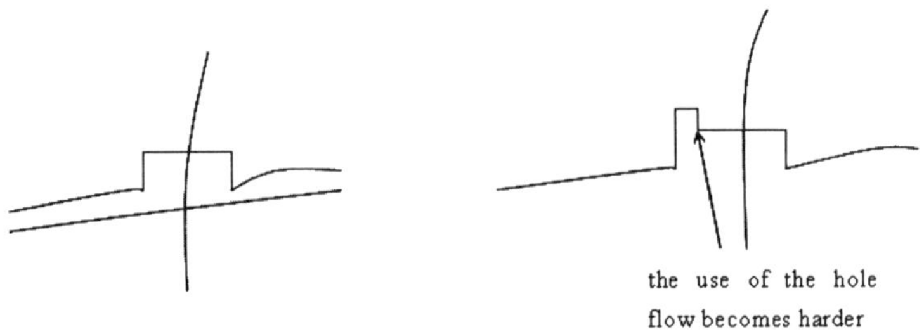

                           the use of the hole
                           flow becomes harder

Thus, as the oscillation is still "wide" and small and is building up, we introduce, at the same point where we introduce the $\pm v$-jump corresponding to the hole flow, a thin oscillation, which we build up.

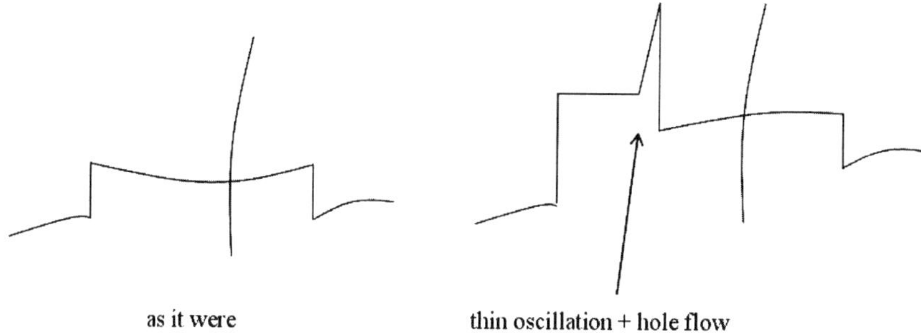

as it were   thin oscillation + hole flow

and progressively cancel the use of the hole flow. We find

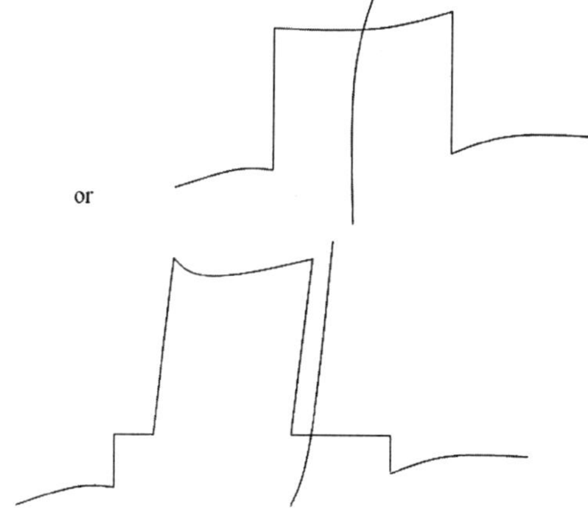

or

All these oscillations in the vicinity, wherever the hole flow has been used, can be brought back to one single wide oscillation or family of wide oscillations, before, after or surrounding the node:

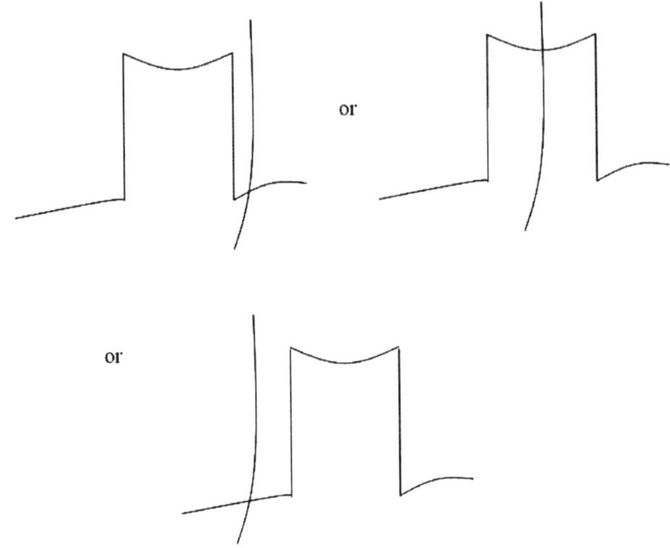

Coming back and looking at our construction, we see that K (after deformation) has been split into two parts P_1 and P_2. $P_1 \cap P_2$ is made of small oscillations and we may assume, because our first deformation (the one used in the compactness part) is defined on it, that $P_1 \cap P_2$ is below x^∞. $P_1 \cap P_2$ is also derived only **with the introduction of companions** in between oscillations. P_2 is also derived only with the introduction of companies and can **all** be moved down, below the level of x^∞. The degree argument can then be applied to $(P_1, P_1 \cap P_2)$. This allows to overcome the Fredholm difficulty.

2.5.11 *Transversality and the compactness argument*

Another technical difficulty comes with the non-Fredholm behavior: the lack of transversality. This implies that we cannot assume anymore that if a critical point (at infinity) dominates another one (i.e. there is a flow-line from the first to the second one), then the second one has a lower index.

We have shown in [Bahri-1 2003] how this difficulty can be overcome, but we have to check here that the associated deformation downwards not only does not increase the number of zeros but also leaves us with the possibility of tracking down the original families in the case of a flow-line from x_{2k} to x_m^∞, $m \geq 2k$ (the case $m \leq 2k-2$ is simple). We want to bypass

x_m^∞ only with the addition of companions, as in the case of the building up of large, then oscillations.

Let us first find precisely where the difficulty lies.

Assume first that all oscillations are small and that all the ξ-pieces of x_m^∞ are non degenerate.

There is then a basic way to patametrize $C_\beta^+ = \{x|\dot{x} = a\xi + bv, a \geq 0\}$ indicated in [Bahri-1 2003]. Roughly speaking, this parametrization is related to the $H_0^1 \oplus T\Gamma_{2s}$ orthogonal decomposition near $x_m^\infty (x_m^\infty \in \Gamma_{2s})$. H_0^1 is the H_0^1-space, see [Bahri-1 2003], related to the verticals of $x_m^\infty, T\Gamma_{2s}$ is the tangent space to Γ_{2s} near x_m^∞. This decomposition is orthogonal and generates a tubular neighborhood of Γ_{2s} near x_m^∞ which is nearly all of C_β^+ near x_m^∞. A part is missing which is related to the "normals" along the verticals of x_m^∞, namely a general curve of C_β^+ near x_m^∞ may be viewed as

$$x_m^\infty + \sigma + h + \tau$$

σ is some sliding in Γ_{2s}, h is some variation in H_0^1

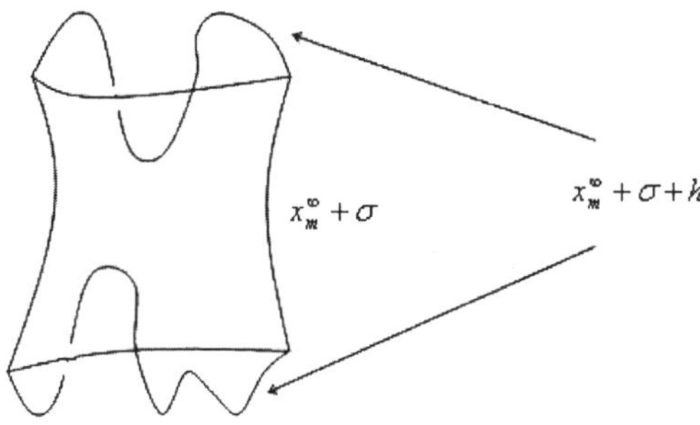

τ yields small ξ-pieces along the verticals of x_m^∞:

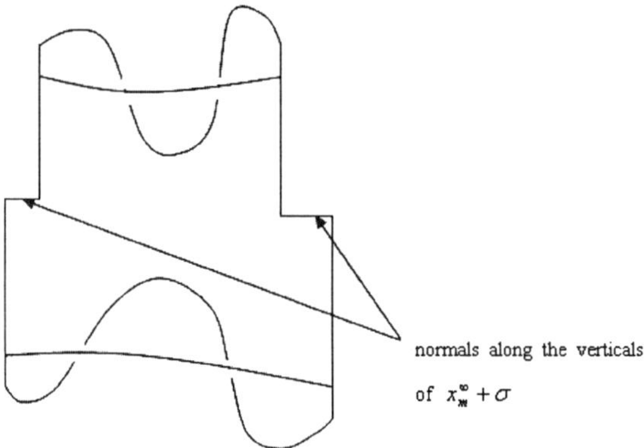

If x_m^∞ is a true critical point at infinity, the introduction of the normals as above along the verticals of x_m^∞ only increases J. Therefore, cancelling progressively these normals, we decrease J and we are brought back to curves parametrized by $H_0^1 + \Gamma_{2s}$.

If x_m^∞ is a false critical point at infinity, we can decrease J by introducing a small normal, i.e. by introducing a small ξ-piece along one of these verticals. This corresponds to the creation of a companion, If x_m^∞ is a true critical point at infinity, then, under the assumption that the oscillations are small, the curves of $W_u(x_{2k})$ may be viewed in the $H_0^1 + \Gamma_{2s}$ parametrization i.e. after a deformation which tracks down the v-jumps (the "cancellation" of the flat small ξ-pieces along the verticals), we find curves which look like:

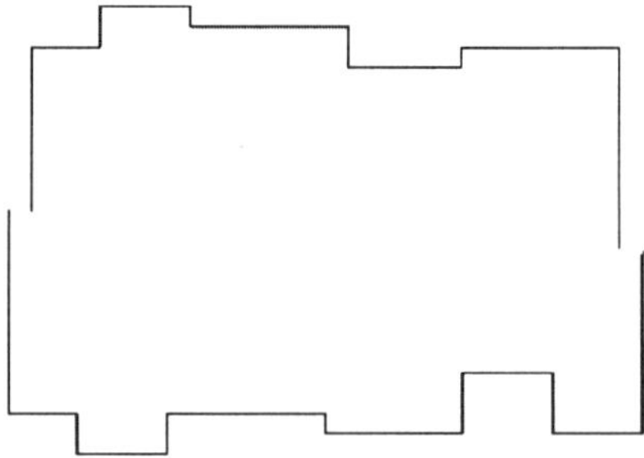

The small lateral breaks are here to indicate the tiny residues of the flat and small ξ-pieces which we might have had along the v-verticals.

We can forget about these small ξ-pieces and think in $H_0^1 + \Gamma_{2s}$, our curves being special though because they have additional small $\pm$-jumps between the large v-verticals. The total dimension of the space where $W_u(x_{2k})$ is represented is $4k$ while the dimension of $W_u(x_{2k})$ is $2k$.

Since we can perturb v near x_{2k} without perturbing $W_s(x_m^\infty)$ near x_m^∞, tranversality should follow easily.

However, x_m^∞ has $2s$ jumps. Assuming for simplicity that $W_u(x_{2k})$ is still made with $2k \pm v$-jumps in the region where we want transversality — i.e. assuming that we have not introduced companions — we have:

$$2s \le 4k.$$

The unstable manifold of x_{2k} could intersect the stable manifold at infinity of x_m^∞, $W_s^\infty(x_m^\infty)$ which is contained in Γ_{2s} because Γ_{2s} is contained in Γ_{4k}.

This happens if $2k - s \pm v$-jumps are of size zero, so that we are left with $s \pm v$-jumps. Thus $W_u(x_{2k}) \cap \Gamma_{2s}$ is a stratified space of top dimension s because the size of the $\pm v$-jumps on $W_u(x_{2k})$ can be taken, at a generic point, to represent each dimension.

$W_u(x_{2k}) \cap \Gamma_{2s}$ intersects $W_s^\infty(x_{2k})$ if (a neccesary condition)

$$s \ge i_\infty + 1.$$

Besides the s basic $\pm v$-jumps in $W_u(x_{2k})$ near x_m^∞, we have $2k - s$ additional $\pm v$-jumps which are "small", in between these basic $\pm v$-jumps. We are assuming that the index of x_m^∞ is $i_0 + i_\infty = m \ge 2k$ so that

$$i_0 + i_\infty \ge 2k$$

i.e.

$$i_0 \ge 2k - i_\infty \ge 2k - s + 1. \tag{*}$$

Observe — and this is **the crucial point** — that these $2k - s \pm v$-jumps can be tracked down on $W_u(x_{2k})$ i.e. their positions can be followed. (*) states that, as we track them down, we are always missing one H_0^1-negative position of x_m^∞.

Let us analyze how these H_0^1-negative positions of x_m^∞ arise. For this, given a ξ-piece of x_m^∞ we need to introduce starting from each edge the solution of

$$\begin{cases} \ddot\eta + a^2 \eta \tau = 0 \\ \eta(\text{edge}) = 0, \dot\eta(\text{edge}) = 1. \end{cases}$$

The above conditions define two functions φ_1 and φ_2:

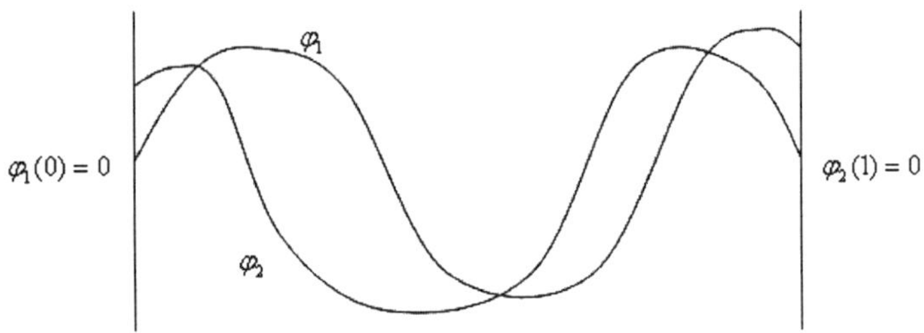

which coincide only if the ξ-piece is characteristic.

The two graphs taken together indicate the zones along the ξ-pieces where the introduction of a small $\pm v$-jump corresponds to a negative direction for $J''_\infty(x^\infty_m)$

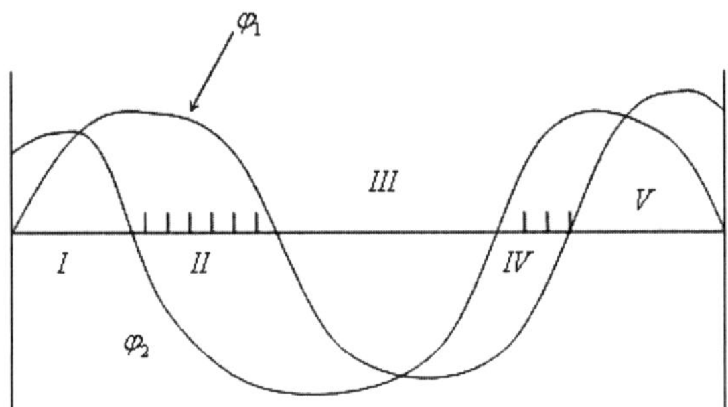

In area I, a small $\pm v$-jump does not provide a negative direction for $J''_\infty(x^\infty_m)$ since the H^1_0-index is totally achieved by the two oscillations of φ_2 to the right. The same claim holds true for area V and φ_1. Each time a zero of φ_1 or φ_2 is crossed, there is a switch: the v-jump generates, after the crossing, a negative direction if we were coming from a positive direction or vice-versa. We thus see that, in regions II and IV, a $\pm v$-jump provides us with a negative direction while the opposite is true in areas I, III and V.

If the ξ-piece is characteristic, φ_1 and φ_2 are confounded. Regions I, III, V reduce to single points while II and IV span the entire region in between nodes.

This repeats with more nodes.

Coming back to the $2k - s \pm v$-jumps whose positions can be tracked down, we assert that, in every occurrence, there will be a ξ-piece where the number of $\pm v$-jumps among these $2k - s$ which live on this ξ-piece is strictly less than the trict H_0^1-index of this ξ-piece.

The positions of these $\pm v$-jumps can be **tracked down**. We can then - it is a technical lemma — slide them so that each of them is in a "negative area", two of them not being in the same negative area (we know how to separate $\pm v$-jumps when they are small while decreasing J).

There is then, in every occurence, one negative area where no $\pm v$-jump lives. If this negative area is the first one after the left edge, we can introduce in it a $\pm v$-jump which has the orientation of this left edge.

This will decrease J and this decrease is completed through the introduction of a companion (to the edge).

If this (there could be several of them) negative area is not the first one, then we can deform our $\pm v$-jump progressively, with the introduction of companions, and make it to be the first negative area.

Indeed, we can fill every empty negative area with a $\pm v$-jump having the same orientation and the same weight than the next existing $\pm v$-jump to the right of this empty negative area. As we create these $\pm v$-jumps (which might be zero if the next — tracked — $\pm v$-jump existing to the right is zero), we can cancel its next companion to the right. Working step by step, we free the first negative area and proceed as above, introducing a $\pm v$-jump in it with the orientation of the right edge. J decreases then and transversality follows.

The definition of companions is slightly extended in this way to include "twin" $\pm v$-jumps because the $\pm v$-jumps which we are replacing may be zero.

The argument as described above assumes that $W_u(x_{2k})$ is constantly achieved through Γ_{4k}, which is of course wrong because companions and twins can be introduced along the way. This builds families of $\pm v$-jumps. Revisiting our argument above, we may reduce a whole family to a single $\pm v$-jump in a negative area and proceed as above. This requires only the addition of a few technical details allowing to slide the $\pm v$-jumps of a family in negative areas and reduce them to a single one. Observe that along these slidings, the $\pm v$-jumps can change orientation. However, families will change orientation together as can be engineered easily.

The transversality argument in this simplified framework thus holds, when x_m^∞ does not have characteristic pieces.

If x_m^∞ has some characteristic pieces, but the oscillations are still assumed to be small, the above argument extends. Indeed, the curve supporting x_m^∞, $\bar{x}^\infty$, supports in fact a cluster of critical points of various indices $\bar{x}^\infty = \bar{x}_0^\infty, \bar{x}_1^\infty, \ldots, x_m^\infty$ etc. corresponding to the various "puzzles" which we can build using the full (half)-unstable manifolds associated to the characteristic pieces of $\bar{x}^\infty$. $\bar{x}^\infty$ itself has an index equal to $i_0 + i_\infty$. If $i_0 + i_\infty \leq 2k - 1$, then some $\bar{x}_i$'s have index $2k - 1$ (see [Bahri-1 2003], the index increases by 1 at most at each step) and the arguments developed for compactness above work. The covering degree of these $\bar{x}_i$'s by x_{2k} is zero.

If $i_0 + i_\infty \geq 2k$, then $\bar{x}^\infty$ has already an index too large for x_{2k} and we would like to establish that $W_u(x_{2k})$ and $W_s(\bar{x}^\infty)$ do not intersect. $\bar{x}^\infty$ does not use any full (half)-unstable manifold of any characteristic piece included in it. Its index is made of i_∞, the index of $\bar{x}^\infty$ in Γ_{2s} and i_0, the **strict** H_0^1-index. The arguments used for x_m^∞ when it did not include any characteristic piece then apply, after being suitably adapted (they require some modifications and extensions though which we do not state here for the sake of conciseness).[1] The case of families which might control several nodes or fill several negative areas at some point over the configuration requires a different approach: the case of H_0^1-index equal to 1 has been solved in 2.5.5.2d. There is nothing to prove for the index 0. In the remaining cases, there is more "room" to move inside whenever there is a hole.We have to use iii) of Proposition 30 and Proposition 31 and we must verify that we can use them and engineer a decrease and we must be able along this decrease to track the families as they evolve. To this aim, we view Proposition 31 differently: instead of switching the orientation of F_2, see Propositions 30 and 31, when its left edge is very near the node x_{m-2} and its right edge is in the middle of the next nodal zone, we use our deformation differently and create a hole at x_{m-2}.This can be done easily through the same deformation which we used in the proof of Proposition 31, only that instead of switching the orientation of F_2, we stop it once a hole has been created at x_{m-2}, before any orientation switch. Some members of F_2 are then canceled, others survive.

---

[1] In the framework of families which can be reduced to a single $\pm v$-jump, the argument goes as follows: a free node allows to build, given a family, a twin family through a non-increasing deformation. In this way, a "twin" — switching is completed family by family, freeing progressively the left extreme node with the use of an inside free node. This process is hindered as a family exits the characteristic piece through the left edge. This can happen if the family is "large" with the same orientation than the left edge. When it is small, we may assume that it is well inside the characteristic piece, past the first node — using the switching described above — As it gets larger, we can make it travel — we are manipulating the original configurations — from inside the characteristic piece through the left extreme node so that it can exit through the left edge. When it is still inside the characteristic piece, after the first node — which is then free — we create a companion to the left edge near the node (past it) so that J decreases. As the family travels and approaches this node, we switch this companion and create it after the family. Only details are now left aside. $\square$

We then have either a hole at x_{m-2} or a hole at x_{m-3}. We can build our deformation using the Hole flow or using companions to the right edge. The Hole flow is of local nature and we can check directly that, throughout its use, we can track our families after a local reordering and redistribution: we need only to compare an alternating sequence of families with our actual sequence together with the normals of the Hole flow which we use. There is then a natural redistribution of the families. The continuity of the transition as a family involved in a repetition changes orientation can be built in through a "splitting" of this family as it becomes small into two small $\pm v$-jumps of small size and opposite orientation which travel back, on each side of the family, to the other "basic" families which are not changing sign. The two families of the edges define "boundary families", with a stable orientation.

The case of non degenerate ξ-pieces and families works essentially in the same way: there is a Hole flow also on these pieces and there is also a kind of Normal (II) flow which corresponds to the cancelation of $\pm v$-jumps which are in the negative areas, see the diagrams above. This Normal (II) flow does not go all the way to the reversal of sign. Using these two features, we reduce the problem of transversality(already solved in 2.5.5.2d if $\bar{\gamma}_j = 1$, if $\bar{\gamma}_j = 0$, there could be a single family missing)to configurations where a family such as F_2-in the degenerate case-occupies the two last nodal positions, with an orientation opposite to the right edge while the family of the right edge is advancing inside the ξ-piece. F_2 then recedes; part of it, let us say after use of the widening process that it is reduced to two $\pm v$-jumps, its right $\pm v$-jump is in the middle of the "cancelation" zone between the two nodal zones. Its left $\pm v$-jump is in the middle of a nodal zone. We are exactly in the same configuration than in the degenerate case, a decreasing Normal N^- can be introduced in the last nodal zone, but as the family of the right edge advances, we need to "switch", we need a Normal (II) flow. We have it partially since we can cancel the right $\pm v$-jump of F_2, but we need this cancelation to decrease J enough so that we can cancel over the same process the left $\pm v$-jump of F_2 and decrease J. Then, we can introduce in the hole created a $\pm v$-jump with an orientation opposite with the orientation of the family, coherent with the assignment of the Hole flow in this hole. Transversality would follow.

A relatively easy computation shows that this reduces to a manipulation of the Green's function of the operator $\ddot{\eta} + \eta\tau$ under Dirichlet boundary conditions on the non degenerate piece. The Green's function reads as the solution η of

$$\ddot{\eta} + \eta\tau = -\delta_x$$

under Dirichlet boundary conditions on the appropriate time interval. This is tantamount to manipulate the transport equation between the right edge and the left edge of F_2 when the right is in the "cancelation" position and the left edge is in the second last nodal zone (counting from the right), it is totally equivalent to the construction of Appendix 2, which is carried out in great detail, together with its application to degenerate ξ-pieces (part of the application is carried out **before** hand in 2.5.5.2c, 2.5.5.2d). The extension to the present framework requires exactly the same result than Appendix 2, with some modifications from the point of view of the application, computations of the second derivative on $\pm v$-jumps located at various times on the ξ-piece have to be carried; one finds out that this quadratic form is controlled by the Green's functions taken at couple of points among these times. Using the results of Appendix 2, one can check on this quadratic form that the Green's function can be manipulated so that the deformation associated to the Normal (II) flow is J-decreasing. The details are left to the reader

The above arguments all assume that all oscillations are small. However, some oscillations can be large and, as explained in [Bahri 1998], [Bahri-1 2003], they might yield decreasing deformation after widening.

We now have two deformations: one, due to transversality, applies when the oscillations are small and moves the curves downwards, past x_m^∞. The other one, due to the non Fredholm character, applies when the oscillations are large and thin and J decreases after "opening and widening" them. We need to glue them together.

The following basic phenomenon shows that this is possible: let us consider some point on $\bar{x}^\infty$ where a thin, large oscillation can be built, then widened so as to decrease J. Once this construction can be carried out for $\bar{x}^\infty$, it can be carried for all nearby curves, the situation being similar (the opening of these oscillations and then widening obey precise equations related to the geometry of α on the v-orbits through the points of $\bar{x}^\infty$ see [Bahri-1 2003], hence extends naturally nearby).

Thus, such a thin large oscillation can be built on all the flow-lines out of the curves of $W_u(x_{2k})$ which are in the vicinity of $\bar{x}^\infty$ and which are represented with **small** oscillations. Once, it is built, it can be opened and widened and this decreases J all along the curves of such flow-lines. We thus have the following diagram:

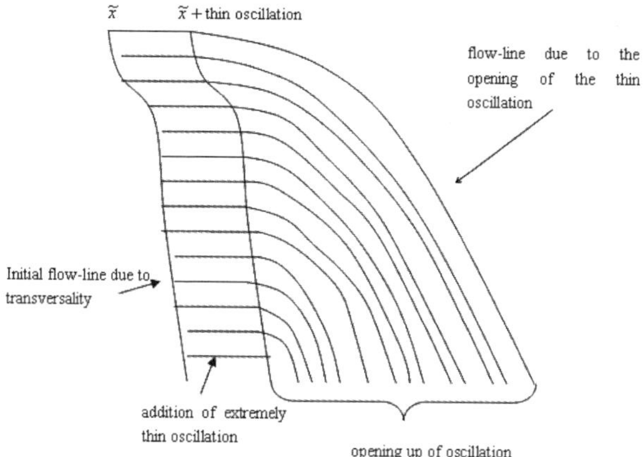

This diagram shows that the two deformations can be patched. We can also reverse it, take a curve near $\bar{x}^\infty$ of $W_u(x_{2k})$ bearing an extremly thin, large oscillation which we could use to decrease J. We can **supress** this oscillation from z, patching then z into $\tilde{z}$. This might increase J but only by a tiny amount related to how thin the oscillation is. Such $\tilde{z}$'s bear still the $\pm v$-jumps at the locations of those of z. they form a $2k$-dimensional manifold to which we can apply the transversality arguments described above. This argument involves sliding of small $\pm v$-jumps, introduction of companions etc. which might seem difficult in this new framework since we have to reintroduce in the end a thin oscillation and patch with the other deformation on thin, large oscillations. However, we may assume that all the negative areas are very close, after redistributing, using the techniques of [Bahri-1 2003], Propositions 27 and 28, Lemma 11, the rotation of v along the ξ-piece so that, in all these negative areas and in between, all thin and large oscillations can be opened and widened while decreasing J.

The deformation on the $\tilde{z}$'s and the widening of the oscillations then patch just as when we were studying the non-Fredholm behavior, in the previous section. $\qquad\square$

Addendum 1

Our statement about the unstable manifold $W_u(\bar{x}), \bar{x}$ a periodic orbit of ξ, includes decreasing deformation i.e. our statement is variational - with control on the maximal number of zeros of b as in [Bahri-1 2003]; this number decreases from $W_u(\bar{y})$ or $W_u(\bar{y}^\infty)$ to $W_u(\bar{x})$ or $W_u(\bar{x}^\infty)$ if $\bar{y}$ or $\bar{y}^\infty$ dominates $\bar{x}$ or $\bar{x}^\infty$.

Indeed, such a variational theory exists if we use near $\bar{x}$ the diffusion flow having $\eta = b$ (see [Bahri-1 2003]). The upperlevel sets $J_{c+\varepsilon}(c = J(\bar{x}))$

deform then on $D^-(\bar{x}) \cup J_{c-\varepsilon}$ where $D^-(\bar{x})$ is a small unstable disk in E^-. E^- is the negative eigenspace of the linearized operator $-(\ddot{\eta} + a^2 \eta \tau)$.

We have defined functions $\eta(c_1, -, c_{2k})$ for $W_u(\bar{x})$ (when the Morse index of $\bar{x}$ is $2k$; there is a similar formula for $2k+1$) such that $J''(\bar{x})$ is negative along these directions. Such functions $\eta(c_1, -, c_{2k})$ split on $E^- \oplus E^+$ into $\eta^-(c_1, -, c_{2k})$ and $\eta^+(c_1, -, c_{2k})$. It is easy to see that $\eta^-(c_1, -, c_{2k})$ is non zero. We can define, using the diffusion flow ($\eta = b$) a decreasing flow-line from $\eta(c_1, -, c_{2k})$ to $\tilde{\eta}^-(c_1, -, c_{2k}) \in E^-$. $\tilde{\eta}^-(c_1, -, c_{2k})$ is non zero: the proof of this fact is equivalent to the statement that $\eta^-(c_1, -, c_{2k})$ is non zero. The map from $D^-(\bar{x})$ to a neighborhood of $\bar{x}$ in $W_u(\bar{x})$ is one to one and onto. The number of zeros of b on such flow lines is at most $2k$.

We thus can define "cylinders" running from disks in our unstable manifold to disks in E^- and the maximal numbers of zeros of b along these "cylinders" (after removal of $\bar{x}$) is $2k$.

There is on the other hand a natural "dilation" on our unstable manifold and J decreases along this dilation. Combining this "dilation" and the flow-lines defined above, we can create reverse cylinders, starting from disks in E^- and ending in (non-standard) disks around $\bar{x}$ in $W_u(\bar{x})$. This family of flow-lines when combined with this dilation allows to define a **decreasing** deformation of $D^-(\bar{x}) \cup J_{c-\varepsilon}$ on $\tilde{D}(\bar{x}) \cup J_{c-\varepsilon'}$ where $\tilde{D}(\bar{x})$ is a "disk" in our unstable manifold. Because we use flow-lines along which b has at most $2k$ zeros and because the "dilation" is only a progressive transversal shift in the flow-lines, the maximal numbers of zero of b does not go beyond $2k$ (including multiplicity). This extends to a small C^∞-neighborhood of these flow-lines. The fibrations of $W_u(\bar{y})$ or $W_u(\bar{y}^\infty)$ (above $W_u(\bar{x})$) can be assumed to be defined using the C^∞-topology (or C^m for m very large) since we basically use the diffusion flow (+ dilation).

The claim follows.

The deformation result of "Compactness" is variational. It goes beyond properties of flow-lines connecting periodic orbits.	$\square$

Addendum 2

Considering the first $*$ before the first node on a characteristic piece, we claim that there is a small constant $c > 0$ such that if the associated $\pm v$-jump does not have the orientation of the left edge (there is a similar statement for the right edge) or is of size less than c, then we may move this $*$ to the middle of the first nodal zone while decreasing J. Thereafter, we can use the normal II-flow on it. No use of the hole flow is needed on such configurations.

The size of the associated $\pm v$-jump might change then. But this deformation can be extended to a global decreasing deformation where the orientation of the $\pm v$-jump associated to this $*$ does not change at least if it was initially of size c^2 or more. In addition, the $\pm v$-jumps among these which have the orientation of the left edge and which are of size $\geq 2c$ are left untouched and those of size between $\frac{5c}{4}$ and $2c$ might be translated and changed but they remain of size $\frac{c}{4}$ to the least.

Finally, picking c_1 small, positive but very large with respect to c (the choice of c follows the choice of c_1), all $\pm v$-jumps associated to this $*$ or not, of size $\geq c_1$ having the orientation of the left edge are only very little perturbed through this process if the $\pm v$-jumps associated to this $*$ have the orientation of the left edge (or are tiny, less than $2c$ in size with the reverse orientation).

This last piece of our claims allows us to stabilize through this process - the aim of which is to push "in the middle" such a $*$ which has an orientation opposite to the left edge or has the same orientation, but is small - the various possible representatives of the left edge for the hole flow.

Indeed, in order to engineer a decrease, if the left edge is positively oriented for example, it suffices to decrease an existing positive v-jump or to increase in size a negative v-jump, see below for the proof of this fact in general.

As we engineer this decrease, we may translate the position of $*$, the position of this $\pm v$-jump. If this v-jump is zero, we may translate it without changing its size. Similarly, we can move a tiny v-jump inside the characteristic piece, inside the first nodal zone while decreasing its size and decreasing the functional. We proceed as follows to achieve this goal: if this tiny v-jump is the first one after the left edge, it is straightforward to move it away from this edge while decreasing the functional. If it is not the first one, we can "widen", as in Proposition 30, the oscillations where it is involved, being careful though to move this v-jump always more inside the characteristic piece. This procedure contrasts in some ways with the one of Proposition 30 because we do not complete the widening of the oscillation necessarily on this side of this oscillation where the $\pm v$-jump is smallest. Since we are basically - we will see that we cannot always do that - widening always on the side of our v-jump, we might cancel in this process $\pm v$-jumps which are intermediate between this v-jump and the left edge. The process can be continued though with the next $\pm v$-jump in the direction of the left edge and it can be made into a continuous process if, when a $\pm v$-jump becomes tiny, we move progressively the position of the left vertical of the oscillation towards the next $\pm v$-jump in the direction of the left edge. This last move might affect the orientation of the $\pm v$-jump associated to our $*$. However, we observe that initially we were deforming a compact subset K of $W_u(x_m)$ in $J^{-1}(c_\infty + \varepsilon), \varepsilon > 0$ a fixed number, $c_\infty = J(x_\infty)$. The $\pm v$-jumps of K are separated by a ξ-distance $\delta(\varepsilon) > 0$ to the least. Thus, as we widen the oscillations, the $\pm v$-jumps associated to our $*$ of size (absolute size) c^2 or more will remain of an absolute size lowerbounded by $\delta_1(\varepsilon) > 0$. As we move the left side of the oscillation (whose right side is our $*$) to make our process continuous, this left side is tiny as we please. We may impose that it is less than $\delta_1(\varepsilon)^2$. Then, the orientation of our $\pm v$-jumps (those associated to this $*$, also the other v-jumps of size $\geq \delta_1(\varepsilon)$) is not affected; they remain of size $\delta_1(\varepsilon)/2$ to the least.

In the end, the $\pm v$-jump which we are pushing inside will have to go to the transition zone and beyond; this process cannot stop since the left edge is "large". We can then adjust this process to the size of this $\pm v$-jump before deformation. As the size of this v-jump increases (once it is positive), we move it less and less inside the first nodal zone. We can scale the changes so that once these v-jumps are positive of size more than $2c$, they are untouched and if they are of size $\frac{5c}{4}$ or more, they are of size $\frac{c}{4}$ to the least in the end. All other v-jumps, positively oriented, of size c_1 very large with respect to c will remain - it is easy to see - of size $\geq c_1 - Cc = c_1(1 + o(1))$ through this process provided the $\pm v$-jumps associated to the $*$ are either positively oriented or are of negative size c to the most, i.e. representatives of the left edge are stabilized in this process if a hole flow needs to be used.

Let us prove now, a fact that we already know if there is only one $\pm v$-jump after the left edge which we decrease (algebraically) in size, that if we decrease (algebraically) a $\pm v$-jump on an almost characteristic ξ-piece (the $\pm v$-jump being in the first nodal zone, "away" from the node) J decreases.

The addition at the $\pm v$-jump is $\mu_0 v, \mu_0 < 0$, which we transport back to the left edge following the history of the curve. We are going to show that the transported vector z at the left edge reads:

$$z = o(\eta_0)\xi - \eta_0(1 + o(1))[\xi, v] + (\mu_0 + o(\eta_0))v$$

where $\eta_0 = -\mu_0 \overline{\Delta t}, \overline{\Delta t}$ being the ξ-length between the left edge and the $\pm v$-jump which we are considering. It is then easy, using z, to construct a decreasing normal N^- as in 4.2. Let us prove the above formula; the transport equations read

$$\begin{cases} \dot{\mu} + a\eta\tau - b\eta\bar{\mu}_\xi = 0 \\ \dot{\eta} = \mu a - \lambda b \qquad\qquad t \in [0, t_0] \\ \overline{\dot{\lambda} + \bar{\mu}\eta} = b\eta \end{cases}$$

0 is the time of our $\pm v$-jump; t_0 is the (negative) time of the left edge. b is equal to a sum of Dirac masses $\sum_{i=1}^m c_i \delta_{t_i}$ with $\sum |c_i| = o(1), t_0 = -\overline{\Delta t}$.

Integrating we find:

$$\eta + \int_0^t b\left(\int_0^s b\eta - \bar{\mu}\eta\right) = -\mu_0 at - a\int_0^t \int_0^s (a\eta\tau - b\eta\bar{\mu}_\xi).$$

A simple argument shows then that

$$\eta = -\mu_0 at(1 + o(1))$$

$\square$

Addendum 3

Curves nearby x^∞ can be split between nearly v-verticals (see [Bahri-1 2003] for the definition of v-verticals) and nearly ξ-pieces. The nearly v-verticals can be modelized ([Bahri 1988], [Bahri 1998], [Bahri-1 2003]) by large pieces of $\pm v$-orbits alternated with small ξ-pieces.

Because x^∞ is assumed not to be false (see [Bahri-1 2003]; false critical points at infinity can be bypassed through the introduction of companions, see 2.5.1 above for the definition of companion), the small ξ-pieces can be absorbed through a J-decreasing deformation into the nearly ξ-pieces.

Thus, any deformation class can be represented near x^∞ by curves made of large $\pm v$-verticals joined by nearly ξ-pieces. The analysis of [Bahri-1 2003, pp. 113–141] tells us then that any chain of $(J_{c_\infty} + \varepsilon, J_{c_\infty} - \varepsilon)$ in $\mathcal{V}(\S_{\updownarrow}^\infty) - \mathcal{V}(\S_{\updownarrow}^\infty)$ is a neighborhood of x_m^∞ in $C_\beta^+ = \{x; \dot{x} = a\xi + bv, a \geq 0\}, c_\infty = J(x^\infty)$ - can be represented as a product of the unstable manifold **infinity** of x^∞ (of dimension i_∞) with the H_0^1-unstable manifold of the non degenerate ξ-pieces of x^∞ and with the H_0^1-unstable manifold (strict or full; the full one is half of the usual unstable manifold because it corresponds to a degenerate critical point) of the characteristic pieces (each piece in this product is taken a number of times). In our situation - i.e. in 2.5.4.5; the statement has to be adapted for 6.- the maximal number of zeros of b on $W_u(x_{m-1}^\infty)$ is $2k$ and $m-1 = 2k-1$. The number of characteristic ξ-pieces needed to define the cycle(s) of x_{m-1}^∞ is then imposed; this is the number of characteristic ξ-pieces whose full unstable H_0^1-manifolds are used in the definition of the cycles of x_{m-1}^∞ see [Bahri-1 2003, p. 126].

Furthermore this is a consequence of the above representation, denoting $C_\beta^{+,2k} = \{x, \dot{x} = a\xi + bv, a \geq 0, b \text{ has at most } 2k \text{ zeros }\}$, the group

$$W = H_{2k}(J_{c_\infty + \varepsilon} \cap C_\beta^{+,2k} \cap \mathcal{V}(\S_{\updownarrow - \infty}^\infty), J_{]_\infty - \varepsilon} \cap C_\beta^{+,\in\|} \cap \mathcal{V}(\S_{\updownarrow - \infty}^\infty), \mathbb{Z})$$

is a free group. Let $\varphi_1, -, \varphi_s$ be a basis of this group. $W_u(x_{2k})$ generates near x_{m-1}^∞ an element ψ of W which therefore reads

$$\psi = \sum n_i \varphi_i, n_i \in \mathbb{Z}.$$

Since $W_u(x_m) \cap \mathcal{V}(\S_{\updownarrow - \infty}^\infty)$ can be moved in $C_\beta^{+,2k}$ below x_{m-1}^∞, each n_i is zero.

$\square$

Appendix 1

We now take care of the ξ-pieces of zero H_0^1-index. Those which are non degenerate do not interfere with our argument, therefore we will not discuss them. For the degenerate ones, the key point relies again into the fact that we are able to track the $*'s$ of the families as they travel.

Our cycles are all taken modulo $J_{c_\infty-\epsilon}$, we are going to prove that we can rewrite any given cycle into a sum of cycles such that for any given cycle taken from this sum, the number of $*'s$ associated to a given degenerate ξ-piece of H_0^1-index 0 is constant equal either to 0 or to 2 all over the cycle. In the former case, the full (half) unstable manifold of the associated ξ-piece is not part of the definition of the cycle while in the latter case, it is part of this definition, ℓ (see Section 2.5.6) is increased by 1, but n also in the argument of Section 2.5.6 can be increased by 1 since we do have two $*'s$ associated to this ξ-piece which thus behaves as if it were of non zero H_0^1-index.

Assume that for a given configuration in our cycle, three $*'s$ are living on a given degenerate ξ-piece of index zero, two for each edge and one in between. Let us track these three $*'s$ as the configuration changes in our cycle. We obviously can follow them as their respective locations change over the basic curve $\bar{x}$ which is close to the critical point at infinity x_∞. As they travel outside of the characteristic piece of index zero and go inside its edges, their orientations might change but we can still relate two of them to this characteristic piece in our global count and the counting argument of Section 2.5.6 can proceed at least as far as this characteristic piece is involved. As a $*$ travels away, it can enter non-characteristic pieces, engineering then a reordering of the $\pm v$-jumps generating their H_0^1-unstable manifolds, a $*$, maybe modified, can be still tracked and attributed to our initial characteristic ξ-piece of index zero, so there is no meaningful modification in this case. If we use the results of [Bahri-2 2003], modify the rotation of v on non-degenerate ξ-pieces until either a ξ-piece becomes characteristic, or it becomes of H_0^1-index zero, so that i_0^f of Section 2.5.6 becomes zero, then we can even assume that this $*$, after decreasing deformation, has the same orientation all over these transitions.

The count could change only as such a $*$ enters another characteristic piece. If it is a single v-jump as it enters, then the Normal(II)-flow can be applied upon it and a related splitting of the cycle occurs. If it enters as a family, a problem might arise if it is a family of an edge, typically of the right edge. We have discussed configurations such as these in 2.5.5.2b, 2.5.5.2c, 2.5.5.2d, in Propositions 30 and 31 although these configurations were more specific in 2.5.5.2. Forgetting about the use of the hole flow in the last nodal zone which we introduced in 2.5.5.2 (in our present configuration, flows are provided by Normal(II)-flows or the use of an appropriate alternating sequence), we can, once the family of our $*$ has penetrated a little bit inside the last nodal zone and there is a hole in the vicinity of x_{m-1}, engineer a decrease through the introduction of a companion to the right edge, also repeating the construction of 2.5.5.2c, 2.5.5.2d and using if needed (for a transition with the use of the alternating sequence) the Normal (II)-flow on the family preceding the family of our incoming $*$, or if this family (with an orientation opposite to the orientation of the right edge) starts overextending, use the previous node x_{m-3}. At this x_{m-3}, the use of the Hole flow of the left edge (which we assume to provide an orientation opposite to the one of the right edge just before the incoming $*$) and the introduction of a companion of this family are compatible because the companion to this family near x_{m-3} comes **after** the $\pm v$-jump due to the Hole flow, nearest to the family itself. We thus can use the Normal due to this Hole flow(recall that Hole flows can be inserted in alternating sequences)in combination with the Normal (II)-flow and thereafter, once we are not using the Normal (II)-flow anymore, introduce the companion to the family, cancel the use of the Normal associated to the Hole flow and keep the use of this companion as the family overextends, past x_{m-3}. In all, our cycle splits into two pieces at such a transition and there is a natural way to extend the flow on each piece of the cycle because of this compatibility at x_{m-3}.

We are thus left with maximal sequences of characteristic pieces of H_0^1-index zero. We count the number of $*'s$ associated to them, which are all the $*'s$ between the $*$ associated to the right edge of the first characteristic piece preceding the sequence (included) and the $*$ associated to the left edge of the first characteristic piece following the sequence (not included). If there are repetitions in these edges, we include the $*'s$ involved in these repetitions in the count except for the last $*$ for the left edge of the first characteristic piece following the sequence . Again, we can split the cycle in sub-cycles along which this number does not change for each given sequence. It can be even equal to $2s$ or odd equal to $2s + 1$. In both cases, the sequence contributes at most s to the number ℓ of Section 2.5.6, but the sequence contributes also at least $2s$ to the number of $*'s$. The argument of Section 2.5.6 can therefore proceed for each of these sub-cycles. Would the number of characteristic pieces of index zero which contribute to a given cycle through their full(half)-unstable manifold change over the span of the cycle (very close to the top level c_∞)without a corresponding change for the number of $*'s$, we could not derive the same conclusion. Observe that near this top level we can assume after Section 2.5.6 that all characteristic pieces of non zero H_0^1 are "filled" up, but maybe for one of them which bears a single simple hole, together with a repetition in its left edge.

We can conclude now along the lines of the arguments used in Section 2.5.6. Either ℓ, see Section 2.5.6, tends to infinity with k; we use Hypothesis (A) , argue as in Section 2.5.6 (with variants) and conclude. Or ℓ is bounded , then n, the number of characteristic pieces of non zero H_0^1-index≥ 1, is bounded (using the arguments of Section 2.5.6). If the H_0^1-index of anyone of these pieces tends infinity or if the number or the H_0^1-index of the non degenerate ξ-pieces tends to infinity, we can argue again as in section 6. Otherwise, the total H_0^1-index is bounded ,ℓ is bounded and so is $\bar{\gamma}_1$. Then k is also bounded since $i_0^f + \bar{\gamma}_1 + 2\ell = 2k$ $\square$

Appendix 2

Let us consider a piece of ξ orbit of length $\frac{\pi}{4}$. We modify ξ into $\tilde{\xi}$, α into $\tilde{\alpha} = (1 + o(1))\alpha$ near the piece of ξ-orbit as follows.

α is $x\,dy + dz$, $\xi = \frac{\partial}{\partial z}$. We replace α by

$$(1 + 2\nu\omega(\frac{x^2 + y^2}{\epsilon_1})\omega_2(z)C_1 xy)^{-1}\alpha$$

The piece of v-orbit is provided by $x = 0, y = 0$, z belongs to $[0, \gamma]$, γ small. ϵ_1 is a small parameter, $\omega_1'(t)t$ and $\omega_1''(t)t^2$ are small; ω_1 is 1 near zero and is zero for $t \geq M$, $\omega_1 = \omega_1(t)$. ω_2 is 1 on a small z-interval inside the ξ-piece, close to its starting point and is zero outside an interval twice as large. The first and second derivatives of ω_2 are bounded in function of the size of these intervals. This size is large when compared to ϵ_1. ν is an appropriate constant, the value of which is provided below. We use the formulae for $\tilde{\xi}$ available in [Bahri-1 2003, p. 83] and we find that

$$\tilde{\xi} = \xi - 2\nu\omega_1\omega_2 C_1 x\frac{\partial}{\partial x} + 2\nu\omega_1\omega_2 C_1 y\frac{\partial}{\partial y} + o(|x| + |y|)$$

$o(|x| + |y|)$ is small in the C^1-sense if ϵ_1 is chosen small enough once ω_1 and ω_2 are chosen.

Differentiating, we find for the $\tilde{\xi}$-transport matrix:

$$\dot{\delta x} = -2\nu\omega_1\omega_2 C_1\delta x + o(|\delta x| + |\delta y|)$$

$$\dot{\delta y} = 2\nu\omega_1\omega_2 C_1\delta y + o(|\delta x| + |\delta y|)$$

with $\omega_1 = 1$ on the piece of $\tilde{\xi}$ or ξ-orbit.

We thus see that the transport matrix of ξ has been modified into the transport matrix of $\tilde{\xi}$ which is:

$$A = \begin{pmatrix} C & 0 \\ 0 & \frac{1}{C} \end{pmatrix}$$

C can be chosen as we please after an appropriate choice of C_1 and ν in function of ω_2 and C.

With this new $\tilde{\xi}$, $\beta = d\tilde{\alpha}(v, .)$ is not necessarily a contact form anymore. It is easy however to modify v into $\tilde{h}v$ so that $\tilde{\beta} = d\tilde{\alpha}(\tilde{v}, .)$ is a contact form with the same orientation than α. Indeed, the transport map of $\tilde{\xi}$ is essentially a contraction or dilation along $\frac{\partial}{\partial x}$ and a dilation or contraction along $\frac{\partial}{\partial y}$.

We first, before any modification of α into $\tilde{\alpha}$, unwind v along the piece of ξ-orbit where we introduce the modification, redistributing the rotation of v so that v builds with ξ on the piece of ξ-orbit a foliation. We take v to be $\frac{\partial}{\partial x}$. The transport map of $\tilde{\xi}$ maps $\frac{\partial}{\partial x}$ onto $\gamma\frac{\partial}{\partial x}$, γ positive. We thus can reintroduce some missing rotation and turn β into $\tilde{\beta}$ a contact form.

In the case when $C \geq 1$, i.e the modification has introduced a dilation along $\frac{\partial}{\partial x} = "v"$ (this is v at the beginning of the ξ-piece), we can take $\tilde{v} = v$ i.e keep v unchanged. Indeed, we compute:

$$d\tilde{\alpha}(v, [\tilde{\xi}, v]) = d\alpha(v, [\frac{\partial}{\partial z} - 2\nu\omega_1\omega_2 C_1 x\frac{\partial}{\partial x} + 2\nu\omega_1\omega_2 C_1 y\frac{\partial}{\partial y}, v]) + o(1)$$

v reads as

$$v = \frac{\partial}{\partial x} - zb(x, y)(\frac{\partial}{\partial y} - x\frac{\partial}{\partial z})$$

if we set $z = 0$ at the beginning of the ξ-piece where v equals $\frac{\partial}{\partial x}$. $b(x, y)$ is a positive smooth function of y and x. The above formula is obtained using the equation $d\alpha(v, [\xi, v]) = -1$; the Darboux coordinates are built so that $v = \frac{\partial}{\partial x}$ on a section to $\frac{\partial}{\partial z}$ at the starting point of the ξ-piece. We find:

$$[\tilde{\xi}, v] = o(1) + [\frac{\partial}{\partial z} - 2\nu\omega_1\omega_2 C_1(x\frac{\partial}{\partial x}), \frac{\partial}{\partial x} - zb(x, y)(\frac{\partial}{\partial y} - x\frac{\partial}{\partial z})]$$

$$= o(1) - b(x, y)\frac{\partial}{\partial y} + 2\nu\omega_1\omega_2 C_1 \times (\frac{\partial}{\partial x} + b(x, y)\frac{\partial}{\partial y})$$

and

$$d\alpha(\frac{\partial}{\partial x} - b(x, y)z\frac{\partial}{\partial y}, -b(x, y)\frac{\partial}{\partial y} + 2\nu\omega_1\omega_2 C_1(\frac{\partial}{\partial x} + b(x, y)z\frac{\partial}{\partial y}))$$

$$= -b(x, y) + 2\nu\omega_1\omega_2 zC_1(1 + b).$$

Since C_1 is negative when $C \geq 1$, $\tilde{\beta} = d\tilde{\alpha}(v, .)$ is indeed a contact form as claimed.

Modifications of the above type, the first one with $\frac{1}{C}$ and the second one with C, are introduced on the first nodal zone, in the vicinity of the left edge, and on the second last nodal zone (x_{m-2}, x_{m-1}), in the vicinity of x_{m-2}, see Sections 2.5.4 and 2.5.5 for the related uses and notations. v is mapped onto $\bar{\theta}v$ from x_{m-2} to the left edge. We can use a first couple of modifications of this type and modify $\bar{\theta}$ into 1. It is easy to see that we do not create new critical points at infinity over this process. Thus we assume in what follows that $\bar{\theta} = 1$. Then, $\frac{\partial}{\partial x}$ can be viewed as a ξ-transported vector from the left edge until x_{m-2} that coincides with v at the left edge and at x_{m-2}. The x, y, z-Darboux coordinates extend also from the left edge until x_{m-1}. The two modifications are thus written in the same set of coordinates. For $x^2 + y^2$ small with respect to ϵ_1, using the formulae of [Bahri-1 2003, p. 83], we find that $\dot{z}$ for $\tilde{\xi}$ is equal to 1 and that the maps of $\tilde{\xi}$ on each interval are linear given by matrices such as A, see above. The error in this approximation is precisely $O(|x|^3 + |y|^3)$. It is simple then to arrange the values of the constants such as C_1, ν and the functions such as ω_2 on each interval modification so that the total $\tilde{\xi} - \xi$-transport map from the middle of the nodal zone (x_{m-2}, x_{m-1}) , or points close to this middle point on curves close to the critical point at infinity, to the left edge of these curves is the identity map in the (x, y)-coordinates, i.e is simply z-translation, of the same amount than the ξ-translation on the original curve, up to an error of size $O(|x|^3 + |y|^3)$. This holds of course under the assumption that there are no $\pm v$-jump on the way.

Assuming now that there are $\pm v$-jumps on the way, but that they are all located in the vicinity of nodes and of total size Σc_i, assuming also for simplicity that we have completed a family of previous modifications so that all the related coefficients $\bar{\theta}_i$ from all nodes to the left of x_{m-2}, x_{m-2} included, are equal to 1, these $\pm v$-jumps are then translations in the x-direction of c_i at the nodes which we can define on each ξ-piece corresponding to the characteristic piece, up to $o(c_i)$. We are assuming that there are no $\pm v$-jumps in the areas where the modifications are completed. The matrices of type A "commute" with such translations, i.e the total transport map is then the ξ-transport map composed with a translation along v at the left edge of size $C\Sigma\theta_i$, up to $o(\Sigma|\theta_i|)$.

We can also, instead of replacing the ξ-piece of curve, with the $\pm v$-jumps inserted in it, by a $\tilde{\xi}$-piece of curve, with all $\pm v$-jumps gathered at the left edge (up to $o(\sigma c_i)$), insert back $\pm v$-jumps at the similar locations on the piece of $\tilde{\xi} - \xi$-curve. A $\pm v$-jump of size c should then be replaced by a $\pm v$-jump of size $\frac{1}{C}$. We can also complete the process in the reverse way, starting with a $\tilde{\xi} - \xi$-piece of curve.

In all, a curve x, close to x_∞ is replaced by a curve $\tilde{x}$. $\tilde{x}$ might not be closed, up to $o(\Sigma c_i)$, but we can close it, using the left edge for example where we have the freedom of using variations along v, variations along ξ and also variations along the transport of ξ from the bottom to the top of the left edge.

There are two functionals involved, J and $\tilde{J}$, corresponding respectively to α and ξ on one hand and $\tilde{\alpha}$ and $\tilde{\xi}$ on the other hand. As we change from x to $\tilde{x}$, our estimates above indicate that:

$$J(x) = \tilde{J}(\tilde{x}) + o(\Sigma c_i)$$

The $c_i's$ involved are the sizes of the $\pm v$-jumps to the left of the middle of the nodal zone (x_{m-2}, x_{m-1}), they are all supposed to be close to the location of the "nodal" points in our estimate above. $\qquad\square$

2.6 Transmutations

In this section we study in more details transmutations of critical points at infinity and the related changes of the indexes at infinity.

Transmutations of a critical point at infinity x^∞ have been introduced in [Bahri-1 2003] following two distinct procedures. They correspond to the following basic fact: given a critical point at infinity x^∞, of index $i_0 + i_\infty = m$ (i_0 is the H_0^1-index, i_∞ is the index at infinity), the maximal number of zeros of b on the unstable manifold of x^∞ is $i_0 + \gamma$ see [Bahri-1 2003]. $i_0 + \gamma$ and $i_0 + i_\infty$ are different. Therefore, one can hope to change $i_0 + \gamma$ through a modification of α into $\lambda\alpha$ near x_m^∞ while $i_0 + i_\infty$ remains unchanged i.e. x_m^∞ remains isolated in its species.

This has been carried out in [Bahri-1 2003] either by introducing a suitable "Hamiltonian" λ in the neighborhood of a non-degenerate ξ-piece of x_m^∞ or by redistributing the v-rotation along the non-degenerate ξ-pieces without changing the Poincaré-return map of x_m^∞, C. The phenomenon is subtle because as explained in [Bahri-1 2003], as this transmutation occurs, x_m^∞ remains isolated in its species but an additional critical point at infinity y^∞ (in fact a couple or more) is created which has an additional characteristic ξ-piece with respect to x_m^∞ (corresponding to the non degenerate ξ-piece where the transmutation is occuring).

We will be particularly interested in this section to x_{2k}^∞'s i.e. critical points at infinity of index $2k$ having an index $i_\infty = k$ as these are, see [Bahri-1 2003], the critical points at infinity which interfere with the homology defined in [Bahri-1 2003].

Assuming hypothesis (A), either x_m^∞ has no characteristic piece. If the number of its ξ-pieces is large, we can redistribute the v-rotation on one of them. We change i_∞ to something different as we will see. When $m = 2k, x_{2k}^\infty$ does not interfere with our homology anymore. We have created through this process other critical points at infinity y^∞'s, but they all have at least one characteristic ξ-piece and we may apply to them our compactness results (after full generalization, we are providing here our program of work).

If x_m^∞ has a characteristic piece, then the compactness results can be directly applied to it.

Our compactness results, as they stand, require that one of the characteristic pieces of x_m^∞ or y^∞ be of strict index different from 1. We believe that they hold in fact in full generality ([Bahri-2 2003]).

Our focus in this section is to understand in details the changes of indexes as a transmutation takes place. We will assume, for the sake of the precision in the details, that $m = 2k, i_\infty = k$; however the results which we establish hold in full generality, for all generic transmutations.

We consider such a transmutation which necessarily involves the collapse (see [Bahri-1 2003]) of x_{2k}^∞ with y^∞ another critical point at infinity having the same number of ξ and $\pm v$-pieces than x_{2k}^∞.

We start with:

Proposition 34 *As x_{2k}^∞ and y^∞ collapse, we may assume that $da_c(v) \neq 0$ on the ξ-piece where the change of H_0^1-index takes place. $d\theta(\xi)$, the differential along ξ of the collinearity coefficient θ of $d\varphi_{a_c}(v)$ on v, can also be perturbed as we please at the time of the collapse at x_0^-, while $\theta(x_0^-) = \bar{\theta}$ is kept unchanged.*

Observation 6. As we pile up rotation on a ξ-piece, bringing it from other free ξ-pieces see [Bahri-1 2003] pp 96-102, the full Poincaré-return map does not change. Let $\bar{a}$ be the length of the ξ-piece. $d\phi_{\bar{a}}(v)$ and $-v$ are closer and closer in direction as we pile up rotation (keeping it less than π). Recalling now the other way of transmuting x_{2k}^∞ i.e. of changing the H_0^1-index of this ξ-piece from 0 to 1, we may complete this transmutation once the rotation is almost π through a change of the Hamiltonian see [Bahri-1 2003] pp. 81-102. Since C has not changed in the first step, since $d\phi_{\bar{a}}(v)$ and $-v$ are then very close, we may assume by genericity that the forbidden direction(s), see [Bahri-1 2003] pp 85-90, do not lie between $d\phi_{\bar{a}}(v)$ and $-v$. Thus, we may assume that, at the time of the collapse, α is unchanged outside of a small neighborhood of the ξ-piece which changes H_0^1-index. We assume in the sequel that the transmutation takes place in this way.

Corollary 8 *Either $i_\infty \geq k + 1$ or the direction $\tilde{v}_k$ on the characteristic piece of y^∞ corresponding to the ξ-piece of x_{2k}^∞ which changes H_0^1-index is in the negative eigenspace of $J_\infty''(y_\infty)$ after the collapse while the index of $J_\infty''(y_\infty)$ on the $J_\infty''(y_\infty)$-orthogonal of $\tilde{v}_k$ does not change.*

Observation 7. As the collapse, $J_\infty''(y_\infty) \cdot \tilde{v}_k \cdot \tilde{v}_k = 0$ as we will see. Thus, $\tilde{v}_k$ is orthogonal to itself. However, we will prove that $\tilde{v}_k$ has a codimension one orthogonal (for $J_\infty''(y_\infty)$) which can be followed continuously through the collapse.

Proof. [Proof Corollary 8 and of Proposition 34] As we add rotation, the index of y^∞ will change from $i_\infty - 1$ to i_∞ through the collapse, at least if y^∞ remains isolated in its type. This will follow from the behavior of $J_\infty''(y_\infty)$ transversally to $\tilde{v}_k$.

If we are able to define continuously an orthogonal G to $\tilde{v}_k$, through the collapse, transversally to $\tilde{v}_k$, and if $J''(y_\infty)|_G$ does not degenerate, then

y_∞ will be isolated in its species. Indeed, since $da_c(v) \neq 0, \tilde{v}_k$ which is the only direction of degeneracy, is transverse to the characteristic manifold and $y_\infty + \varepsilon \tilde{v}_k$ has not the corresponding ξ-piece as characteristic.

We therefore have now to understand the behavior of the tangent space at y_∞ and of $J''(y_\infty)$ on this tangent space.

We draw the degenerate ξ-piece of y^∞ which is collapsing with the free piece of x^∞_{2k}.

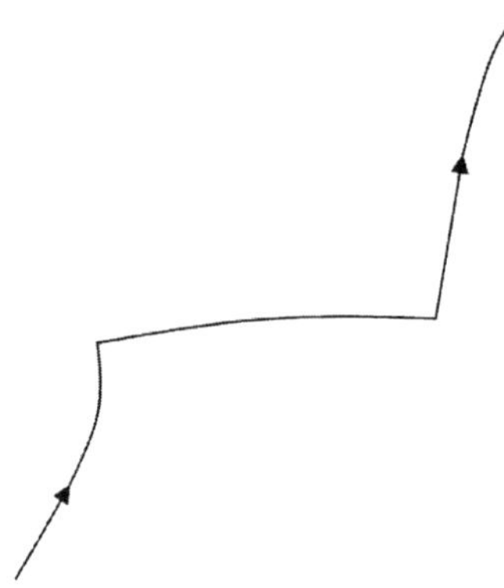

We define, aside of $\tilde{v}_k$, three additional tangent vectors having support on this ξ-piece

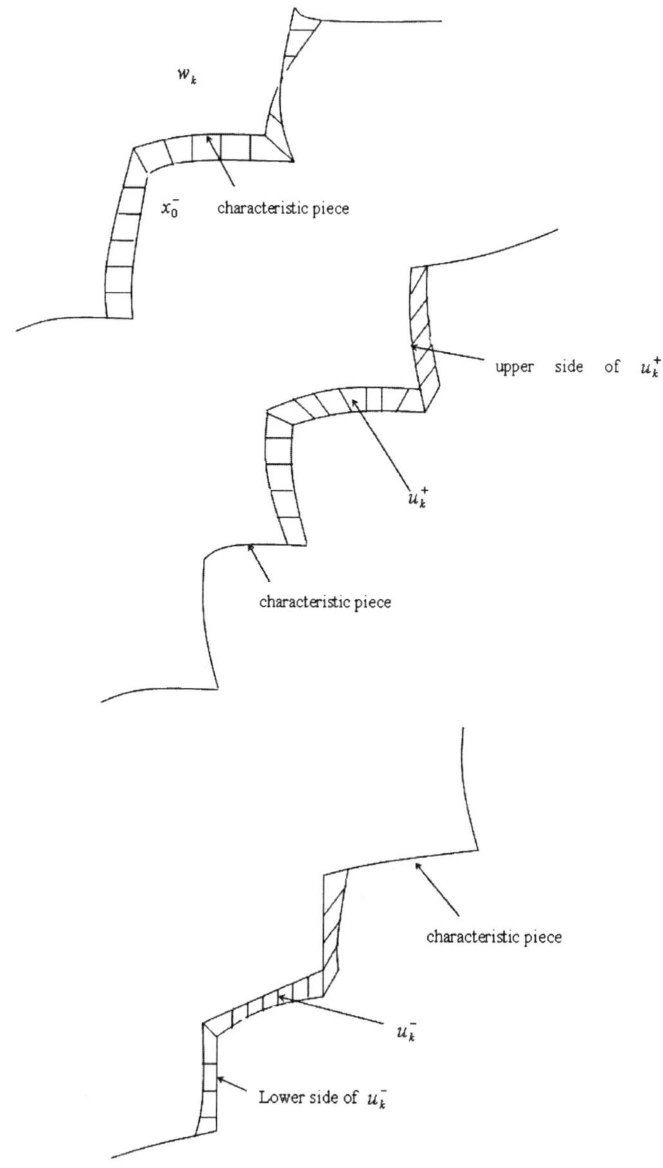

We continue building up tangent vectors using each ξ-piece and its neighboring ξ-pieces. If it is characteristic, it has a vector such as u_k^+, u_k^- and a $\tilde{v}$. If it is free, the upperside of a vector such as u_k^+ and the lower side of a vector such as u_k^- can be removed and we derive two vectors u^+, u^-. This builds a base of $T_{y\infty}(\Gamma_{2s})(y_\infty \in \Gamma_{2s})$. Clearly, $\tilde{v}_k$ interacts with w_k, u_k^+, u_k^-. The other vectors do not have support on the characteristic piece of y_∞. Orthogonality for $J''_\infty(y_\infty)$ follows. Let us draw curves of Γ_{2s} nearby y_∞:

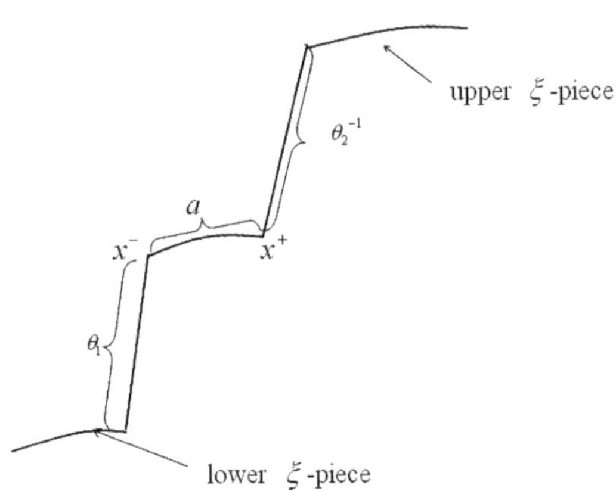

θ_1 and θ_2^{-1} are the v-transport maps corresponding to the edge v-jumps of the curve. We write:

$$d\theta_1(\xi) = (1 + A_1)\xi + B_1[\xi, v] + \mu_1 v$$
$$d\phi_a(v) = -\theta v + c(x^-)(a - a_c(x^{-1})[\xi, v]$$

with $c(x^-)$ bounded away from zero.

$$d\theta_2^{-1}(\xi) = (1 + A_2)\xi + B_2[\xi, v] + \mu_2 v \qquad d\phi_{\bar{a}_c}(v(x_0^-)) = -\bar{\theta}v(x_0^+).$$

Accordingly, $\tilde{v}_k, w_k, u_k^+, u_k^-$ find natural extensions to nearby curves. In order to extend $\tilde{v}_k$, we take v at x^-, transport it via ξ into $-\theta v + c(x^-)(a - a_c(x^-))[\xi, v]$ at x^+ and we compensate the component on $[\xi, v]$ of $d\varphi_a(v)$ by $d\theta_2^{-1}\left(\frac{c(x^-)(a - a_c(x^-))}{B_2}\xi\right)$. We thus have at a nearby curve $x(\tilde{v}_k$ is also the notation for the extension of $\tilde{v}_k$):

$$J'_\infty(x) \cdot \tilde{v}_k = \frac{A_2}{B_2}c(x^-)(a - a_c(x^-)).$$

Observation 8. At the collapse, $A_2 = A_1 = 0$ on y^∞ because x_{2k}^∞ which is then confounded with y^∞ has this ξ-piece free.

On $y^\infty, a = a_c$. Therefore, any tangent vector $\varphi \in T_{y_\infty}\Gamma_{2s}$

$$J''_\infty(y_\infty) \cdot \tilde{v}_k \cdot \varphi = \frac{A_2}{B_2} c(x_0^-)(\delta a - \delta a_c)(\varphi)$$

and the formula holds through the collapse.

We thus may define the $J''_\infty(y_\infty)$ - orthogonal of $\tilde{v}_k$ to be:

$$G = \ker(\delta a - \delta a_c).$$

Observe that

$$(\delta a - \delta a_c)(\tilde{v}_k) = -da_c(v) \neq 0.$$

It is therefore easy to project w_k, u_k^+, u_k^- orthogonally on G; we use the formula:

$$\varphi + \frac{(\delta a(\varphi) - \delta a_c(\varphi))\tilde{v}_k}{da_c(v)}.$$

We denote these vectors $\tilde{w}_k, \tilde{u}_k^+, \tilde{u}_k^-$. they are independent and we add to them $\tilde{v}_k$, we find the same span for this family of vectors.

$J''_\infty(y_\infty)$ takes then the simple form through the collapse:

$$\begin{matrix} & \tilde{v}_k & & \\ & \begin{pmatrix} c_k & 0 \dots & & 0 \\ 0 & & & \\ \vdots & & B & \\ 0 & & & \end{pmatrix} \end{matrix}$$

c_k is equal to $-\frac{A_2}{B_2} c(x_0^-)da_c(v)$.

Since $\tilde{v}_k$ is orthogonal to itself at the collapse, we might as well compute B using w_k, u_k^+, u_k^- and the other vectors having no support on the ξ-piece where the tranmutation takes place.

B reads:

$$\begin{pmatrix} J''_\infty(y_\infty) \cdot w_k \cdot w_k & J''_\infty(y_\infty) \cdot w_k \cdot u_k^+ & J''_\infty(y_\infty) \cdot w_k \cdot u_k^- & \\ J''_\infty(y_\infty) \cdot w_k \cdot u_k^+ & J''_\infty(y_\infty) \cdot u_k^+ \cdot u_k^+ & J''_\infty(y_\infty) \cdot u_k^+ \cdot u_k^- & C \\ J''_\infty(y_\infty) \cdot w_k \cdot u_k^- & J''_\infty(y_\infty) \cdot u_k^+ \cdot u_k^- & J''_\infty(y_\infty) \cdot u_k^- \cdot u_k^- & \\ & C & & B_1 \end{pmatrix}.$$

We may compute B at the collapse, when $y_\infty = x_{2k}^\infty$. Outside of the ξ-piece where the transmutation takes place, y_∞ and x_{2k}^∞ and their partial Poincaré-return maps (those not involving this ξ-piece) coincide.

The second step of the transmutation can be assumed to take place away from the edges of the ξ-piece of x_{2k}^{∞}. Thus, if we think of all the tangent vectors, outside of w_k, as defined for x_{2k}^{∞} at the collapse and also slightly before and after the collapse, we derive that B_1, C and also $J_{\infty}''(x_{2k}^{\infty}) \cdot u_k^{\pm} \cdot u_k^{\pm}$ and $J_{\infty}''(x_{2k}^{\infty}) \cdot u_k^{+} \cdot u_k^{-}$ are constant through the collapse. At the collapse, they identify with the corresponding part of $J_{\infty}''(y_{\infty})$. We are going to compute $J_{\infty}''(y_{\infty}) \cdot w_k \cdot \varphi$ and show that we can stabilize $J_{\infty}''(y_{\infty}) \cdot w_k \cdot w_k$ and $J_{\infty}''(y_{\infty}) \cdot w_k \cdot u_k^{\pm}$ near the collapse. The entire matrix B will therefore change very little through the collapse. Because we can perturb a little bit $\bar{\theta}$ at the collapse, we will see that we may assume that $\det B \neq 0$ at the collapse and nearby, which is our claim.

We compute now $J_{\infty}''(y_{\infty}) \cdot w_k \cdot \varphi$:

We need to know the value of $d\phi_{a_c}([\xi, v])(x^-)$. We have:

$$d\phi_a(v) = -\theta v + c(x^-)(a - a_c(x^-))[\xi, v]$$

which we differentiate along ξ to get:

$$d\phi_a([\xi, v]) = -d\theta(\xi)v + \gamma[\xi, v] + (a - a_c)A.$$

Thus,

$$d\phi_{a_c}([\xi, v]) = -d\theta(\xi)v + \gamma[\xi, v].$$

Since

$$d\alpha(d\phi_a(v), d\phi_a([\xi, v])) = -d\alpha(v, [\xi, v]),$$

γ is equal to $-\frac{1}{\theta}$ and

$$d\phi_{a_c}([\xi, v]) = -\frac{1}{\theta}[\xi, v] - d\theta(\xi)v.$$

We use the following figure.

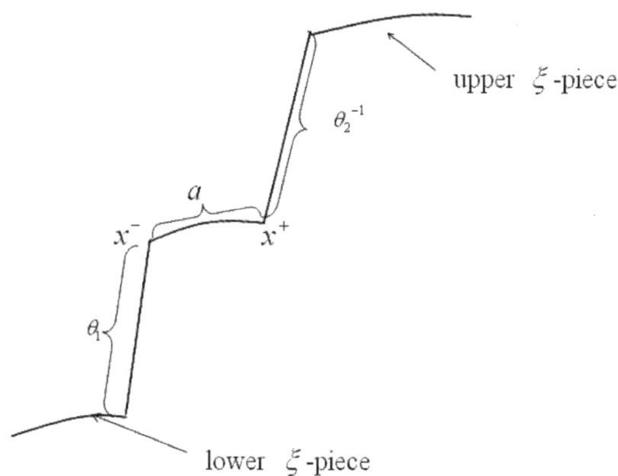

to support our construction.

In order to build w_k, we transport ξ using $d\theta_1$ from the lower ξ-piece. At x^- we derive:

$$(1 + A_1)\xi + B_1[\xi, v] + \mu v.$$

We use $d\phi_a$ to transport this vector to x^+. We derive at x^+ a vector equal to:

$$(1 + A_1)\xi - B_1 \left(\frac{1}{\theta}[\xi, v] + d\theta(\xi)v \right) - \mu_1 \theta v.$$

We pull back $-\frac{B_1}{B_2\theta}\xi$ from the upper ξ-branch using $d\theta_2^{-1}$. We derive at x^+ the vector:

$$-\frac{B_1}{B_2\theta}(1 + A_2)\xi - \frac{B_1}{B_2\theta}[\xi, v] - \frac{\mu_2 B_1}{B_2\theta} v.$$

The $[\xi, v]$- components of the above two vectors match. Using v-transport, we can adjust the v-component of $d\theta_2^{-1}\left(-\frac{B_1}{B_2\theta}\xi\right)$ to be $-\mu_1\theta$. Using ξ-transport, we can adjust the ξ-component of $d\phi_a \circ d\theta_1(\xi)$ to be $-\frac{B_1}{B_2\theta}(1 + A_2)$. We then have ($w_k$ is the tangent vector which we have built):

$$J'_\infty(x) \cdot w_k = A_1 - \frac{A_2 B_1}{B_2\theta}.$$

Since $A_1 = A_2 = 0$ when $y_\infty = x_\infty^{2k}$, we have:

$$J_\infty''(y_\infty) \cdot w_k \cdot \varphi = dA_1(\varphi(x_0^-)) - \frac{\bar{B}_1}{\bar{B}_2 \bar{\theta}} dA_2(\varphi(x_0^+)).$$

If we take φ to be $u_k^\pm$, $\varphi(x_0^\pm)$ is independent of the transmutation process and so is therefore $J_\infty''(x_{2k}^\infty) \cdot w_k \cdot u_k^\pm$ because dA_1 and dA_2 are given on x_{2k}^∞. Since $J_\infty''(y_\infty) \cdot w_k \cdot u_k^\pm$ equals $J_\infty''(x_{2k}^\infty) \cdot w_k \cdot u_k^\pm$ at the collapse, we see that the only term in B which depends on the transmutation (i.e. cannot be computed on x_{2k}^∞ also nearby the transmutation) is $J_\infty''(y_\infty) \cdot w_k \cdot w_k$, which is equal to:

$$dA_1(w_k(x_0^-)) - \frac{\bar{B}_1}{\bar{B}_2 \bar{\theta}} dA_2(w_k(x_0^+))$$

$w_k(x_0^-)$ is independent of the transmutation process. But $w_k(x_0^+)$ is equal, according to our computations above, to:

$$-\frac{\bar{B}_1}{\bar{B}_2 \bar{\theta}} \xi - \frac{\bar{B}_1}{\bar{\theta}} [\xi, v] - v \left(\bar{B}_1 d\theta(\xi) + \mu, \bar{\theta} \right)$$

which depends on the transmutation process via the two quantities $\bar{\theta}$ and $d\theta(\xi)$. dA_1 and dA_2 are independent of the transmutation process. Computing $\det B$, we find that it reads:

$$\det B = A_0 d\theta(\xi) + \frac{B_0}{\bar{\theta}} + C_0 \bar{\theta} + D_0,$$

where A_0, B_0, C_0, D_0 are independent of the transmutation process near the collapse i.e. can be computed on x_{2k}^∞ at the collapse and nearby. They are equal to the corresponding quantities for y_∞ at the collapse. Using the results of Part 6 below, which provide us with a large set of free parameters, at least $2(s-1)$ if x_{2k}^∞ and y_∞ are in Γ_{2s}, we may assume that A_0, B_0, C_0 and D_0 are non zero.

Let us prove that we can change $s\theta(\xi)$ while $\bar{\theta}$ stays unchanged. θ depends on the ξ-transport and of the values $v(x_0^\pm)$. If these values are unchanged, $\bar{\theta}$ is unchanged if $d\alpha(v, [\xi, v])$ remains equal to -1. To track the parameters which we can modify, we follow the framework introduced in Part 6 (below). We pick up Darboux coordinates near x_0^+ where α reads $ydz + dx$ and we transport them along $-\xi$ to x_0^- at the time of the collapse. ξ is therefore $\frac{\partial}{\partial x}$.

We can use at x_0^+ special Darboux coordinates where v reads:

$$v = \lambda \left(\frac{\partial}{\partial z} - y \frac{\partial}{\partial x} + (a_1 x + b_1 y + p_2) \frac{\partial}{\partial y} \right)$$

p_2 being of order 2 or more in (x, y, z) v near x_0^- then reads:

$$v = \mu \left(\frac{\partial}{\partial z} - y \frac{\partial}{\partial x} + (ax + b y_c z + q_2) \frac{\partial}{\partial y} \right).$$

Again, q_2 is of order 2 or more in (x, y, z).

It is easy to see that

$$da_c(v)(x_0^-) = \frac{c}{a_1}$$

a_1 is certainly non zero because

$$-1 = da(v, [\xi, v])$$

$$= \lambda^2 dy \wedge dz \left(\frac{\partial}{\partial z} - y \frac{\partial}{\partial x} + (a_1 x + b_2 y + p_2) \frac{\partial}{\partial y}, \left(a_1 + \frac{\partial p_2}{\partial x} \right) \frac{\partial}{\partial y} \right)$$

$$= -\lambda^2 \left(a_1 + \frac{\partial p_2}{\partial x} \right).$$

Also

$$-1 = -\mu^2 \left(a + \frac{\partial q_2}{\partial x} \right).$$

We may clearly assume, after a C^1-small perturbation of v, with no change in $v(x_0^\pm)$ that $c \neq 0$. Thus, we may assume as claimed in the statement of Proposition 34 that $da_c(v)(x_0^-)$ is non zero at the collapse.

We need to compute $\theta(x, 0, 0)$ at first order, with $\theta(0, 0, 0,) = \theta(x_0^-) = \bar{\theta}$. At $(\delta x, 0, 0)$, v is collinear to:

$$\frac{\partial}{\partial z} + (a \delta x + c_2 \delta x^2) \frac{\partial}{\partial y}.$$

We transport it along $\frac{\partial}{\partial x}$ near $x_0^+ = (0, 0, 0)$. It is collinear to $v(\delta x', 0, 0)$ if

$$a_1 \delta x' + \bar{c}_2 \delta x'^2 = a \delta x + c_2 \delta x^2.$$

Thus,

$$\delta x' = \frac{a}{a_1} \delta x (1 + O(\delta x))$$

and

$$\theta(\delta x, 0, 0) = \frac{\lambda(\delta x, 0, 0)}{\mu(\delta x', 0, 0)} = \frac{\bar{\lambda} + \frac{\partial \lambda}{\partial x}(0, 0, 0) \delta x + O(\delta x^2)}{\bar{\mu} + \frac{\partial \mu}{\partial x}(0, 0, 0) \frac{a}{a_1} \delta x + O(\delta x^2))}.$$

Thus,

$$\theta(\delta x, 0, 0) = \bar{\theta}\left(1 + \left(\frac{1}{\lambda}\frac{\partial\lambda}{\partial x}(0) - \frac{1}{\bar{\mu}}\frac{\partial\mu}{\partial x}(0)\right)\delta x + O(\delta x^2)\right).$$

It follows that

$$\frac{d\theta(\xi)(0)}{\theta} = \frac{1}{\bar{\lambda}}\frac{d\lambda}{\partial x}(0) - \frac{1}{\mu}\frac{\partial\mu}{\partial x}(0).$$

Since $\lambda = \dfrac{1}{\sqrt{a_1 + \frac{\partial p_2}{\partial x}}}$ and $\mu = \dfrac{1}{\sqrt{a + \frac{\partial q_2}{\partial x}}}$,

$$\frac{d\theta(\xi)}{\theta}(0) = -\frac{1}{2a_1}\frac{\partial^2 p_3}{\partial x^2}(0,0,0) + \frac{1}{2a}\frac{\partial^2 q_2}{\partial x^2}(0,0,0).$$

Changing $\frac{\partial^2 p_2}{\partial x^2}(0,0,0)$ or $\frac{\partial^2 q_2}{\partial x^2}(0,0,0)$ is a C^1-small perturbation of v which does not affect $v(x_0^{\pm})$. Proposition 34 and Corollary 8 follow. $\Box$

Let us now consider x_{2k}^{∞} made of a cycle which uses k (half)-full unstable of characteristic pieces in its definition.

Using the results of Part 6, we may assume since $i_{\infty} = k$ and since x_{2k}^{∞} has at least $k+1$ characteristic pieces $(\ell = k)$, that $\tilde{v}_1, \ldots, \tilde{v}_{k-1}$ are in the negative eigenspace of $J_{\infty}''(x_{2k}^{\infty})$. At the collapse, $\tilde{v}_1, \ldots, \tilde{v}_{k-1}$ are therefore in the negative eigenspace of $J_{\infty}''(y_{\infty})$ and after the collapse, $\tilde{v}_k$ is also in this negative eigenspace for y^{∞}, see Proposition 35, below. We then have:

Corollary 9   *Assume that hypothesis (A) holds on y^{∞}. Then $i_{\infty} \geq k+1$.*

Proof.   If hypothesis (A) holds on y^{∞}, we can carry rotation on an additional free piece (of y^{∞}, this time) and change the H_0^1-index by 1. We create then another critical point at infinity z^{∞}, of index $i_{\infty} \cdot z^{\infty}$ has $\tilde{v}_1, ..., \tilde{v}_k$ in its negative eigenspace (because it collapses with y_{∞}) and also, by Proposition 35 below, after the collapse, an additional $\tilde{v}_{k+1}$. The result follows.

$\Box$

We now establish that, as we pile up rotation, the index at infinity of y_{∞} increases. We prove this result under the assumption that the H_0^1-index of the ξ piece of x_{2k}^{∞} to which v-rotation is added is initially zero. This assumption can be easily removed.

Proposition 35   *If the H_0^1-index on a non degenerate ξ-piece of x_{2k}^{∞} changes from zero to 1, then x_{2k}^{∞} collapses with another critical point at infinity y^{∞} having the corresponding piece as a degenerate ξ-piece of index zero. The index at infinity of x_{2k}^{∞} changes from i_{∞} to $i_{\infty} - 1$ and the index at infinity of y^{∞} changes from $i_{\infty} - 1$ to i_{∞}.*

Proof. The phenomenon has been described and discussed in [Bahri-1 2003] pp 129—130 and pp 133—136.

Our proof here sets the description of [Bahri-1 2003] on a rigorous standing. The isolation of x_{2k}^{∞} has been proved in [Bahri-1 2003, Proposition 16]. The existence of at least one y^{∞} follows. We need however to show that the index at infinity of x_{2k}^{∞} changes from i_{∞} to $i_{\infty} - 1$, that y_{∞} is unique (collapsing with x_{2k}^{∞}) and that its index at infinity changes from $i_{\infty} - 1$ to i_{∞}.

We are going to define a space $\mathcal{W}_{2s+1}$, a manifold of dimension $2s + 1$ to which x^{∞} belongs because Γ_{2s} near x^{∞} is embedded in $\mathcal{W}_{2s+1}$ in a natural way.

In $\mathcal{W}_{2s+1}, x^{\infty}$ which is the critical point at infinity with an additional degenerate ξ-piece collapsing with x_{∞}, will be degenerate.

x_{∞} has index i_{∞} before the collapse. Through the collapse, all other critical points at infinity created in the process are degenerate in C_{β} and will be so in $\mathcal{W}_{2s+1}$. Since an H_0^1-index equal to 1 has been created on x^{∞}, one of the negative directions of x^{∞} is tranverse to Γ_{2s}. Thus, the index of x^{∞} in Γ_{2s} is $i_{\infty} - 1$ after the collapse.

The space $\mathcal{W}_{2s+1}$ is defined as follows: x^{∞} belongs to Γ_{2s}, hence has $s \pm v$-verticals. Two consecutive v-verticals are special here because it is on the ξ-piece lying between them that the deformation of α will be completed. As it stands i.e. before this process is started, the ξ-piece between these two verticals is non degenerate, of H_0^1-index zero. Let us denote $\bar{O}_1, \bar{O}_2$ these two verticals and let us consider all neighboring v-verticals O_1, O_2. If these are close enough to $\bar{O}_1, \bar{O}_2$, there is a unique ξ-piece joining them in the neighborhood of the ξ-piece of x^{∞}. Let us denote $y(t, O_1, O_2) = y(t)$ this ξ-piece. Assume that $y(t)$ is parametrized by $[0,1]$, pick up a time t_0, for example $t_0 = \frac{1}{2}$. Consider the orbit $z_s, s \in (-\varepsilon, \varepsilon), \varepsilon$ uniform and small, of the vector-field w through $y(t_0)$. For each s, we can consider the v-orbit through z_s, which we denote O_3^s.

We then have $2s + 1$ verticals which are the $2s$ former v-verticals close to the v-orbits of x^{∞}, to which O_3^s has been added.

O_3^s is not as free to move as are O_1 and O_2 since it is constrained to move along the prescribed orbit of w.

We then join all these v-verticals by ξ-pieces close to those of x^{∞}.

If we remove O_3^s, we get a neighborhood of x_{∞} in Γ_{2s}. If we now replace, for s given, the ξ-piece between O_1 and O_2 with a ξ-piece between O_1 and O_2, we derive a manifold of dimension $2s$. If s varies now, we find a manifold of dimension $2s + 1$.

Indeed, the ξ-piece between O_1 and O_2 is unique. Thus, the dimension of the space of curves made of ξ and v-pieces between the assigned verticals even after the ξ-piece between O_1 and O_2 is removed is $2s$. O_1, O_2 being given, z^s is unique because the ξ-piece between O_1 and O_2 is unique, thus the family O_3^s is given and we have an additional dimension.

The space thus defined is a manifold of dimension $2s + 1$.

Along our process, ξ and α change into $\xi_\tau, \alpha_\tau, \tau \in [0,1]$. The construction of z^s, O_3^s does not change: it is performed for ξ_0, α_0, w_0. However $\mathcal{W}_{2s+1}$ is in fact $\mathcal{W}_{2s+1}^\tau$ i.e. is built using ξ_τ. Since the ξ_τ-piece of x^∞ has, all along the process, an index at most equal to 1, with a rotation of v only slightly larger than π at most, there are always two unique ξ_τ-pieces joining O_1 and O_3^s on one hand and O_3^s and O_2 on the other hand. Thus, $\mathcal{W}_{\in f + \infty}^\tau$ is a manifold near x^∞, of dimension $2s + 1$, all along the process. We remove in the sequel the index τ for convenience.

All the critical points of J_∞ in $\mathcal{W}_{2s+1}$ are in fact in Γ_{2s}. Indeed, suppose that we have a small v-jump along $O_3^{s_0}$ on a critical point z^∞ of J_∞ near x^∞:

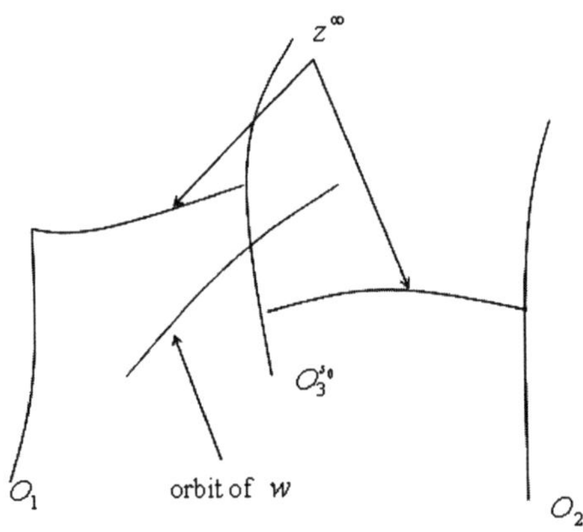

Let us consider variations s near s_0 and let us consider all curves y which have the same graph than z^∞ outside of O_1 and O_2 and are built between O_1 and O_2 by moving O_3^s.

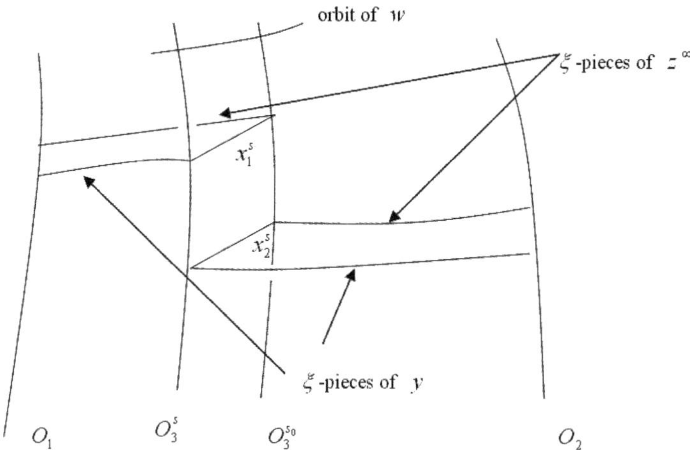

x_2^s is derived from x_1^s by v-transport since both curves join the v-orbits $O_3^{s_0}$ and O_3^s as s varies continuously near s_0.

Clearly,

$$J_\infty(y) - J_\infty(z^\infty) = \int_{x_2^s} \alpha - \int_{x_1^s} \alpha$$

using Stokes formula (as s varies, we span surfaces to which ξ is tangent and which are bounded by the ξ-pieces of y and z^∞, v-pieces along O_1 and O_2, and x_1^s, x_2^s. x_1^3 is derived after ξ-transport from a piece of v-orbit along O_1). We may assume that α is not changed on the first portion of the ξ-piece of x^∞ so that x_1^s has a non-zero component along w. The transport equations along v then read:

$$\overline{\dot{\lambda} + \bar{\mu}\eta} = \eta.$$

Since x_2^s is derived from x_1^s by v-transport and η is non zero along $x_1^s(s - s_0$ is very small) initially, since the v-jump s genuine, $J_\infty(y) - J_\infty(z^\infty)$ is non zero and our claim follows.

We thus can follow the critical points of J_∞ near x^∞ in $\mathcal{W}_{2s+1}$ in restriction to Γ_{2s}. We know that x^∞ survives the collapse and that any other critical point y^∞ has a characteristic piece bewteen its O_1, O_2 verticals.

We now prove that the index of x^∞ along the additional direction of $\mathcal{W}_{2s+1}$ transverse to Γ_{2s} changes from 0 to 1 and that y^∞ is degenerate in $\mathcal{W}_{2s+1}$.

Since $\mathcal{W}_{2s+1}$ is a manifold, it follows that x^∞ is the only critical point of J_∞ which induces a difference of topology in its level sets, the index of

x^∞ has to remain unchanged, equal to i_∞. Since its normal index changes from 0 to 1, its index along Γ_{2s} has to change from i_∞ to $i_\infty - 1$.

We will later prove our claim about y^∞. Let us prove for the moment that the normal index of i^∞ changes from 0 to 1. We draw a normal direction near x^∞:

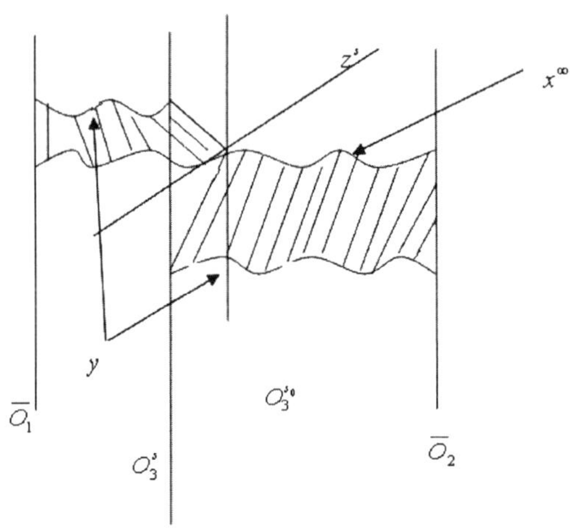

Let η_1 be the first eigenfunction of the linearized operator under Dirichlet boundary conditions. η_1 satisfies:

$$\begin{cases} -(\ddot{\eta}_1 + a^2\eta_1\tau) = \lambda_1\eta_1 \\ \eta_1(0) = \eta_1(1) = 0 \end{cases}.$$

We will follow this equation along the deformation process, as λ_1 crosses the value zero (being positive and becoming negative).

A little bit before and a little bit after this crossing, the operator $-(\ddot{\eta} + a^2\eta\tau)$ is invertible. Therefore, it is easy to take the limit of $(y - x^\infty)(t)$, for $t \in [0,1]$. We find, point by point, a vector $z(t)$ tangent to M. Its w - component η satisfies:

$$\begin{cases} -(\ddot{\eta} + a^2\eta\tau) = \delta_{t_0} \\ \eta(0) = \eta(1) = 0 \end{cases},$$

where δ_{t_0} is the Dirac mass at t_0, the time of the v-jump. η_1 is the first eigenfunction, with $\int_0^1 \eta_1^2 = 1$. We then have:

$$\eta = \frac{\eta_1(t_0)}{\lambda_1}\eta_1 + U$$

with U bounded H^1 and orthogonal to η_1. Thus,

$$\int_0^1 \dot{\eta}^2 - a^2 \eta^2 \tau = O(1) + \frac{\eta_1(t_0)^2}{\lambda_1}.$$

This implies that, at the crossing of $\lambda_1 = 0$, the normal index switches from zero to 1.

We now turn to y^∞. First, we establish that Γ_{2s}, near x^∞, is embedded, **all along the deformation**, in a natural way in $\mathcal{W}_{2s+1}$. This is obvious as the deformation starts, but less obvious for later times of the deformation.

Given a curve y of Γ_{2s} near x^∞, we have two well defined associated verticals O_1 and O_2. O_1 and O_2 are close to the verticals $\bar{O}_1$ and $\bar{O}_2$ of x^∞. O_1 and O_2 are then joined by a unique orbit of ξ, here $\xi = \xi_0$ the Reeb vector-field before the deformation starts. This ξ-piece is close to the ξ-orbit of x^∞ which lies between $\bar{O}_1$ and $\bar{O}_2$. This defines a curve y_0 of Γ_{2s}^0 i.e. of the initial Γ_{2s}^0, near x^∞. Thus, we have a well defined surface $\underset{s \in (\varepsilon, \varepsilon)}{U} O_3^s$. This surface is spanned by v-orbits along a small path defined by an orbit of w, z^s, going through $y_0(t_0)$. There is a basic strip around z^s where the tangent plane to this surface is close to Span (v, w):

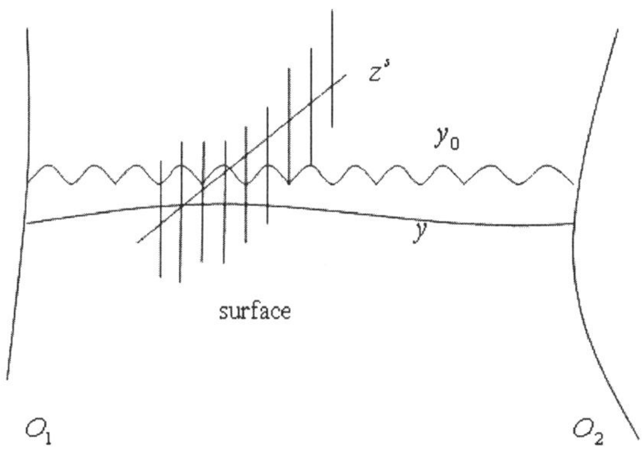

y is very close to y_0 in the C^1-sense since ξ is very close to ξ_0 in the C^0-sence (the deformation is C^0-bounded, C^1-small in α). Since (ξ_0, v, w) span TM and since ξ is C^0-close to ξ_0, the intersection between y and the surface is transverse and is a unique point. Thus, y is defined unambiguously as an element of $\mathcal{W}_{2s+1}$.

Next, we consider y^∞ in Γ_{2s} having a characteristic ξ-piece between its verticals O_1, O_2:

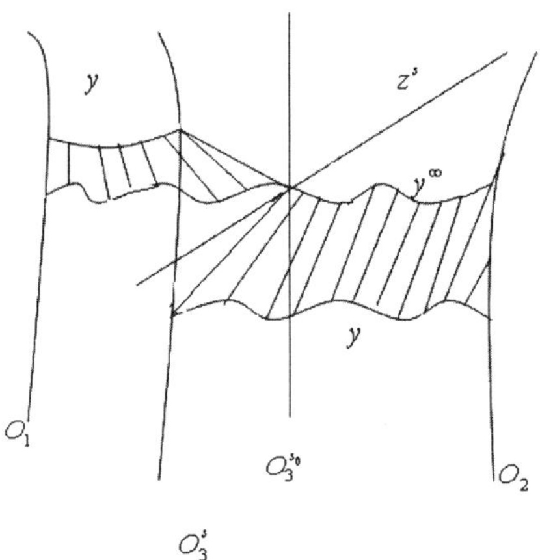

y is a curve of $\mathcal{W}_{2s+1}$ in the same family.

y^∞ is defined by $O_3^{s_0}, y$ by O_3^s. By uniqueness and minimality, the distance between y and y_∞ i.e. between the two pieces of y and the two pieces of y^∞ (the ones between O_1 and $O_3^s, O_3^{s_0}$ and the ones between O_3^s or $O_3^{s_0}$ and O_2) is bounded by the distance d^s between O_3^s and $O_3^{s_0}$. It is easy to see, arguing as before, that $d^s, (y - x^\infty)(0)$ and $(y - x^\infty)(1)$ and all of the same order. We thus can take the limit of $\frac{|y-x_\infty|}{|y-x_\infty|(0)}$ for $t \in [0,1]$. We find a vector $z(t)$ tangent to M along y^∞. Its component η along w satisfies then:

$$\begin{cases} -\ddot{\eta} + a^2 \eta \tau = c\delta_{t_0} \\ \eta(0) = \eta(1) = 0 \end{cases}.$$

Because the H_0^1-problem on y^0 is degenerate of index 1, i.e. there is a positive solution η_1 for:

$$\begin{cases} -(\ddot{\eta}_1 + a^2 \eta_1 \tau) = 0 \\ \eta_1(0) = \eta_1(1) = 0 \end{cases}.$$

c must be zero so that η is collinear to η_1 and the size δ of the v-jump along O_3^s is $o(d^s)$. This implies, after expansion, that

$$J_\infty(y) - J_\infty(y_\infty) = O(\delta d_s) = o(d^{s2}) = 0(|y - y_\infty|^2)$$

which indicates the degeneracy along the normal.

We now prove that $\delta = O(d^{s^2}) = O(|y - y_\infty|^2)$, which implies that

$$J_\infty(y) - J_\infty(y_\infty) = O(d_s^3).$$

Indeed, coming back to our drawing, we label:

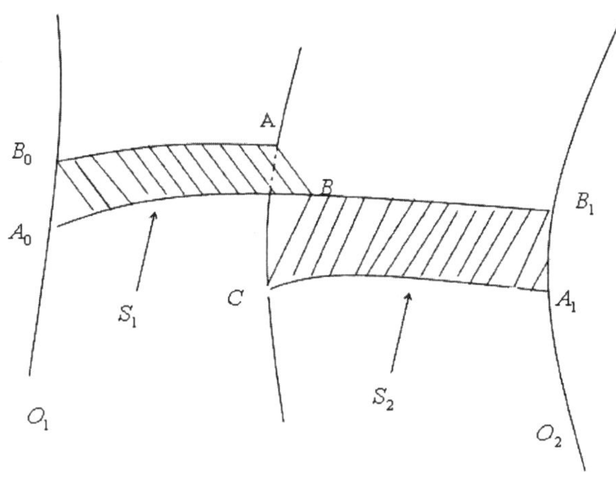

$$J_\infty(y) - J_\infty(y_\infty) = \int_{[B,C]} \alpha - \int_{[A,B]} \alpha$$

since the surfaces S_1, S_2 are spanned by ξ.

$[A, C]$ has size δ, $[A, B]$ has size $O(d^s)$, has a component along w at first order equal to $\bar{c}_1 d^s, \bar{c}_1 \neq 0$ and $[B, C]$ is derived from $[A, B]$ by v-transport during a time $O(\delta)$ so that

$$J_\infty(y) - J_\infty(y_\infty) = O(\delta d^s) = O(d^{s3}).$$

If we establish that $\delta \sim cd_s^2, c \neq 0$, then

$$J_\infty(y) - J_\infty(y_\infty) = c_1 d^{s3} + o(d^{s3})$$

and our statement about y^∞ and its unstable manifold follows.

In order to prove that $\delta \sim cd^{s2}$, let us follow the ξ-trajectory from A_0 past A until it arrives near A_1 at a point A_2. The ξ-length $[A_0, A_2]$ equals the ξ-length $[B_0, B]$. Since the ξ-piece $[B_0, B_1]$ is characteristic and $[A_0, B_0]$ is a piece of v-orbit of length $\ell^s \sim c_1 d^s$, this ξ-trajectory reaches a point A_2 with $d(A_2, O_2) \sim c_2 d^{s2}, c_2 \neq 0$, because $[A_2, B_1]$ is at first order tangent to

v ($[B_0, B_1]$ is characteristic) and the situation is generic (no vanishing at third order):

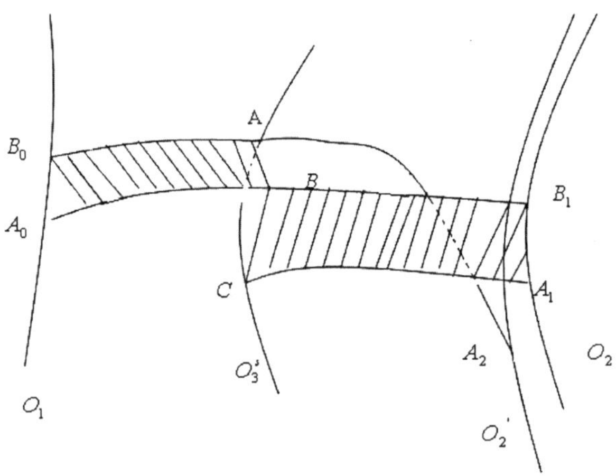

Thus, there is a minimal ξ-orbit from O_3^s to O_2^1 from A to A_2 and $[C, A_1]$ is a minimal ξ-orbit from O_3^s to O_2 with $d(O_2, O_2') \sim c_2 d^{s2}$. The conclusion follows for d^s small enough, since the distance from A to C is of the same order than the distance between the two curves $[A_0, A]$ and $[C, A_1]$. This distance is equivalent past any given positive time along $[C, A_1]$, to any of the distances between the two v-verticals at two corresponding points, one taken on $[C, A_1]$ and the other along $[A, A_2]$. $\square$

2.6.1 *Study of the Poincaré-return maps*

We prove in this section that the Poincaré-return map of a general critical point at infinity preserve $d\alpha$. This generalizes the result of [Bahri-1 2003] IIIb.9

Lemma 22 *Assume that (a part of) the Poincaré-return map $g : \sigma_1 \rightarrow \sigma_2$, where σ_1 and σ_2 are sections to ξ, satisfies $g^*\alpha = \alpha + df$.*

We can then find a section σ_3 to ξ as close as we please to σ_1 so that the Poincaré-return map $\tilde{g}: \sigma_3 \rightarrow \sigma_2$ satisfies $\tilde{g}^*\alpha = \alpha$.

Proof. Let o be the center of σ_1 and σ_2 so that $g(o) = 0$. Let $f_1(x) = f(x) - f(o) = df_o(x) + O(|x|^2)$. $\square$

Observe that

$$(g^{-1})^*(\alpha + df) = \alpha.$$

Let ψ_s be the one parameter group of ξ. Define

$$\gamma : \sigma_1 \to \sigma_3$$

$$x \to \psi_{f_1(x)}(x).$$

Then,

$$\gamma^*\alpha = \alpha(d\psi_{f_1(x)} + df_1\xi) = \alpha + df.$$

Thus,

$$(g^{-1})^* o\gamma^*\alpha = \alpha \quad \text{i.e.} (\gamma o g^{-1})^*\alpha = \alpha.$$

Let

$$\tilde{g} = g o \gamma^{-1}.$$

The conclusion follows.

The next Lemma generalizes [Bahri-1 2003, III B.9, pp 134–136]. As we carry out rotation from (non characteristic) pieces of x^∞ to other (non characteristic) pieces, we can view this transport as a composition of maps. Each map bypasses a sequence of consecutive characteristic pieces, from one side of the family to the other side.

When the family is reduced to one characteristic piece, we have proved in [Bahri-1 2003] that there are sections σ_1 and σ_2 which can be defined in a natural way such that the (partial) Poincaré-return map $g : \sigma_1 \to \sigma_2$ satisfies $g^*\alpha = \alpha + df$. Using the previous lemma, we may assume that $df = 0$. We may then use Proposition 16 of [Bahri-1 2003] and transport the v-rotation from one to the other side of the characteristic piece.

We extend III B. 9 of [Bahri-1 2003] as follows:

Consider a (full) sequence of characteristic ξ-piece of x^∞.

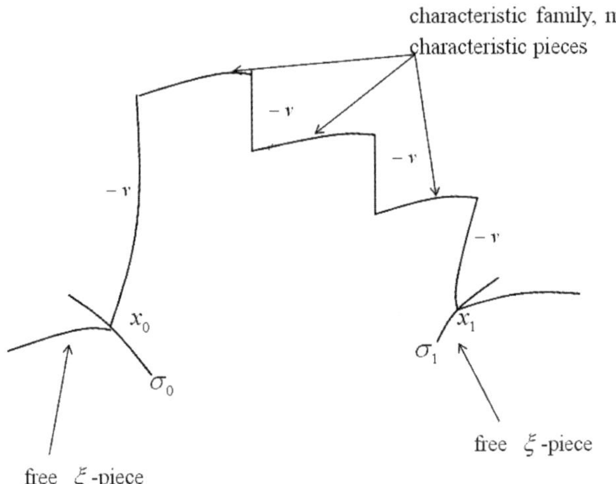

We define σ_0 and σ_1 as follows:
Given a characteristic piece and the two edge v-orbits,

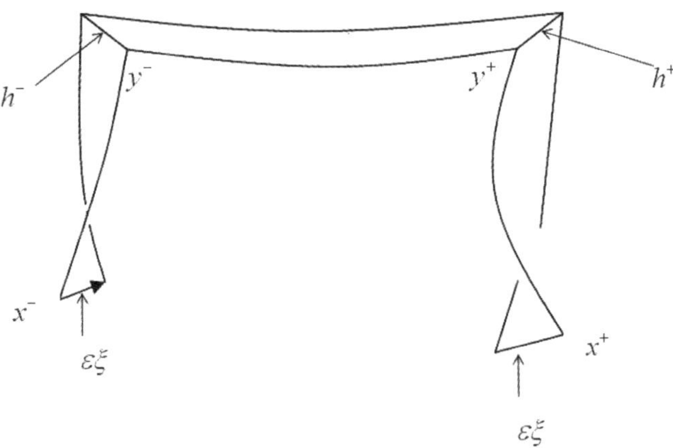

condition (*) of [Bahri-1 2003] is satisfied between x^- and x^+ if the vector h defined by transport of ξ-appropriately scaled, with different scales $\varepsilon, \varepsilon'$ on each side - from x^- to y^-, x^+ to y^+ respectively and by matching the two transported vectors along the ξ-piece satisfies:

$$\varepsilon' - \alpha(h^+) + (h^-) - \varepsilon = 0 \qquad (*)$$

x^+ is defined by (*) as a function of x^-.

σ_0 and σ_1 are then defined by the following property that they should satisfy:

From $z \in \sigma_0$, we move along v a time s_0 until we reach a point y^- such that $-\gamma_s$ is the one parameter group of $v\text{-}\alpha_z(D\gamma_{-s_0}(\xi(y))) = 1$. We then move along ξ from y^- to y^+ a characteristic length (of strict H_0^1- index zero) and from y^+ we move along $-v$ until we reach a point z' so that (*) holds between z and z'. From z', we move a characteristic ξ-length to z'' and from z'' we move along $-v$ a time s_2 until we reach a point z'' so that (*) holds between y^+ and z'''; and so on until we reach the last v-jump of the family. Then, in addition to (*), we require that, with s_m denoting the time along the last v-jump:

$$\alpha(D\gamma_{-s_m}(\xi)) = 1.$$

This imposes a extra-condition and provides a restriction which defines σ_0 and σ_1.

$$g : \sigma_0 \to \sigma_1$$

is defined accordingly and we claim that:

Lemma 23

$$g^*\alpha = \alpha + df.$$

Proof. Observe that (*) allow us to define a variation $\tilde{u}$ (a tangent vector) along each characteristic ξ-piece with edge v-pieces so that x^+ is derived from x^- through the use of (*):

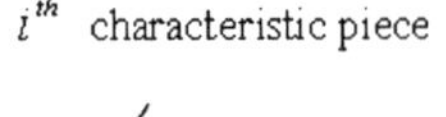

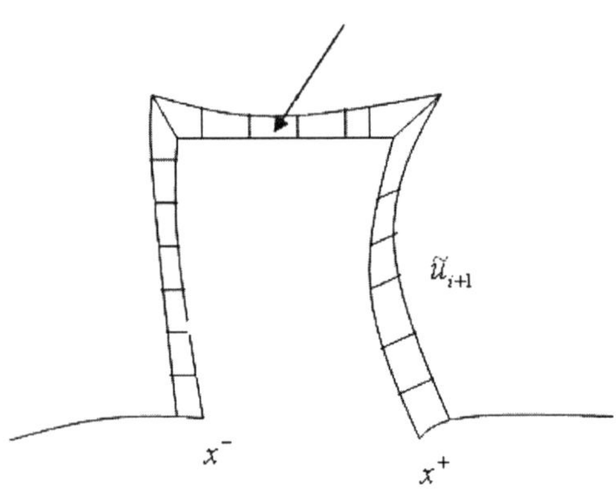

We thus have m vectors $\tilde{u}_2, \ldots, \tilde{u}_{m+1}$ along each trajectory from σ_0 to σ_1. Considering such a trajectory, it need not be part of a curve of $\Gamma_{2\ell}$, let alone a critical curve. However, we can partially complete it with the addition (the continuous addition) of a initial and of a final non degenerate ξ-piece:

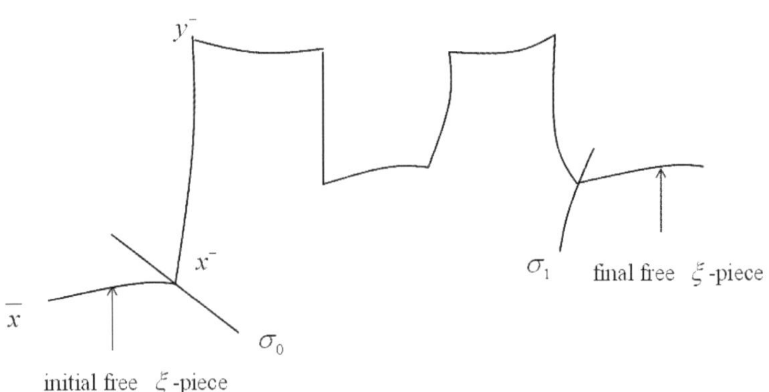

A "tangent vector" $\tilde{u}_1$ and another "tangent vector" $\tilde{u}_{m+2}$ can be defined using the free ξ-pieces as follows:

We pull back $\varepsilon\xi$ from y^- to x^- and adjust using $\varepsilon'v$ at $\bar{x}$ transported along the initial ξ-piece (similarly for the final ξ-piece):

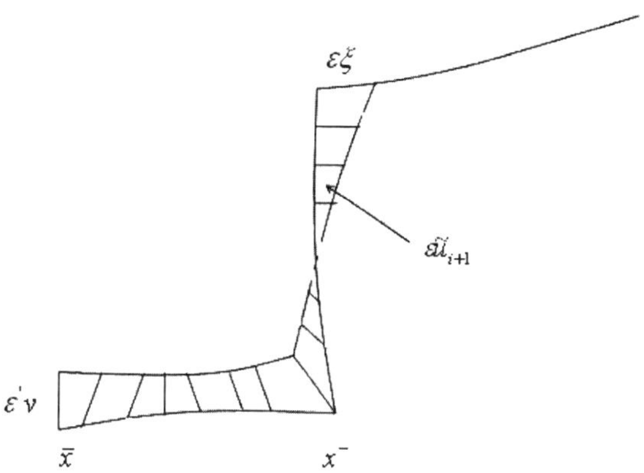

We now have $m+2$ vectors $\tilde{u}_1,\ldots,\tilde{u}_{m+2}$. Using the vectors $\tilde{v}_1,\ldots,\tilde{v}_m$, we can modify them into $u_1,\ldots,u_{m+2}$ which satisfy:

$$(\delta a^i - \delta a_c^i)(u_j) = 0 \quad \forall i = 1,\ldots,m, \forall j = 1,\ldots,m+2.$$

Condition (*) as well as the conditions on the initial and final v-jump $\alpha(D\gamma_{-s}(\xi)) = 1$ imply that the variation of the ξ-length of the trajectory along $u_1,\ldots,u_{m+2}$ is zero.

Observe that each edge y_i^- the vectors u_i and u_{i+1} generate, tranversally to v, $\mathrm{Span}(\xi, w)$. Given a tangent vector z_i at y_i^-, we can thus substract to z_i a combination of u_i and u_{i+1} so that it becomes collinear to v.

Let us now consider a vector h tangent to σ_0 and transport it through (partial) Poincaré-return maps until σ_1. We derive in this way a "tangent vector" $\tilde{h}$ along a trajectory.

At y_1^-, we can substract to $\tilde{h}_1 = \tilde{h}(y_1^-)$ a combination of u_1, u_2 so that the difference is along v. Then $\tilde{h} - \alpha_1 u_1 - \alpha_2 u_2$ at the other edge y_1^+ is collinear to $D\psi(v) = \theta_1 v + da_c'(v)\xi$. If we subtract $\alpha_3 u_3$ from $\tilde{h} - \alpha_1 u_1 - \alpha_2 u_2$, we eliminate the ξ-component. $\alpha_3 u_3$ is by construction collinear to v at y^-. Thus,

$$\tilde{h} - \alpha_1 u_1 - \alpha_2 u_2 - \alpha_3 u_3$$

leaves the first ξ-piece characteristic and is collinear to v at both edges. It is therefore zero on this ξ-piece.

At y_2^-, $\tilde{h} - \alpha_1 u_1 - \alpha_2 u_2 - \alpha_3 u_3$ might not be zero because h and the u_i's are transported differently along the v-jump. However, because it is zero at y_1^+, it is collinear to v at y_2^-, thus collinear to $d\psi(x) = \theta_2 v + da_c^2(v)\xi$ at y_2^+. Subtracting $\alpha_4 u_4$, $\tilde{h} - \alpha_1 u_1 - \alpha_2 u_2 - \alpha_3 u_3 - \alpha_4 u_4$ is zero on the second ξ-piece as well.

Iterating, we find that

$$\tilde{h} = \sum_{i=1}^{m+2} \alpha_i u_i$$

between σ and σ_1. Observe that the u_i's are parallel to v at the edges of the initial and final non degenerate ξ-pieces. Since the variation of the ξ-length along each u_i is zero, we can write:

$$\alpha(h(x_1)) - \alpha(h(x_0)) = \sum_{i=1}^{m+2} \alpha_i \sum_{j=1}^{m} \left(\alpha(u_i(y_j^+)) - \alpha(u_i(y_j^-)) \right)$$

$$= \sum_{j=1}^{m} \left(\alpha(\tilde{h}(y_j^+)) - \alpha(\tilde{h}(y_j^-)) \right)$$

$$= \sum_{j=1}^{m} \left(\alpha(D\psi(\tilde{h}(y_j^-))) - \alpha(\tilde{h}(y_j^-)) \right) = \sum_{j=1}^{m} da_c^j(\tilde{h}(y_j^-))$$

$\tilde{h}(y_j^-)$ is the image of $h(x_0)$ through the (partial) Poincaré-return map dg_j so that

$$\alpha \left(dg(h(x_0)) - \alpha(h(x_0)) = \sum_{j=1}^{m} da_c^j(dg_j(h(x_0))) \right)$$

hence

$$g^*\alpha = \alpha + df \text{ on } \sigma_0.$$

$\square$

We now state a result (Lemma 8) which may be useful in the verification of hypothesis (A):

Let z be a critical point at infinity with all its v-jumps having the same orientation, for example positive and having all its ξ-pieces of H_0^1-index zero. Assume that z has index at infinity $i_\infty = k$. Assume that z supports a critical point at infinity x_{2k}^∞ which corresponds to a non trivial cycle of dimension $2k$. Assume that this cycle has a non zero component on $\partial(w_1 \otimes \ldots w_{k+1} \otimes W^{i_\infty})$. W^{i_∞} is the unstable manifold of z and each w_i is the (half) — full unstable manifold of a degenerate ξ-piece of z (they are of H_0^1-index zero). We label the w_i's in the order in which they follow each other on z, as we describe the full circle of time.

There might be sequences of other degenerate ξ-piece spaced with free ξ-pieces between w_i and w_{i+1}. We still label them w_i, w_{i+1} for the sake of simplicity. From Proposition 33, we know that each w_i belongs to a different sequence of degenerate ξ-pieces. We then claim:

Lemma 24 *For each i, there are times $t_{-i} < \bar{t}_i, z(\underline{t}_i)$ and $z(\bar{t}_i)$ belonging to non degenerate ξ-pieces between w_i and w_{i+1} such that $z(\underline{t}_i)$ belong to some characteristic surface $\sum_{-j}$ of α and $z(\bar{t}_i)$ belongs to some other characteristic surface $\sum_\ell$ of α i.e. $z(\underline{t}_i)$ has a conjugate point after $2\pi j$ revolutions of $\ker\alpha$ along $-v$ and $z(\bar{t}_i)$ has a conjugate point after $2\pi\ell$ revolutions of $\ker\alpha$ along $+v$.*

Proof. We consider one edge of a sequence of degenerate ξ-pieces, the end one for example:

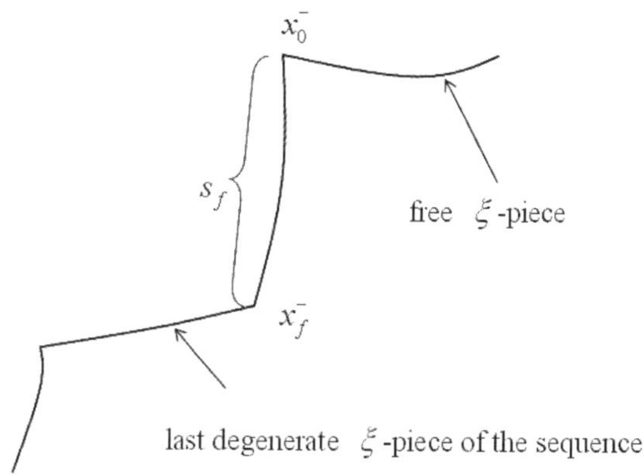

We know that

$$\alpha_{x_0^-}(D\gamma_{s_f}(\xi(x_f^+))) = 1$$

and that (by genericity) x_f^+ and x_0^- are not conjugate. Thus,

$$a(s_f) = \frac{\partial}{\partial s}\alpha_{x_0^-}(D\gamma_s(\xi(\gamma_{-s}(x_0^-))))\Big|_{s=s_f} \neq 0.$$

It is easy to see that for $s > s_f, s < s_f + c$, we have:

$$\alpha_{x_0^-}(D\gamma_s(\xi(\gamma_{-s}(x_0^-)))) > 1.$$

Indeed, would $a(s_f)$ be negative, there would be times between x_f^+ and x_0^- such that
$\alpha_{x_0^-}(D\gamma_s(\xi(\gamma_{-s}(x_0^-)))) > 1$ and z would have a decreasing normal which would not increase the number of zeros of b see [Bahri-1 2003] pp 109-112:

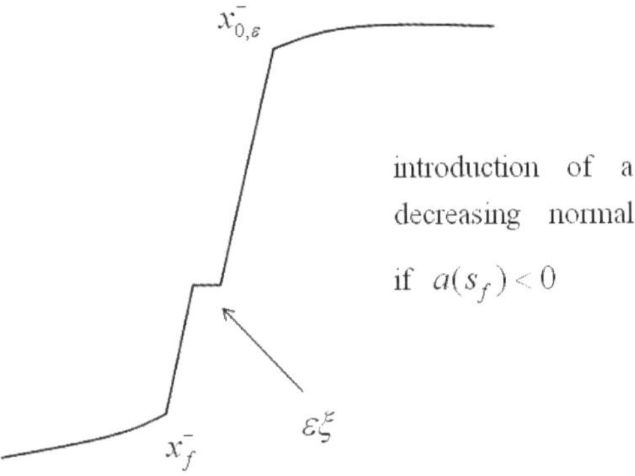

Thus, we may assume that $a(s_f) > 0$ and we can decrease z and x^∞ if we can use a negative v-jump:

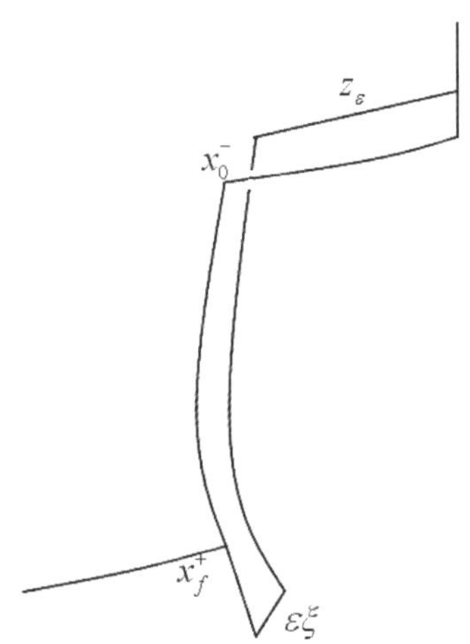

This can be viewed as part of

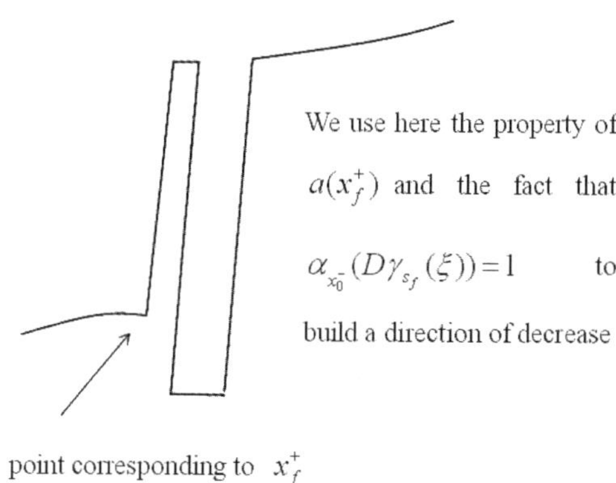

point corresponding to x_f^+

or part of

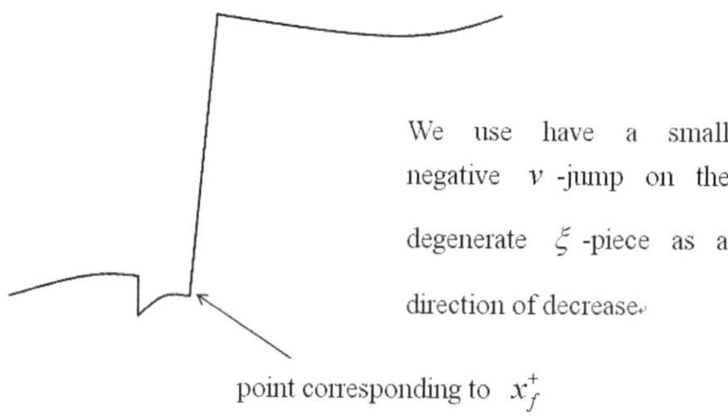

point corresponding to x_f^+

We use here a small negative v-jump on the degenerate ξ-piece as a direction of decrease.

The two decreasing deformations can be made part of the same global decreasing deformation, just as in the proof of Proposition 33.

When is this deformation hindered after x_0^-? : We may define such a deformation at $z(t)$ as long as we can find $s > 0$ such that

$$\alpha_{z(t)}(D\gamma_s(\xi)) > 1.$$

If we cross a v-jump between conjugate points after x_0^-:

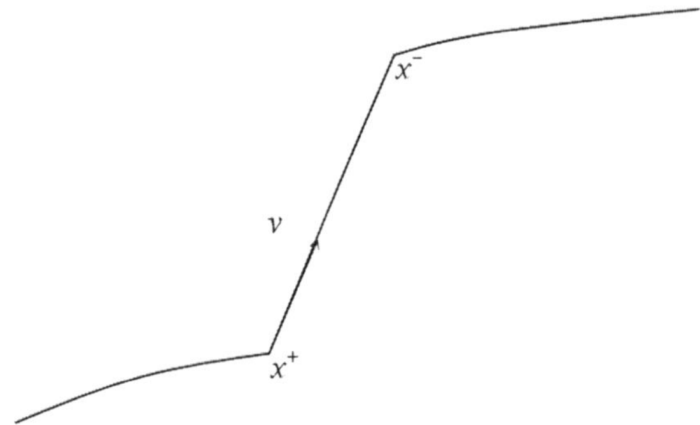

and the condition holds for x^+, it holds at x^- because the form α is transported from x^+ to x^-, hence we can continue on the upper ξ-branch.

Therefore, this decreasing deformation stops expanding after x_0^- only if, for some time $\underline{t}_i$, we have

$$\begin{cases} \alpha_{z(\underline{t}_i)}(D\gamma_{\bar{s}}(\xi)) = 1 \\ \frac{\partial}{\partial s}(\alpha_{z(\underline{t}_i)}(D\gamma_s(\xi)))\Big|_{s=\bar{s}} = 0 \end{cases}$$

which implies that $z(\underline{t}_i)$ and $\gamma_{-\bar{s}}(\underline{t}_i))$ are conjugate.

We may apply the same reasoning to the starting edge of next sequence thereby deriving $z(\bar{t}_i)$ and $\bar{s}^+$ such that $z(\bar{t}_i)$ and $\gamma_{\bar{s}+}(z(\bar{t}_i))$ are conjugate.

If such a $z(\underline{t}_i)$ and such a $z(\bar{t}_i)$ did not exist between a degenerate sequence and the next one, we could consider both of them to give the same w because, denoting w_i the (half) full unstable manifold of the first sequence and w_f the one of the second sequence, we would be able to write:

$$w_i - w_f = \partial\psi$$

with no increase in the number of zeros for ψ with respect to w_i, w_f (2 zeros). From that second sequence, we would then go to the third one and so forth. But w_i and w_{i+1} are not homologous, this is our assumption, otherwise $\partial(w_1 \otimes \cdots \otimes q_{k+1} \otimes W^{i\infty}) = \partial\sigma, \sigma$ carrying $2k$ zeros for b at most, see the proof of Proposition 33.

Thus, $z(\underline{t}_i)$ must exist, on the same sequence of **non-degenerate** ξ-pieces. Furthermore, $z(\underline{t}_i)$ must be **before** $z(\bar{t}_i)$ otherwise, if they overlap, we can build in the overlap upon a transition between the two decreasing deformations using:

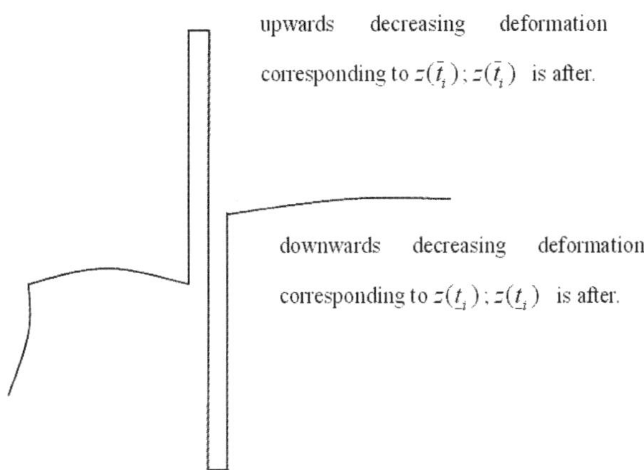

As we combine the two oscillations, the functional is still decreasing and the number of zeros is still only increased by 2(as allowed by the use of w_i).

We can thin down and cancel one of these oscillations as we evolve to one or to the other edge. $\qquad\qquad\qquad\qquad\qquad\qquad\qquad\qquad\square$

Before ending Part V, we define a suitable basis at $\bar{x}_\infty \in \Gamma_{2s}$ of $T_{\bar{x}_\infty}\Gamma_{2k}$ for the reduction of $d^2J(\bar{x}_\infty)$ and we also establish that β remains a contact form through all our perturbations above (and below in Part VI).

2.6.2 Definition of a basis of $T_{\bar{x}_\infty}\Gamma_{2s}$ for the reduction of $d^2J(\bar{x}_\infty)$

We define, for a general x_∞ belonging to Γ_{2s} a basis of $T_{\bar{x}_\infty}\Gamma_{2s}$ suitable for the reading of $d^2J(\bar{x}_\infty)$. This basis will be redefined in section VI when we will be studying more specifically critical points at infinity having only characteristic ξ-pieces.

We need first to define, given a ξ-piece, free or characteristic, what is the associated vector $\tilde{v}_i$. We draw this ξ-piece and the next one. We assume in a first step that both are free ξ-pieces.

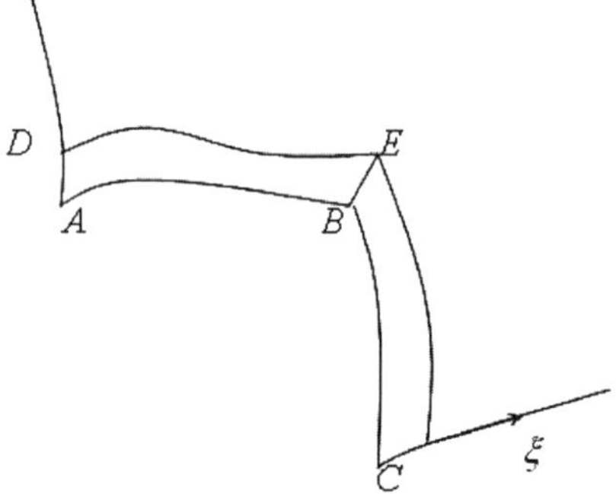

We build $\tilde{v}_i$ by transporting v from A to B, then compensating the $[\xi, v]$-component of the transported vector at B using the transport of ξ, with the proper scale, back from C to B, then adding or subtracting a little of ξ at B to the transport of v in order to adjust the ξ-component. The v-component can also be adjusted easily. This defines $\tilde{v}_i$. This definition coincides with the definition of $\tilde{v}_i$ for a characteristic piece see Section 2.7 below, since the transport of v from A to B is collinear to v under such a circumstance.

If now a free piece is followed by a characteristic piece, then this characteristic piece has an associated vector $\tilde{v}_{i+1}$, with

$$\delta a^{i+1}(\tilde{v}_{i+1}) = 0; \quad \delta a_c^{i+1}(\tilde{v}_{i+1}) \neq 0.$$

We can add to $\tilde{v}_i$ a component along $\tilde{v}_{i+1}$ so that the two vectors $\tilde{v}_i$ and $\tilde{v}_{i+1}$ become orthogonal to each other. We achieve with the appropriate choice of γ:

$$(\delta a^{i+1} - \delta a_c^{i+1})(\tilde{v}_i + \gamma \tilde{v}_{i+1}) = 0.$$

This defines $\tilde{v}_i$ under such a circumstance.

We now establish:

Lemma 25   *In all cases, $J_\infty''(\bar{x}) \cdot \tilde{v}_i \cdot \tilde{v}_j = 0$ for $i \neq j$.*

Proof.   We need to provide the proof in the case of two consecutive free ξ-pieces with $j = i + 1$.

If a ξ-piece is free, we can pick up $\varepsilon\xi$ on the next ξ-piece and transport it back along v

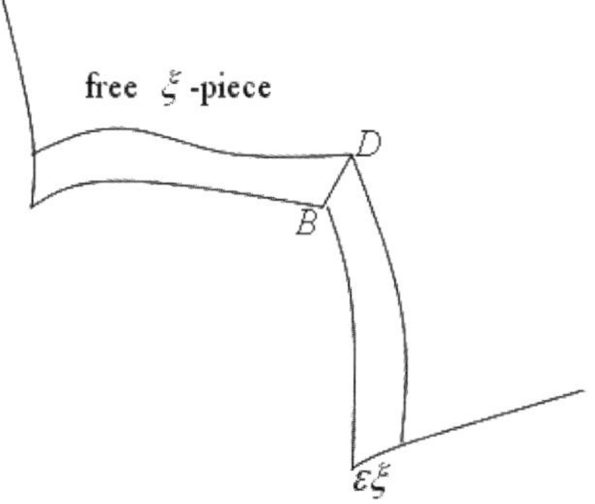

Then,

$$\overrightarrow{BD} = \varepsilon(1 + A_1)\xi + B_1[\xi, v] + \theta v.$$

At a critical curve, $A_1 = 0$.

This fact is not changed on the second ξ-piece after $\bar{x}_\infty$ has been displaced along $\tilde{v}_i$ since $\tilde{v}_i$ has no support near the second edge of the second free ξ-piece:

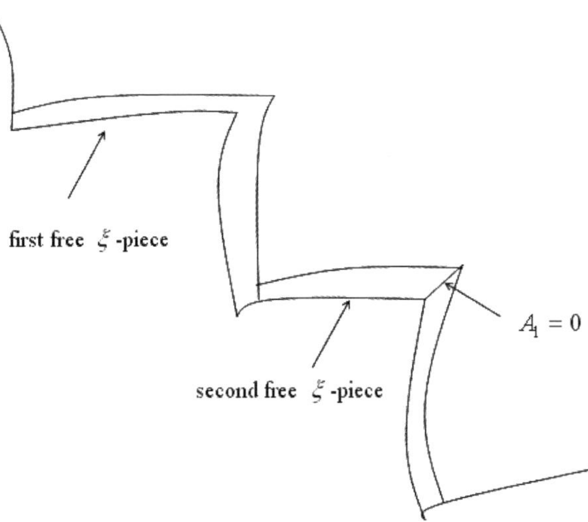

Thus, the natural extension of $\tilde{v}_{i+1}$ along $\bar{x}_\infty + \varepsilon\tilde{v}_i$ still satisfies $J'_\infty \cdot \tilde{v}_{i+1} = 0$ and we must have:

$$J''_\infty(\bar{x}_\infty) \cdot \tilde{v}_i \cdot \tilde{v}_{i+1} = 0.$$

$\square$

We now extend to a free piece the definition of the vectors u_i provided above (in the proof of Proposition 34):

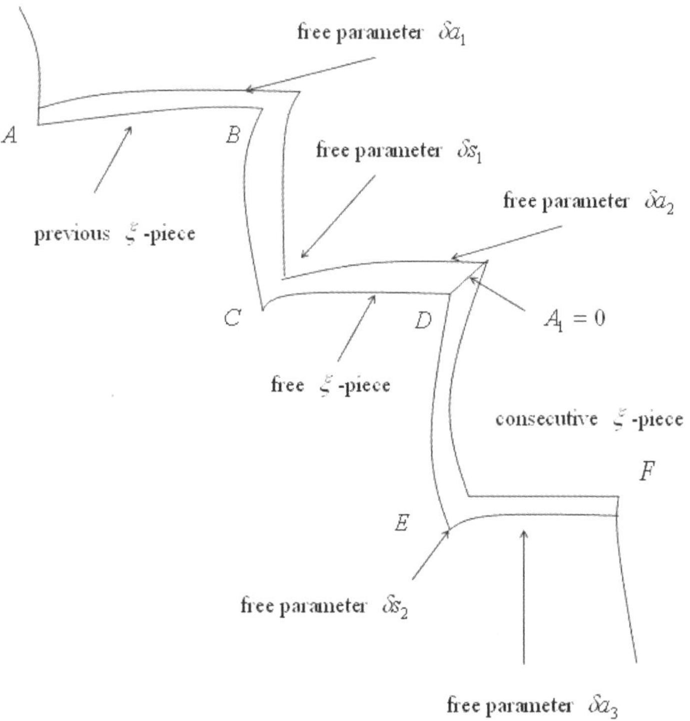

We take v at A, we transport it along ξ at B, then along v to C, then along ξ to D, then again along v to E and finally along ξ to F and we expect that it glues up with μv at F.

The free parameters are $\delta a_1, \delta s_1, \delta a_2, \delta s_2, \delta a_3$ since we are free to change slightly the length of the ξ- and of the v-pieces along our transports.

At F, we have two equations to satisfy since we need u_i at F to be equal to μv.

Furthermore, we want u_i to be orthogonal to $\tilde{v}_{i-1}, \tilde{v}_i$ and $\tilde{v}_{i+1}$. This requirement introduces three additional conditions. We have five parameters for five conditions. We will see later that the system is compatible.

We now have a basis for $T_{\bar{x}_\infty}\Gamma_{2s}$ made of $\tilde{v}_1, \ldots, \tilde{v}_s, u_1, \ldots, u_s$ where $J''_\infty(\bar{x}_\infty)$ takes the form:

$$
\begin{pmatrix}
c_1 & & 0 & \\
& \ddots & & \\
0 & & c_s & 0 \\
& & & \\
0 & & & A
\end{pmatrix} \cdot
$$

Inspecting A, we find that as in [Bahri-1 2003], also Section 2.7 below, A is as follows:

$$
\begin{pmatrix}
\ddots & \ddots & 0 \\
\ddots & \ddots & \ddots \\
0 & \ddots & \ddots
\end{pmatrix}
$$

2.6.3 *Compatibility*

We now prove that the C^2-bounded perturbations of v which we have introduced do not destroy the requirement on β to be a contact form. Since the arguments establishing that $i_\infty \geq k+1$ (Corollary 9) are of local type (local near $\bar{x}_{2k}^\infty$), we will not worry about establishing global statements about the perturbed α, v, β etc, we will worry only about local properties near $\bar{x}_{2k}^\infty$. Our v-jump is between x^- and x_0^-:

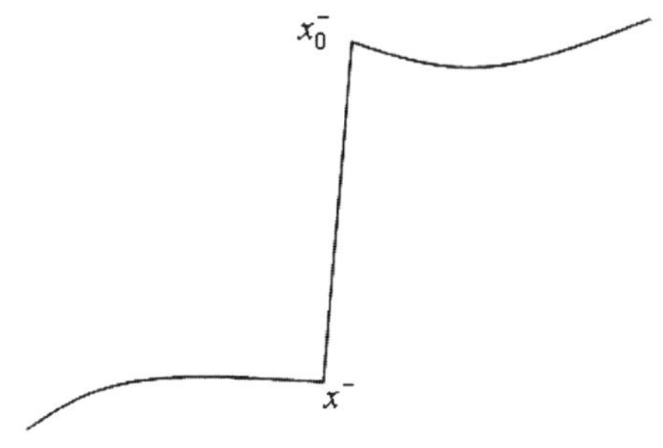

We denote v_0 the unperturbed $v = \frac{\partial}{\partial z}$ (rescaled). We introduce transverse coordinates x and y and perturb v_0 into:

$$v = \frac{\partial}{\partial z} + a(x, y, z)\frac{\partial}{\partial x} + b(x, y, z)\frac{\partial}{\partial y}$$

where both a and b are second order in x, y and are zero near x^-, x_0^-.

For such a general perturbation of v_0, the new v is not anymore in $\ker \alpha$. It is in $\ker(\alpha + df)$, with $df(v) = -\alpha(v)$ of order 2. Such an f can easily be built on all of M^3 as we will see later, although we only need it locally, near $\bar{x}_{2k}^\infty$.

Replacing α with $\alpha' = \alpha + df$, we find:

$$d\alpha'\left(v, \left[\frac{\xi}{1 + df(\xi)}, v\right]\right) = d\alpha(v_0, [\xi, v_0]) + o(1) \leq -\frac{1}{2}.$$

Hence, $\beta' = d\alpha'(v, \cdot)$ is again a contact form.

In order to build f, we use the equation:

$$df(v) = -\alpha(v).$$

Let us assume first that the v_0-orbit through x^-, x_0^- is not closed. Let us assume that the jump x^- to x_0^- is along $+v$.

We build a section σ to v_0 and v at x_0, a section $\bar{\sigma}$ at x^- and another section σ' on the negative v-orbit through x_0^-, far past x^-:

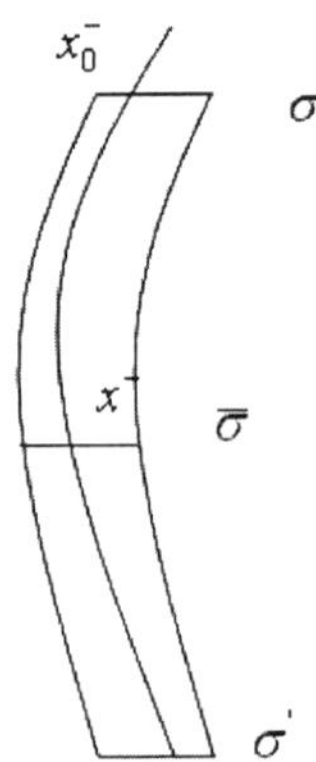

We would like f to vanish after σ'. For this, we consider the v-orbits originating in σ after they have passed x^-. We perturb α into $\tilde{\alpha} = \alpha + \gamma, \gamma =$

$o(x^2 + y^2)$ (after $\bar{\sigma}$). σ' is chosen very far from $\bar{\sigma}$ and

$$\int_{\bar{\sigma}}^{\sigma'} \gamma(v) = -\int_{\sigma}^{\bar{\sigma}} \alpha(v).$$

Setting then $\tilde{\alpha} = \alpha + \gamma$ all over M and

$$f(x_s) = -\int_{o \in \sigma}^{s} \tilde{\alpha}(v)$$

f is zero after $\bar{\sigma}$ (on the negative v-orbit) and is defined globally.

If the support of a and b in (x, y) is chosen small enough, independently from the length of the v-orbits from σ to σ', df will be small, as needed $(\tilde{\alpha} + df)(v) = 0$. Furthermore,

$$d(\tilde{\alpha} + df)(v, [\tilde{\xi}, v]) = d\tilde{\alpha}(v, [\tilde{\xi}, v]).$$

Since

$$d\tilde{\alpha} = d\alpha + o(|x| + |y|),$$

$$\tilde{\xi} = \xi + o(|x| + |y|)$$

and

$$d\tilde{\alpha}(v, [\tilde{\xi}, v]) = d\alpha(v, [\xi, v]) + o(1)$$

$\tilde{\beta}$ is a contact form and we can proceed.

If the v_0-orbit through x_0^-, x^- is closed, we can also build a global f (which we do not need though for our arguments) after introducing from x^- to x_0^- the same perturbation than between x_0^- and x^-, only that it will now be taken with the negative sign.

In all, as claimed, v can be perturbed freely at second order and we can set

$$v(\text{ rescaled}) = \frac{\partial}{\partial z} + \omega(x, y)\bar{\omega}(z)$$
$$\left((h_1 x^2 + \ell_1 xy + k_1 y^2)\frac{\partial}{\partial x} + (h_2 x^2 + \ell_2 xy + k_2 y^2)\frac{\partial}{\partial y} \right).$$

$\square$

We now conclude this book with a thorough study of $d^2 J(\bar{x}_\infty)$ in the case when all the ξ-pieces of $\bar{x}_\infty$ are characteristic.

2.7   On the Morse Index of a Functional Arising in Contact Form Geometry

2.7.1   *Introduction*

Let (M^3, α) be a three dimensional compact manifold without boundary. Let α be the contact form on M. Let ξ be the contact vector field of α, and let v be a non-singular vector field in $\ker \alpha$ such that

$$\beta = d\alpha(v, \cdot)$$

is also a contact form with the same orientation than α.

The existence of such v and β is discussed in other papers [Bahri 1998] and [Bahri-1 2003]. It is not needed for our results here but is used for the sake of simplicity.

Let $k \in N^*$ be a non-zero integer and let Γ_{2k} be the space of closed curves on M made of k pieces of ξ-orbits alternated by k-pieces of $\pm v$-orbits. The orientation ε_i of the ith v-orbit is pre-assigned so that we should rigorously work with $\Gamma_{2k}(\varepsilon_1, \cdots, \varepsilon_k)$, but we work with Γ_{2k} for the sake of simplicity.

Γ_{2k} is, under a generic perturbation on v if needed [Bahri 1998], a $2k$ dimensional manifold. It is natural to ask, a question which we will not address here, whether Γ_{2k} is a symplectic or complex manifold.

A curve of Γ_{2k} $\overline{x}$ is typically drawn as follows:

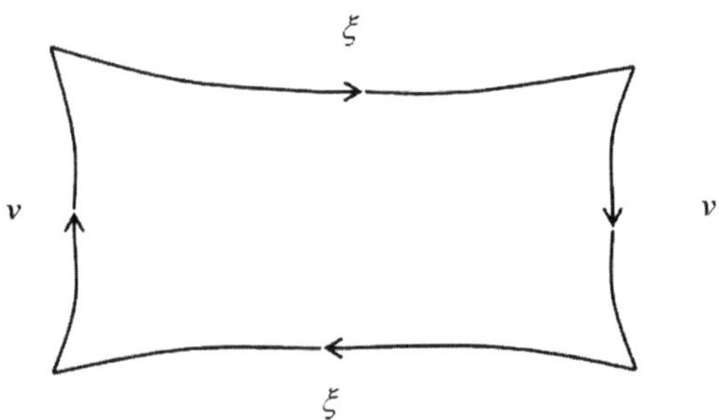

And there is a natural functional defined on Γ_{2k},

$$J_\infty(\overline{x}) = \sum_{i=1}^{k} a_i,$$

where a_i is the length of the ith ξ-piece.

The critical points of J_∞ on Γ_{2k} have been stated in [Bahri 1998], they are of several types. The first type are the so-called "true critical points at infinity" where the v-jumps occur between conjugate points, i.e. referring to the drawing above and denoting Φ_s the one-parameter group of v, we have

$$\left(\Phi_{s_i}^* \alpha\right)_{x_{i-1}^+} = \alpha_{x_{i-1}^+}$$

at each base point x_{i-1}^+ of a v-jump of algebraic length s_i. We will not be interested by these critical points here.

The second type are curves $\bar{x}$ such that all the ξ-piece of $\bar{x}$ are characteristic, i.e. in the ξ-transport along a ξ-piece from x_i^- to x_i^+, $v(x_i^-)$ is mapped onto $\theta_i v(x_i^+)$, i.e.

$$D\psi_{a_i}(v(x_i^-)) = \theta_i v(x_i^+)$$

where ψ_a is the one-parameter group of ξ and a_i is the ξ-length of the ith ξ-piece.

$\bar{x}$ has to satisfy other conditions to be a critical point of J_∞ which are discussed below.

There are other types of critical points which are intermediate between the first type (the true ones) and the second type which we partially described above. We will not, for sake of simplicity, consider them here.

It is interesting to study the second derivative of J_∞, $J_\infty''(\bar{x}^\infty)$ at a critical point at infinity of the second type, i.e. having all its ξ-pieces characteristic.

Let us denote
$$C_k = \{\bar{x} \text{ of } \Gamma_{2k} \text{ which have all their } \xi\text{-pieces characteristic}\}.$$
At a curve $\bar{x}$ of C_k, let us consider its $i^{th} - \xi$ piece,

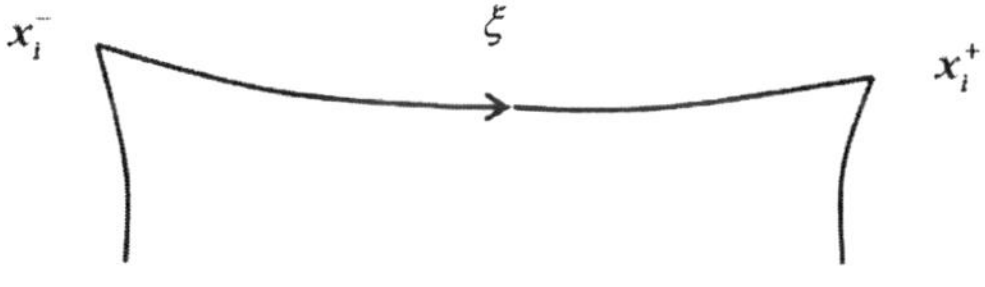

Here x_i^- is the starting point of the ith ξ-piece, x_i^+ is its ending point.

We can define along this piece a tangent vector to Γ_{2k} (not to C_k) which we denote $\widetilde{v}_i$. It is generated by the transport of v along ξ from x_i^- to x_i^+.

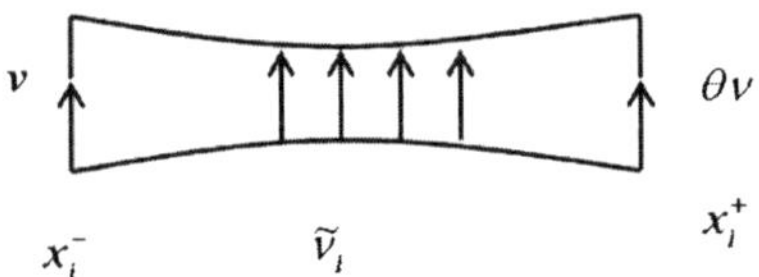

At a generic point of C_k, in particular at $\overline{x}^\infty$, we have

$$T_{\overline{x}}\Gamma_{2k} = \overline{T_{\overline{x}}C_k} \oplus \left(\underset{i=1}{\overset{k}{\to}} \oplus R\widetilde{v}_i \right).$$

This decomposition might not be a direct sum along a $k-1$ dimensional sub-manifold of C_k. At a critical point at infinity x^∞, it will be generically a direct sum.

$J_\infty''(\overline{x}^\infty)$ reads in the decomposition as follows:

$$\left(\begin{pmatrix} c_1 & 0 & 0 \\ 0 & \dots & 0 \\ 0 & 0 & c_k \end{pmatrix} \quad 0 \\ \quad\quad 0 \quad\quad\quad A \right)$$

here,

$$c_i = J_\infty''(\overline{x}_\infty).\widetilde{v}_i.\widetilde{v}_i,$$

$$J_\infty''(\overline{x}_\infty).\widetilde{v}_i.\widetilde{v}_j = 0 \text{ for } i \neq j.$$

Thus, the index i_∞ of $J_\infty''(\overline{x}^\infty)$ splits into

$$i_\infty = i_\infty^1 + i_\infty^2$$

with $i_\infty^1 =$ number of negative $c_i's$ and $i_\infty^2 =$ index A. i_∞^1 is what we call the normal index.

Usually, at a non-degenerate critical point y_0 of a function $f(x)$, a negative direction can be adjusted to be any prescribed direction through rotation.

We prove in this paper that we can adjust these negative directions as we please between i^1_∞ and i^2_∞, switching them from one side to the other and vice-versa, not through rotations in the parameter spaces but through local deformation of α and v around $\overline{x}^\infty$.

Namely, we prove,

Theorem 2 (Main Theorem) *Given i^1_∞ and i^2_∞ and another way of writing $i^1_\infty + i^2_\infty$ as $j^1_\infty + j^2_\infty$, $j^1_\infty \leq k, j^2_\infty \leq k$, as long as $i^2_\infty \geq 2$ and $j^2_\infty \geq 1$, there is a C^2-bounded and C^1-small deformation of α and v keeping $\overline{x}^\infty$ isolated and non-degenerate such that, for the new α and v, $J''_\infty(\overline{x}^\infty)$ has normal index j^1_∞ and j^2_∞ as index along C_k.*

Corollary 10 *Assume $i^1_\infty < k$ and $i^2_\infty \geq 2$, let i be such that $c_i = J''_\infty(\overline{x}_\infty).\widetilde{v}_i.\widetilde{v}_i > 0$, we can complete a C^2-bounded and C^1-small deformation of α and v which keeps $\overline{x}^\infty$ isolated and non-degenerate and changes c_i from $c_i > 0$ to $c_i < 0$.*

Results of this kind are directly related to the homology defined in [Bahri-1 2003] and which we studied more thoroughly in [Bahri-2 2003]. They give us some freedom in the definition of this homology as decreasing flow-lines which go to $\overline{x}_\infty$'s as above can be assumed to leave later neighborhoods of such $\overline{x}_\infty$'s using decreasing directions corresponding to the normal (i^1_∞) or the tangential (i^2_∞) as we please.

The Γ_{2k}'s and J_∞ arise in the variational problem at infinity associated with the variational problem

$$J(x) = \int_0^1 \alpha_x(\dot{x})dt$$

on loops $x = x(t)$ of $C_\beta = \{x \in H^1(S^1, M)$ s.t. $\beta_x(\dot{x}) = 0, \alpha_x(\cdot) = const > 0\}$.

Such curves have a tangent vector which reads $\dot{x} = a\xi + bv$. There are decreasing flow trajectories for a suitable pseudo-gradient which do not converge to the periodic orbits of ξ and end up in the Γ_{2k}'s, more precisely at the $\overline{x}_\infty$'s. Hence, the usual ∂-operator of Morse theory when related to the critical points only (periodic orbits of ξ) does not work and we have to include the $\overline{x}_\infty$'s. The $\overline{x}_\infty$'s have two indices, an H^1_0- index i_0 and an index at infinity i_∞, see [Bahri-1 2003]. i_0, if we use an outstretched analogy with piece-wise geodesics between prescribed points, is the index of the piece-wise geodesics between the prescribed points once these prescribed points are fixed. i_∞ is the index when the curves between the prescribed points are geodesics and the preassigned points are free to move.

Studying flow lines more closely out of or reaching periodic orbits, we derive that we may assume for the $\overline{x}_\infty$'s which interfere with the homology that $i_0 = 0$, see [Bahri-2 2003].

This implies that the total index of such $\overline{x}_\infty$'s can be computed in restriction to the slice of Γ_{2k} to which these $\overline{x}_\infty$'s belong.

Understanding their unstable manifold is useful and our theorem provides some preliminary information on this index. A more thorough study of the C_k's and of their injection in the Γ_{2k}'s should provide some insight about this index. Other related questions may be asked such as, are the C_k's for example real sub-manifolds in Γ_{2k}? Are they lagrangian?

Another interesting question is to study the part of Γ_{2k} where $i_0 = 0$, see [Bahri-2 2003]. This part corresponds to curves such that their ξ- pieces, if thought as geodesics (this can always be done locally), are minima in a variational problem $J(x) = \int_0^1 \alpha_x(\dot{x})dt$ defined on curves whose ends are not completely prescribed but constrained to move along prescribed v-orbits.

Such a part of Γ_{2k} can be thought of as a covering of the configuration space of pieces of surfaces transverse to v (S^2 in the case of Hopf fibration $S^3 \to S^2$ and the standard contact structure). A natural question which arises then is the question of understanding the relationship between i_∞^1, i_∞^2 and the topology of the configuration spaces which are underlying the set of curves having $i_0 = 0$.

We proceed now with the proof of the theorem as follows.

2.7.2   *The Case of* Γ_2

Theorem 3   *Let x^∞ be a false critical point at infinity in Γ_2, with its characteristic ξ-piece. Assume that the index at infinity of x^∞ is 1. We can then modify the vector field v in the vicinity of x^∞ so that the vector field $\tilde{v}$ at x^∞ defines the negative direction of the second derivative at infinity. After this modification, β remains a contact form and no new critical point at infinity is created.*

The proof starts with the choice of suitable coordinates in order to describe Γ_2 near x^∞. Once these coordinates are found and the equations for Γ_2 become explicit, the vector-field will be modified so that $\tilde{v}$ becomes the negative direction for J''_∞.

x_∞ has a base point x_0. The characteristic ξ-piece of x^∞ starts at x_0 and ends at x_1:

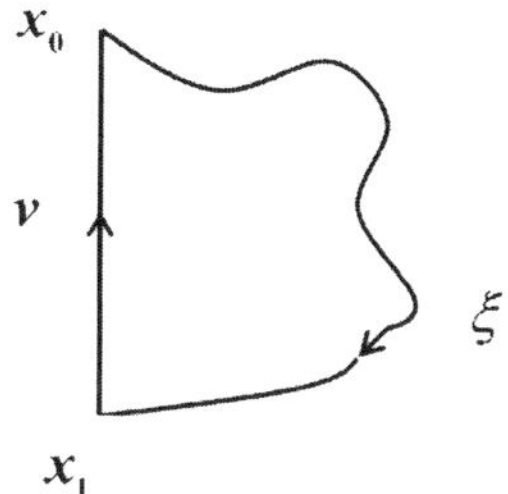

We will use Darboux coordinates at x_0 which we transport at x_1. v will have a special reduction in these coordinates near x_1. Since ξ is reduced in these coordinates to be the constant vector $\frac{\partial}{\partial x}$ ($\alpha = ydz + dx$), its one-parameter group ϕ_s is x-translation by s. We need to reduce v at each edge to $\frac{\partial}{\partial z}$.

This will be completed as follows:

We have special Darboux coordinates at x_1 and we derive from these Darboux coordinates near x_0. In the Darboux coordinates at $x_1, v(x_1) = \frac{\partial}{\partial z}, \left(\frac{\partial}{\partial z}v\right)(x_1) = \gamma v(x_1)$.

Near x_1, we choose an (almost) explicit change of coordinates, from the special Darboux coordinates to other coordinates, which reduces v to $\frac{\partial}{\partial z}$ (up to a multiplication factor). We propagate these coordinates along the v-piece from x_1 to x_0. This provides us with a set of coordinates near x_0 where v is reduced to $\frac{\partial}{\partial z}$ (up to a multiplication factor). We thus derive a change of coordinates near x_0 in order to come back to the Darboux coordinates.

Two changes of coordinates are therefore needed, one near x_0, the other one near x_1; both relate coordinates where $v = \frac{\partial}{\partial z}$ (up to a multiplication factor) to Darboux coordinates.

We give in what follows the explicit form of all these coordinates.

2.7.3 *Darboux Coordinates*

We start with the **Darboux coordinates**.

Lemma 26   *There exist near $\bar{x}_1$ Darboux coordinates where α reads $ydz + dx$ and $v(\bar{x}_1)$ is collinear to $\frac{\partial}{\partial z}$ while $\frac{\partial v}{\partial z}(\bar{x}_1)$ is also collinear to $\frac{\partial}{\partial z}$.*

Proof.   Choose first arbitrary Darboux coordinates near $\bar{x}_1$, where α reads $ydz + dx$. Then v reads:

$$v = A_0\left(\frac{\partial}{\partial z} - y\frac{\partial}{\partial x}\right) + B_0\frac{\partial}{\partial y}.$$

$\square$

Let

$$x' = x - \theta z^2/2$$
$$y' = y + \theta z$$
$$z' = z.$$

This provides new Darboux coordinates where

$$v(0,0,0) = A_0 \frac{\partial}{\partial z} + B_0 \frac{\partial}{\partial y}$$

$$= A_0 \left(\frac{\partial}{\partial z'} + \theta \frac{\partial}{\partial y'} \right) + B_0 \frac{\partial}{\partial y'}$$

$$= A_0 \frac{\partial}{\partial z'} + (A_0\theta + B_0) \frac{\partial}{\partial y'}.$$

A_0 can be assumed not to be zero since we can use if needed the new set of Darboux coordinates:

$$\tilde{x}' = x - \frac{y^2}{2}$$
$$\tilde{y}' = y$$
$$\tilde{z}' = z + y$$

where $\frac{\partial}{\partial y} = \frac{\partial}{\partial \tilde{y}'} + \frac{\partial}{\partial \tilde{z}'}$ at $(0,0,0)$.

Choosing $\theta = -\frac{B_0}{A_0}$, v reads as:

$$v = \mu_1 \left(\frac{\partial}{\partial z} - y\frac{\partial}{\partial x} + (a_1 x + b_1 y + g(z) + O(x^2 + y^2 + xz + yz)) \frac{\partial}{\partial y} \right)$$

a_1 is not zero since $d\alpha(v, \frac{\partial v}{\partial x}) \neq 0$ at $\bar{x}_1$.

We need to get rid of $g(z)\frac{\partial}{\partial y}$. We set:

$$x' = x + h(z)$$
$$y' = y - h'(z)$$
$$z' = z.$$

Then,

$$y\,dz + dx = (y - h'(z'))dz' + dx + h'(z')dz' = y'dz' + dx'.$$

In these new coordinates,

$$\frac{\partial}{\partial z} = \frac{\partial}{\partial z'} - h''(z')\frac{\partial}{\partial y'} + h'(z')\frac{\partial}{\partial x'}$$

$$\frac{\partial}{\partial x'} = \frac{\partial}{\partial x}$$

$$\frac{\partial}{\partial y'} = \frac{\partial}{\partial y}.$$

So that,

$$\frac{\partial}{\partial z} - y\frac{\partial}{\partial x} + (a_1 x + b_1 y + g(z) + O(x^2 + y^2 + xz + yz))\frac{\partial}{\partial y}$$

$$= \frac{\partial}{\partial z'} - y'\frac{\partial}{\partial x'} + a_1 x' + b_1 y' + g(z') - a_1 h(z') + b_1 h'(z') - h''(z')$$

$$+ O\left(x'^2 + y'^2 + x'z' + +y'z' + |z'h(z')| + |z'h'(z')| + h^2(z') + h'^2(z')\right)\frac{\partial}{\partial y'}.$$

We choose h so that

$$b_1 h' - h'' = g(z) \quad \text{with} \quad h = O(z^3), h' = O(z^2).$$

The claim follows.

Next, we reduce v to $\frac{\partial}{\partial z}$ (up to collinearly) at each edge $\bar{x}_0, \bar{x}_1$. Near $\bar{x}_0$ and $\bar{x}_1$, v is now collinear to $\frac{\partial}{\partial z} - y\frac{\partial}{\partial x} + \left(\tilde{a}x + \tilde{b}y + \tilde{c}z + q_2(x, y, z)\right)\frac{\partial}{\partial y}$, where q_2 stands for a function which is $O(x^2 + y^2 + z^2)$ at the origin. $\bar{q}_2$ is its second order term.

Let ψ_1 be a local diffeomorphism such that $D_0\psi_1 = Id$, which maps the above vector-field onto $\frac{\partial}{\partial z}$. Writing:

$$\psi_1(x, y, z) = (x + P_2^1 + P_3^1 + R)\frac{\partial}{\partial x'} + (y + P_2^2 + P_3^2 + S)\frac{\partial}{\partial y'} + (z + P_2^3 + P_3^3 + T)\frac{\partial}{\partial z'}$$

where P_2^i and P_3^i are of degree 2 and 3 respectively and where R, S and T are remainders of order four or more, we seek that

$$D\psi_1\left(\frac{\partial}{\partial z} - y\frac{\partial}{\partial x} + (\tilde{a}x + \tilde{b}y + \tilde{c}z + q_2(x, y, z))\frac{\partial}{\partial y}\right) = \frac{\partial}{\partial z'}.$$

This yields:

$$\frac{\partial P_2^1}{\partial z} = y \qquad \frac{\partial P_2^2}{\partial z} + \tilde{a}x + \tilde{b}y + \tilde{c}z = 0 \qquad \frac{\partial P_2^3}{\partial z} = 0.$$

$$\frac{\partial P_3^1}{\partial z} = y\frac{\partial P_2^1}{\partial x} - (\tilde{a}x + \tilde{b}y + \tilde{c}z)\frac{\partial P_2^1}{\partial y}$$

$$\frac{\partial P_3^2}{\partial z} = y\frac{\partial P_2^2}{\partial x} - (\tilde{a}x + \tilde{b}y + \tilde{c}z)\frac{\partial P_2^2}{\partial y} - \bar{q}_2(x, y, z)$$

$$\frac{\partial P_3^3}{\partial z} = y\frac{\partial P_2^3}{\partial x} - (\tilde{a}x + \tilde{b}y + \tilde{c}z)\frac{\partial P_2^3}{\partial y}$$

R, S, T satisfy other equations. We thus have:

Proposition 36　*If ψ_1 satisfies the above system, then $D\psi_1(\frac{\partial}{\partial z} - y\frac{\partial}{\partial x} + (\tilde{a}x + \tilde{b}y + \tilde{c}z + q_2(x, y, z))\frac{\partial}{\partial y}) = \frac{\partial}{\partial z'} + O_3$, where O_3 is of order 3 at $(0, 0, 0)$.*

A particular solution to the above system is (set $\bar{q}_3 = \int_0^z \bar{q}_2(x, y, \tau)d\tau$) :

$$\bar{P}_2^1 = yz \qquad \bar{P}_2^2 = -(\tilde{a}x + \tilde{b}y + \frac{\tilde{c}z}{2})z \qquad \bar{P}_2^3 = 0$$

$$\bar{P}_3^1 = -\frac{\tilde{a}xz^2}{2} - \frac{\tilde{b}yz^2}{2} - \frac{\tilde{c}z^3}{3}$$

$$\bar{P}_3^2 = -\frac{\tilde{a}yz^2}{2} + (\tilde{a}x + \tilde{b}y)\frac{\tilde{b}z^2}{2} + \tilde{c}\frac{\tilde{b}z^3}{3} - \bar{q}_3(x, y, z)$$

$$\bar{P}_3^3 = 0.$$

Let now $\psi(x, y, z) = A\frac{\partial}{\partial x} + B\frac{\partial}{\partial y} + C\frac{\partial}{\partial z}$ be a diffeomorphism achieving

$$D\psi\left(\frac{\partial}{\partial z}\right) = \frac{\partial}{\partial z} - B\frac{\partial}{\partial x} + (aA + bB + cC + p_2)\frac{\partial}{\partial y} \quad \text{(collinear to } v\text{)}$$

with $A = x + \tilde{A}, B = y + \tilde{B}, C = z + \tilde{C}, \tilde{A}, \tilde{B}, \tilde{C}$ of order 2 and higher. In our framework, the image $\frac{\partial}{\partial x}, \frac{\partial}{\partial y}, \frac{\partial}{\partial z}$ will be the Darboux coordinates for α near $\bar{x}_0$. We then have:

$$\frac{\partial A}{\partial z} = -B, \frac{\partial B}{\partial z} = aA + bB + cC + p_2, \frac{\partial C}{\partial z} = 1.$$

Thus, $C = z + h(x, y)$, with h of order 2 and higher and

$$\frac{\partial A}{\partial z} = -B$$

$$\frac{\partial B}{\partial z} = aA + bB + c(z + h(x, y)) + p_2(x, y, z) + O_3.$$

Hence,

$$
\begin{pmatrix} A \\ B \end{pmatrix}(x,y,z)
$$

$$
= e^{\begin{pmatrix} 0 & -1 \\ a & b \end{pmatrix} z} \begin{pmatrix} A \\ B \end{pmatrix}(x,y,0) + \begin{pmatrix} 0 \\ ch(x,y)z + \int_0^3 p_2(x,y,\tau)d\tau \end{pmatrix}
$$

$$
+ c e^{\begin{pmatrix} 0 & -1 \\ a & b \end{pmatrix} z} \int_0^z \left(\tau \begin{pmatrix} 0 \\ 1 \end{pmatrix} - \tau^2 \begin{pmatrix} 0 & -1 \\ a & b \end{pmatrix} \begin{pmatrix} 0 \\ 1 \end{pmatrix} \right) d\tau + O_4
$$

$$
= e^{\begin{pmatrix} 0 & -1 \\ a & b \end{pmatrix} z} \begin{pmatrix} x + \widetilde{A}(x,y,0) \\ y + \widetilde{B}(x,y,0) \end{pmatrix}
$$

$$
+ \begin{pmatrix} -\frac{cz^3}{6} \\ czh + \int_0^z p_2(x,y,\tau)d\tau + \frac{cz^2}{2} + \frac{z^3}{6} + cb \end{pmatrix} + O_4
$$

$$
= \left(Id + \begin{pmatrix} 0 & -1 \\ a & b \end{pmatrix} z + \begin{pmatrix} -a & -b \\ ab & b^2 - a \end{pmatrix} \frac{z^2}{2} \right) \begin{pmatrix} x \\ y \end{pmatrix}
$$

$$
+ \left(Id + \begin{pmatrix} 0 & -1 \\ a & b \end{pmatrix} z \right) \begin{pmatrix} \widetilde{A}(x,y,0) \\ \widetilde{B}(x,y,0) \end{pmatrix}
$$

$$
+ \begin{pmatrix} -\frac{cz^3}{6} \\ czh + \int_0^z p_2(x,y,\tau)d\tau + \frac{cz^2}{2} + \frac{cbz^3}{6} \end{pmatrix} + O_4.
$$

We thus have:

Lemma 27

$$
\psi(x,y,z) = \left(x - yz + \widetilde{A}(x,y,0) - (ax+by)\frac{z^2}{2} - z\widetilde{B}(x,y,0) - \frac{cz^3}{6} \right) \frac{\partial}{\partial x}
$$

$$
+ \begin{pmatrix} y + (ax+by)z + \frac{cz^2}{2} + \widetilde{B}(x,y,0) + (abx + (b^2 - a)y)\frac{z^2}{2} \\ +(a\widetilde{A}(x,y,0) + b\widetilde{B}(x,y,0))z + czh + \int_0^z p_2(x,y,\tau)d\tau + cb\frac{z^3}{6} \end{pmatrix} \frac{\partial}{\partial y}
$$

$$
+ (z + h_2(x,y))\frac{\partial}{\partial z} + O_4.
$$

We now compute the differential of ψ with respect to a variation $(\delta x, \delta y, 0)$.

Differentiating, we find:

Lemma 28

$$D\psi\left(\delta x\frac{\partial}{\partial x}+\delta y\frac{\partial}{\partial y}\right)$$

$$=\left(\begin{array}{l}\delta x-\delta yz-(a\delta x+b\delta y)\frac{z^2}{2}+\frac{\partial\widetilde{A}}{\partial x}(x,y,0)\delta x+\frac{\partial\widetilde{A}}{\partial y}(x,y,0)\delta y\\[2mm]-z\frac{\partial\widetilde{B}}{\partial x}(x,y,0)\delta x-z\frac{\partial B}{\partial y}(x,y,0)\delta y+y\left(\frac{\partial h_2}{\partial x}\delta x+\frac{\partial h_2}{\partial y}\delta y\right)\end{array}\right)\frac{\partial}{\partial x}$$

$$+\left(\delta y+(a\delta x+b\delta y)z+\frac{\partial\widetilde{B}}{\partial x}(x,y,0)\delta x+\frac{\partial\widetilde{B}}{\partial y}(x,y,0)\delta y\right.$$

$$+(ab\delta x+(b^2-a)\delta y)\frac{z^2}{2}$$

$$+a\left(\frac{\partial\widetilde{A}}{\partial x}(x,y,0)\delta x+\frac{\partial\widetilde{A}}{\partial y}(x,y,0)\delta y\right)$$

$$+b\left(\frac{\partial\widetilde{B}}{\partial x}(x,y,0)\delta x+\frac{\partial\widetilde{B}}{\partial y}(x,y,0)\delta yz\right)$$

$$+c\left(\frac{\partial h_2}{\partial x}\delta x+\frac{\partial h_2}{\partial y}\delta y\right)z+\int_0^z\left(\frac{\partial p_2}{\partial x}\delta x+\frac{\partial p_2}{\partial y}\delta y\right)(x,y,\tau)d\tau$$

$$\left.-\left(\frac{\partial h_2}{\partial x}\delta x+\frac{\partial h_2}{\partial y}\delta y\right)(ax+by+cz)\right)\frac{\partial}{\partial y}+\gamma_1 v+h.o.$$

Proof.   Differentiating in restriction to $\delta z=0$ and observing that:

$$\left(\frac{\partial h_2}{\partial x}\delta x+\frac{\partial h_2}{\partial y}\right)\frac{\partial}{\partial z}$$

$$=\left(\frac{\partial h_2}{\partial x}\delta x+\frac{\partial h_2}{\partial y}\delta y\right)\left(\frac{\partial}{\partial z}-\left(y+\widetilde{B}(x,y,z)\right)\frac{\partial}{\partial x}\right.$$

$$+(aA+bB+cC+p_2)\frac{\partial}{\partial y}\Bigg)+(y+\widetilde{B}(x,y,z))\left(\frac{\partial h_2}{\partial x}\delta x+\frac{\partial h_2}{\partial y}\delta y\right)\frac{\partial}{\partial x}$$

$$-\left(\frac{\partial h_2}{\partial x}\delta x+\frac{\partial h_2}{\partial y}\delta y\right)(ax+by+cz+a\widetilde{A}+b\widetilde{B}+c\widetilde{C}+p_2)\frac{\partial}{\partial y}+h.o.$$

$$\square$$

We derive the claim.

ψ will be used to reduce v near $\bar{x}_0$. Near $\bar{x}_1$, we use ψ_1 given by (P), to reduce v. Then, $\psi_1|_{z=0}$ and $D\psi_1|_{z=0,\delta z=0}$ are identity maps.

2.7.4   *The v-transport maps*

We now study the map along the v-side of x^∞.

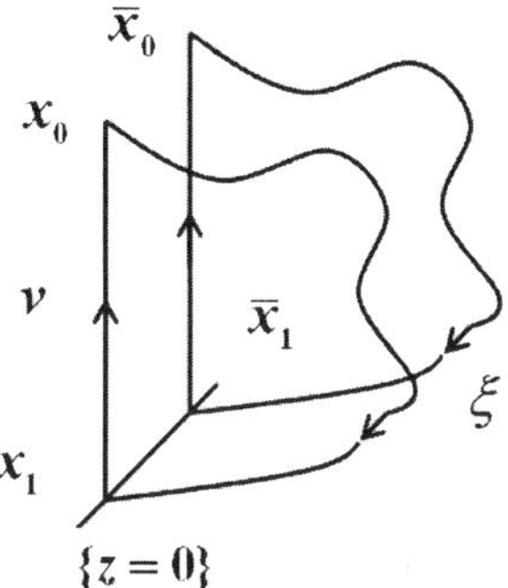

This map can be seen as a composition of two maps: a v-transport map into a section to v near $\bar{x}_1$ defined by $\{z = 0\}$, starting from a neighborhood of $\bar{x}_0$, composed with a transport along v during a small time $\tilde{s}$, starting from this section.

$\tilde{s}$ can be thought of in two ways: it is the gap needed to close up the curve along the z-variable. $\tilde{s}$ can also be thought of as the complement displacement which warrants a condition satisfied at a false critical point at infinity of the second kind — that $\det(d\ell_{\bar{x}} - Id) = 0$. $d\ell_{\bar{x}}$ is the differential of the transport map associated to a curve $\bar{x}$ of Γ_2. $\tilde{s}$, when thought of in these two ways, is different. But both ways of thinking of this displacement coincide at a false critical point at infinity of the second kind.

We first write the second part of the map, that is the v-transport during a small time $\tilde{s}$, starting from $z = 0$, near the $\bar{x}_1$-edge. Our map will denoted $\Gamma_{\tilde{s}}$. v, in these coordinates, is equal, up to a multiplication factor, to

$$\frac{\partial}{\partial z} - y\frac{\partial}{\partial x} + (a_1 x + b_1 y + \bar{p}_2)\frac{\partial}{\partial y}.$$

We then have:

Lemma 29 *The transport equation during the time $\tilde{s}$ reads:*

$$x = x_0 - y_0\tilde{s}_0 - (a_1 x_0 + b_1 y_0)\frac{\tilde{s}^2}{2} + h.o.$$

$$y = y_0 + (a_1 x_0 + b_1 y_0)\tilde{s} + (b_1(a_1 x_0 + b_1 y_0) - a_1 y_0)\frac{\tilde{s}^2}{2}$$

$$+ \int_{z_0}^{\tilde{s}+z_0} \bar{p}_2(x_0, y_0, \tau)d\tau + h.o.$$

$$z = z_0 + \tilde{s}.$$

Differentiating, we have (starting from $z_0 = 0$):

Lemma 30

$$\delta x = \delta x_0 - \delta y_0 \tilde{s} - (a_1 \delta x_0 + b_1 \delta y_0)\frac{\tilde{s}^2}{2} + h.o.$$

$$\delta y = \delta y_0(1 + b_1 \tilde{s} + (b_1^2 - a_1)\frac{\tilde{s}^2}{2}) + \delta x_0(a_1 \tilde{s} + a_1 b_1 \frac{\tilde{s}^2}{2}) + \int_0^{\tilde{s}} \delta \bar{p}_2 + h.o.$$

$$\delta z = \delta z_0 = 0.$$

Proof. The transport equations read:

$$\dot{x} = -y$$
$$\dot{y} = a_1 x + b_1 y + \bar{p}_2$$
$$\dot{z} = 1$$

from which the two lemmas follow.   □

Next, we write the v-map into $\{z = 0\}$. In $\{z = 0\}$, near the end-point $\bar{x}_1$, we pick up commuting vector-fields $\frac{\partial}{\partial x'}$ and $\frac{\partial}{\partial y'}$, with $\frac{\partial}{\partial x'} = \xi(x', y', 0)$. We pull them back along v and we obtain in this way three commuting vector-fields $\frac{\partial}{\partial x}, \frac{\partial}{\partial y}$ and v collinear to $\frac{\partial}{\partial z}$.

We can arrange so that $\frac{\partial}{\partial x}(0), \frac{\partial}{\partial y}(0)$ and $\frac{\partial}{\partial z}$ are the Darboux coordinates at $\bar{x}_0$ and that $\frac{\partial}{\partial x'}, \frac{\partial}{\partial y'}, v(\bar{x}_1)$ are the Darboux coordinates at $\bar{x}_1$. We can also arrange so that $\left(\frac{\partial}{\partial x}, \frac{\partial}{\partial y}\right)$ is derived from $\left(\frac{\partial}{\partial x'}, \frac{\partial}{\partial y'}\right)$ using a constant matrix. This involves a reparametrization of v (for the transport equations only) so that $\{z = 0\}$ at $\bar{x}_1$ is mapped into $\{z = 0\}$ at $\bar{x}_0$. Then, the transport along v from (x, y) to (x', y') and cutting into the v-section $\{z = 0\}$ reads through a constant matrix $\bar{D}$ which satisfies $\det(\bar{D} - Id) = 0$ and thus reads:

$$\bar{D} = Id + \begin{pmatrix} \bar{A}_1 & \lambda \bar{A}_1 \\ \bar{B}_1 & \lambda \bar{B}_1 \end{pmatrix}.$$

Therefore, the map on the v-side reads as:

$$\Gamma_{\bar{z}}\left(Id + \begin{pmatrix} \bar{A}_1 & \lambda \bar{A}_1 \\ \bar{B}_1 & \lambda \bar{B}_1 \end{pmatrix}\right)\begin{pmatrix} x \\ y \end{pmatrix} = \Gamma_{\bar{z}}\begin{pmatrix} x + \bar{A}_1(x + \lambda y) \\ y + \bar{B}_1(x + \lambda y) \end{pmatrix}.$$

Composing, we find:

Lemma 31

$$\Gamma_{\tilde{z}}\begin{pmatrix} x + \bar{A}_1(x + \lambda y) \\ y + \bar{B}_1(x + \lambda y) \end{pmatrix}$$

$$= \begin{pmatrix} x + \bar{A}_1(x + \lambda y) - (y + \bar{B}_1(x + \lambda y))\tilde{z} + O(|X|^3) \\ y + \bar{B}_1(x + \lambda y) + (a_1(x + \bar{A}_1(x + \lambda y)) + b_1(y + \bar{B}_1(x + \lambda y)))\tilde{z} \\ +O(|X|^3) \\ \tilde{z} \end{pmatrix}.$$

Differentiating, we introduce new variables:

$$\delta\hat{x}(0) = \delta x(0) + \bar{A}_1(\delta x(0) + \lambda\delta y(0))$$

$$\delta\hat{y}(0) = \delta y(0) + \bar{B}_1(\delta x(0) + \lambda\delta y(0))$$

and we have:

Lemma 32

$$\delta\hat{x} = \delta\hat{x}(0)(1 - a_1\frac{\tilde{z}^2}{2}) - \delta\hat{y}(0)(\tilde{z} + b_1\frac{\tilde{z}^2}{2}) + h.o$$

$$\delta\hat{y} = \delta\hat{y}(0)(1 + b_1\tilde{z} + (b_1^2 - a_1)\frac{\tilde{z}^2}{2}) + a_1\delta\hat{x}(0)(\tilde{z} + b_1\frac{\tilde{z}^2}{2})$$

$$+ \int_0^{\tilde{z}}\left(\frac{\partial\bar{p}_2}{\partial x}\delta\hat{x}(0) + \frac{\partial\bar{p}_2}{\partial y}\delta\hat{y}(0)\right)d\tau + h.o.$$

Ultimately, $\tilde{z}$ is a function of (x, y, z), but no differentiation with respect to $\tilde{z}$ is involved in the computation of $\det(d\ell_{\tilde{x}} - Id)$.

Using the above lemmas, with ψ as in Lemma 27 and ϕ_s being the one-parameter group of ξ, a_c the characteristic length ($a_c = a_c(x, y, z)$), we derive:

Lemma 33

$$\phi_{a_c} \circ \psi(x, y, z)$$

$$= \left(x - yz - (ax + by)\frac{z^2}{2} + \tilde{A}(x, y, 0) - z\tilde{B}(x, y, 0) - c\frac{z^3}{6} + a_c \right) \frac{\partial}{\partial x}$$

$$+ \left(y + (ax + by)z + \frac{cz^2}{2} + \tilde{B}(x, y, 0) + (abx + (b^2 - a)y)\frac{z^2}{2} + (a\tilde{A} + b\tilde{B})z \right.$$

$$+ czh_2 + \int_0^z p_2 + \frac{cbz^3}{6})\frac{\partial}{\partial y} + (z + h_2 + \frac{cz^2}{2} + \tilde{B}(x, y, 0))\frac{\partial}{\partial z} + O(|X|^3)$$

$$D\phi_{a_c} \circ D\psi(\delta x \frac{\partial}{\partial x} + \delta y \frac{\partial}{\partial y})$$

$$= (\delta x - z\delta y - (a\delta x + b\delta y)\frac{z^2}{2} + \frac{\partial \tilde{A}}{\partial x}(x, y, 0)\delta x + \frac{\partial \tilde{A}}{\partial y}(x, y, 0)\delta y$$

$$- z\frac{\partial \tilde{B}}{\partial x}(x, y, 0)\delta x - z\frac{\partial \tilde{B}}{\partial y}(x, y, 0)\delta y + (y\frac{\partial h_2}{\partial x}\delta x + y\frac{\partial h_2}{\partial y}\delta y) + h.o)\frac{\partial}{\partial x}$$

$$+ (\delta y + (a\delta x + b\delta y)z + \frac{\partial \tilde{B}}{\partial x}(x, y, 0)\delta x + \frac{\partial \tilde{B}}{\partial y}(x, y, 0)\delta y$$

$$+ (ab\delta x + (b^2 - a)\delta y)\frac{z^2}{2}$$

$$+ \left(a(\frac{\partial \tilde{A}}{\partial x}\delta x + \frac{\partial \tilde{A}}{\partial y}(x, y, 0)\delta y + b\left(\frac{\partial \tilde{B}}{\partial x}(x, y, 0)\delta x + \frac{\partial \tilde{B}}{\partial y}(x, y, 0)\delta y \right) \right) z$$

$$+ c\left(\frac{\partial h_2}{\partial x}\delta x + \frac{\partial h_2}{\partial y}\delta y \right) z + \int_0^z \left(\frac{\partial p_2}{\partial x}\delta x + \frac{\partial p_2}{\partial y}\delta y \right)(x, y, \tau)d\tau -$$

$$\left(\frac{\partial h_2}{\partial x}\delta x + \frac{\partial h_2}{\partial y}\delta y \right)(ax + by + cz))\frac{\partial}{\partial y} + \gamma_2 v + h.o.$$

Observe that $\tilde{z}$, on the v-side of the curve, should be equal to the third component of $\phi_{a_c} \circ \psi(x, y, z)$ so that the curve closes along the z-axis. We denote in what follows $\bar{a}_0$ the quantity $a_c(0, 0, 0)$.

2.7.5   *The equations of the characteristic manifold near x^∞; the equations of a critical point*

2.7.5.1   *The characteristic manifold for the unperturbed problem*

The characteristic manifold near x^∞ is the one-dimensional sub-manifold of Γ_2 of curves made of one ξ-piece and one v-piece, the ξ-piece being of characteristic length. We have:

Proposition 37 *Let (x, y, z) be the base point of a curve of the characteristic manifold near x^∞. ((x, y, z) track a point similar to x_0). Then (x, y, z) satisfies*

$$\bar{A}_1(x+\lambda y)-(y+\bar{B}_1(x+\lambda y))\tilde{z} = \tilde{A}(x, y, 0)+a_c(x, y, z)-\bar{a}_0-yz+O(|X|^3)$$

$$\bar{B}_1(x + \lambda y) + (a_1(x + \bar{A}_1(x + \lambda y)) + b_1(y + \bar{B}_1(x + \lambda y))\tilde{z} = (ax + by)z + \frac{cz^2}{2} + \tilde{B}(x, y, 0) + O(|X|^3)$$

$$\tilde{z} = z + h_2 + \frac{cz^2}{2} + \tilde{B}(x, y, 0) + O(|X|^3)$$

The first system of equations is equivalent to:

$$z = a_2 x + b_2 y + \frac{1}{da_c(v)}\tilde{A}_1(x + \lambda y) + O(|X|^2)\,(a_2, b_2 \text{ independent of } \bar{A}_1)$$

$$\bar{B}_1(x + \lambda y) = (x + \lambda y)(\theta x + \mu y) + y^2 \tilde{B}_2(-\lambda, 1, 0) + ((a - a_1)x + (b - b_1)y$$
$$+ \frac{c}{2}(a_2 x + b_2 y))(a_2 x + b_2 y) + O(|X|^3)$$

$\tilde{B}_2$ is the (initial) term of order 2 in $\tilde{B}(x, y, 0)$.

Proof. We write that

$$\phi_{a_c} \circ \psi(x, y, z) = \begin{pmatrix} \Gamma_z(Id + \begin{pmatrix} \bar{A}_1 & \lambda\bar{A}_1 \\ \bar{B}_1 & \lambda\bar{B}_1 \end{pmatrix})\binom{x}{y}) \\ \tilde{z} \end{pmatrix}.$$

This yields using Lemmas 32 and 33:

$$x + \bar{A}_1(x + \lambda y) - (y + \bar{B}_1(x + \lambda y))\tilde{z}$$
$$= x - yz + \tilde{A}(x, y, 0) + a_c - a_0 + O(|X|^3)$$
$$y + \bar{B}_1(x + \lambda y) + (a_1(x + \tilde{A}(x + \lambda y)) + b_1(y + \tilde{B}_1(x + \lambda y)))\tilde{z}$$
$$= y + (ax + by)z + \frac{cz^2}{2} + \tilde{B}(x, y, 0) + O(|X|^3)$$
$$\tilde{z} = z + h_2 + \frac{cz^2}{2} + \tilde{B}(x, y, 0) + O(|X|^3).$$

The first equation can be rewritten in the form:

$$da_{c,0}(x) = x\frac{\partial a_c}{\partial x}(0) + y\frac{\partial a_c}{\partial y}(0) + z\frac{\partial a_c}{\partial z}(0) = \tilde{A}_1(x + \lambda y) + O(|X|^2).$$

Since we assume that $\frac{\partial a_c}{\partial z}(0) = a'_c(v)(0)$ is non zero, this yields:

$$z = a_2 x + b_2 y + \frac{1}{a'_c(v)(0)}\tilde{A}_1(x + \lambda y) + O(|X|^2).$$

The second equation then yields:

$$\tilde{B}_1(x + \lambda y) = (x + \lambda y)(\theta x + \mu y) + y^2 \tilde{B}_2(-\lambda, 1, 0) + (a_2 x + b_2 y)((a - a_1)x$$

$$+ (b - b_1)y + \frac{c}{2}(a_2 x + b_2 y)) + 0(|x|^3)$$

as stated. $\qquad\qquad\qquad\qquad\qquad\qquad\qquad\qquad\qquad\qquad\qquad\qquad\qquad\qquad\square$

2.7.6   *Critical points, vanishing of the determinant*

Let γ_s be the one-parameter group of $v, \tilde{x}$ be a curve of Γ_2 near x^∞, with base point $\tilde{x}_0$ and v-length $\tilde{s}_0$.

$\tilde{x}$ is a critical point of the functional $a = $ length along ξ near x^∞ if the length of the ξ-piece of x^∞ is characteristic and if $\det(D\phi_{a_c} \circ D\psi - D\gamma_{\tilde{s}_0}) = 0$.

$D\gamma_{\tilde{s}_0}$ is taken here from the coordinates which reduce v near x_0 to the Darboux coordinates near x_1. Since a_c is characteristic,

$$D\phi_{a_c} \circ D\psi \left(\frac{\partial}{\partial z}\right) = \theta_1 v.$$

Also,

$$D\gamma_{\tilde{s}_0} \left(\frac{\partial}{\partial z}\right) = \theta_2 v \quad \text{with } \theta_1 \neq \theta_2 \text{ generically.}$$

Thus, denoting

$$\pi_2 : \ \text{Span} \left(\frac{\partial}{\partial x}, \frac{\partial}{\partial y}, v\right) \longrightarrow \text{Span} \left(\frac{\partial}{\partial x}, \frac{\partial}{\partial y}\right)$$

the projection onto the space generated by the two first vectors of the Darboux reduction near x_1, parallel to v, we must have:

$$\pi_2 \circ D\phi_{a_c} \circ D\psi \begin{pmatrix} \delta x \\ \delta y \\ 0 \end{pmatrix} = D\gamma_{\tilde{s}_0} \begin{pmatrix} \delta x \\ \delta y \\ 0 \end{pmatrix} = \begin{pmatrix} \delta x \\ \delta y \end{pmatrix}$$

for some non zero $(\delta x, \delta y)$.

We then have:

Lemma 34   *The coordinates (x, y, z) of the base point $\tilde{x}_0$ of the critical curve $\tilde{x}$ satisfy:*

$$\bar{A}_1 \left(b_1 \tilde{z} - \frac{\partial \tilde{B}(x, y, 0)}{\partial y} - bz + \lambda \left(\frac{\partial \tilde{B}}{\partial x}(x, y, 0) + az - a_1 \tilde{z}\right)\right)$$

$$+ \bar{B}_1 \left(\frac{\partial \widetilde{A}}{\partial y}(x, y, 0) + \tilde{z} - z - \lambda \frac{\partial \widetilde{A}}{\partial x}(x, y, 0) \right) + O(|X|^2) = 0.$$

Proof. Using Lemma 32, the condition on the vanishing of the determinant reads:

$$O(|X|^2)$$

$$= \begin{vmatrix} \widetilde{A}_1 - \dfrac{\partial \bar{A}}{\partial x} - \bar{B}_1 \tilde{z} & z - \dfrac{\partial \widetilde{A}}{\partial y} + \lambda \bar{A}_1 - \lambda \bar{B}_1 \tilde{z} - \tilde{z} \\[2mm] b_1 \bar{B}_1 \tilde{z} + \bar{B}_1 - \dfrac{\partial \widetilde{B}}{\partial x} - az + a_1 \tilde{z} + a_1 \bar{A}_1 \tilde{z} & b_1(1 + \lambda \bar{B}_1)\tilde{z} + \lambda B_1 - \dfrac{\partial \widetilde{B}}{\partial y} - bz + \lambda a_1 \bar{A}_1 \tilde{z} \end{vmatrix}.$$

This yields: $\bar{A}_1 (b_1(1 + \lambda \bar{B}_1)\tilde{z} - \frac{\partial \widetilde{B}}{\partial y} - bz + \lambda a_1 \bar{A}_1 \tilde{z}) - \lambda \bar{B}_1 \left(\frac{\partial \widetilde{A}}{\partial x} + \bar{B}_1 \tilde{z} \right) -$
$\lambda \bar{A}_1 (b_1 \bar{B}_1 \tilde{z} - \frac{\partial \widetilde{B}}{\partial x} - az + a_1 \tilde{z} + a_1 \bar{A}_1 \tilde{z}) - \bar{B}_1 (z - \frac{\partial \widetilde{A}}{\partial y} - \lambda \bar{B}_1 \tilde{z} - \tilde{z}) + O(|X|^2) = 0.$

$\square$

Lemma 35 follows.

2.7.7 *Introducing the perturbation*

To start with, we observe that the formulae for the characteristic manifold, for γ_2 and for the false critical points at infinity involve p_2, but only at a higher order. a_c involves p_2 at second order but we have:

Lemma 35 *Using a cut-off function ω near x_0, we may suppress p_2 from v. This does not create new critical points at infinity.*

Lemma 35 will follow from the same arguments (easier form) than the ones used in order to prove that the sign of $a'_c(v) = da_c(v)$ can be changed without creating new critical points at infinity.

Assuming that p_2 is zero, v is, in the Darboux coordinates near x_0, equal up to a collinearity coefficient to $\frac{\partial}{\partial z} - y \frac{\partial}{\partial x} + (x + by + cz)\frac{\partial}{\partial y}$. We want to replace c with $c_\tau(x, y, z), \tau \in [0, 1]$, a modification which will take place on a small neighborhood of x_0 with $c_0 = c$ and $c_1(0, 0, 0) \cdot c$ negative.
c_τ will be of the form:

$$c + \tau \omega(x, y, z)(\tilde{c}(z) - c)$$

$\tilde{c}(z)$ is equal to c for $|z| \geq z_0$, where z_0 is small. $\omega(x, y, z)$ is a cut-off function which is constructed in a non-standard way. In fact, the modified v is defined in a first stage as a non-autonomous modification of v which we prove later to be autonomous. For this, we consider the plane $\{z = 0\}$ at

x_0. The plane is a section to v. The flow of the modified v can be written up to re-parametrization ($|z| \leq z_0$):

$$\dot{z} = 1$$
$$\dot{y} = ax + by + c_\tau(x, y, z)z$$
$$\dot{x} = -y.$$

We want to define $c_\tau(x, y, z)$ and for this purpose, we start with the section $\{z = 0\}$. On this section, $c_\tau(x, y, 0)$ is defined to be

$$c_\tau(x, y, 0) = c + \tau w(x, y)(\tilde{c}(0) - c)$$

$w(x, y)$ is a cut-off function equal to 1 on a small disk of radius η and equal to zero outside of the disk of radius 2η. $\tilde{c}$ will be defined later. At each $(x^0, y^0, 0)$,

We define a differential equation:

$$\dot{x} = -y \qquad\qquad x(0) = x^0, y(0) = y^0$$
$$\dot{y} = ax + by + (c + \tau w(x^0, y^0)(\tilde{c}(s) - c))s \tag{*}$$

This defines $(x(s), y(s))$ which we view in three dimensions:

$$(x(s), y(s), s)$$

which satisfies:

$$\dot{x} = -y$$
$$\dot{y} = ax + by + (c + w(x^0, y^0) \cdot \tau(\tilde{c}(s) - c))s$$
$$x(0) = x^0, y(0) = y^0, z(0) = 0 \tag{*}$$
$$\dot{z} = 1$$

and we have:

Lemma 36 *For each s, (*) defines a diffemorphism from $\{z = 0\}$ to $\{z = s\}$. Thus, (*) defines a vector-field v in* ker α.

We differentiate (*) and we find:

$$\overline{\dot{\delta x}} = -\delta y$$
$$\dot{\delta y} = a\delta x + b\delta y + \tau \delta w s(\tilde{c}(s) - c)$$

δw is taken with respect to variations $(\delta x^0, \delta y^0)$ in the initial conditions.

Denoting A the matrix $\begin{pmatrix} 0 & -1 \\ a & b \end{pmatrix}$, we have,

$$\begin{pmatrix} \delta x \\ \delta y \end{pmatrix} = A \begin{pmatrix} \delta x \\ \delta y \end{pmatrix} + O\left(\frac{z}{\eta}\right) \begin{pmatrix} \delta x^0 \\ \delta y^0 \end{pmatrix}.$$

Thus,

$$\begin{pmatrix} \delta x \\ \delta y \end{pmatrix} = e^{sA} \left[\begin{pmatrix} \delta x^0 \\ \delta y^0 \end{pmatrix} + \int_0^s e^{-\xi A} O\left(\frac{z}{\eta}\right) \begin{pmatrix} \delta x^0 \\ \delta y^0 \end{pmatrix} \right]$$

$$= e^{sA} \left(Id + O\left(\frac{z^2}{\eta}\right) \right) \begin{pmatrix} \delta x^0 \\ \delta y^0 \end{pmatrix} = e^{sA} \left(Id + O\left(\frac{\inf(z,z_0)^2}{\eta}\right) \right) \begin{pmatrix} \delta x^0 \\ \delta y^0 \end{pmatrix}.$$

Next, we have:

Lemma 37 $\tilde{c}(z)$ *can be chosen so that:*

(i) $\tilde{c}(0) = -c$ *and for the time 1 of the homotopy,* $a'_c(v)(x_0)$ *has changed sign*

(ii) The time z_0-map of v, starting from $\{z = 0\}$ to $\{z = z_0\}$, does not change along the homotopy.

(iii) x^∞ *has the same transport map in a section to v such as $\{z = 0\}$ at x_0.*

(iv) $\det\left(d\ell_{x_0} - Id|_{\{z=0\}}\right) = 0$ *and* x^∞ *remains a false critical point at infinity of the same nature along the homotopy.*

Proof. We use (*) the time z_0-map:

$$\begin{pmatrix} x \\ y \end{pmatrix}(z_0) = e^{z_0 A} \left(\begin{matrix} \begin{pmatrix} x^0 \\ y^0 \end{pmatrix} + \left(c\int_0^{z_0} e^{-sA} s\,ds\right)\begin{pmatrix} 0 \\ 1 \end{pmatrix} \\ +\omega(x^0,y^0)\tau \int_0^{z_0} e^{-sA} s(\tilde{c}(s) - c)\,ds \begin{pmatrix} 0 \\ 1 \end{pmatrix} \end{matrix} \right). \qquad \square$$

We thus want to have:

$$\int_0^{z_0} e^{-sA} s(\tilde{c}(s) - c)\,ds01 = \begin{pmatrix} 0 \\ 0 \end{pmatrix}. \qquad (**)$$

We also ask that $\tilde{c}(s) = c$ after $\frac{3z_0}{4}$ so that our new vector-field glues up smoothly with the original v. We prove later that (**) can be satisfied, with $\tilde{c}(0) = -c, \tilde{c}(s) = c$ for $|s| \geq \frac{3z_0}{4}$. Lemma 37 follows.

We then have:

Lemma 38   *Under (**), the one-parameter group $\tilde{\gamma}_s$ of the new vector-field $\tilde{v}$ reads for $|s| \geq \frac{3z_0}{4}$.*

$$\tilde{\gamma}_s(x, y, z) = \gamma_s(x, y, z) + O\left(\inf(z^2, (s + z_0)^2)\right).$$

Proof. $\tilde{\gamma}_s$ and γ_s differ only if z or $s + z$ is in $[-z_0, z_0]$. Assume that $z \in [-z_0, z_0]$. Then $|z + s| \geq z_0$ and we write:

$$\tilde{\gamma}_s(x, y, z) = \tilde{\gamma}_{z+s}(\tilde{\gamma}_{-z}(x, y, z)) = \gamma_s(x, y, z) + O(|\tilde{\gamma}_{-z}(x, y, z) - \gamma_{-z}(x, y, z)|)$$
$$= \gamma_s(x, y, z) + O(|\tilde{\gamma}_{-z}(x, y, 0) - \gamma_{-z}(x, y, 0)|).$$

Using (*), we derive that $|\tilde{\gamma}_{-z}(x, y, 0) - \gamma_{-z}(x, y, 0)| = O(z^2)$, thus $\tilde{\gamma}_s(x, y, z) = \gamma_s(x, y, z) + O(z^2)$ as claimed. If $z + s \in [-z_0, z_0]$, we find $O((z + s)^2)$, again as claimed. Lastly, we claim that: $\qquad\qquad\square$

Lemma 39   *Let $P = \{dz = 0\}$ at x_0, which we transport by $\xi = \frac{\partial}{\partial x}$ to x_1. Let $\pi_2 : \mathbb{R}^3 \to \mathbb{P}(x, y, z)$ be the projection along v.*
     Then, the solution $\tilde{s}(x, y, z)$ of the equation

$$\det\left(\left(\pi_2 \circ d\phi_{a_c} \circ d\psi - \pi_2 \circ d\gamma_{\tilde{s}}\right)\big|_{p(x,y,z)}\right) = 0$$

is perturbated by $o(z)$ when v is modified into $\tilde{v}$ and γ_s into $\tilde{v}_s$.

Observation 1. The above condition is equivalent for a closed curve having a characteristic ξ-piece to the vanishing of the determinant.

Observation 2. $\tilde{s}$ does not represent the time of transport along v or $\tilde{v}$, rather the section $z = \tilde{s}$ where the flow-lines of v or $\tilde{v}$ are made to abut. Hence, since P identifies after $\frac{\partial}{\partial x}$ transport with the tangent space to those sections, we do not need π_2 in front of $d\gamma_{\tilde{s}}$ once we restrict to variations $(\delta x, \delta y, 0)$. π_2, in front of $d\phi_{a_c} \circ d\psi$ or $d\phi_{\tilde{a}_c} \circ d\psi$ is taken at $\gamma_{\tilde{s}}(x, y, z)$ or $\tilde{\gamma}_{\tilde{s}}(x, y, z)$. These two points are $O(z^2)$ apart (we need only to go back to the section $\{z = 0\}$). Hence π_2 is changed by $O(z^2)$.

Proof.   Since γ_s is perturbed into $\tilde{\gamma}_s$ which is $O(z^2)$ away from it, $d\gamma_s$ is close to $d\tilde{\gamma}_s$ only at $O(z)$, so that we would expect $\tilde{s}$ to perturbed $0(z)$, not $o(z)$. This gain from $O(z)$ to $o(z)$ is due to the fact that the differentials are considered with respect to $\left(\frac{\partial}{\partial x}, \frac{\partial}{\partial y}\right)$, that is with respect to variations $\delta x^0, \delta y^0$ in the $(x^0, y^0, 0)$-plane. We have shown, in the proof of Lemma 38, that the perturbation in the $d\gamma_s$-differential was then $O\left(\frac{\inf(z_0, z_0)^2}{\eta}\right)$, which is $o(z)$ if z_0 is chosen to be $o(\eta)$ (observe that then $\frac{\partial v}{\partial x} = a + 0(\frac{z_0}{\eta}) = a + o(1)$ and β remains a contact form). On the other hand, $d\phi_{a_c} \circ d\psi$ changes only at second order as v is perturbed because ψ in the above statement is related to the original $v(D\psi = \bar{A}\delta x + \bar{B}\delta y + \delta h_2 \frac{\partial}{\partial z} = \bar{A}\frac{\partial}{\partial x} + \bar{B}\frac{\partial}{\partial y} + \delta h_2 v + O_2 =$

$\bar{A}\frac{\partial}{\partial x} + \bar{B}\frac{\partial}{\partial y} + \delta h_2 \tilde{v} + O_2)$ and $d\phi_{a_c}$ is not sensitive to the modification in a_c. Lemma 2.14 follows. $\qquad\square$

2.7.8 *The characteristic manifold for the perturbed problem; the determinant equations*

As we perturb v to change the sign of $da_c(v)(x_0)$, we modify the one-parameter group of v. We have estimated this perturbation previously. We found that, for $s \geq 2z_0$

$$\tilde{\gamma}_s(x, y, z) = \gamma_s(x, y, z) + D\gamma_{z+s}(\tilde{\gamma}_{-z}(x, y, z) - \gamma_{-z}(x, y, z) + 0(z^4).$$

On the other hand,

$$\overline{\dot{\overline{\tilde{\gamma}_{-s}(x, y, z) - \gamma_{-s}(x, y, z)}}}$$

$$= \begin{pmatrix} \dot{X} \\ \dot{Y} \\ 0 \end{pmatrix} \xrightarrow{B} \overbrace{\begin{pmatrix} A & \begin{matrix}0\\0\end{matrix} \\ 0\,0\,0 \end{pmatrix}} \begin{pmatrix} X \\ Y \\ 0 \end{pmatrix} - \tau\omega(\tilde{c}(z-s) - c)(z-s)\begin{pmatrix} 0 \\ 1 \\ 0 \end{pmatrix}.$$

Thus,

$$\tilde{\gamma}_{-z}(x, y, z) - \gamma_{-z}(x, y, z) = -e^{zB}\tau\omega\int_0^z e^{-sB}(\tilde{c}(z-s) - c)(z-s)ds \begin{pmatrix} 0 \\ 1 \\ 0 \end{pmatrix}$$

$$= -\tau\omega\int_0^z e^{sB}(\tilde{c}(s) - c)sds \begin{pmatrix} 0 \\ 1 \\ 0 \end{pmatrix}.$$

Since $D_{x_0}\psi = Id$, the map on the v-side is changed into (observe that $D\gamma_{z+s} = Id$ in coordinates where $v = \frac{\partial}{\partial z}$): Set $\tilde{X} = (x, y, z)$

$$\Gamma_z\left(Id + \begin{pmatrix} \bar{A}_1 & \lambda\bar{A}_1 \\ \bar{B}_1 & \lambda\bar{B}_1 \end{pmatrix}\right)\left\{\begin{pmatrix} x \\ y \end{pmatrix} - \tau\omega\int_0^z (\tilde{c}(s) - c)sds\begin{pmatrix} 0 \\ 1 \end{pmatrix}\right\} + O(|\tilde{X}|^3)$$

$$= \begin{pmatrix} x + \bar{A}_1(x + \lambda y) - (y + \bar{B}_1(x + \lambda y))\tilde{z} + O(|\tilde{X}|^2) \\ y + \bar{B}_1(x + \lambda y) + (a_1(x + \bar{A}_1(x + \bar{A}_1(x + \lambda y)) + b_1(x + \lambda y)))\tilde{z} + O(|\tilde{X}|^2) \end{pmatrix}.$$

On the other side, that is the side of ξ, we have:

$$\phi_{a_c} \circ \psi(x,y,z)$$

$$= (x - yz + (ax+by)\frac{z^2}{2} + \tilde{A}(x,y,0) - z\tilde{B}(x,y,0) - c\frac{z^3}{6} + a_c)\frac{\partial}{\partial x}$$

$$+ (y + (ax+by)z + c\frac{z^2}{2} + \tilde{B}(x,y,0) + (abx + (b^2-a)y)\frac{z^2}{2} + (a\tilde{A}+b\tilde{B})z$$

$$+ czh_2 + \int_0^z p_2 + +cb\frac{z^3}{6})\frac{\partial}{\partial y} + (z + h_2 + c\frac{z^2}{2} + \tilde{B}(x,y,0))\frac{\partial}{\partial z} + 0(|X|^3).$$

This formula does not involve $\tau\omega\tilde{c}$ because ψ is the map such that ψ^{-1} reduces the original v, before perturbation. It is however modified with respect to the unperturbed v because a_c is modified.

A companion of the above formula is the formula for $d\phi_{a_c} \circ d\psi$ which is unperturbed when v is modified. We have seen how $\gamma_{\tilde{s}}$ and $\tilde{s}(x,y,z)$ were modified with our perturbation (only with $o(x,y,z)$). This argument extends to a modification of v which gets rid of p_2 and $\bar{p}_2$ at each of x_0, x_1 : $d\phi_{a_c} \circ d\psi$ is modified only at second order; $\gamma_{\tilde{s}}$ is modified at the $o(|X|^2)$ order if the support of the modification is small, so that the differential is modified only at the $o(|X|)$-order. So is $\tilde{s}(x,y,z)$. Therefore, Lemma 39 holds with $0(|X|^2)$ replaced by $o(|X|)$.

We thus have:

Proposition 38   *1) The equations of the characteristic manifold read:*

$$a_c(x,y,z) - \bar{a}_c = \bar{A}_1(x+\lambda y) + O(|X|^2)$$

$$\bar{B}_1(x+\lambda y) = O(|X|^2)$$

$$\tilde{z} = z + O(|X|^2).$$

2) The function $a_c(x,y,z)$ has the expansion:

$$a_c = \bar{a}_c + \frac{(a-a_1)(x+\lambda y) + ((b-b_1) - \lambda(a-a_1))y + c_\tau z}{a_1} + O(|X|^2).$$

3) The determinant equation reads:

$$\bar{A}_1\left((a-a_1) - (b-b))z + \lambda\frac{\partial\tilde{B}}{\partial x}(x,y,0) - \frac{\partial\tilde{B}}{\partial y}(x,y,0)\right)$$

$$+ \bar{B}_1\left(\frac{\partial\tilde{A}}{\partial y}(x,y,0) - \lambda\frac{\partial\tilde{A}}{\partial x}(x,y,0)\right) + o(|X|) = 0.$$

Proof. The proof of 3) follows readily from Lemma 34 and our remarks above. The proof of 2) follows from the definition of a_c. Observe that $\frac{\partial}{\partial y}, \frac{\partial}{\partial x}$ and $\xi = \frac{\partial}{\partial x}$ are ξ-transported. For v to come back collinear to itself in the ξ-transport, we need:

$$a_1(x + s - \bar{a}_c)b_1 y + \tilde{\bar{p}}_2(x + s - \bar{a}_c, y, z) = ax + by + c_\tau z + \tilde{p}_2(x, y, z).$$

This happens for $s = a_c(x, y, z)$ and yields 2) immediately. $\qquad\square$

The proof of 1) is straightforward using the expansions provided above. We proceed now with the **the proof of the theorem.** Combining 1) and 2), we derive that:

$$((b - b_1) - \lambda(a - a_1))\, y + c_\tau z = O(y^2 + z^2).$$

Let $\tilde{A}_2(x, y, 0), \tilde{B}_2(x, y, o)$ the first terms (of order 2) of $\tilde{A}(x, y, 0)$, $\tilde{B}(x, y, 0)$. After using the fact that $x + \lambda y = O(|X|^2)$, the determinant equation yields:

$$\bar{A}_1 \left((b - b_1) - \lambda(a - a_1)\right) z + 2 \left(\bar{A}_1 \tilde{B}_2(-\lambda, 1, 0) - \bar{B}_1 \tilde{A}_2(-\lambda, 1, 0)\right) y$$

$$= o(|y| + |z|).$$

We thus have at a critical point:

$$((b - b_1) - \lambda(a - a_1))\, y + c_\tau z$$
$$= o(|y| + |z|)$$
$$\bar{A}_1 \left((b - b_1) - \lambda(a - a_1)\right) z + 2(\bar{A}_1 \tilde{B}_2(-\lambda, 1, 0) - \bar{B}_1 \tilde{A}_2(-\lambda, 1, 0)) y$$
$$= o(|y| + |z|).$$

This yields $y = z = 0$ if the determinant is far from zero. Since $c_\tau = c$ when we want to rid of $p_2, \bar{p}_2$, we see that such a homotopy will not create new critical points at infinity. This proves Lemma 28.

The determinant of the above "linear" system is:

$$\Delta = \bar{A}_1 \left((b - b_1) - \lambda(a - a_1)\right)^2 - 2c_\tau \left(\bar{A}_1 \tilde{B}_2(-\lambda, 1, 0) - \bar{B}_1 \tilde{A}_2(-\lambda, 1, 0)\right).$$

We now claim that:

Lemma 40 *x^∞ is of index 1 at infinity if and only if*

$$2\bar{A}_1 c \left(\bar{A}_1 \tilde{B}_2(-\lambda, 1, 0) - \bar{B}_1 \tilde{A}_2(-\lambda, 1, 0)\right) < \bar{A}_1^2 \left((b - b_1) - \lambda(a_1 - a)\right)^2.$$

Assuming this lemma, we see that we can modify the sign of c, i.e. introduce c_τ and keep Δ away from zero. c_τ must satisfy (**) above, i.e.

$$\int_0^z e^{-sA} s(\tilde{c}(s) - c)ds \begin{pmatrix} 0 \\ 1 \end{pmatrix} = \begin{pmatrix} 0 \\ 0 \end{pmatrix} \text{ for } z_0 > 0 \text{ and } z_0 < 0 \quad (**)$$

$$\text{with } A = \begin{pmatrix} 0 & -1 \\ a & b \end{pmatrix}.$$

$\tilde{c}$, assuming that c is positive, behaves as in the drawing on next page.

Clearly, for ε small enough, Δ will be away from zero on $[-z_0, -\gamma_1], [\gamma_1, z_0]$. On $[-\gamma_0, \gamma_0]$, the statement follows from Lemma 39 above. We need now to show that we can choose γ_0, γ_1, z_0 so that (**) holds. We expand (**) at first order i.e. replace e^{-sA} by $1 - sA$. We derive:

$$\int_0^z (\tilde{c} - c)s \begin{pmatrix} 1 & s \\ -sa & 1 - sb \end{pmatrix} \begin{pmatrix} 0 \\ 1 \end{pmatrix} ds = O(z_0^4)$$

that is

$$\int_0^{z_0} (\tilde{c} - c)s^2 ds = O(z_0^4)$$

$$\int_0^{z_0} (\tilde{c} - c)s(1 - sb) = O(z_0^4).$$

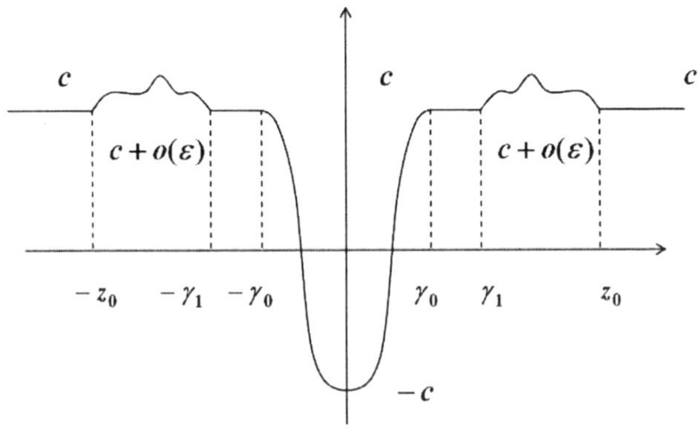

Take $\gamma_0 = O(e^{-1/z_0})$. Impose $\tilde{c}$ as above, on the graph, on $[-\gamma_0, \gamma_0]$. The above conditions then reread:

$$\int_{\gamma_1}^{z_0} (\tilde{c} - c)s^2 ds = O(z_0^4) \qquad \text{(also on } [-z_0, -\gamma_1])$$

$$\int_{\gamma_1}^{z_0} (\tilde{c} - c)s(1 - sb)ds = O(z_0^4).$$

Choose $\gamma_1 = \frac{1}{2}z_0$. This rereads:

$$\int_{1/2}^{1} (\tilde{c} - c)x^2 dx = O(z_0)$$

$$\int_{1/2}^{1} (\tilde{c} - c)x \, dx = O(z_0^2)$$

$\tilde{c} - c$ are re-scaled. The above conditions are easy to fulfill with small perturbations of zero. The argument clearly extends to the full e^{-sA}.

Proof. [Proof of Lemma 40]

First, let us observe that, since $v = \gamma \left(\frac{\partial}{\partial z} - y \frac{\partial}{\partial x} + (\tilde{a}x + \tilde{b}y + \tilde{c}z) \frac{\partial}{\partial y}, \right)$ where $\tilde{a}, \tilde{b}, \tilde{c}$ are constants (we are looking here at the unperturbed problem after suppressing $p_2, \bar{p}_2$), we derive from the normalization $d\alpha(v, [\xi, v]) = -1$ that $\gamma = \pm \frac{1}{\sqrt{\tilde{a}}}$.

Assuming $\gamma = \frac{1}{\sqrt{a}}$ near x_0 (the other case is similar), then since the rotation of v along the ξ-piece of x^∞ is 1, $\gamma = -\frac{1}{\sqrt{a_1}}$ near x_1. Observe that $[\xi, v](x_0) = \sqrt{a} \frac{\partial}{\partial y}$ and $[\xi, v](x_1) = -\sqrt{a_1} \frac{\partial}{\partial y}$.

Using the notations of [Bahri-1 2003], let θ_1 be the transport map from x_1 to x_0 along v.

Then,

$$d\theta_1^{-1}(\xi) - \xi = \bar{A}_1 \xi + \bar{B}_1 \frac{\partial}{\partial y} = \bar{A}_1 \xi - \frac{\bar{B}_1}{\sqrt{a_1}}[\xi, v](x_1).$$

According to [Bahri-1 2003], the vector-field $\tilde{v}$ generates then an eigenvalue for $J_\infty''(x^\infty)$ equal to $\frac{\bar{A}_1}{\bar{B}_1} \times \frac{c}{a}$. $\left(da_c(v)(x_0) = \frac{1}{\sqrt{a}} \times \frac{c}{a_1} \right)$.

Since a is positive $(d\alpha(v, [\xi, v]) < 0)$, this has the sign of $\frac{\bar{A}_1}{\bar{B}_1} \times c$.

We use the equation of the characteristic manifold to find the other eigenvalue. By Proposition 38, along the characteristic manifold:

$$a_c - \bar{a}_0$$

$$= -\tilde{A}(x,y,0) + \frac{\bar{A}_1}{\bar{B}_1}\left(\tilde{B}(x,y,0) + c\frac{z^2}{2} + ((b-b_1) - \lambda(a-a_1))\,yz\right)$$

$$+ O(|X|^3) = \left(\frac{\bar{A}_1}{\bar{B}_1}\tilde{B}_2(-\lambda,1,0) - \tilde{A}_2(-\lambda,1,0)\right)y^2$$

$$+ \frac{\bar{A}_1}{\bar{B}_1}\left(c\frac{z^2}{2} + ((b-b_1) - \lambda(a-a_1))\,yz\right) + O(|X|^3).$$

On the other hand, a direct expansion shows that:

$$a_c - \bar{a}_0 = \frac{\partial a_c}{\partial x}x + \frac{\partial a_c}{\partial y}y + \frac{\partial a_c}{\partial z}z + O(|X|^2) = O(|X|^2)$$

$(= O(|X|^2)$ by the expansion above).

Thus, since $x + \lambda y = O(|X|^2)$,

$$\frac{\partial a_c}{\partial z}(0)z = \left(-\frac{\partial a_c}{\partial y}(0) + \lambda\frac{\partial a_c}{\partial x}(0)\right)y + O(|X|^2).$$

By Proposition 38,

$$\frac{\partial a_c}{\partial x}(0) = \frac{a-a_1}{a_1}; \quad \frac{\partial a_c}{\partial y} = \frac{b-b_1}{a_1}; \quad \frac{\partial a_c}{\partial z} = \frac{c}{a_1}$$

so that

$$z = \frac{y}{c}\,(b_1 - b - \lambda(a_1 - a)) + O(y^2).$$

and

$$a_c - \bar{a}_0$$

$$= \left(\frac{\bar{A}_1}{\bar{B}_1}\tilde{B}_2(-\lambda,1,0) - \tilde{A}_2(-\lambda,1,0) - \frac{\bar{A}_1}{2\bar{B}_1 c}\,((b_1-b) - \lambda(a_1-a))^2\right)y^2$$

$$+ O(y^3).$$

Since the other eigenvalue is $\frac{\bar{A}_1}{\bar{B}_1} \times \frac{c}{a}$ with a positive, x^∞ is of index 1 if

$$\bar{A}_1 c\left(\bar{A}_1\tilde{B}_2(-\lambda,1,0) - \tilde{A}_2(-\lambda,1,0)\bar{B}_1 - \frac{\bar{A}_1}{2c}\,((b_1-b) - \lambda(a_1-a))^2\right) < 0.$$

i.e.

$$2\bar{A}_1 c\left(\bar{A}_1\tilde{B}_2(-\lambda,1,0) - \bar{B}_1\tilde{A}_2(-\lambda,1,0)\right) < \bar{A}_1^2\,((b_1-b) - \lambda(a_1-a))^2. \qquad \square$$

2.7.9 *Reduction to the Case $k = 1$*

We show in this section how to reduce the general case to the case of Γ_2 with minor changes after a result of Y.Xu about the deformation, using the deformation of the Hamiltonians, of the second derivative $J''_\infty(\overline{x}_\infty)$ or $\overline{x}_\infty$ of Γ_{2k}. $\overline{x}_\infty$ is a critical point of J_∞ and has k (all) of its ξ- pieces characteristic.

We need to think of the criticality conditions in a non-standard way since for Γ_2, they are expressed (see above) in the vanishing of a determinant and the fact that the ξ-piece is characteristic.

Since here the variations are many more, we cannot expect to have such a simple situation. We will then need the following definition:

Definition 6 A ξ-piece of an orbit of ξ is characteristic if it transports the v-vector onto itself, i.e. $D\psi_a(v(x^-)) = \theta v(x^+)$, here ψ_a is the one-parameter group of ξ, and a is the length of this ξ-piece.

We then denote $a_c = a_c^k(x^-)$ the characteristic length, i.e. the length along ξ of a piece of ξ-orbit starting at x^- which is characteristic with a rotation $k\pi$ for v.

Definition 7 We define the characteristic manifold $C_k(\overline{x}_\infty)$ to be the manifold of the curves of Γ_{2k} having all their ξ-pieces characteristic.

It is easy to see that $C_k(\overline{x}_\infty)$ is a manifold after a generic assumption of v [Bahri-2 2003]. $C_k(\overline{x}_\infty)$ has codimension k in Γ_{2k} and a complement subspace to the tangent space at a curve $\overline{x}$ near $\overline{x}_\infty$ is generated by the vectors $\widetilde{v}_1, \cdots, \widetilde{v}_k$ which are defined by transport of v along ξ, along each ξ- piece, from one edge of the ξ piece to the other one (from x_i^- to x_i^+ for $\widetilde{v}_i$).

This holds for each curve $\overline{x}$ of $C_k(\overline{x}_\infty)$ such that the differential $da_c^i(v(x_i^-)) \neq 0$, for $i = 1, \cdots, k.a_c^i$ is the characteristic length of the point x_i.

Considering a curve $\overline{x}$ of Γ_{2k}, we may relate to this curve k ξ-lengths, $a^1, \cdots, a^k$, and k characteristic lengths $a_c^1(x_1^-), \cdots, a_c^k(x_k^-)$. Accordingly, $\delta a^1, \cdots, \delta a^k$ are the differentials of the $a^i{}'s$ on $T_{\overline{x}}\Gamma_{2k}$ and $\delta a_c^1, \cdots, \delta a_c^k$ are the differentials of the characteristic lengths $a_c^i{}'s$ on $T_{\overline{x}}\Gamma_{2k}$.

At $\overline{x}_\infty$, we have,

Proposition 39 *There exists a constant $c_i \neq 0$ such that $J''_\infty(\overline{x}_\infty).\widetilde{v}_i.z = c_i(\delta a^i - \delta a_c^i)(z)$ for any $z \epsilon T_{\overline{x}_\infty}\Gamma_{2k}$ where $\delta a^i(z)$ is the variation along z of the ith ξ-piece of $\overline{x}_\infty$.*

See [Bahri-1 2003], [Bahri-2 2003].

For all $z \epsilon T_{\overline{x}_\infty} C_k$, in order to keep every ξ-pieces characteristic, we must have $(\delta a^i - \delta a^i_c)(z) = 0$. Therefore we have,

Corollary 11 $T_{\overline{x}_\infty} C_k(\overline{x}_\infty)$ *and* $Span\{\tilde{v}_1, \cdots, \tilde{v}_k\}$ *are* $J''_\infty(\overline{x}_\infty)$ *-orthogonal.*

Given a quadratic form q on a space E and $G \subset E$ a subspace of E, we denote G^o the orthogonal for q of G.

As in the case of Γ_2, we can find a suitable Darboux coordinates near the first ξ-piece, such that in the small neighborhood of x_1^-, v reads $\dfrac{\partial}{\partial z} - y\dfrac{\partial}{\partial x} + (ax + by + cz + P(x,y,z))\dfrac{\partial}{\partial y}$, and in the small neighborhood of x_1^+, v reads $\dfrac{\partial}{\partial z} - y\dfrac{\partial}{\partial x} + (a_1 x + b_1 y + Q(x,y,z))\dfrac{\partial}{\partial y}$. We assume that

1) $J''_\infty(\overline{x}_\infty)\,|_{T_{\overline{x}_\infty} C_k(\overline{x}_\infty)}$ is non-degenerate,

2) $J''_\infty(\overline{x}_\infty)$ has two negative directions in

$$L = G^o \cap \{\delta a^2 - \delta a_c^2 = \cdots = \delta a^k - \delta a_c^k = \delta z^1 = 0\},$$

where G is the subspace

$$Span\{z \epsilon T_{\overline{x}_\infty} C_k(\overline{x}_\infty), z \equiv 0 \text{ on the first} \xi \text{piece}\}.$$

It is easy to see that G is of codimension three, and L is a three dimensional subspace of $T_{\overline{x}_\infty} \Gamma_{2k}$.

Let $L_1 = L \cap \{\delta a^1 - \delta a_c^1 = 0\}$, it is a two dimensional subspace of $T_{\overline{x}_\infty} C_k(\overline{x}_\infty)$. Now, $dJ^2_\infty(\overline{x}_\infty)$ has the following form,

$$\left(\begin{pmatrix} C & a_1 & a_2 & a_3 \\ a_1 & & & \\ a_2 & & D & \\ a_3 & & & \\ & 0 & & \end{pmatrix} \quad \begin{matrix} 0 \\ \\ C \end{matrix} \right)$$

here

$$c = d^2 J_\infty(\overline{x}_\infty).\tilde{v}_1.\tilde{v}_1,$$

$$a_i = c = d^2 J_\infty(\overline{x}_\infty).\tilde{v}_1.u_i, \qquad \text{for} \qquad 1 \leq i \leq 3,$$

$$D = d^2 J_\infty(\overline{x}_\infty)\,|_L,$$

$$C = d^2 J_\infty(\overline{x}_\infty)\,|_{(L+R\tilde{v}_1)^o}.$$

We now prove the following result,

Lemma 41 *There is a deformation of L_τ, such that either c changes sign or D changes index.*

Proof. [Proof of Lemma 41] We are going to use the following lemma in the proof. $\square$

Lemma 42 *In a suitable fixed connection in Γ_{2k}, and after having extended $\widetilde{v}_1$ to a neighborhood of $\overline{x}_\infty$, for $\overline{x}$ near $\overline{x}_\infty$, while modifying a, a_1, b, b_1 and c of v near x_1^- and x_1^+ along $a_\tau, a_{1\tau}, b_\tau, b_{1\tau}$ and c_τ, we have*

$$d^2 J_\infty^\tau(\overline{x}).\widetilde{v}_1.h = d^2 J_\infty^0(\overline{x}).\widetilde{v}_1.h + o(|h|),$$

$$d^2 J_\infty^\tau(\overline{x}).h.k = d^2 J_\infty^0(\overline{x}).h.k + o(|h| + |k|)$$

$$\text{and } d^2 J_\infty^\tau(\overline{x}).\widetilde{v}_1.\widetilde{v}_1 = c_\tau + o(1)$$

$$\text{for all } h, k \in T_{\overline{x}_\infty} C_k(\overline{x}_\infty) \text{ such that } dz_1(h) = dz_1(k) = 0.$$

The proof the Lemma 42 is deferred to the appendix.

Assume $c = d^2 J_\infty(\overline{x}).\widetilde{v}_1.\widetilde{v}_1$ is positive. We know how to change c (see the proof of Γ_2). This does not change C. Using Lemma42, a_1, a_2, a_3 and D change very little in a neighborhood. The determinant of L is a linear form in c. Either we can decrease c and make it cross zero while $\det L$ remains non zero, we are done or we cannot do that. But then we can increase c to a very large value without ever crossing the value 0 for $\det L$.

If we change index of C while the total matrix does not degenerate, then the upper matrix changes index. Clearly if $|c|$ is large enough with respect to a_1, a_2, a_3, this implies that D changes index.

We will establish the following theorem in Section 2.7.10, which states,

Theorem 4 *Under the assumption $i_\infty \geq 2$, there is a deformation of α and v such that*

1) $d^2 J_\infty^\tau(\overline{x})$ does not degenerate.

2) Index of C changes by 2.

A corollary of this theorem is that either we can modify $c = d^2 J_\infty(\overline{x}).\widetilde{v}_1.\widetilde{v}_1$ from positive to negative or we can adjust α and v near $\overline{x}_\infty$ such that there are two negative directions in

$$L = G^o \cap \{\delta a^2 - \delta a_c^2 = \cdots = \delta a^k - \delta a_c^k = \delta z^1 = 0\}$$

since we can create two more positive directions in $span\{u_3, \ldots, u_{k-1}\}$.

Thus we established Lemma 41.

Using the notation in the proof of Section 2.7.2 (the case of Γ_2), we claim that,

Lemma 43　　*a and b of v at x_1^+ and a_1, b_1 of v at x_1^- can be adjusted so that there is a negative direction u for $J_\infty''(\overline{x}_\infty)$ in the $J_\infty''(\overline{x}_\infty)$-orthogonal of L_1 in L.*

Proof.　　[Proof of Lemma 43]

A tangent vector on the first ξ piece has at x_1^- components $\delta x_1, \delta y_1, \delta z_1$ and we denote $\delta x_1'$ the variation $\delta a + \delta x_1'$ along this vector.　　$\square$

We are asserting that we have two negative directions in

$$L = G^o \cap \{\delta z_1 = \delta a^2 - \delta a_c^2 = \cdots = \delta a^k - \delta a_c^k = 0\}.$$

We want to create a negative direction in $L \cap \{\delta z_1 = \delta a^1 - \delta a_c^1 = 0\}$. We work in the sequel in L.

On $\{\delta z_1 = 0\}$, $\delta a^1 - \delta a_c^1$ reduces to

$$\delta x_1' - \delta x_1 - (\frac{a - a_1}{a_1})\delta x_1 - (\frac{b - b_1}{a_1})\delta y_1 = \delta x_1' - \frac{a}{a_1}\delta x_1 - (\frac{b - b_1}{a_1})\delta y_1.$$

$\delta x_1', \delta x_1, \delta y_1$ are independent linear forms since $\delta x_1, \delta y_1, \delta a^1$ are linearly independent , we have to restrict the value of $\dfrac{a}{a_1}$. Thus $\dfrac{a}{a_1}, \dfrac{b - b_1}{a_1}$ are free to choose in a half plane (see [Bahri-2 2003]), because $-\dfrac{a}{a_1}$ needs to remain negative if β has to remain a contact form.

According to Lemma 42, $d^2 J_\infty(\overline{x}).h.k$ does not change as we change a, a_1, b, b_1 if $\delta z_1(h) = \delta z_1(k) = 0$.

Since c_1 is so large, $d^2 J_\infty$ does not change sign and $\overline{x}_\infty$ remains isolated. Actually, the proof of Lemma 42 implies that we do not create any new critical point at infinity.

Let now

$$\widetilde{A} = dJ_\infty^2(\overline{x}_\infty) \, |_{\{\delta z_1 = 0\}} \text{ in } L.$$

Let P be a generic plane which is negative for $\widetilde{A}^{-1}$. We may assume that $(0, 1, 0)$ (for $(\delta x_1, \delta y_1, \delta x_1')$) is not in $\widetilde{A}P$. $\widetilde{A}P$ has then the equation $\underline{a}\delta x_1 + \underline{c}\delta y_1 + \underline{b}\delta x_1' = 0$ with $\underline{c} \neq 0$.

Let u in $\widetilde{A}P$ be such that $\delta x_1(u) < 0, \delta x_1'(u) > 0$. Choose u with $\delta x_1'(u) = 1$, we set a, a_1, b, b_1 so that

$$\delta x_1(u) = -\frac{a}{a_1}$$

$$\delta y_1(u) = -\frac{b - b_1}{a_1}$$

$$\delta x_1'(u) = 1$$

then

$$\widetilde{A}^{-1}u \in P.$$

If x is in $\{\delta z_1 = \delta a^1 - \delta a_c^1 = 0\}$, then ${}^t x \widetilde{A}(\widetilde{A}^{-1}u) = {}^t xu = \delta x_1' - \frac{a}{a_1}\delta x_1 - \left(\frac{b - b_1}{a_1}\right)\delta y_1 = 0$.

Thus $\widetilde{A}^{-1}u$ is $d^2 J_\infty(\overline{x})$- orthogonal to $\{\delta z_1 = \delta a^1 - \delta a_c^1 = 0\}$ in L. Furthermore,
${}^t u\widetilde{A}^{-1}\widetilde{A}\widetilde{A}^{-1}u = {}^t u\widetilde{A}^{-1}u < 0$, thus $\widetilde{A}^{-1}u$ satisfies our purpose.

After this lemma, $T_{\overline{x}_\infty}C_k(\overline{x}_\infty)$ reads

$$G \oplus L_1 \oplus R\widetilde{u} = G \oplus L_1 \oplus R\left(u - \frac{J_\infty''(\overline{x}_\infty).\widetilde{v_1}.u}{J_\infty''(\overline{x}_\infty).\widetilde{v_1}.\widetilde{v_1}}\widetilde{v_1}\right).$$

Since L_1 and $\widetilde{v_1}$ are $J_\infty''(\overline{x}_\infty)$-orthogonal and $u - \dfrac{J_\infty''(\overline{x}_\infty).\widetilde{v_1}.u}{J_\infty''(\overline{x}_\infty).\widetilde{v_1}.\widetilde{v_1}}\widetilde{v_1}$ is independent of $\widetilde{v_1}(\delta z^1(u) = 0)$ and satisfies $\delta a^i - \delta a_c^i = 0$ for every $i \geq 1$, we can pick up a co-dimensional one sub-manifold C_{k-1} of $C_k(\overline{x}_\infty)$ which is tangent to $G \oplus L_1$ at $\overline{x}_\infty$. We then have $R\widetilde{v_1} \oplus \left(\overset{k}{\underset{i=2}{\to}} \oplus R\widetilde{v_i}\right) = J_\infty''(\overline{x}_\infty)^o T_{\overline{x}_\infty}C_k$.

We foliate $C_k(\overline{x}_\infty)$ by manifolds C_{k-1}^s with $C_{k-1}^0 = C_{k-1}$, here C_{k-1}^s are the co-dimensional 1 sub-manifolds of C_k which are tangent to $G \oplus L_1$ at some base points x_∞^s. x_∞^s, $s \in (-\varepsilon, \varepsilon)$ is a curve tangent to $\widetilde{u}$, with $x_\infty^0 = \overline{x}_\infty$. We prove below that the C_{k-1}^s's are basically unchanged along the homotopy which we complete.

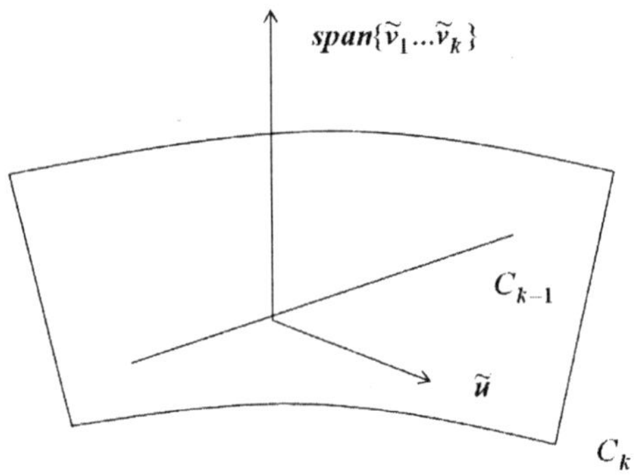

We then have,

Lemma 44 $\overline{x}$ *near* $\overline{x}_\infty$ *is a critical point of* J_∞ *if and only if*
  i) $\overline{x}$ *is in some* C^s_{k-1},
  ii) $\overline{x}$ *is critical of* $J_\infty |_{C^s_{k-1}}$,
  iii) $\det(dl_{\overline{x}} - Id) = 0,$ *where* $dl_{\overline{x}}$ *is the differential of the usual transport map of* $\overline{x}$.

Proof of Lemma 44: Once i) is established, ii) and iii) are clearly necessary . Together with the requirement that $\overline{x}_\infty$ has k characteristic pieces, this builds 2k independent conditions (generically to v) and criticality follows.

Condition i) is equivalent to the statement that $\overline{x}$ is critical (near $\overline{x}_\infty$). Then all its ξ pieces are characteristic. This follows from the fact that near C_k each v-jump offers, by v-transport along it, a decreasing variation for J_∞, only that the fact that a ξ-piece is characteristic does not allow to transform this variation into a tangent vector.

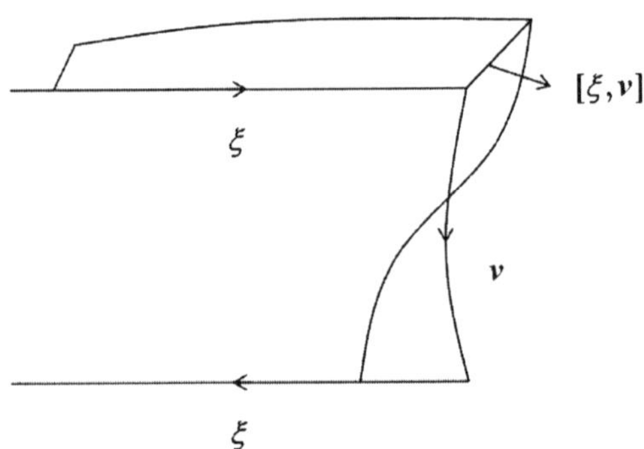

If ξ is non-characteristic, the variation can be closed and is J_∞-decreasing.

Hence, if $\overline{x}$ is in the neighborhood of C_k and has a non-characteristic ξ-piece, it is not critical for J_∞. i) follows, as well as ii) and iii) and the lemma.

We now establish that the proof of theorem 2 follows from the proof of Γ_2.

We will think of $\mathrm{Span}\{\widetilde{v}_1, \widetilde{u}\}$ as the extension of $T_{\overline{x}_\infty}\Gamma_2$ and complete our modification of $\widetilde{v}_1$ so that,

(i) $\mathrm{Span}\{\widetilde{v}_1, \widetilde{u}\} = \mathrm{Span}\{\ \widetilde{v}_1, u\}$ is basically unperturbed.

(ii) L_1, G and the $C^{s'}_{k-1}s$ is basically unperturbed.

(iii) The orthogonality relations of $\mathrm{Span}\{\widetilde{v}_1, \widetilde{u}\}$ and L_1 and the matrix $J''_\infty(\overline{x}_\infty)|_{L_1}$ are unperturbed.

(iv) $J''_\infty(\overline{x}_\infty)|_{Span\{\widetilde{v}_1, \widetilde{u}\}}$ does not degenerate.

(v) Along the deformation, $J''_\infty(\overline{x}_\infty).\widetilde{v}_1.\widetilde{v}_1$ changes sign.

We think of the whole Poincare-return map as the composition of three maps, the transport along $\widetilde{u}$, the transport along $\widetilde{v}_1$ and the remaining part of the Poincare-return map. According to Lemma 30, the remaining part is basically unperturbed.

Now, $dJ^2_\infty(\overline{x}_\infty)$ has the following form,

$$\begin{pmatrix} c & \overline{d} & 0 \\ \overline{d} & \overline{a} & \\ 0 & & B \end{pmatrix}$$

with

$$E_1 = Span\{\widetilde{v}_1, u\}$$
$$E_2 = T_{\overline{x}_\infty}C_{k-1} \oplus Span\{\widetilde{v}_2, \ldots, \widetilde{v}_k\} \subset \{\delta z_1 = 0\}.$$

We also know that

$$\overline{a} = dJ^2_\infty(\overline{x}_\infty).u.u < 0, \qquad c = dJ^2_\infty(\overline{x}_\infty).\widetilde{v}_1.\widetilde{v}_1 > 0.$$

Hence,

$$c\overline{a} - \overline{d}^2 < 0.$$

According to Lemma 41, we have,

Lemma 45 *As we change c along $c_\tau, \tau\epsilon[0,1]$ from positive to negative, $B, \overline{a}, \overline{d}$ do not change.*

Then, the isolation of $\overline{x}_\infty$ reduces to the condition

$$c_\tau \overline{a} - \overline{d}^2 < 0$$

(i.e. does not change sign).

This is equivalent to our determinant condition in the case of Γ_2 and the proof are completely parallel. The vanishing of the determinant in the case of Γ_2 being equivalent to $c\overline{a} - \overline{d}^2 = 0$. The two are linear forms on c_τ which vanish when the other one vanishes.

Obviously, since $\overline{a}$ is positive, unchanged and $c\overline{a}$ is negative, we can change c_τ, make it cross zero, using the same construction than the one for Γ_2 and never have $c_\tau \overline{a} - \overline{d}^2 = 0$.

We now prove that i), ii), iii), iv) and v) stated above hold.

iv) and v) are clear. iii) follows from Lemma 31. Along G and L_1, δz_1 equals 0. The base vectors at x_1^- have no component on $v(x_1^-)$. This also holds for u, since $u \in L$. The proof of theorem 6 implies then that their natural extension to a neighborhood of $\overline{x}_\infty$ in Γ_{2k} changes very little.

2.7.10   *Modification of $d^2 J_\infty^\tau(\overline{x}_\infty)|_{span\{u_2,\cdots,u_{k-1}\}}$*

In this section, we will prove Theorem 4, which states as,

Theorem 4   *There is a deformation of α and v such that*

    1) $d^2 J_\infty^\tau(\overline{x}_\infty)$ does not degenerate.

    2) Index of $d^2 J_\infty^\tau(\overline{x}_\infty)|_{span\{u_2,\cdots,u_{k-1})}$ changes by 2

First we need to describe a basis for the tangent space at $\overline{x}$. Given three consecutive characteristic ξ-pieces:

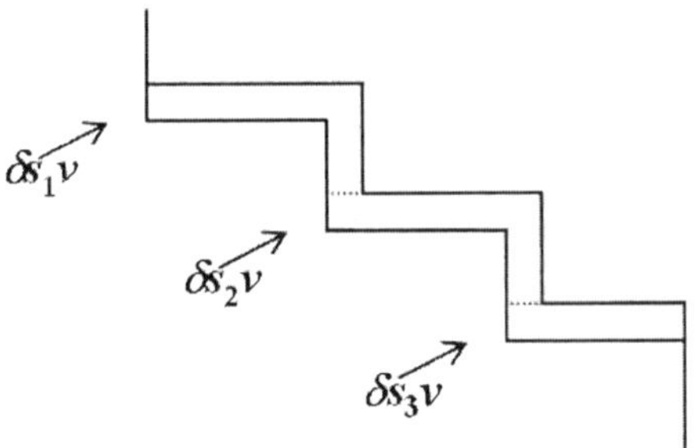

We create a tangent vector at $\overline{x}_\infty$ after picking up at x_1^- a vector $\delta s_1 v$, transporting it a characteristic length to x_1^+, then along v from x_2 to x_3 with an addition of $\delta s_2 v$, transporting it then a characteristic length again from x_2^- to x_2^+ and along v from x_2^+ to x_3^-, with an addition of $\delta s_3 v$, transporting it finally a characteristic length from x_3^- to x_3^+. Using the freedom of δs_2 and δs_3, we may arrange so that the vector obtained at x_3^+ is parallel to v.

The vector will be denoted u_2, where 2 is the index of the middle characteristic piece. In this way, we build k vectors $\{u_1, \ldots, u_k\}$. It is clear that $\partial^2 J(\overline{x}_\infty).u_i.u_j$ is zero when $|i - j| > 2$.

As in the case of Γ_2, we work with Darboux coordinates $\alpha = dx + ydz$, so we can follow the curves in a small neighborhood of the base curve explicitly. We will perturb all the ξ−pieces and all the $v-$ pieces, so we can change $\partial^2 J_\infty(\overline{x}_\infty).u_i.u_i$ and $\partial^2 J_\infty(\overline{x}_\infty).u_i.u_{i+1}$ freely and independently, for $1 \leq i \leq k$, without changing $da_c^i(v)$.

After the modification, in a small neighborhood of the ith v-piece, v becomes (up to a multiplication factor)

$$\frac{\partial}{\partial z} + \omega_{3i}(z)\omega_{4i}(x, y)(h_{1i}x^2 + l_{1i}xy)\frac{\partial}{\partial x} + \omega_{3i}(z)\omega_{4i}(x, y)(h_{2i}x^2 + l_{2i}xy)\frac{\partial}{\partial y}.$$

Integrating along the v-pieces, the modification occurs only at the second order and higher. Here, $\omega_{3i}(z)$ and $\omega_{4i}(x, y)$ are the cut-off functions in a small neighborhood of the v-pieces.

And α on the ith ξ−piece becomes $\lambda\alpha$, where $\dfrac{1}{\lambda} = 1 + \omega_{1i}(x)\omega_{2i}(y, z)d_i y^2 z$, here, $\omega_{1i}(x)$ and $\omega_{2i}(y, z)$ are the cut-off functions in a small neighborhood of the ξ−pieces. The corresponding Reeb vector field becomes

$$\xi = \frac{1}{\lambda}\left(\frac{\partial}{\partial x} + \frac{\lambda_z - y\lambda_x}{\lambda^2}\frac{\partial}{\partial y} - \frac{\lambda_y}{\lambda^2}\left(\frac{\partial}{\partial z} - y\frac{\partial}{\partial x}\right)\right).$$

We will prove,

Lemma 46 *Using the perturbation we described above, we can modify the second derivatives*

$$\partial^2 J_\infty(\overline{x}_\infty).u_i.u_i, \partial^2 J_\infty(\overline{x}_\infty).u_i.u_{i+1}$$

for $1 \leq i \leq k$ freely and independently, i.e. we can prescribe any set of $2k$ real values for

$$\partial^2 J_\infty(\overline{x}_\infty).u_i.u_i, \partial^2 J_\infty(\overline{x}_\infty).u_i.u_{i+1}, 1 \leq i \leq k.$$

A corollary of Lemma 46 is,

Corollary 12   *With a C^1-small, and C^2-bounded modification, we can change the index of $J''_\infty(\bar{x}_\infty)\,|_G$ by 2 while keeping $J''_\infty(\bar{x}_\infty)\,|_{TC_k(\bar{x}_\infty)}$ non-degenerate.*

Proof of Corollary 12: We reorder A so that the first coefficient is u_k because the supplement of G in the space $\mathrm{Span}\{u_i\}$ is $\mathrm{Span}\{u_1, u_2, u_k\}$. A reads

$$
\begin{pmatrix}
\partial^2 J_\infty(\bar{x}_\infty).u_k.u_k & \cdot\ \cdot\ 0\ 0\ \cdot & \partial^2 J_\infty(\bar{x}_\infty).u_k.u_{k-1} \\
\partial^2 J_\infty(\bar{x}_\infty).u_1.u_k & \cdot\ \cdot\ \cdot\ 0\ 0 & \partial^2 J_\infty(\bar{x}_\infty).u_1.u_{k-1} \\
\partial^2 J_\infty(\bar{x}_\infty).u_2.u_k & \cdot\ \cdot\ \cdot\ \cdot\ 0 & 0 \\
0 & \cdot\ \cdot\ \cdot\ \cdot\ \cdot & 0 \\
0 & 0\ \cdot\ \cdot\ \cdot\ \cdot & \cdot \\
\partial^2 J_\infty(\bar{x}_\infty).u_1.u_{k-1}\ 0\ 0 & \cdot\ \cdot\ \cdot & \partial^2 J_\infty(\bar{x}_\infty).u_{k-2}.u_{k-1} \\
\partial^2 J_\infty(\bar{x}_\infty).u_{k-1}.u_k\ \cdot\ 0\ 0 & \cdot\ \cdot & \partial^2 J_\infty(\bar{x}_\infty).u_{k-1}.u_{k-1}
\end{pmatrix}.
$$

If we look at the matrix A, it has on each line 5 non zero terms at most. For all lines besides the first two ones and the two last ones, these terms are the diagonal term and the two next terms on each side. For the first line, we have the diagonal, two terms on the immediate right and two other terms on the extreme right of the matrix. For the second line, we have four terms, one to the left of the diagonal, two to the immediate right and one at the extreme right. The situation is similar but with right replaced by left for the two last lines. Our lemma above establishes that the diagonal of A and the two next "diagonals" on the right and on the left are made of free parameters which we can choose as we please. The next two diagonals on the right and on the left are untouched through these changes. The extreme top coefficients on the right and its symmetric on the left are free to choose. The two remaining coefficients are untouched through these changes.

det A is linear in the diagonal terms. We can therefore send the diagonal coefficients to $+\infty$ or $-\infty$ so that det A is never zero. The other terms are untouched. Now A has a very large diagonal, all other terms are bounded. Let us take the other coefficients which are free to choose and make them very large (still small with respect to the absolute value of the diagonal terms).

Setting the untouched coefficients to be zero, including the top right and the bottom left coefficients, we derive a matrix A_0 whose coefficients are totally free to choose. A_0 reads

$$\begin{pmatrix} \cdot & \cdot & 0 & 0 & 0 & 0 & \cdot \\ \cdot & \cdot & \cdot & 0 & 0 & 0 & 0 \\ 0 & \cdot & \cdot & \cdot & 0 & 0 & 0 \\ 0 & 0 & \cdot & \cdot & \cdot & 0 & 0 \\ 0 & 0 & 0 & \cdot & \cdot & \cdot & 0 \\ 0 & 0 & 0 & 0 & \cdot & \cdot & \cdot \\ \cdot & 0 & 0 & 0 & 0 & \cdot & \cdot \end{pmatrix}.$$

In this form our claim becomes a simple algebraic statement involving pathes and connectedness of the set of non degenerate matrices of the above type as long as they have the same index. We do not include the proof here.

Now we proceed with the proof of Lemma 46. Since the proof is somehow long and technical, we will only outline the main steps here and leave the details in the appendix.

First, let us look at what happens if we modify only the third ξ-piece and the second v-piece, the result extends to the general case. For the sake of simplicity, we drop the subscript i in the proof. After the modification, in a small neighborhood of the second v-piece, v becomes (up to a multiplication factor)

$$\frac{\partial}{\partial z} + \omega_3(z)\omega_4(x,y)(h_1 x^2 + l_1 xy)\frac{\partial}{\partial x} + \omega_3(z)\omega_4(x,y)(h_2 x^2 + l_2 xy)\frac{\partial}{\partial y}.$$

Here, $\omega_3(z)$ and $\omega_4(x,y)$ are the cut-off functions in a small neighborhood of the second v-piece.

And α on the third $\xi-$piece becomes $\lambda\alpha$, where

$$\frac{1}{\lambda} = 1 + \omega_1(x)\omega_2(y,z)dy^2 z.$$

Here, $\omega_1(x)$ and $\omega_2(y,z)$ are the cut-off functions in a small neighborhood of the third ξ-piece.

Let $\tilde{v}$ be the new vector field introduced. We build a function ψ using the orbits of $\tilde{v}$ such that

$$d\psi(\tilde{v}) = -\alpha(\tilde{v}).$$

Setting then

$$\tilde{\alpha} = \alpha - d\psi,$$

we find

$$\tilde{\alpha}(\tilde{v}) = 0.$$

Furthermore,

$$\widetilde{\alpha} \wedge d\widetilde{\alpha} = \alpha \wedge d\alpha - d\psi \wedge d\alpha.$$

If $d\psi$ is small, $\widetilde{\alpha}$ is still a contact form.

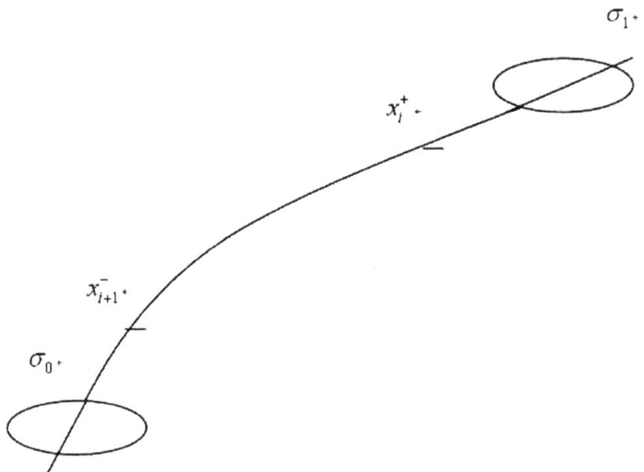

Finally,

$$\widetilde{\beta} = d\alpha(\widetilde{v}, \cdot) = d\alpha(v, \cdot) + O(x^2 + y^2),$$

so that $\widetilde{\beta}$ is still a contact form.

To construct ψ, we pick up the v-orbit through the edge and the two sectors σ_0, σ_1, one before the starting point of the edge, another one after.

We build up one perturbation between x_{i+1}^- and x_i^+ and build a compensating perturbation, still $O(x^2 + y^2)$, between x_i^+ and σ_1.

We solve

$$d\Psi(\widetilde{v}) = -\alpha(\widetilde{v})$$

by integration along the flow-lines of $\widetilde{v}$ starting with $\Psi = 0$ on σ_0. We get

$$\Psi(s, x) = -\int_0^s \alpha(\widetilde{v})dt \text{ for } x \in \sigma_0,$$

t is the time along $\widetilde{v}$, Ψ is $O(x^2 + y^2)$, $d\Psi$ is $O(|x| + |y|)$.

Following the notation in the case of Γ_2, in a small neighborhood of point x_i^+, the $v-$vector after re-normalization reads $\frac{\partial}{\partial z} - y\frac{\partial}{\partial x} + (a_i^+ x + b_i^+ y + Q_i(x,y,z))\frac{\partial}{\partial y}$, and in the small neighborhood of x_i^-, the $v-$ vector after re-normalization reads $\frac{\partial}{\partial z} - y\frac{\partial}{\partial x} + (a_i^- x + b_i^- y + cz + P_i(x,y,z))\frac{\partial}{\partial y}$.

As we modify the third ξ-piece, δa_c^3 changes. For the sake of simplicity, we omit the superscript below. We will need the following lemma:

Lemma 47 *After performing the modification on the third ξ-piece, δa_c becomes*

$$\delta a_c = \frac{(a_3^- - a_3^+)x_0 + (b_3^- - b_3^+)y_0 + c_3 z_0 - \bar{a}2dy_0}{a_3^+}$$

$$- \frac{(a_3^- - a_3^+)x_0 + (b_3^- - b_3^+)y_0 + c_3 z_0 - \bar{a}2dy_0}{a_3^+} \cdot \frac{2dy_0 + Q_{2,x}(x_0,y_0,z_0)}{a_3^+}$$

$$+ \{P_2^3(x_0,y_0,z_0) - Q_2^3(x_0,y_0,z_0) + \bar{a}b_3^+ 2dy_0 z_0$$

$$- \bar{a}2dz_0(a_3^- x_0 + b_3^- y_0 + c_3 z_0) - \bar{a}^2 2dz_0(a_3^- x_0 + b_3^- y_0 + c_3 z_0)\}/a_3^+ .$$

Here, the first line is the first order terms, and the rest are the second order terms.

The proof of Lemma 47 is deferred to the Appendix.

We proceed with the calculation of $\partial^2 J_\infty(\bar{x}_\infty).u_2.u_2$, $\partial^2 J_\infty(\bar{x}_\infty).u_2.u_3$, $\partial^2 J_\infty(\bar{x}_\infty).u_2.u_4$ now.

2.7.11 *Calculation of* $\partial^2 J_\infty(\bar{x}_\infty).u_2.u_3$

First let us focus on the u_3-piece, the bottom variation. We start at x_1 with a small variation δs_1 along v, we transport this small variation along the second ξ-piece a characteristic length to x_2. From x_2 we transport it along the second v-piece to x_3, then from x_3 with a small variation δs_2 along v, we transport it along the third ξ-piece a characteristic length to x_4. From x_4 we transport it along the third v-piece to x_5, then from x_5 with a small variation δs_3 along v, we transport it along the fourth ξ-piece a characteristic length to x_6 . The z-coordinates of the 1st, 2nd and 3rd $\xi-$piece of u_2 are respectively $\delta s_1, \delta s_2, \delta s_3$. δs_1 and δs_3 are constant along the first and third ξ-pieces of u_2. Because the third $\xi-$piece is perturbed,

δs_2 varies at the second order (refer to Lemma 47) along this piece, but it stays the same at $\overline{x}_\infty$. We have:

Lemma 48 $\delta s_1, \delta s_2, \delta s_3$ *satisfy the following relation,*

$$\delta s_2 = p_2 \delta s_1 + O\left(\delta s_1^2\right), \qquad \delta s_3 = q_2 \delta s_1 + O\left(\delta s_1^2\right)$$

where

$$p_2 = -\frac{c_2}{a_2^+}\frac{a_3^+}{c_3}\left(\frac{a_3^- B_1 + (\Delta b_3 - 2d\overline{a})B_2}{a_3^+} + \frac{B_2}{C_2}C_4\right),$$

$$q_2 = \frac{c_2}{a_2^+}\frac{a_4^-}{c_4}\frac{B_2}{C_2}\det C$$

$$\text{and } A = \begin{pmatrix} A_1 & A_3 \\ A_2 & A_4 \end{pmatrix}, B = \begin{pmatrix} B_1 & B_3 \\ B_2 & B_4 \end{pmatrix} \text{ and } C = \begin{pmatrix} C_1 & C_3 \\ C_2 & C_4 \end{pmatrix}$$

are respectively the transport matrices of the 1st, 2nd, and 3rd v-pieces.

The proof of this lemma is deferred to the Appendix.

On top of u_3, now we construct $u_2(u_3)$, assume that the z-coordinates of the 1st, 2nd and 3rd ξ−pieces of $u_2(u_2)$ are respectively $\overline{\delta s_1}$, $\overline{\delta s_2} + \delta s_1$ and $\delta s_2 + \overline{\delta s_3}$. We start at x_7 with a small variation $\overline{\delta s_1}$ along v, then we transport this small variation along the third ξ-piece a characteristic length to x_0. From x_0 we transport along the first v-piece to x_1', then from x_1' with a small variation $\delta s_1 + \overline{\delta s_2}$ along v, we transport it along the second ξ-piece a characteristic length to x_2'. From x_2' we transport it along the second v-piece to x_3', then from x_3' with a small variation $\delta s_2 + \overline{\delta s_3}$ along v, we transport it along the third ξ-piece a characteristic length to x_4'.

Lemma 49   *The x, y coordinates of point x_4' reads*

$$\begin{pmatrix} -\dfrac{c_2}{a_2^+}\dfrac{B_2}{C_2}C_4\delta s_1 - \dfrac{c_2}{a_2^+}B_2\delta s_1\overline{\delta s_3} \\[2ex] \dfrac{c_2}{a_2^+}B_2\delta s_1 + \dfrac{c_2}{a_2^+}(b_3^+ B_2 - a_3^+\dfrac{B_2}{C_2}C_4)\delta s_1\overline{\delta s_3} \end{pmatrix} + U\delta s_1^2 + V\overline{\delta s_1}^2 + O(\delta s_1^3 + \overline{\delta s_1}^3),$$

here U and and V are constants which we will not track.

The proof of Lemma 49 is in the Appendix.

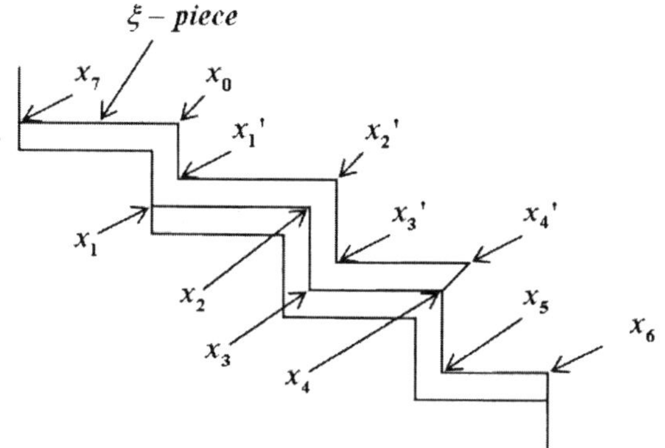

Assume

$$\overline{\delta s_2} = p\overline{\delta s_1} + R\delta s_1\overline{\delta s_1} + \widetilde{U}\overline{\delta s_1}^2 + \widetilde{V}\delta s_1^2 + O(\overline{\delta s_1}^3 + \delta s_1^3),$$

$$\overline{\delta s_3} = q\overline{\delta s_1} + T\delta s_1\overline{\delta s_1} + \widetilde{U}_1\overline{\delta s_1}^2 + \widetilde{V}_1\delta s_1^2 + O(\overline{\delta s_1}^3 + \delta s_1^3).$$

Here, we will track only R and T, and we will not track $\widetilde{U}, \widetilde{V}, \widetilde{U}_1$ and $\widetilde{V}_1$.

First, we track only first order terms along $u_2(u_3)$, we get

Lemma 50 $\overline{\delta s_1}, \overline{\delta s_2}$ *and* $\overline{\delta s_3}$ *satisfy the following relation at the first order,*

$$\overline{\delta s_2} = p_1\overline{\delta s_1} + O\left(\delta s_1^2, \overline{\delta s_1}^2, \delta s_1\overline{\delta s_1}\right),$$

$$\overline{\delta s_3} = q_1\overline{\delta s_1} + O\left(\delta s_1^2, \overline{\delta s_1}^2, \delta s_1\overline{\delta s_1}\right),$$

where

$$p_1 = -\frac{c_1}{a_1^+} \cdot \frac{a_2^+}{c_2} \cdot \left(\frac{a_2^- A_1 + \Delta b_2 A_2}{a_3^+} + \frac{A_2}{B_2}B_4\right),$$

$$q_1 = \frac{c_1}{a_1^+} \frac{a_3^-}{c_3} \frac{A_2}{B_2} \det B.$$

Hence we know how u_3 and $u_2(u_3)$ constitute at the first order. The proof of the Lemma 50 is in the Appendix.

462   *Recent Progress in Conformal Geometry*

Let us look at

$$J(\overline{x}_\infty + \overline{\delta s_1} u_2(u_3) + \delta s_1 u_3) - J(\overline{x}_\infty + \delta s_1 u_3)$$

$$= \overline{\delta s_1} \partial J(\overline{x}_\infty + \delta s_1 u_3).u_2(u_3) + O(\overline{\delta s_1}^2)$$

$$= \overline{\delta s_1} \delta s_1 \partial^2 J(\overline{x}_\infty).u_2.u_3 + O(\delta s_1^2, \overline{\delta s_1}^2).$$

Since $\overline{x}_\infty$ is critical, $\partial J(\overline{x}_\infty).(\cdot)$ is 0. If we take δs_1 very small compared with $\overline{\delta s_1}$, then we can ignore the δs_1^2 terms. Similarly if we take $\overline{\delta s_1}$ very small compared with δs_1, then we can ignore the $\overline{\delta s_1}^2$ terms. Since our aim is to compute $\partial^2 J.u_2.u_3$ we need only to concentrate on the $\delta s_1 \overline{\delta s_1}$ terms, so from now on in all our computation, we keep only the $\delta s_1 \overline{\delta s_1}$ terms, and drop the $\delta s_1^2, \overline{\delta s_1}^2$ terms and terms with order higher than 2.

Claim 1 Along $u_2(u_3)$, the transport matrix in x, y- coordinates along the first v-piece is

$$\left(Id + \overline{(\delta s_2} + \delta s_1) \begin{pmatrix} 0 & -1 \\ a_2^- & b_2^- \end{pmatrix} \right) A \left(Id - \overline{\delta s_1} \begin{pmatrix} 0 & -1 \\ a_1^+ & b_1^+ \end{pmatrix} \right) \cdot x_0$$

$$+ \begin{pmatrix} 0 \\ c_2 \end{pmatrix} \overline{\delta s_2} \delta s_1 + O\left(\delta s_1^2, \overline{\delta s_1}^2 \right) + O_3.$$

As before, the proof of the claim is deferred to the Appendix. Therefore we can write the x, y-coordinates of x_1' as

$$\left(Id + \overline{(\delta s_2} + \delta s_1) \begin{pmatrix} 0 & -1 \\ a_2^- & b_2^- \end{pmatrix} \right) A \left(Id - \overline{\delta s_1} \begin{pmatrix} 0 & -1 \\ a_1^+ & b_1^+ \end{pmatrix} \right) \begin{pmatrix} \frac{c_1}{a_1^+} \\ 0 \end{pmatrix} \overline{\delta s_1} + \begin{pmatrix} 0 \\ c_2 \end{pmatrix} \overline{\delta s_2} \delta s_1$$

$$= A \begin{pmatrix} \frac{c_1}{a_1^+} \\ 0 \end{pmatrix} \overline{\delta s_1} + \begin{pmatrix} 0 & -1 \\ a_2^- & b_2^- \end{pmatrix} A \begin{pmatrix} \frac{c_1}{a_1^+} \\ 0 \end{pmatrix} \overline{\delta s_1} \delta s_1 + \begin{pmatrix} 0 \\ c_2 p_1 \end{pmatrix} \overline{\delta s_1} \delta s_1.$$

So the x, y-coordinates of x_2' read

$$\begin{pmatrix} \frac{a_2^-}{a_2^+} & \frac{\Delta b_2}{a_2^+} \\ 0 & 1 \end{pmatrix} \cdot x_1' + \begin{pmatrix} \frac{c_2}{a_2^+} \\ 0 \end{pmatrix} (\overline{\delta s_2} + \delta s_1) + \begin{pmatrix} N_2 \\ 0 \end{pmatrix}$$

$$= \begin{pmatrix} \frac{a_2^-}{a_2^+} & \frac{\Delta b_2}{a_2^+} \\ 0 & 1 \end{pmatrix} A \begin{pmatrix} \frac{c_1}{a_1^+} \\ 0 \end{pmatrix} \overline{\delta s_1} + \begin{pmatrix} \frac{c_2}{a_2^+} \\ 0 \end{pmatrix} (\delta s_1 + p_1 \overline{\delta s_1}) + R \begin{pmatrix} \frac{c_2}{a_2^+} \\ 0 \end{pmatrix} \overline{\delta s_1} \delta s_1$$

$$+ \begin{pmatrix} \frac{a_2^-}{a_2^+} & \frac{\Delta b_2}{a_2^+} \\ 0 & 1 \end{pmatrix} \left(\begin{pmatrix} 0 & -1 \\ a_2^- & b_2^- \end{pmatrix} A \begin{pmatrix} \frac{c_1}{a_1^+} \\ 0 \end{pmatrix} + \begin{pmatrix} 0 \\ c_2 p_1 \end{pmatrix} \right) \overline{\delta s_1} \delta s_1 + \begin{pmatrix} N_2 \\ 0 \end{pmatrix}$$

here N_2 is the second order term of $da_c(x_1')$.

Claim 2 The transport matrix along the second v- piece is

$$\left(Id + (\overline{\delta s_3} + \delta s_2)\begin{pmatrix} 0 & -1 \\ a_3^- & b_3^- \end{pmatrix}\right) B\left(Id - (\overline{\delta s_2} + \delta s_1)\begin{pmatrix} 0 & -1 \\ a_2^+ & b_2^+ \end{pmatrix}\right) \cdot x_2'$$

$$+ \begin{pmatrix} 0 \\ c_3 \end{pmatrix}\overline{\delta s_3}\delta s_2 + U\delta s_1^2 + V\overline{\delta s_1}^2 + O(\delta s_1^3 + \overline{\delta s_1}^3).$$

Here, the values of U and V has changed. However, we will not track these values. The proof of Claim 2 is similar with the proof of the Claim 1, so we omit it here.

The first order term of x_2' is

$$\begin{pmatrix} \dfrac{a_2^-}{a_2^+} & \dfrac{\Delta b_2}{a_2^+} \\ 0 & 1 \end{pmatrix} \cdot \text{point}1' + \begin{pmatrix} \dfrac{c_2}{a_2^+} \\ 0 \end{pmatrix}(\overline{\delta s_2} + \delta s_1)$$

$$= \frac{c_1}{a_1^+}\begin{pmatrix} -\dfrac{A_2}{B_2}B_4 \\ A_2 \end{pmatrix}\overline{\delta s_1} + \begin{pmatrix} \dfrac{c_2}{a_2^+} \\ 0 \end{pmatrix}\delta s_1.$$

Integrating along the second $v-$piece, we get

$$\begin{pmatrix} h_1(-\dfrac{c_1}{a_1^+}\dfrac{A_2}{B_2}B_4\overline{\delta s_1} + \dfrac{c_2}{a_2^+}\delta s_1)^2 + l_1(-\dfrac{c_1}{a_1^+}\dfrac{A_2}{B_2}B_4\overline{\delta s_1} + \dfrac{c_2}{a_2^+}\delta s_1)\dfrac{c_1}{a_1^+}A_2\overline{\delta s_1} \\ h_2(-\dfrac{c_1}{a_1^+}\dfrac{A_2}{B_2}B_4\overline{\delta s_1} + \dfrac{c_2}{a_2^+}\delta s_1)^2 + l_2(-\dfrac{c_1}{a_1^+}\dfrac{A_2}{B_2}B_4\overline{\delta s_1} + \dfrac{c_2}{a_2^+}\delta s_1)\dfrac{c_1}{a_1^+}A_2\overline{\delta s_1} \end{pmatrix}$$

$$= \frac{c_1}{a_1^+}\frac{c_2}{a_2^+}A_2\begin{pmatrix} -2h_1\dfrac{B_4}{B_2} + l_1 \\ -2h_2\dfrac{B_4}{B_2} + l_2 \end{pmatrix}\overline{\delta s_1}\delta s_1 + O(\delta s_1^2, \overline{\delta s_1}^2).$$

We are able to write the crossing terms of x_3' now.
Thus x_3' can be written as

$$\left(Id + (\overline{\delta s_3} + \delta s_2)\begin{pmatrix} 0 - 1 \\ a_3^- & b_3^- \end{pmatrix}\right) B\left(Id - (\overline{\delta s_2} + \delta s_1)\begin{pmatrix} 0 & -1 \\ a_2^+ & b_2^+ \end{pmatrix}\right) \cdot x_2'$$

$$+ \begin{pmatrix} 0 \\ c_3 \end{pmatrix}\overline{\delta s_3}\delta s_2 + O(\delta s_1^2, \overline{\delta s_1}^2) + \frac{c_1}{a_1^+}\frac{c_2}{a_2^+}A_2\begin{pmatrix} -2h_1\dfrac{B_4}{B_2} + l_1 \\ -2h_2\dfrac{B_4}{B_2} + l_2 \end{pmatrix}\overline{\delta s_1}\delta s_1 + O_3.$$

Then we transport it along third ξ-piece, we are able then to get an expression of x_4'. Comparing the crossing terms, i.e. $\overline{\delta s_1}\delta s_1$ terms of point x_4' with the expression in Lemma 49, we get

Claim 3 R satisfies the following relation,

$$-\frac{c_1}{a_1^+}A_2B_2\left(\frac{a_2^- - a_2^+}{a_2^+} + \frac{\Delta b_2}{a_2^+}\left(a_2^+\frac{B_4}{B_2} - b_2^+\right)\right) - c_3p_2q_1 - c_2p_1B_4$$

$$-2d\bar{a}\frac{c_2}{a_2^+}q_1B_2 + \frac{c_2}{a_2^+}RB_2 + NB_2 + \frac{c_1}{a_1^+}\frac{c_2}{a_2^+}A_2\left(-2h_2\frac{B_4}{B_2} + l_2\right) = 0.$$

We leave the details of the proof of this claim in the Appendix.

It is clear that there is no $\overline{\delta s_1}\delta s_1$ term which is contributed by the first and the fourth $\xi-$pieces. So in order to calculate the $\overline{\delta s_1}\delta s_1$ terms in $J_\infty(\overline{x}_\infty + \overline{\delta s_1}u_1(u_2) + \delta s_1 u_2) - J_\infty(\overline{x}_\infty + \delta s_1 u_2)$, we need only to count the $\overline{\delta s_1}\delta s_1$ terms contributed by the second and the third ξ-pieces.

For the second $\xi-$piece, the $\overline{\delta s_1}\delta s_1$ terms come from three sources. Since $\overline{\delta s_2} = p_1\overline{\delta s_1} + R\overline{\delta s_1}\delta s_1$, so R is one of the sources. Also the second order terms of $da_c(x_1')$, N_2 contributes $\overline{\delta s_1}\delta s_1$ terms, and all the other $\overline{\delta s_1}\delta s_1$ terms come from the v-transport.

For the third ξ-piece, the $\overline{\delta s_1}\delta s_1$ terms also come from three sources. Since $\overline{\delta s_3} = q_1\overline{\delta s_1} + T\overline{\delta s_1}\delta s_1$, so T is one of the sources. Also the second order terms of $da_c(x_3')$ N_3 contributes $\overline{\delta s_1}\delta s_1$ terms, and all the other $\overline{\delta s_1}\delta s_1$ terms come from the v-transport.

Summing them up, we get,

$$\partial^2 J_\infty(\overline{x}_\infty).u_2.u_3 = \underbrace{q_1 a_3^- B_1 + (\Delta b_3 - 4d\bar{a}_c)B_2 + a_3^+\frac{B_2}{C_2}C_4}_{(I)}$$

$$\underbrace{+ c_2p_1\frac{B_4}{B_2} - c_2p_1\frac{\det B}{B_2} c_2p_1\frac{\det B}{B_2}}_{(II)} + \underbrace{\frac{c_1}{a_1^+}\frac{c_2}{a_2^+}A_2((2h_1\frac{B_1}{B_2} - 2h_2)\frac{B_4}{B_2} + l_2 - l_1)}_{(III)}.$$

(I) involves the perturbation parameter d.

(II) involves no perturbation parameter, since $p_1 = -\frac{c_1}{a_1^+}\cdot\frac{a_2^+}{c_2}\cdot$

$\left(\frac{a_2^- A_1 + \Delta b_2 A_2}{a_3^+} + \frac{A_2}{B_2}B_4\right)$ involves no perturbation parameter.

And since (III) involves parameters h_1, l_1, h_2, l_2 , we know $\partial^2 J_\infty(\overline{x}_\infty).u_1.u_2$ involves perturbation parameters h_1, l_1, h_2, l_2 and d. As we modify the contact form, we can change it freely without changing $da_c(v)$.

The details of the calculation are in the Appendix.

2.8   Calculation of $\partial^2 J_\infty(\overline{x}_\infty).u_2.u_2$

The aim of this section is to perturb $\partial^2 J_\infty(\overline{x}_\infty).u_2.u_2$, without changing $da_c(v)$.

First let us focus on u_2 , the bottom variation. We start at x_1 with a small variation δs_1 along v, we transport this small variation along the first ξ-piece a characteristic length to x_2. From x_2 we transport it along the first v-piece to x_3, then from x_3 with a small variation δs_2 along v, we transport it along the second ξ-piece a characteristic length to x_4. From x_4 we transport it along the second v-piece to x_5, then from x_5 with a small variation δs_3 along v, we transport it along the third ξ-piece a characteristic length to x_6. The $z-$coordinates of the 1st, 2nd and 3rd $\xi-$piece of u_1 are respectively $\delta s_1, \delta s_2, \delta s_3$. δs_1 and δs_2 are constant along the first and second ξ-pieces. Because the third $\xi-$piece is perturbed, δs_3 varies at the second order (refer to Lemma 47) along this piece.

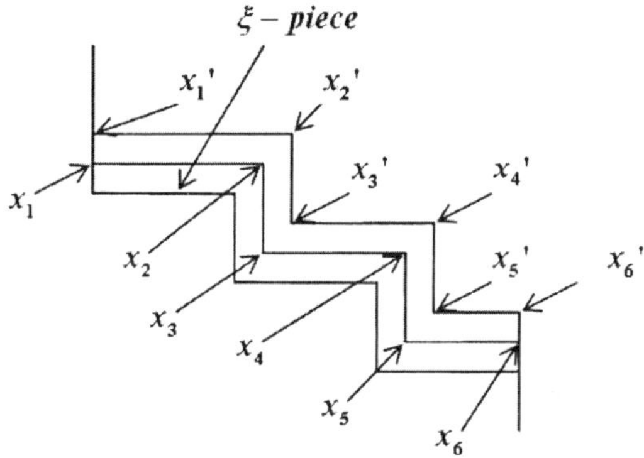

Lemma 51 $\delta s_1, \delta s_2, \delta s_3$ *satisfy the following relation,*

$$\delta s_2 = p\delta s_1 + O\left(\delta s_1^2\right) \qquad \delta s_3 = q\delta s_1 + O\left(\delta s_1^2\right)$$

where

$$p = -\frac{c_1}{a_1^+}\frac{a_2^+}{c_2}\left(\frac{a_2^- A_1 + \Delta b_2 A_2}{a_3^+} + \frac{A_2}{B_2}B_4\right),$$

$$q = \frac{c_1}{a_1^+}\frac{a_3^-}{c_3}\frac{A_2}{B_2}\det B,$$

and

$$A = \begin{pmatrix} A_1 & A_3 \\ A_2 & A_4 \end{pmatrix}, B = \begin{pmatrix} B_1 & B_3 \\ B_2 & B_4 \end{pmatrix}$$

are the transport matrices of the 1st and 2nd v–pieces.

The proof of this lemma is similar to the proof of Lemma 48, so we omit it here.

Lemma 52   *The x, y coordinates of x_6' reads $\begin{pmatrix} 0 \\ 0 \end{pmatrix} + U\delta s_1^2 + V\overline{\delta s_1}^2 + O(\delta s_1^3 + \overline{\delta s_1}^3)$, and the z coordinate is $\delta s_3 + \overline{\delta s_3}$, here U and V are constants which we will not track.*

The proof is similar with the proof of Lemma 49 which we defer to the Appendix.

Now on top of u_2, we construct $u_2(u_2)$. We start at x_1 with a small variation $\delta s_1 + \overline{\delta s_1}$ along v, we transport this small variation along the first ξ-piece a characteristic length to x_2. From x_2 we transport it along the first v-piece to x_3, then from x_3 with a small variation $\delta s_2 + \overline{\delta s_2}$ along v, we transport it along the second ξ-piece a characteristic length to x_4. From x_4 we transport it along the second v-piece to x_5, then from x_5 with a small variation $\delta s_3 + \overline{\delta s_3}$ along v, we transport it along the third ξ-piece a characteristic length to x_6. The z-coordinates of the 1st, 2nd and 3rd ξ-pieces of $u_1(u_2)$ are respectively $\delta s_1 + \overline{\delta s_1}$, $\delta s_2 + \overline{\delta s_2}$ and $\delta s_3 + \overline{\delta s_3}$, since v has a constant component equal to 1 on $\dfrac{\partial}{\partial z}$. As before, first we concentrate only on the first order terms, then we go to the $\overline{\delta s_1}\delta s_1$ terms.

Considering only the first order terms, we get:

Lemma 53 $\overline{\delta s_1}, \overline{\delta s_2}$ *and* $\overline{\delta s_3}$ *satisfy the following relation at the first order,*

$$\overline{\delta s_2} = p\overline{\delta s_1} + O\left(\delta s_1^2, \overline{\delta s_1}^2, \delta s_1\overline{\delta s_1}\right),$$

$$\overline{\delta s_3} = q\overline{\delta s_1} + O\left(\delta s_1^2, \overline{\delta s_1}^2, \delta s_1\overline{\delta s_1}\right),$$

where

$$p = -\frac{c_1}{a_1^+}\frac{a_2^+}{c_2}\left(\frac{a_2^- A_1 + \Delta b_2 A_2}{a_2^+} + \frac{A_2}{B_2}B_4\right),$$

$$q = \frac{c_1}{a_1^+}\frac{a_3^-}{c_3}\frac{A_2}{B_2}\det B.$$

The proof of this lemma is along the same line with the proof of Lemma 50, we defer it to the Appendix.

As before, since our goal is to compute $\partial^2 J_\infty(\bar{x}_\infty).u_2.u_2$, we need only to concentrate on the $\delta s_1 \overline{\delta s_1}$ terms, so from now on in all our computation, we keep only the $\delta s_1 \overline{\delta s_1}$ terms, and drop the $\delta s_1^2, \overline{\delta s_1}^2$ terms and terms with order higher than 2.

Assume

$$\overline{\delta s_2} = p\overline{\delta s_1} + R\delta s_1\overline{\delta s_1} + \tilde{U}\overline{\delta s_1}^2 + \tilde{V}\delta s_1^2 + O(\overline{\delta s_1}^3 + \delta s_1^3),$$

$$\overline{\delta s_3} = q\overline{\delta s_1} + T\delta s_1\overline{\delta s_1} + \tilde{U}_1\overline{\delta s_1}^2 + \tilde{V}_1\delta s_1^2 + O(\overline{\delta s_1}^3 + \delta s_1^3).$$

Now let us track the $\delta s_1\overline{\delta s_1}$ terms along $u_1(u_1)$.

Claim 4 x_1' reads $\begin{pmatrix} 0 \\ c\delta s_1\overline{\delta s_1} \\ \delta s_1 + \overline{\delta s_1} \end{pmatrix} + \tilde{V}\delta s_1^2 + O(\overline{\delta s_1}^3 + \delta s_1^3)$

The proof is deferred to the Appendix.

x_2' reads

$$\left(\begin{pmatrix} c_1 \\ a_1^+ \\ 0 \end{pmatrix}(\delta s_1 + \overline{\delta s_1}) + \begin{pmatrix} a_1^- & b_1^- - b_1^+ \\ a_1^+ & a_1^+ \\ & 01 \\ \hline & \overline{\delta s_1} + \delta s_1 \end{pmatrix}\begin{pmatrix} 0 \\ c_1 \end{pmatrix}\delta s_1\overline{\delta s_1} + \begin{pmatrix} N_1 \\ 0 \end{pmatrix} \right),$$

where N_1 is the second order term of $a_c(x_1') - a_c(x_2')$.

As in Section 2.7.11, we can prove that the transport matrix for $x, y-$ coordinates along the first v-piece is

$$\left(Id + \overline{(\delta s_2} + \delta s_2)\begin{pmatrix} 0 & -1 \\ a_2^- & b_2^- \end{pmatrix} \right) A \left(Id - (\delta s_1 + \overline{\delta s_1})\begin{pmatrix} 0 & -1 \\ a_1^+ & b_1^+ \end{pmatrix} \right) \cdot x_2'$$

$$+ \begin{pmatrix} 0 \\ c_2 \end{pmatrix}\overline{\delta s_2}\delta s_2 + \tilde{U}\overline{\delta s_1}^2 + \tilde{V}\delta s_1^2 + O(\overline{\delta s_1}^3 + \delta s_1^3),$$

here $\tilde{U}$ and $\tilde{V}$ are changed than before, which we will not track exactly.

Writing only the first order and the $\delta s_1 \overline{\delta s_1}$ terms, the x, y-coordinates of x_3' read

$$A\begin{pmatrix}\dfrac{c_1}{a_1^+}\\[4pt]0\end{pmatrix}(\delta s_1 + \overline{\delta s_1}) + A\begin{pmatrix}\dfrac{a_1^-}{a_1^+} & \dfrac{b_1^- - b_1^+}{a_1^+}\\[4pt]0 & 1\end{pmatrix}\begin{pmatrix}0\\c_1\end{pmatrix}\delta s_1\overline{\delta s_1}$$

$$+ 2p\begin{pmatrix}0 & -1\\a_2^- & b_2^-\end{pmatrix}A\begin{pmatrix}\dfrac{c_1}{a_1^+}\\[4pt]0\end{pmatrix}\delta s_1\overline{\delta s_1}$$

$$- 2A\begin{pmatrix}0 & -1\\a_1^+ & b_1^+\end{pmatrix}\begin{pmatrix}\dfrac{c_1}{a_1^+}\\[4pt]0\end{pmatrix}\delta s_1\overline{\delta s_1} + \begin{pmatrix}0\\c_2\end{pmatrix}p^2\overline{\delta s_1}\delta s_1 + A\begin{pmatrix}N_1\\0\end{pmatrix},$$

and the z-coordinate remains $\overline{\delta s_2} + \delta s_2$.

The x, y-coordinates of x_4' read

$$\begin{pmatrix}\dfrac{a_2^-}{a_2^+} & \dfrac{\Delta b_2}{a_2^+}\\[4pt]0 & 1\end{pmatrix}\cdot x_3' + \begin{pmatrix}\dfrac{c_2}{a_2^+}\\[4pt]0\end{pmatrix}(\delta s_2 + \overline{\delta s_2}),$$

and the z-coordinate remains $\overline{\delta s_2} + \delta s_2$, here, $\Delta b_i = b_i^- - b_i^+$.

The transport matrix along the second v-piece is

$$\left(Id + \overline{(\delta s_3} + \delta s_3)\begin{pmatrix}0 & -1\\a_3^- & b_3^-\end{pmatrix}\right)B\left(Id - (\delta s_2 + \overline{\delta s_2})\begin{pmatrix}0 & -1\\a_2^+ & b_2^+\end{pmatrix}\right)\cdot x_4'$$

$$+ \begin{pmatrix}0\\c_3\end{pmatrix}\overline{\delta s_3}\delta s_3 + \widetilde{U}\overline{\delta s_1}^2 + \widetilde{V}\delta s_1^2 + O(\overline{\delta s_1}^3 + \delta s_1^3)$$

here $\widetilde{U}$ and $\widetilde{V}$ are changed than before, which we won't track exactly. Therefore the first order term of x_4' is

$$\begin{pmatrix}\dfrac{a_2^-}{a_2^+} & \dfrac{\Delta b_2}{a_2^+}\\[4pt]0 & 1\end{pmatrix}A\begin{pmatrix}\dfrac{c_1}{a_1^+}\\[4pt]0\end{pmatrix}(\delta s_1 + \overline{\delta s_1}) + \begin{pmatrix}\dfrac{c_2}{a_2^+}\\[4pt]0\end{pmatrix}(\delta s_2 + \overline{\delta s_2})$$

$$= \begin{pmatrix}-\dfrac{c_1}{a_1^+}\dfrac{A_2}{B_2}B_4\\[8pt]\dfrac{c_1}{a_1^+}A_2\end{pmatrix}(\delta s_1 + \overline{\delta s_1}).$$

Thus the effect of the modification on v is

$$\begin{pmatrix} h_1\left(-\dfrac{c_1}{a_1^+}\dfrac{A_2}{B_2}B_4\right)^2 + l_1\left(-\dfrac{c_1}{a_1^+}\dfrac{A_2}{B_2}B_4\right)\dfrac{c_1}{a_1^+}A_2 \\[2ex] h_1\left(-\dfrac{c_1}{a_1^+}\dfrac{A_2}{B_2}B_4\right)^2 + l_1\left(-\dfrac{c_1}{a_1^+}\dfrac{A_2}{B_2}B_4\right)\dfrac{c_1}{a_1^+}A_2 \end{pmatrix}(\delta s_1 + \overline{\delta s_1})^2$$

$$= 2\left(\dfrac{c_1}{a_1^+}A_2\right)^2\dfrac{B_4}{B_2}\begin{pmatrix} -l_1 + \dfrac{B_4}{B_2}h_1 \\[2ex] -l_2 + \dfrac{B_4}{B_2}h_2 \end{pmatrix} + \widetilde{U}\overline{\delta s_1}^2 + \widetilde{V}\delta s_1^2.$$

Therefore the x,y–coordinates of x_5' read

$$\left(Id + \overline{(\delta s_3} + \delta s_3)\begin{pmatrix} 0 & -1 \\ a_3^- & b_3^- \end{pmatrix}\right)B\left(Id - (\delta s_2 + \overline{\delta s_2})\begin{pmatrix} 0 & -1 \\ a_2^+ & b_2^+ \end{pmatrix}\right)\cdot x_4'$$

$$+\,(0\ c_3)\,\overline{\delta s_3}\delta s_3 + 2\left(\dfrac{c_1}{a_1^+}A_2\right)^2\dfrac{B_4}{B_2}\begin{pmatrix} -l_1 + \dfrac{B_4}{B_2}h_1 \\[2ex] -l_2 + \dfrac{B_4}{B_2}h_2 \end{pmatrix}$$

$$+\,\widetilde{U}\overline{\delta s_1}^2 + \widetilde{V}\delta s_1^2 + O(\overline{\delta s_1}^3 + \delta s_1^3),$$

and the z–coordinate is $\overline{\delta s_3} + \delta s_3$.

The x and y coordinates of x_5' at the first order are

$$B\begin{pmatrix} \dfrac{a_2^-}{a_2^+} & \dfrac{\Delta b_2}{a_2^+} \\[2ex] 0 & 1 \end{pmatrix}A\begin{pmatrix} \dfrac{c_1}{a_1^+} \\[1ex] 0 \end{pmatrix}(\delta s_1 + \overline{\delta s_1}) + B\begin{pmatrix} \dfrac{c_2}{a_2^+} \\[1ex] 0 \end{pmatrix}(\delta s_2 + \overline{\delta s_2})$$

$$= \begin{pmatrix} A_1\dfrac{c_1}{a_1^+}(\delta s_1 + \overline{\delta s_1}) \\[2ex] 0 \end{pmatrix}.$$

Using Lemma 47, along the third ξ–piece,

$$\delta x = N_3,$$
$$\delta y = 0,$$
$$\delta z = d(\delta s_3 + \overline{\delta s_3})^2\overline{a}_c.$$

Writing only the $\delta s_1\overline{\delta s_1}$ terms,

$$\delta x = N_3,$$
$$\delta y = 0,$$
$$\delta z = 2d\overline{a}_c q^2 \delta s_1\overline{\delta s_1},$$

here, N_3 is the second order term of $a_c(x_5') - a_c(\widetilde{x_5})$.

$$x_6 \text{ is } \begin{pmatrix} \dfrac{a_3^-}{a_3^+} & \dfrac{\Delta b_3 - 2d\bar{a}_c}{a_3^+} \\ 0 & 1 \end{pmatrix} \cdot \text{ Point } x_5' + \begin{pmatrix} \dfrac{c_3}{a_3^+} \\ 0 \end{pmatrix} \overline{(\delta s_3}+\delta s_3) + \begin{pmatrix} \delta x \\ \delta y \end{pmatrix}.$$

Comparing the $x, y-$ coordinates of x_6' with Lemma 52, we get,

Claim 5 R satisfies the following relation,

$$c_1 \frac{\Delta b_1}{a_1^+} B_2 \frac{a_2^- A_1 + \Delta b_2 A_2}{a_2^+} - c_1 B_2 \frac{a_2^- A_3 + \Delta b_2 A_4}{a_2^+} + c_1 B_4 (\frac{\Delta b_1}{a_1^+} A_2 - A_4)$$

$$+2p A_2 B_2 \frac{c_1}{a_1^+}(1 - \frac{a_2^-}{a_2^+}) + 2p \frac{c_1}{a_1^+} B_2 \frac{\Delta b_2}{a_2^+}(a_2^- A_1 + b_2^- A_2) + c_2 p^2 (\frac{\Delta b_2}{a_2^+} B_2 - B_4)$$

$$-c_3 q^2 + \frac{c_2}{a_2^+} B_2 R + B_2 N_2 + (B_4 A_2 + B_2 \frac{a_2^- A_1 + \Delta b_2 A_2}{a_2^+}) N_1$$

$$+2 \left(\frac{c_1}{a_1^+} A_2 \right)^2 \frac{B_4}{B_2} \left(-l_2 + \frac{B_4}{B_2} h_2 \right) = 0.$$

In order to calculate the $\overline{\delta s_1}\delta s_1$ terms in $J_\infty(\bar{x}_\infty + \overline{\delta s_1}u_1(u_2) + \delta s_1 u_2) - J_\infty(\bar{x}_\infty + \delta s_1 u_2)$, we need to count the $\overline{\delta s_1}\delta s_1$ terms contributed by all the three ξ-pieces.

For the second $\xi-$piece, like before, the $\overline{\delta s_1}\delta s_1$ terms come from three sources. Since $\overline{\delta s_2} = p_1 \overline{\delta s_1} + R\overline{\delta s_1}\delta s_1$, so R is one of the sources. The second order terms of $da_c(x_1')$ N_2 contributes $\overline{\delta s_1}\delta s_1$ terms too, and all the other $\overline{\delta s_1}\delta s_1$ terms come from the v-transport.

For the third ξ-piece, the $\overline{\delta s_1}\delta s_1$ terms come from three sources, the R term of $\overline{\delta s_3}$, the second order terms of $da_c(x_5')$ N_3 , and those from the v-transport.

Summing them together, we have

$$\partial^2 J_\infty(\bar{x}_\infty).u_2.u_2$$

$$= \underbrace{(1 - A_1 + \frac{A_2}{B_2}(\det B - B_4))N_1 + c_1(\frac{\Delta b_1}{a_1^+} A_2 - A_4)\frac{1}{B_2}(\det B + B_4)}_{(I)}$$

$$\underbrace{-c_2 p^2 \frac{1}{B_2}(\det B - B_4) + c_1(A_3 - \frac{\Delta b_1}{a_1^+} A_1) + \frac{\Delta b_1}{a_1^+} c_1 + c_3 \frac{1 - B_1}{B_2} q^2}_{(I)}$$

$$+2\left(\frac{c_1}{a_1^+}A_2\right)^2\frac{B_4}{B_2}\left(\frac{B_1}{B_2}\left(l_2-\frac{B_4}{B_2}h_2\right)-\left(l_1-\frac{B_4}{B_2}h_1\right)\right)$$

$$\underbrace{\phantom{+2\left(\frac{c_1}{a_1^+}A_2\right)^2\frac{B_4}{B_2}\left(\frac{B_1}{B_2}\left(l_2-\frac{B_4}{B_2}h_2\right)-\left(l_1-\frac{B_4}{B_2}h_1\right)\right)}}_{(II)}$$

$$-2\left(\frac{c_1}{a_1^+}A_2\right)^2\frac{B_4}{B_2^2}\left(-l_2+\frac{B_4}{B_2}h_2\right).$$

$$\underbrace{\phantom{-2\left(\frac{c_1}{a_1^+}A_2\right)^2\frac{B_4}{B_2^2}\left(-l_2+\frac{B_4}{B_2}h_2\right)}}_{(II)}$$

The details of the calculation are carried out in the Appendix.

We can see that (I) involves no parameter, and (II) involves only $h_1, l_1.h_2, l_2$, Comparing with the expression for $\partial^2 J_\infty(\overline{x}_\infty).u_2.u_3$, it is easy to see that so as we perturb $h_1, l_1.h_2, l_2$, we can modify $\partial^2 J_\infty(\overline{x}_\infty).u_2.u_2$ freely and independent of $\partial^2 J_\infty(\overline{x}_\infty).u_2.u_3$.

2.9 Calculation of $\partial^2 J_\infty(\overline{x}_\infty).u_2.u_4$

First let us focus on u_4, the bottom variation; The $z-$coordinates of the 1st, 2nd and 3rd ξ-piece of u_4 will be denoted $\delta s_1, \delta s_2, \delta s_3$. They are constant along u_2. We have:

Lemma 54 $\delta s_1, \delta s_2, \delta s_3$ *satisfy the following relation,*

$$\delta s_2 = p_3\delta s_1 + O\left(\delta s_1^2\right)\qquad \delta s_3 = q_3\delta s_1 + O\left(\delta s_1^2\right),$$

where

$$p_3 = -\frac{c_3}{a_3^+}\frac{a_4^+}{c_4}\left(\frac{a_4^-C_1+\Delta b_4 C_2}{a_3^+}+\frac{C_2}{D_2}D_4\right),$$

$$q_3 = \frac{c_3}{a_3^+}\frac{a_5^-}{c_5}\frac{C_2}{D_2}\det D,$$

and

$$C = \begin{pmatrix} C_1 & C_3 \\ C_2 & C_4 \end{pmatrix}\quad and\quad D = \begin{pmatrix} D_1 & D_3 \\ D_2 & D_4 \end{pmatrix}$$

are respectively the transport matrices of the 3rd and 4th $v-$pieces.

The proof is the same than the proof of Lemma 48, we will omit it here.

On top of u_4, now we construct $u_2(u_4)$, assume that the z-coordinates of the 1st, 2nd and 3rd $\xi-$pieces of $u_2(u_4)$ are respectively $\overline{\delta s_1}, \overline{\delta s_2}+\delta s_1$ and $\overline{\delta s_2}+\overline{\delta s_3}$.

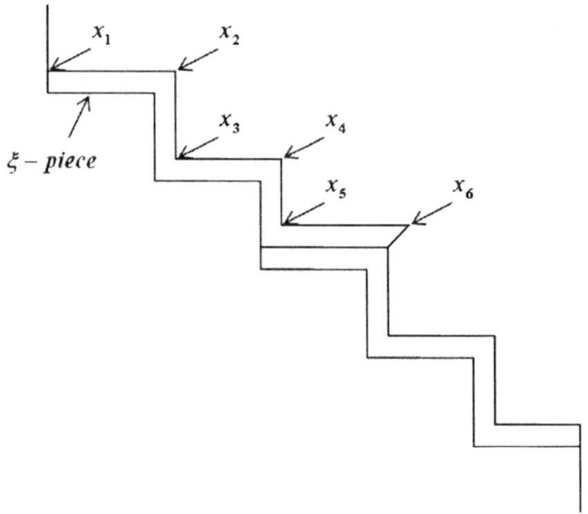

Lemma 55   *The x, y coordinates of x_6 read*

$$\begin{pmatrix} \dfrac{c_3}{a_3^+}\delta s_1 \\ c_3\delta s_1\overline{\delta s_3} \end{pmatrix} + U\delta s_1^2 + V\delta s_1^2 + O(\delta s_1^3 + \overline{\delta s_1}^3),$$

here U and and V are constants which we will not track.

The proof is similar with Lemma 49
Assume

$$\overline{\delta s_2} = p\overline{\delta s_1} + R\delta s_1\overline{\delta s_1} + \widetilde{U}\overline{\delta s_1}^2 + \widetilde{V}\delta s_1^2 + O(\overline{\delta s_1}^3 + \delta s_1^3),$$

$$\overline{\delta s_3} = q\overline{\delta s_1} + T\delta s_1\overline{\delta s_1} + \widetilde{U}_1\overline{\delta s_1}^2 + \widetilde{V}_1\delta s_1^2 + O(\overline{\delta s_1}^3 + \delta s_1^3).$$

Here, we will track only R and T, and we will not track $\widetilde{U}, \widetilde{V}, \widetilde{U}_1$ and $\widetilde{V}_1$.

First, we track only the first order terms along $u_2(u_4)$; we find:

Lemma 56 $\overline{\delta s_1}, \overline{\delta s_2}$ *and* $\overline{\delta s_3}$ *satisfy the following relation at the first order,*

$$\overline{\delta s_2} = p_1\overline{\delta s_1} + O\left(\delta s_1^2, \overline{\delta s_1}^2, \delta s_1\overline{\delta s_1}\right),$$

$$\overline{\delta s_3} = q_1\overline{\delta s_1} + O\left(\delta s_1^2, \overline{\delta s_1}^2, \delta s_1\overline{\delta s_1}\right),$$

where

$$p_1 = -\frac{c_1}{a_1^+} \cdot \frac{a_2^+}{c_2} \cdot \left(\frac{a_2^- A_1 + \Delta b_2 A_2}{a_3^+} + \frac{A_2}{B_2} B_4 \right),$$

$$q_1 = \frac{c_1}{a_1^+} \frac{a_3^-}{c_3} \frac{A_2}{B_2} \det B.$$

Hence we know how u_4 and $u_2(u_4)$ are built at the first order.

As before, since our aim is to compute $\partial^2 J_\infty(\overline{x}_\infty).u_2.u_4$ we need only to concentrate on the $\delta s_1 \overline{\delta s_1}$ terms, so from now on in all our computations, we keep only the $\delta s_1 \overline{\delta s_1}$ terms, and drop the $\delta s_1^2, \overline{\delta s_1}^2$ terms and terms with order higher than 2.

Let us track $u_2(u_4)$, the x, y coordinates of x_1 is $\begin{pmatrix} 0 \\ 0 \end{pmatrix}$, transport them to the point x_2, we get $\begin{pmatrix} \frac{c_1}{a_1^+} \overline{\delta s_1} \\ 0 \end{pmatrix}$; the z-coordinate along the first ξ-piece is constant $\overline{\delta s_1}$.

The x, y coordinates of x_3 read $A \begin{pmatrix} \frac{c_1}{a_1^+} \\ 0 \end{pmatrix} \overline{\delta s_1}$, the transport along the second ξ-piece reads

$$\begin{pmatrix} \frac{a_2^-}{a_2^+} & \frac{\Delta b_2}{a_2^+} \\ 0 & 1 \end{pmatrix} \cdot \text{point } x_3 + \begin{pmatrix} \frac{c_2}{a_2^+} \\ 0 \end{pmatrix} \overline{\delta s_2}.$$

The first order term of x_4 is

$$\begin{pmatrix} \frac{a_2^-}{a_2^+} & \frac{\Delta b_2}{a_2^+} \\ 0 & 1 \end{pmatrix} \cdot \text{point } x_3 + \begin{pmatrix} \frac{c_2}{a_2^+} \\ 0 \end{pmatrix} \overline{\delta s_2} = \frac{c_1}{a_1^+} \begin{pmatrix} -\frac{A_2}{B_2} B_4 \\ A_2 \end{pmatrix} \overline{\delta s_1}.$$

Integrating along the second $v-$piece, we get only $\overline{\delta s_1}^2$ terms. These terms would not influence the second derivative we are calculating.

We are able to write the crossing terms of x_5 now, they read,

$$B \begin{pmatrix} \frac{a_2^-}{a_2^+} & \frac{\Delta b_2}{a_2^+} \\ 0 & 1 \end{pmatrix} A \begin{pmatrix} \frac{c_1}{a_1^+} \\ 0 \end{pmatrix} \overline{\delta s_1} + B \begin{pmatrix} \frac{c_2}{a_2^+} \\ 0 \end{pmatrix} \overline{\delta s_2} + \begin{pmatrix} 0 \\ c_3 \end{pmatrix} \overline{\delta s_3} \delta s_1.$$

The first order terms are

$$B\begin{pmatrix} \dfrac{a_2^-}{a_2^+} & \dfrac{\Delta b_2}{a_2^+} \\ 0 & 1 \end{pmatrix} A\begin{pmatrix} \dfrac{c_1}{a_1^+} \\ 0 \end{pmatrix}\overline{\delta s_1} + B\begin{pmatrix} \dfrac{c_2}{a_2^+} \\ 0 \end{pmatrix}\overline{\delta s_2} = \begin{pmatrix} -\dfrac{a_3^-}{c_3}q_1 \\ 0 \end{pmatrix}\overline{\delta s_1}.$$

According to Lemma 47, along the third ξ-piece, $(\delta x, \delta y) = \begin{pmatrix} N_3 \\ 0 \end{pmatrix}$.

Then transporting along the third ξ-piece, we are able to derive an expression of x_6. Comparing the crossing terms, i.e. $\overline{\delta s_1}\delta s_1$ terms of x_4' with the expression in Lemma 55, we get

$$\frac{c_2}{a_2^+}B_2R = c_3q_1.$$

It is clear the only ξ-pieces that contribute $\overline{\delta s_1}\delta s_1$ terms are the second and the third one, so we need only to count the $\overline{\delta s_1}\delta s_1$ terms contributed by them. The second ξ-piece contributes $R\dfrac{c_2}{a_2^+}\overline{\delta s_1}\delta s_1$. The third ξ-piece contributes

$$(1,0)\begin{pmatrix} \dfrac{a_3^-}{a_3^+} & \dfrac{\Delta b_3 - 2d\bar{a}}{a_3^+} \\ 0 & 1 \end{pmatrix} B\begin{pmatrix} \dfrac{c_2}{a_2^+} \\ 0 \end{pmatrix} R + \frac{c_3}{a_3^+}T + N_3.$$

Summing them up, we get

$$\partial^2 J_\infty(\overline{x}_\infty).u_2.u_4 = \frac{1 - B_1}{B_2}c_3q_1.$$

Since q_1 contains no perturbation parameters, we know the modification which we built does not change the value of $\partial^2 J(\overline{x}_\infty).u_2.u_4$.

2.10   Other Second Order Derivatives

The modification which we performed on the third ξ-piece and the second v-piece will also change the following second derivatives: $\partial^2 J_\infty(\overline{x}_\infty).u_3.u_3$, $\partial^2 J_\infty(\overline{x}_\infty).u_4.u_4$, $\partial^2 J_\infty(\overline{x}_\infty).u_3.u_4$. Since the computation of these second derivatives is very similar to the one which we carried out for $\partial^2 J_\infty(\overline{x}_\infty).u_2.u_2$, $\partial^2 J_\infty(\overline{x}_\infty).u_2.u_3$, $\partial^2 J_\infty(\overline{x}_\infty).u_2.u_4$ in Sections 2.7.11, 2.8 and 2.9, we will only give the results without going into the details of the calculation here.

$$\partial^2 J_\infty(\overline{x}_\infty).u_3.u_3$$

$$= \left(1 - B_1 + \frac{B_2}{C_2}(\det C - C_4)\right)N_1 + c_2\left(\frac{\Delta b_2}{a_2^+}B_2 - B_4\right)\frac{1}{C_2}(\det C + C_4) + \frac{\Delta b_2}{a_2^+}c_2$$

$$\underbrace{\phantom{= \left(1 - B_1 + \frac{B_2}{C_2}(\det C - C_4)\right)N_1 + c_2\left(\frac{\Delta b_2}{a_2^+}B_2 - B_4\right)\frac{1}{C_2}(\det C + C_4) + \frac{\Delta b_2}{a_2^+}c_2}}_{(I)}$$

$$-c_3 p^2 \frac{1}{C_2}(\det C - C_4) + c_2\left(B_3 - \frac{\Delta b_2}{a_2^+}B_1\right) + c_4\frac{1 - C_1}{C_2}q^2 + 4d\frac{c_2}{a_2^+}B_2 p\frac{1}{C_2}(\det C + C_4)$$

$$\underbrace{\phantom{-c_3 p^2 \frac{1}{C_2}(\det C - C_4) + c_2\left(B_3 - \frac{\Delta b_2}{a_2^+}B_1\right) + c_4\frac{1 - C_1}{C_2}q^2 + 4d\frac{c_2}{a_2^+}B_2 p\frac{1}{C_2}(\det C + C_4)}}_{(I)}$$

$$\underbrace{-\frac{c_2}{a_2^+}h_2\left(C_4 + \frac{C_4}{C_2}(1 - C_1)\right)}_{(II)},$$

here,

$$p = -\frac{c_2}{a_2^+}\frac{a_3^+}{c_3}\left(\frac{a_3^- B_1 + (\Delta b_3 - 2d\overline{a}_c)B_2}{a_3^+} + \frac{B_2}{C_2}C_4\right),$$

$$q = \frac{c_2}{a_2^+}\frac{a_4^-}{c_4}\frac{B_2}{C_2}\det C.$$

$$\partial^2 J_\infty(\overline{x}_\infty).u_4.u_4 =$$

$$\left(1 - C_1 + \frac{C_2}{D_2}(\det D - D_4)\right)N_1 + c_3\left(\frac{\Delta b_3 - 2b\overline{a}_c}{a_3^+}C_2 - C_4\right)\frac{1}{D_2}(\det D + D_4)$$

$$\underbrace{\phantom{\left(1 - C_1 + \frac{C_2}{D_2}(\det D - D_4)\right)N_1 + c_3\left(\frac{\Delta b_3 - 2b\overline{a}_c}{a_3^+}C_2 - C_4\right)\frac{1}{D_2}(\det D + D_4)}}_{(I)}$$

$$-c_4 p^2 \frac{1}{D_2}(\det D - D_4) + c_3\left(C_3 - \frac{\Delta b_3 - 2b\overline{a}_c}{a_3^+}C_1\right) + \frac{\Delta b_3 - 2b\overline{a}_c}{a_3^+}c_3 + c_5\frac{1 - D_1}{D_2}q^2,$$

$$\underbrace{\phantom{-c_4 p^2 \frac{1}{D_2}(\det D - D_4) + c_3\left(C_3 - \frac{\Delta b_3 - 2b\overline{a}_c}{a_3^+}C_1\right) + \frac{\Delta b_3 - 2b\overline{a}_c}{a_3^+}c_3 + c_5\frac{1 - D_1}{D_2}q^2}}_{(I)}$$

here

$$p = -\frac{c_3}{a_3^+}\frac{a_4^+}{c_4}\left(\frac{a_4^- C_1 + \Delta b_4 C_2}{a_4^+} + \frac{C_2}{D_2}D_4\right),$$

$$q = \frac{c_3}{a_3^+}\frac{a_5^-}{c_5}\frac{C_2}{D_2}\det D.$$

And we also have,

$$\partial^2 J_\infty(\overline{x}_\infty).u_3.u_4$$

$$= -\frac{c_2}{a_2^+}(a_3^- B_1 + b_3^- B_2) + \frac{c_2}{a_2^+}\frac{B_2}{a_3^+} + 2c_4 q_2 p_3 \frac{1}{C_2} - c_4 q_2 \frac{C_4}{C_2}$$

$$+ p_2 \left(\frac{c_3}{a_3^+} a_4^+ \frac{C_4 - \det C}{C_2} - 2dB_2 \frac{c_2}{a_2^+} \frac{C_4 + \det C}{C_2} \right),$$

here,

$$p_2 = -\frac{c_2}{a_2^+}\frac{a_3^+}{c_3}\left(\frac{a_3^- B_1 + (\Delta b_3 - 2d\overline{a_c})B_2}{a_3^+} + \frac{B_2}{C_2}C_4 \right),$$

$$q_2 = \frac{c_2}{a_2^+}\frac{a_4^-}{c_4}\frac{B_2}{C_2}\det C,$$

and

$$p_3 = -\frac{c_3}{a_3^+}\frac{a_4^+}{c_4}\left(\frac{a_4^- C_1 + \Delta b_4 C_2}{a_4^+} + \frac{C_2}{D_2}D_4 \right),$$

$$q_3 = \frac{c_3}{a_3^+}\frac{a_5^-}{c_5}\frac{C_2}{D_2}\det D.$$

Comparing the expressions for $\partial^2 J_\infty(\overline{x}_\infty).u_2.u_2$, $\partial^2 J_\infty(\overline{x}_\infty).u_2.u_3$, $\partial^2 J_\infty(\overline{x}_\infty).u_2.u_4$, $\partial^2 J_\infty(\overline{x}_\infty).u_3.u_3$, $\partial^2 J_\infty(\overline{x}_\infty).u_4.u_4$ and $\partial^2 J_\infty(\overline{x}_\infty).u_3.u_4$, we can see that, if we let $d = 0$, then only $\partial^2 J_\infty(\overline{x}_\infty).u_2.u_3$, $\partial^2 J_\infty(\overline{x}_\infty).u_3.u_3$, $\partial^2 J_\infty(\overline{x}_\infty).u_4.u_4$ are modified, and they change independently. Therefore, we can prescribe any three real values for $\partial^2 J_\infty(\overline{x}_\infty).u_2.u_3$, $\partial^2 J_\infty(\overline{x}_\infty).u_3.u_3$, $\partial^2 J_\infty(\overline{x}_\infty).u_4.u_4$. In this way, we can prescribe any set of $2k$ real values for $\partial^2 J_\infty(\overline{x}_\infty).u_i.u_i$, $\partial^2 J_\infty(\overline{x}_\infty).u_i.u_{i+1}$ for $1 \leq i \leq k$. Thus we proved Lemma 46.

2.11   Appendix

2.11.1   *The Proof of Lemma 42*

Let us consider first two directions h and k in $T_{\overline{x}_\infty}\Gamma_{2k}$ such that $\delta z_1(h) = \delta z_1(k) = 0$. We study $dJ_{\overline{x}_\infty}^t(h,k)$ along the homotopy of v into v_t.

As we change c into c_t, a, b into a_t, b_t, we have to study the change in the v_t transport matrix. If the second derivative of this v_t transport along (h,k) varies little, then $dJ_{\overline{x}_\infty}^t(h,k)$ varies also very little.

Writing the equation of the tangent space, we obtain,

$$w = dl_{\overline{x}}^t(w) + \sum \delta a_i \xi_i^t + \sum \delta s_i v_i^t.$$

Observe that,

$$dz_1(w) = 0 \text{ for both } h \text{ and } k.$$

Now we compute the variation of the equation defining h along k,

$$\delta_h w = (\delta_h dl_{\overline{x}}^t)(w) + \sum \delta a_i \delta_h \xi_i^t + \sum \delta s_i \delta_h v_i^t$$

$$+ dl_{\overline{x}}^t(\delta_h w) + \sum \delta_h(\delta a_i)\xi_i^t + \sum \delta_h(\delta s_i)v_i^t.$$

Once $(\delta_h dl_{\overline{x}}^t)(w) + \sum \delta a_i \delta_h \xi_i^t + \sum \delta s_i \delta_h v_i^t$ is computed, $\delta_h w, \delta_h(\delta a_i)$ and $\delta_h(\delta s_i)$ can be found since $Span(Id - dl_{\overline{x}}) + Span_{i=1}^k(\xi_i^t, v_i^t) = R^3$ since Γ_{2k} is a manifold.

$(\delta_h dl_{\overline{x}}^t)(w) + \sum \delta a_i \delta_h \xi_i^t + \sum \delta s_i \delta_h v_i^t$ can be computed as we modify a, b, c into a_t, b_t, c_t. Each involves the transport of a vector $(\widetilde{w}, \widetilde{\xi}_i, \widetilde{v}_i)$ which is (w, ξ, v) after suitable transport along part of the curve. These vectors do not change until the last v jump where

$$v = \frac{\partial}{\partial z} - y\frac{\partial}{\partial x} + (ax + by + cz + h.o.)\frac{\partial}{\partial y}.$$

The tangent equation along v reads,

$$\dot{z} = 1,$$

$$\dot{y} = ax + by + cz + h.o.,$$

$$\dot{x} = -y.$$

Its differential reads

$$\dot{\widehat{\delta z}} = 0,$$

$$\dot{\widehat{\delta y}} = a\delta x + b\delta y + c\delta z + \delta ax + \delta by + \delta cz + \delta(h.o.),$$

$$\dot{\widehat{\delta x}} = -\delta y.$$

The modification which we introduce occurs along $\frac{\partial}{\partial y}$. We make the modification explicit.

The space of variation h such that $\delta z_1(h) = 0$ can be seen as the tangent space to the sub-manifold of curves $\overline{x}$ near $\overline{x}_\infty$ such that $\overline{x}_{\overline{1}}$ has its z_1-coordinate equal to zero if $\overline{x_{1\infty}}$ has its z_1 coordinate equal to zero. h can be seen as a variation of μ.

The tangent equation along v takes then the form

$$\dot{R} = A(t, \mu)R \qquad R(0, \mu) = Id$$

$$\widehat{\frac{\partial R}{\partial \mu}} = \frac{\partial A}{\partial \mu}R + A\frac{\partial R}{\partial \mu}.$$

Here A comes from the differential of the differential equation

$$\dot{X} = v(X),$$

which reads

$$\widehat{\delta \dot{X}} = Dv(\delta X),$$

A corresponds to Dv, i.e. to the linearization of v along the v-orbit. We compute this linearization in the region where

$$v = \frac{\partial}{\partial z} - y\frac{\partial}{\partial x} + (ax + by + cz + h.o.)\frac{\partial}{\partial y}.$$

Therefore, $A(x, y, z)$ is explicit, and

$$A = \begin{pmatrix} 0 & -1 & 0 \\ a + xa_x + yb_x + zc_x & b + xa_y + yb_y + zc_y & c + xa_z + yb_z + zc_z \\ 0 & 0 & 0 \end{pmatrix}.$$

The variation map of the transport map dl_t as we move along εh reads as $dA(h)(\delta x)$.

We then have to study

$$Q = \int_I d^2(xa + yb + cz)(h, dl_{\bar{x}}^t(w) + \sum \delta a_i \xi_i^t + \sum \delta s_i v_i^t),$$

where I is the z-interval of modification and this second derivative $(d^2(xa + yb + cz))$ is taken along the value of h and the value of $dl_{\bar{x}}^t(w) + \sum \delta a_i \xi_i^t + \sum \delta s_i v_i^t$ at the precise points of the interval I.

We observe that

$$\delta z_1(h) = 0 \quad (\text{ at } x_{\bar{1}}).$$

But also, since

$$dl_{\overline{x}}^t(w) + \sum \delta a_i \xi_i^t + \sum \delta s_i v_i^t = w(k) \text{ at } x_{\overline{1}},$$

$$\delta z_1(dl_{\overline{x}}^t(w) + \sum \delta a_i \xi_i^t + \sum \delta s_i v_i^t) = 0 \text{ at } x_{\overline{1}},$$

if $\delta z(x_{\overline{1}}) = 0$, δz is zero throughout and if the interval of integration over z along which a, b, c are modified is small with respect to the x and y interval of modification, the Poincaré-return map is very little changed.

We look now at $dl_{\overline{x}}^t(\delta_h w) + \sum \delta_h(\delta a_i)\xi_i^t + \sum \delta_h(\delta s_i)v_i^t$, with $\delta z_1(h) = \delta z_1(k) = 0$ through out I.

Computing, we find terms of the order of

$$|I| \left(|x| \left|\nabla^2_{(x,y)} a\right| + |y| \left|\nabla^2_{(x,y)} b\right| + |z| \left|\nabla^2_{(x,y)} c\right| + \left|\nabla_{(x,y)} a\right| + \left|\nabla_{(x,y)} b\right| + \left|\nabla_{(x,y)} c\right|\right).$$

Assuming $|I|$ is small with respect to the size of the (x, y) domain of the variations, we obtain that Q is small.

Hence, the contribution of the term

$$(\delta_h dl_{\overline{x}}^t)(w) + \sum \delta a_i \delta_h \xi_i^t + \sum \delta s_i \delta_h v_i^t$$

over the interval I of the modification is $o(|h|, |k|)$.

The remainder corresponds to the computation of the second derivative before perturbation.

This shows that

$$d^2 J_{\overline{x}_\infty}^t(h, k) = d^2 J_{\overline{x}_\infty}(h, k) + o(|h|, |k|),$$

if $\delta z_1(h) = \delta z_1(k) = 0$.

Computing $d^2 J_{\overline{x}_\infty}(\widetilde{v}_1, k)$, we find as usual

$$\overline{c}(\delta a - \delta a_c^1)(h).$$

If $\delta z_1(h) = 0$ and we are modifying only c along c_t, this does not change much.

Lastly, we compute

$$d^2 J_{\overline{x}_\infty}(\widetilde{v}_1, \widetilde{v}_1) = \delta a_c^1(\widetilde{v}_1) = c_\tau$$

as usual, which changes as much as c_τ changes.

Lemma 46 follows.

2.11.2　*The proof of Lemma 47*

We will prove it in a more general setting. After the modification, α on the third ξ-piece becomes $\lambda\alpha$, where $\dfrac{1}{\lambda} = 1 + \omega_1(x)\omega_2(y,z)(az^3 + byz^2 + dy^2z)$, here, $\omega_1(x)$ and $\omega_2(y,z)$ are the cut-off function in a small neighborhood of the third ξ-piece. The corresponding Reeb vector field becomes $\xi = \dfrac{1}{\lambda}\dfrac{\partial}{\partial x} + \dfrac{\lambda_z - y\lambda_x}{\lambda^2}\dfrac{\partial}{\partial y} - \dfrac{\lambda_y}{\lambda^2}\left(\dfrac{\partial}{\partial z} - y\dfrac{\partial}{\partial x}\right)$.

Thus

$$\dot{x} = \frac{1}{\lambda} + y\frac{\lambda_y}{\lambda^2} = 1 - az^3 - by^2z + O_4(y,z),$$

$$\dot{y} = \frac{\lambda_z - y\lambda_x}{\lambda^2} = -3az^2 - by^2 - 2dyz + O_3(y,z),$$

$$\dot{z} = -\frac{\lambda_y}{\lambda^2} = 2byz + dz^2 + O_3(y,z).$$

Integrating it at first order, we have,

$$x = x_0 + (az_0^3 + by_0^2z_0)t + O_4,$$

$$y = y_0 - (3az_0^2 + by_0^2 + 2dy_0z_0)t + O_3,$$

$$z = z_0 + (2by_0z_0 + dz_0^2)t + O_3.$$

Now we go to the second order,

$$\dot{x} = 1 + az_0^3 - by_0^2z_0 + O_4(y,z),$$

$$\dot{y} = -(3az_0^2 + by_0^2 + 2dy_0z_0) - 6az_0(2by_0z_0 + dz_0^2)t,$$

$$+ 2by_0(3az_0^2 + by_0^2 + 2dy_0z_0)t - 2dy_0(2by_0z_0 + dz_0^2) + O_4$$

$$= -(3az_0^2 + by_0^2 + 2dy_0z_0) + (2d^2 - 6ab)y_0z_0^2t + 2by_0^2z_0t + 2by_0^3t + O_4,$$

$$\dot{z} = 2by_0z_0 + dz_0^2 + 2by_0(2by_0z_0 + dz_0^2)t - 2bz_0(3az_0^2 + by_0^2 + 2dy_0z_0)t$$

$$+ 2dz_0(2by_0z_0 + dz_0^2)t + O_4$$

$$= 2by_0z_0 + dz_0^2 + (2d^2 - 6ab)z_0^3t + 2b^2y_0^2z_0t + 2bdy_0z_0^2t + O_4.$$

Integrating it, we get

$$x = x_0 + (az_0^3 - by_0^2 z_0)a_c + O_4(y_0, z_0),$$

$$y = y_0 - (3az_0^2 + by_0^2 + 2dy_0 z_0)a_c$$

$$+ \left((2d^2 - 6ab)y_0 z_0^2 + 2bdy_0^2 z_0 + 2b^2 y_0^3\right)\frac{a_c^2}{2} + O_4,$$

$$z = z_0 + \left(2by_0 z_0 + dz_0^2\right)a_c + \left((2d^2 - 6ab)z_0^3 + 2b^2 y_0^2 z_0 + 2bdy_0 z_0^2\right)\frac{a_c^2}{2} + O_4.$$

On one hand, we can find Darboux coordinates such that around x_3^-, the vector v reads $\dfrac{\partial}{\partial z} - y\dfrac{\partial}{\partial x} + (a_3^- x + b_3^- y + c_3 z + P_2^3(x, y, z))\dfrac{\partial}{\partial y}$, and around x_3^+, the vector v reads $\dfrac{\partial}{\partial z} - y\dfrac{\partial}{\partial x} + (a_3^+ x + b_3^+ y + Q_2^3(x_0, y_0, z_0))\dfrac{\partial}{\partial y}$. So

$$dl(v_{initial}) = \left(-y_0 + a_c\left(3az_0^2 - by_0^2 - 2by_0 z_0(a_3^- x_0 + b_3^- y_0 + c_3 z_0)\right)\right)\frac{\partial}{\partial x}$$

$$+ \{a_3^- x_0 + b_3^- y_0 + c_3 z_0 + P_2^3(x_0, y_0, z_0)$$

$$- a_c(6az_0 + by_0(a_3^- x_0 + b_3^- y_0 + c_3 z_0) + 2dy_0$$

$$+ 2dz_0(a_3^- x_0 + b_3^- y_0 + c_3 z_0)) + a_c^2((2d^2 - 6ab)y_0 z_0 + bdy_0^2$$

$$+ 3b^2 y_0^2(a_3^- x_0 + b_3^- y_0 + c_3 z_0) + (2d^2 - 6ab)z_0^2(a_3^- x_0 + b_3^- y_0 + c_3 z_0)$$

$$+ 2bdy_0 z_0(a_3^- x_0 + b_3^- y_0 + c_3 z_0))\}\frac{\partial}{\partial y}$$

$$+ \{1 + a_c(2bz_0(a_3^- x_0 + b_3^- y_0 + c_3 z_0) + 2dz_0 + 2by_0)$$

$$+ a_c^2((3d^2 - 9ab)z_0^2 + b^2 y_0^2 + 2bdy_0 z_0 + b^2 y_0 z_0(a_3^- x_0 + b_3^- y_0 + c_3 z_0)$$

$$+ bdz_0^2(a_3^- x_0 + b_3^- y_0 + c_3 z_0))\}\frac{\partial}{\partial z}.$$

But on the other hand, around x_3^+, the vector v reads

$$\frac{\partial}{\partial z} - y\frac{\partial}{\partial x} + (a_3^+ x + b_3^+ y + Q_2^3(x_0, y_0, z_0))\frac{\partial}{\partial y}.$$

Therefore,

$$v_{final}$$

$$= \frac{\partial}{\partial z} - (y_0 - (3az_0^2 + by_0^2 + 2dy_0z_0)a_c + ((2d^2 - 6ab)y_0z_0^2 + 2bdy_0^2z_0$$

$$+ 2b^2y_0^3)\frac{a_c^2}{2})\frac{\partial}{\partial x} + (a_3^+(x_0 + (az_0^3 - by_0^2z_0)a_c)$$

$$+ b_3^+(y_0 - (3az_0^2 + by_0^2 + 2dy_0z_0)a_c$$

$$+ ((2d^2 - 6ab)y_0z_0^2 + 2bdy_0^2z_0 + 2b^2y_0^3)\frac{a_c^2}{2})$$

$$+ c_3(z_0 + (2by_0z_0 + dz_0^2)\,a_c + ((2d^2 - 6ab)z_0^3 + 2b^2y_0^2z_0 + 2bdy_0z_0^2)\frac{a_c^2}{2})$$

$$+ Q_2^3(x_0 + a_c - \bar{a}, y_0, z_0))\frac{\partial}{\partial y}.$$

Comparing v_{final} and $dl(v_{initial})$ up to second order, we have,

$$a_3^+(x_0 + a_c - \bar{a}) + b_3^+y_0 - a_cb_3^+(3az_0^2 + by_0^2 + 2dy_0z_0)$$

$$+ Q_2^3(x_0 + a_c - \bar{a}, y_0, z_0)$$

$$a_3^-x_0 + b_3^-y_0 + c_3z_0 + P_2^3(x_0, y_0, z_0) - a_c(6az_0 + by_0(a_3^-x_0 + b_3^-y_0 + c_3z_0)$$

$$+ 2dy_0 = \frac{+2dz_0(a_3^-x_0 + b_3^-y_0 + c_3z_0)) + a_c^2((2d^2 - 6ab)y_0z_0 + bdy_0^2)}{1 + 2a_c(2by_0 + 2dz_0)}.$$

Therefore,

$$\delta a_c = \frac{(a_3^- - a_3^+)x_0 + (b_3^- - b_3^+)y_0 + c_3z_0 - \bar{a}(6az_0 + 2dy_0)}{a_3^+}$$

$$- \frac{(a_3^- - a_3^+)x_0 + (b_3^- - b_3^+)y_0 + c_3z_0 - \bar{a}(6az_0 + 2dy_0)}{a_3^+}$$

$$\cdot \frac{6az_0 + 2dy_0 + Q_{2,x}(x_0, y_0, z_0)}{a_3^+}$$

$$+ \{P_2^3(x_0, y_0, z_0) - Q_2^3(x_0, y_0, z_0) + \bar{a}b_3^+(3az_0^2 + by_0^2 + 2dy_0z_0)$$

$$- \bar{a}(2by_0 + 2dz_0)(a_3^-x_0 + b_3^-y_0 + c_3z_0)$$

$$+ \bar{a}(bdy_0^2 + (2d^2 - 6ab)y_0z_0 - \bar{a}(2by_0 + 2dz_0)(a_3^-x_0 + b_3^-y_0 + c_3z_0)\}/a_3^+.$$

In order to keep $da_c(v)$ unchanged, we take $\dfrac{1}{\lambda} = 1 + \varpi_1(x)\varpi_2(y,z)dy^2 z$, therefore,

$$\delta a_c = \frac{(a_3^- - a_3^+)x_0 + (b_3^- - b_3^+)y_0 + c_3 z_0 - \bar{a}2dy_0}{a_3^+}$$

$$- \frac{(a_3^- - a_3^+)x_0 + (b_3^- - b_3^+)y_0 + c_3 z_0 - \bar{a}2dy_0}{a_3^+} \cdot \frac{2dy_0 + Q_{2,x}(x_0, y_0, z_0)}{a_3^+}$$

$$+ \{P_2^3(x_0, y_0, z_0) - Q_2^3(x_0, y_0, z_0) + \bar{a}b_3^+ 2dy_0 z_0 - \bar{a}2dz_0(a_3^- x_0 + b_3^- y_0 + c_3 z_0)$$

$$- \bar{a}^2 2dz_0(a_3^- x_0 + b_3^- y_0 + c_3 z_0)\}/a_3^+.$$

2.11.3 *Proof of the Lemmas in 2.7.11*

2.11.4 *Proof of Lemma 48*

Following u_2, in Darboux coordinates in the small neighborhood of the base curve, the x, y coordinates of x_1 read $\begin{pmatrix} 0 \\ 0 \end{pmatrix}$, x_2 reads $\begin{pmatrix} c_2 \\ a_2^+ \\ 0 \end{pmatrix} \delta s_1$, and the z coordinates of both x_1 and x_2 is δs_1; x_3 reads $\begin{pmatrix} B\begin{pmatrix} c_2 \\ a_2^+ \\ 0 \end{pmatrix} \delta s_1 \\ \delta s_2 \end{pmatrix}$, here B is the transport matrix along the second v-piece.

Since the third ξ–piece is modified, the transport matrix in x, y- coordinates along this piece is $\begin{pmatrix} \dfrac{a_3^- - a_3^+}{a_3^+} & \dfrac{b_3^- - b_3^+ - 2d\bar{a}}{a_3^+} \\ 0 & 0 \end{pmatrix} + Id$, and there is a small increment $\begin{pmatrix} \dfrac{c_3 - 6a \cdot \bar{a}}{a_3^+} \\ 0 \end{pmatrix} \delta s_2$ due to the effect of the z–component (refer to Lemma 47)

Therefore, the x, y–coordinates of x_4 read $\begin{pmatrix} \dfrac{a_3^-}{a_3^+} & \dfrac{\Delta b_3 - 2d\bar{a}}{a_3^+} \\ 0 & 1 \end{pmatrix}$ $B\begin{pmatrix} c_2 \\ a_2^+ \\ 0 \end{pmatrix} \delta s_1 + \begin{pmatrix} c_3 \\ a_3^+ \\ 0 \end{pmatrix} \delta s_2$, here $\Delta b_3 = b_3^- - b_3^+$, and the z–coordinate of x_4 is δs_2.

The x, y–coordinates of x_5 are

$$
C\begin{pmatrix} \dfrac{a_3^-}{a_3^+} & \dfrac{\Delta b_3 - 2d\bar{a}}{a_3^+} \\ 0 & 1 \end{pmatrix} B\begin{pmatrix} \dfrac{c_2}{a_2^+} \\ 0 \end{pmatrix} \delta s_1 + C\begin{pmatrix} \dfrac{c_3 - 6a\cdot\bar{a}}{a_3^+} \\ 0 \end{pmatrix} \delta s_2,
$$

and the z–coordinate of x_5 is δs_3.

Therefore the x, y–coordinates of x_6 read

$$
\begin{pmatrix} \dfrac{a_4^-}{a_4^+} & \dfrac{b_4^- - b_4^+}{a_4^+} \\ 0 & 1 \end{pmatrix} C\begin{pmatrix} \dfrac{a_3^-}{a_3^+} & \dfrac{\Delta b_3 - 2d\bar{a}}{a_3^+} \\ 0 & 1 \end{pmatrix} B\begin{pmatrix} \dfrac{c_2}{a_2^+} \\ 0 \end{pmatrix} \delta s_1
$$

$$
+ \begin{pmatrix} \dfrac{a_4^-}{a_4^+} & \dfrac{b_4^- - b_4^+}{a_4^+} \\ 0 & 1 \end{pmatrix} C\begin{pmatrix} \dfrac{c_3 - 6a\cdot\bar{a}}{a_3^+} \\ 0 \end{pmatrix} \delta s_2 + \begin{pmatrix} \dfrac{c_4}{a_4^+} \\ 0 \end{pmatrix} \delta s_3,
$$

and the z–coordinate of x_6 is δs_3.

On the other hand, u_3 is parallel with v at x_6.

Thus we have

$$
\begin{pmatrix} \dfrac{a_4^-}{a_4^+} & \dfrac{b_4^- - b_4^+}{a_4^+} \\ 0 & 1 \end{pmatrix} C\begin{pmatrix} \dfrac{a_3^-}{a_3^+} & \dfrac{\Delta b_3 - 2d\bar{a}}{a_3^+} \\ 0 & 1 \end{pmatrix} B\begin{pmatrix} \dfrac{c_2}{a_2^+} \\ 0 \end{pmatrix} \delta s_1.
$$

$$
+ \begin{pmatrix} \dfrac{a_4^-}{a_4^+} & \dfrac{b_4^- - b_4^+}{a_4^+} \\ 0 & 1 \end{pmatrix} C\begin{pmatrix} \dfrac{c_3 - 6a\cdot\bar{a}}{a_3^+} \\ 0 \end{pmatrix} \delta s_2 + \begin{pmatrix} \dfrac{c_4}{a_4^+} \\ 0 \end{pmatrix} \delta s_3 = \begin{pmatrix} 0 \\ 0 \end{pmatrix}.
$$

Solving it for $\delta s_2, \delta s_3$, we get

$$
\delta s_2 = -\frac{c_2}{a_2^+}\frac{a_3^+}{c_3 - 6a\cdot\bar{a}}\left(\frac{a_3^- B_1 + (\Delta b_3 - 2d\bar{a}_c)B_2}{a_3^+} + \frac{B_2}{C_2}C_4 \right) \delta s_1,
$$

$$
\delta s_3 = \frac{c_2}{a_2^+}\frac{a_4^-}{c_4}\frac{B_2}{B_4} \det C \delta s_1.
$$

2.11.5   *The Proof of Lemma 49*

Along u_3,

$$
\begin{pmatrix} 0 \\ 0 \\ \delta s_1 \end{pmatrix} \longrightarrow \begin{pmatrix} \dfrac{c_2}{a_2^+} \\ 0 \\ 1 \end{pmatrix} \longrightarrow \begin{pmatrix} B\begin{pmatrix} \dfrac{c_2}{a_2^+} \\ 0 \end{pmatrix} \delta s_1 \\ \delta s_2 \end{pmatrix}
$$

$$\longrightarrow \left(\left(\begin{pmatrix} \dfrac{a_3^- - a_3^+}{a_4^+} & \dfrac{\Delta b_3 - 2d\bar{a}}{a_4^+} \\ 0 & 1 \end{pmatrix} B \begin{pmatrix} \dfrac{c_2}{a_2^+} \\ 0 \end{pmatrix} \delta s_1 + \begin{pmatrix} \dfrac{c_3 - 6a\cdot\bar{a}}{a_3^+} \\ 0 \end{pmatrix} \delta s_2 \right), \; \delta s_2 \right)$$

so point x_4 reads
$$\begin{pmatrix} \dfrac{c_2}{a_2^+}\delta s_1 \begin{pmatrix} -\dfrac{B_2}{C_2}C_4 \\ B_2 \end{pmatrix} \\ \delta s_2 \end{pmatrix}.$$

And since in the small neighborhood of x_6, the vector v reads

$$\frac{\partial}{\partial z} - y\frac{\partial}{\partial x} + (a_3^+ x + b_3^+ y + Q_2^3(x,y,z))\frac{\partial}{\partial y}.$$

So the vector v at x_6 is parallel with

$$\begin{pmatrix} -\dfrac{c_2}{a_2^+}B_2\delta s_1 \\ \dfrac{c_2}{a_2^+}(b_3^+ B_2 - a_3^+ \dfrac{B_2}{C_2}C_4)\delta s_1 \\ 1 \end{pmatrix} + O(\delta s_1^2).$$

Thus the x,y coordinates of x_4' are

$$\frac{c_2}{a_2^+}\delta s_1 \left(-\frac{B_2}{C_2}C_4 B_2 \right)$$

$$+ \begin{pmatrix} -\dfrac{c_2}{a_2^+}B_2\delta s_1 \\ \dfrac{c_2}{a_2^+}\left(b_3^+ B_2 - a_3^+ \dfrac{B_2}{C_2}C_4\right)\delta s_1 \end{pmatrix} \overline{\delta s_3} + U\delta s_1^2 + O(\delta s_1^3 + \overline{\delta s_1}^3).$$

2.11.6 *The proof of Lemma 50*

We track only the first order terms of the base points along $u_2(u_3)$. The x,y coordinates of point x_7 are $\begin{pmatrix} 0 \\ 0 \end{pmatrix}$, and the x,y coordinates of x_0 read $\begin{pmatrix} \dfrac{c_1}{a_1^+} \\ 0 \end{pmatrix} \overline{\delta s_1}$, both of their z coordinates are $\overline{\delta s_1}$; Thus x_1' reads

$$\begin{pmatrix} A \begin{pmatrix} \dfrac{c_1}{a_1^+} \\ 0 \end{pmatrix} \overline{\delta s_1} \\ \overline{\delta s_2} + \overline{\delta s_1} \end{pmatrix}.$$

The transport matrix in x, y- coordinates along the second ξ-piece is

$$
\begin{pmatrix}
\dfrac{a_2^- - a_2^+}{a_2^+} & \dfrac{b_2^- - b_2^+}{a_4^+} \\
0 & 0
\end{pmatrix} + Id,
$$

and there is a small increment $\begin{pmatrix} \dfrac{c_2}{a_2^+} \\ 0 \end{pmatrix} \overline{(\delta s_2 + \delta s_1)}$ due to the effect of the $z-$ component.

Therefore, the x, y- coordinates of x_2' read

$$
\begin{pmatrix}
\dfrac{a_2^- - a_2^+}{a_2^+} & \dfrac{\Delta b_2}{a_4^+} \\
0 & 1
\end{pmatrix}
A \begin{pmatrix} \dfrac{c_1}{a_1^+} \\ 0 \end{pmatrix} \overline{\delta s_1}
+ \begin{pmatrix} \dfrac{c_2}{a_2^+} \\ 0 \end{pmatrix} \overline{(\delta s_2 + \delta s_1)},
$$

and the $z-$coordinate is $\overline{(\delta s_2 + \delta s_1)}$;

The x, y- coordinates of x_3' are

$$
B \begin{pmatrix}
\dfrac{a_2^- - a_2^+}{a_2^+} & \dfrac{\Delta b_2}{a_4^+} \\
0 & 1
\end{pmatrix}
A \begin{pmatrix} \dfrac{c_1}{a_1^+} \\ 0 \end{pmatrix} \overline{\delta s_1}
+ B \begin{pmatrix} \dfrac{c_2}{a_2^+} \\ 0 \end{pmatrix} \overline{(\delta s_2 + \delta s_1)},
$$

and the $z-$coordinate is $\overline{\delta s_3} + \delta s_2$.

Therefore the x, y- coordinates of x_4' read

$$
\begin{pmatrix}
\dfrac{a_3^-}{a_3^+} & \dfrac{\Delta b_3 - 2d\bar{a}}{a_3^+} \\
0 & 1
\end{pmatrix}
B \begin{pmatrix}
\dfrac{a_2^- - a_2^+}{a_2^+} & \dfrac{\Delta b_2}{a_4^+} \\
0 & 1
\end{pmatrix}
A \begin{pmatrix} \dfrac{c_1}{a_1^+} \\ 0 \end{pmatrix} \overline{\delta s_1},
$$

$$
+ \begin{pmatrix}
\dfrac{a_3^-}{a_3^+} & \dfrac{\Delta b_3 - 2d\bar{a}}{a_3^+} \\
0 & 1
\end{pmatrix}
B \begin{pmatrix} \dfrac{c_2}{a_2^+} \\ 0 \end{pmatrix} \overline{(\delta s_2 + \delta s_1)}
+ \begin{pmatrix}
\dfrac{c_3 - 6a \cdot \bar{a}}{a_3^+} \\ 0
\end{pmatrix} \overline{(\delta s_3 + \delta s_2)},
$$

and the $z-$coordinate remains the same.

Comparing with the first order terms of Lemma 49, we have

$$\begin{pmatrix} a_3^- & \dfrac{\Delta b_3 - 2d\bar{a}}{a_3^+} \\ a_3^+ & \\ 0 & 1 \end{pmatrix} B \begin{pmatrix} a_2^- - a_2^+ & \dfrac{\Delta b_2}{a_4^+} \\ a_2^+ & \\ 0 & 1 \end{pmatrix} A \begin{pmatrix} \dfrac{c_1}{a_1^+} \\ 0 \end{pmatrix} \overline{\delta s_1} +$$

$$\begin{pmatrix} a_3^- & \dfrac{\Delta b_3 - 2d\bar{a}}{a_3^+} \\ a_3^+ & \\ 0 & 1 \end{pmatrix} B \begin{pmatrix} \dfrac{c_2}{a_2^+} \\ 0 \end{pmatrix} \overline{(\delta s_2 + \delta s_1)} + \begin{pmatrix} \dfrac{c_3 - 6a \cdot \bar{a}}{a_3^+} \\ 0 \end{pmatrix} \overline{(\delta s_3 + \delta s_2)}$$

$$= \dfrac{c_2}{a_2^+} \delta s_1 \begin{pmatrix} -\dfrac{B_2}{C_2} C_4 \\ B_2 \end{pmatrix}.$$

Solving it for $\overline{\delta s_2}, \overline{\delta s_3}$, we know that $\overline{\delta s_1}, \overline{\delta s_2}$ and $\overline{\delta s_3}$ satisfy the following relation at the first order,

$$\overline{\delta s_2} = p_1 \overline{\delta s_1} + O\left(\delta s_1^2, \overline{\delta s_1}^2, \delta s_1 \overline{\delta s_1}\right),$$

$$\overline{\delta s_3} = q_1 \overline{\delta s_1} + O\left(\delta s_1^2, \overline{\delta s_1}^2, \delta s_1 \overline{\delta s_1}\right),$$

with

$$p_1 = -\dfrac{c_1}{a_1^+} \cdot \dfrac{a_2^+}{c_2} \cdot \left(\dfrac{a_2^- A_1 + \Delta b_2 A_2}{a_3^+} + \dfrac{A_2}{B_2} B_4\right),$$

$$q_1 = \dfrac{c_1}{a_1^+} \dfrac{a_3^-}{c_3} \dfrac{A_2}{B_2} \det B.$$

2.11.7 *Proof of Claim 1*

In a neighborhood of x_0, the vector v reads up to a multiplicative factor $\dfrac{\partial}{\partial z} - y\dfrac{\partial}{\partial x} + (a_1^+ x + b_1^+ y + Q_2^1(x, y, z))\dfrac{\partial}{\partial y}$, so the v trajectory satisfies

$$\dot{x} = -y,$$

$$\dot{y} = a_1^+ x + b_1^+ y + Q_2^1(x, y, z),$$

$$\dot{z} = 1.$$

Integrating it, we get

$$x = x_0 - y_0 t - \left(a_1^+ x_0 + b_1^+ y_0\right)\frac{t^2}{2} + O_2(x_0, y_0, z_0)t + O(t^3),$$

$$y = y_0 + \left(a_1^+ x_0 + b_1^+ y_0\right)t + \left(-a_1^+ y_0 + b_1^+ \left(a_1^+ x_0 + b_1^+ y_0\right)\right)\frac{t^2}{2}$$
$$+ O_2(x_0, y_0, z_0)t + O(t^3),$$

$$z = z_0 + t.$$

Here $t = \overline{\delta s_1}$, so the transport equation along the vector v is

$$x = x_0 - y_0 t + O(x_0, y_0, z_0)t^2 + O_2(x_0, y_0, z_0)t + O(t^3),$$

$$y = y_0 + (a_1^+ x_0 + b_1^+ y_0)t + O(x_0, y_0, z_0)t^2 + O_2(x_0, y_0, z_0)t + O(t^3),$$

$$z = z_0 + t.$$

If we only look at the x, y−coordinates, and write them in terms of a matrix, we find

$$\left(Id - t\begin{pmatrix} 0 & -1 \\ a_1^+ & b_1^+ \end{pmatrix}\right) \cdot (xy) + O(x, y, z)t^2 + O_2(x, y, z)t + O(t^3).$$

Similarly, in the neighborhood of x_1', up to a multiplicative factor, v reads

$$\frac{\partial}{\partial z} - y\frac{\partial}{\partial x} + \left(a_2^- x + b_2^- y + c_2 z + P_2^2(x, y, z)\right)\frac{\partial}{\partial y}.$$

So the v-trajectory satisfies

$$\dot{x} = -y,$$

$$\dot{y} = a_2^- x + b_2^- y + c_2 z + P_2^2(x, y, z),$$

$$\dot{z} = 1.$$

Integrating, we get

$$x = x_0 - y_0 t - \left(a_2^- x_0 + b_2^- y_0 + c_2 z_0\right)\frac{t^2}{2} + O_2(x_0, y_0, z_0)t + O(t^3),$$

$$y = y_0 + \left(a_2^- x_0 + b_2^- y_0 + c_2 z_0\right)t$$

$$+ \left(-a_2^- y_0 + c_2 + b_2^- \left(a_2^- x_0 + b_2^- y_0 + c_2 z_0\right)\right)\frac{t^2}{2} + O_2(x_0, y_0, z_0)t + O(t^3),$$

$$z = z_0 + t.$$

If we only look at the x, y-coordinates, and write them in terms of a matrix, we find

$$\left(Id - t \begin{pmatrix} 0 & -1 \\ a_1^+ & b_1^+ \end{pmatrix}\right) \cdot \begin{pmatrix} x \\ y \end{pmatrix} + c_2 \begin{pmatrix} 0 \\ t^2 \\ \frac{2}{2} \end{pmatrix} O(x, y, z)t^2 + O_2(x, y, z)t + O(t^3).$$

We can think of the movement from point x_0 to point x_1' as made of three sub-steps: first, we pull point 0 back to $\{\delta z = 0\}$, then along the v-piece, we transport it to $\{\delta z = 0\}$ near the point x_1'; the transport matrix is A; from there we pull it onto $\{\delta z = \overline{\delta s_2} + \delta s_1\}$. Putting them together, the transport matrix is

$$\left(Id + (\overline{\delta s_2} + \delta s_1) \begin{pmatrix} 0 & -1 \\ a_2^- & b_2^- \end{pmatrix}\right) A \left(Id - \overline{\delta s_1} \begin{pmatrix} 0 & -1 \\ a_1^+ & b_1^+ \end{pmatrix}\right) \cdot x_0$$

$$+ \begin{pmatrix} 0 \\ c_2 \end{pmatrix} \delta s_1 \overline{\delta s_2} + U \delta s_1^2 + V \overline{\delta s_1^2} + O(\delta s_1^3 + \overline{\delta s_1^3}).$$

Here, the values of U and V have changed but we will not track these values.

2.11.8 *Proof of Claim 3*

x_3' can be written as

$$\left(Id + (\overline{\delta s_3} + \delta s_2) \begin{pmatrix} 0 & -1 \\ a_3^- & b_3^- \end{pmatrix}\right) B \left(Id - (\overline{\delta s_2} + \delta s_1) \begin{pmatrix} 0 & -1 \\ a_2^+ & b_2^+ \end{pmatrix}\right) \cdot x_2'$$

$$+ \begin{pmatrix} 0 & c_3 \end{pmatrix} \overline{\delta s_3} \delta s_2 + O(\delta s_1^2, \overline{\delta s_1}^2) + O_3 + \frac{c_1}{a_1^+} \frac{c_2}{a_2^+} A_2 \begin{pmatrix} -2h_1 \dfrac{B_4}{B_2} + l_1 \\ -2h_2 \dfrac{B_4}{B_2} + l_2 \end{pmatrix} \overline{\delta s_1} \delta s_1$$

$$= B \begin{pmatrix} a_2^- & \Delta b_2 \\ a_2^+ & a_4^+ \\ 0 & 1 \end{pmatrix} A \begin{pmatrix} c_1 \\ a_1^+ \\ 0 \end{pmatrix} \overline{\delta s_1} + B \begin{pmatrix} c_2 \\ a_2^+ \\ 0 \end{pmatrix} (\delta s_1 + p_1 \overline{\delta s_1}) + BR \begin{pmatrix} c_2 \\ a_2^+ \\ 0 \end{pmatrix} \overline{\delta s_1} \delta s_1$$

$$+ B \begin{pmatrix} a_2^- & \Delta b_2 \\ a_2^+ & a_4^+ \\ 0 & 1 \end{pmatrix} \left(\begin{pmatrix} 0 & -1 \\ a_2^- & b_2^- \end{pmatrix} A \begin{pmatrix} c_1 \\ a_1^+ \\ 0 \end{pmatrix} + \begin{pmatrix} 0 \\ c_2 p_1 \end{pmatrix}\right) \overline{\delta s_1} \delta s_1$$

$$+ \begin{pmatrix} 0 & -1 \\ a_3^- & b_3^- \end{pmatrix} B \begin{pmatrix} a_2^- & \Delta b_2 \\ a_2^+ & a_4^+ \\ 0 & 1 \end{pmatrix} A \begin{pmatrix} c_1 \\ a_1^+ \\ 0 \end{pmatrix} p_2 \overline{\delta s_1} \delta s_1 + B \begin{pmatrix} N_2 \\ 0 \end{pmatrix}$$

$$-B\begin{pmatrix}0 & -1\\ a_2^+ & b_2^+\end{pmatrix}\begin{pmatrix}\dfrac{a_2^-}{a_2^+} & \dfrac{\Delta b_2}{a_4^+}\\[2mm] 0 & 1\end{pmatrix}A\begin{pmatrix}\dfrac{c_1}{a_1^+}\\[2mm] 0\end{pmatrix}\overline{\delta s_1}\delta s_1+\begin{pmatrix}0\\ c_3\end{pmatrix}p_2 q_1\overline{\delta s_1}\delta s_1$$

$$+\begin{pmatrix}0 & -1\\ a_3^- & b_3^-\end{pmatrix}B\begin{pmatrix}\dfrac{c_2}{a_2^+}\\[2mm] 0\end{pmatrix}(q_1+p_1 p_2)\overline{\delta s_1}\delta s_1-B\begin{pmatrix}0 & -1\\ a_2^+ & b_2^+\end{pmatrix}2p_1\overline{\delta s_1}\delta s_1$$

$$+\frac{c_1}{a_1^+}\frac{c_2}{a_2^+}A_2\begin{pmatrix}-2h_1\dfrac{B_4}{B_2}+l_1\\[3mm] -2h_2\dfrac{B_4}{B_2}+l_2\end{pmatrix}\overline{\delta s_1}\delta s_1.$$

Therefore, the first order terms of x_3' are

$$B\begin{pmatrix}\dfrac{a_2^-}{a_2^+} & \dfrac{\Delta b_2}{a_4^+}\\[2mm] 0 & 1\end{pmatrix}A\begin{pmatrix}\dfrac{c_1}{a_1^+}\\[2mm] 0\end{pmatrix}\overline{\delta s_1}+B\begin{pmatrix}\dfrac{c_2}{a_2^+}\\[2mm] 0\end{pmatrix}p_1\overline{\delta s_1}$$

$$=\begin{pmatrix}-\dfrac{c_3}{a_3^-}q_1\\[2mm] 0\end{pmatrix}\overline{\delta s_1}+\frac{c_2}{a_2^+}\begin{pmatrix}B_1\\ B_2\end{pmatrix}\delta s_1.$$

Using Lemma 47, along the 3rd ξ−piece of $u_2(u_3)$, the second order terms of the displacement in x, y and z are

$$\delta x=N_3(\overline{\delta s_1},\delta s_1)=a_c(x_3')-a_c(\tilde x_2),$$

$$\delta y=-2d\frac{c_2}{a_2^+}B_2\delta s_1\overline{(\delta s_3+\delta s_2)}\bar a,$$

$$\delta z=d\overline{(\delta s_3+\delta s_2)}^2\bar a.$$

As we explained before, drop the $\delta s_1^2, \overline{\delta s_1}^2$ terms, we have,

$$\delta x=N_3(\overline{\delta s_1},\delta s_1),$$

$$\delta y=-2d\frac{c_2}{a_2^+}B_2 q_1\bar a\overline{\delta s_1}\delta s_1,$$

$$\delta z=2dp_2 q_1\bar a\overline{\delta s_1}\delta s_1.$$

Thus the x, y−coordinates of x_4' are

$$\begin{pmatrix}\dfrac{a_3^-}{a_3^+} & \dfrac{\Delta b_3-2d\bar a}{a_3^+}\\[2mm] 0 & 1\end{pmatrix}\cdot x_3'+\begin{pmatrix}\dfrac{c_3}{a_3^+}\\[2mm] 0\end{pmatrix}\overline{(\delta s_3+\delta s_2)}+(\delta x\delta y.)$$

Comparing the $\overline{\delta s_1}\delta s_1$ terms of point x_4' with Lemma 49, we get

$$\begin{pmatrix} \dfrac{a_3^-}{a_3^+} & \dfrac{\Delta b_3 - 2d\overline{a}}{a_3^+} \\ 0 & 1 \end{pmatrix} \left(BR\begin{pmatrix} \dfrac{c_2}{a_2^+} \\ 0 \end{pmatrix}\overline{\delta s_1}\delta s_1 + B\begin{pmatrix} N_2 \\ 0 \end{pmatrix} + \begin{pmatrix} 0 \\ c_3 \end{pmatrix}p_2 q_1\overline{\delta s_1}\delta s_1 \right)$$

$$+ \begin{pmatrix} N_3(\overline{\delta s_1}\delta s_1) \\ -2d\dfrac{c_2}{a_2^+}B_2 q_1\overline{a}\overline{\delta s_1}\delta s_1 \end{pmatrix}$$

$$+ \begin{pmatrix} \dfrac{a_3^-}{a_3^+} & \dfrac{\Delta b_3 - 2d\overline{a}}{a_3^+} \\ 0 & 1 \end{pmatrix} B\begin{pmatrix} \dfrac{a_2^-}{a_2^+} & \dfrac{\Delta b_2}{a_2^+} \\ 0 & 1 \end{pmatrix}\left(\begin{pmatrix} 0 & -1 \\ a_2^- & b_2^- \end{pmatrix}A\begin{pmatrix} \dfrac{c_1}{a_1^+} \\ 0 \end{pmatrix} + \begin{pmatrix} 0 \\ c_2 p_1 \end{pmatrix} \right)\overline{\delta s_1}\delta s_1$$

$$+ \begin{pmatrix} \dfrac{a_3^-}{a_3^+} & \dfrac{\Delta b_3 - 2d\overline{a}}{a_3^+} \\ 0 & 1 \end{pmatrix}\left(\begin{pmatrix} 0 & -1 \\ a_3^- & b_3^- \end{pmatrix}B\begin{pmatrix} \dfrac{a_2^-}{a_2^+} & \dfrac{\Delta b_2}{a_2^+} \\ 0 & 1 \end{pmatrix}A\begin{pmatrix} \dfrac{c_1}{a_1^+} \\ 0 \end{pmatrix}p_2 + \begin{pmatrix} 0 & -1 \\ a_3^- & b_3^- \end{pmatrix}B\begin{pmatrix} \dfrac{c_2}{a_2^+} \\ 0 \end{pmatrix}(q_1 + p_1 p_2) \right)\overline{\delta s_1}\delta s_1$$

$$- \begin{pmatrix} \dfrac{a_3^-}{a_3^+} & \dfrac{\Delta b_3 - 2d\overline{a}}{a_3^+} \\ 0 & 1 \end{pmatrix}\left(B\begin{pmatrix} 0 & -1 \\ a_2^+ & b_2^+ \end{pmatrix}\begin{pmatrix} \dfrac{a_2^-}{a_2^+} & \dfrac{\Delta b_2}{a_2^+} \\ 0 & 1 \end{pmatrix}A\begin{pmatrix} \dfrac{c_1}{a_1^+} \\ 0 \end{pmatrix} + B\begin{pmatrix} 0 & -1 \\ a_2^+ & b_2^+ \end{pmatrix}\begin{pmatrix} \dfrac{c_2}{a_2^+} \\ 0 \end{pmatrix}2p_1 \right)\overline{\delta s_1}\delta s_1$$

$$+ T\begin{pmatrix} \dfrac{c_3}{a_3^+} \\ 0 \end{pmatrix}\overline{\delta s_1}\delta s_1$$

$$+ \dfrac{c_1}{a_1^+}\dfrac{c_2}{a_2^+}A_2\begin{pmatrix} \dfrac{a_3^- - a_3^+}{a_3^+} & \dfrac{\Delta b_3 - 2d\overline{a}}{a_3^+} \\ 0 & 1 \end{pmatrix}\begin{pmatrix} -2h_1\dfrac{B_4}{B_2} + l_1 \\ -2h_2\dfrac{B_4}{B_2} + l_2 \end{pmatrix}\overline{\delta s_1}\delta s_1$$

$$= \begin{pmatrix} -\dfrac{c_2}{a_2^+}B_2 \\ \dfrac{c_2}{a_2^+}\left(b_3^+ B_2 - a_3^+\dfrac{B_2}{C_2}C_4 \right) \end{pmatrix}q_1\overline{\delta s_1}\delta s_1. \qquad (*)$$

Computing the y-coordinate, we get

$$-\dfrac{c_1}{a_1^+}A_2 B_2\left(\dfrac{a_2^- - a_2^+}{a_2^+} + \dfrac{\Delta b_2}{a_2^+}\left(a_2^+\dfrac{B_4}{B_2} - b_2^+ \right) \right) - c_3 p_2 q_1 - c_2 p_1 B_4$$

$$-2d\bar{a}\frac{c_2}{a_2^+}q_1 B_2 + \frac{c_2}{a_2^+}RB_2 + NB_2 + \frac{c_1}{a_1^+}\frac{c_2}{a_2^+}A_2\left(-2h_2\frac{B_4}{B_2}+l_2\right) = 0.$$

2.11.9 *The Final Details of the Calculation of* $\partial^2 J(\overline{x}_\infty).u_2.u_3$

For the second ξ–piece, the $\overline{\delta s_1}\delta s_1$ terms come from three sources. Since $\overline{\delta s_2} = p_1\overline{\delta s_1} + R\overline{\delta s_1}\delta s_1$, R is one of the sources. Also the second order terms of $da_c(x_1')$, N_2 contribute $\overline{\delta s_1}\delta s_1$ terms too, and all the other $\overline{\delta s_1}\delta s_1$ terms come from the v-transport. Now we write it explicitly,

$$(1,0)\cdot\left(\begin{array}{c} R\begin{pmatrix}\frac{c_2}{a_2^+}\\0\end{pmatrix}\overline{\delta s_1}\delta s_1 + \begin{pmatrix}N_2\\0\end{pmatrix} + \begin{pmatrix}0\\c_2p_1\end{pmatrix}\\[2em] +\begin{pmatrix}\frac{a_2^- - a_2^+}{a_2^+} & \frac{\Delta b_2}{a_2^+}\\0 & 1\end{pmatrix}\left(\begin{pmatrix}0 & -1\\a_2^- & b_2^-\end{pmatrix}A\begin{pmatrix}\frac{c_1}{a_1^+}\\0\end{pmatrix}\right)\end{array}\right)$$

$$= \left(\frac{c_3}{B_2}p_2q_1 + c_2p_1\frac{B_4}{B_2} + 2d\bar{a}\frac{c_2}{a_2^+}q_1\right)\overline{\delta s_1}\delta s_1.$$

For the third ξ–piece, the $\overline{\delta s_1}\delta s_1$ terms also come from three sources. Since $\overline{\delta s_3} = q_1\overline{\delta s_1} + T\overline{\delta s_1}\delta s_1$, T is one of the sources. Also the second order terms of $da_c(point3')$, N_3 contributes to the $\overline{\delta s_1}\delta s_1$ terms too. All the other $\overline{\delta s_1}\delta s_1$ terms come from the v-transport. Computing, we find:

$$(1,0)\cdot\begin{pmatrix}\frac{a_3^- - a_3^+}{a_3^+} & \frac{\Delta b_3 - 2d\bar{a}}{a_3^+}\\0 & 1\end{pmatrix}$$

$$\times\left(BR\begin{pmatrix}\frac{c_2}{a_2^+}\\0\end{pmatrix}\overline{\delta s_1}\delta s_1 + B\begin{pmatrix}N_2\\0\end{pmatrix} + \begin{pmatrix}0\\c_3\end{pmatrix}p_2q_1\overline{\delta s_1}\delta s_1\right)$$

$$+(1,0)\cdot\begin{pmatrix}\frac{a_3^- - a_3^+}{a_3^+} & \frac{\Delta b_3 - 2d\bar{a}}{a_3^+}\\0 & 1\end{pmatrix}B\begin{pmatrix}\frac{a_2^-}{a_2^+} & \frac{\Delta b_2}{a_2^+}\\0 & 1\end{pmatrix}$$

$$\times\left(\begin{pmatrix}0 & -1\\a_2^- & b_2^-\end{pmatrix}A\begin{pmatrix}\frac{c_1}{a_1^+}\\0\end{pmatrix} + \begin{pmatrix}0\\c_2p_1\end{pmatrix}\right)\overline{\delta s_1}\delta s_1$$

$$+(1,0)\cdot\begin{pmatrix}\dfrac{a_3^- - a_3^+}{a_3^+} & \dfrac{\Delta b_3 - 2d\bar{a}}{a_3^+}\\[2mm] 0 & 1\end{pmatrix}$$

$$\times\left(\begin{pmatrix}0 & -1\\ a_3^- & b_3^-\end{pmatrix} B \begin{pmatrix}\dfrac{a_2^-}{a_2^+} & \dfrac{\Delta b_2}{a_2^+}\\[2mm] 0 & 1\end{pmatrix} A \begin{pmatrix}\dfrac{c_1}{a_1^+}\\[1mm] 0\end{pmatrix} p_2 \atop +\begin{pmatrix}0 & -1\\ a_3^- & b_3^-\end{pmatrix} B \begin{pmatrix}\dfrac{c_2}{a_2^+}\\[1mm] 0\end{pmatrix}(^2q_1 + p_1 p_2)\right)\overline{\delta s_1}\delta s_1$$

$$-(1,0)\cdot\begin{pmatrix}\dfrac{a_3^- - a_3^+}{a_3^+} & \dfrac{\Delta b_3 - 2d\bar{a}}{a_3^+}\\[2mm] 0 & 1\end{pmatrix}\left(B\begin{pmatrix}0 & -1\\ a_2^+ & b_2^+\end{pmatrix}\begin{pmatrix}\dfrac{a_2^-}{a_2^+} & \dfrac{\Delta b_2}{a_2^+}\\[2mm] 0 & 1\end{pmatrix} A \begin{pmatrix}\dfrac{c_1}{a_1^+}\\[1mm] 0\end{pmatrix} \atop +B\begin{pmatrix}0 & -1\\ a_2^+ & b_2^+\end{pmatrix}\begin{pmatrix}\dfrac{c_2}{a_2^+}\\[1mm] 0\end{pmatrix} 2p_1\right)\overline{\delta s_1}\delta s_1$$

$$+T(1,0)\cdot\begin{pmatrix}\dfrac{c_3}{a_3^+}\\[1mm] 0\end{pmatrix}\overline{\delta s_1}\delta s_1 + N_3(\overline{\delta s_1},\delta s_1)$$

$$+(1,0)\frac{c_1}{a_1^+}\frac{c_2}{a_2^+} A_2 \begin{pmatrix}\dfrac{a_3^- - a_3^+}{a_3^+} & \dfrac{\Delta b_3 - 2d\bar{a}}{a_3^+}\\[2mm] 0 & 1\end{pmatrix}\begin{pmatrix}-2h_1\dfrac{B_4}{B_2} + l_1\\[2mm] 2h_2\dfrac{B_4}{B_2} + l_2\end{pmatrix}\overline{\delta s_1}\delta s_1.$$

Comparing with $(*)$, we find that this is equal to:

$$-(1,0)\cdot\left(BR\begin{pmatrix}\dfrac{c_2}{a_2^+}\\[1mm] 0\end{pmatrix}\overline{\delta s_1}\delta s_1 + B\begin{pmatrix}N_2\\ 0\end{pmatrix} + \begin{pmatrix}0\\ c_3\end{pmatrix} p_2 q_1 \overline{\delta s_1}\delta s_1\right)$$

$$-(1,0)\cdot B\begin{pmatrix}\dfrac{a_2^-}{a_2^+} & \dfrac{\Delta b_2}{a_2^+}\\[2mm] 0 & 1\end{pmatrix}\left(\begin{pmatrix}0 & -1\\ a_2^- & b_2^-\end{pmatrix} A \begin{pmatrix}\dfrac{c_1}{a_1^+}\\[1mm] 0\end{pmatrix} + \begin{pmatrix}0\\ c_2 p_1\end{pmatrix}\right)\delta s_1 \delta s_1$$

$$-(1,0)\cdot\left(\begin{pmatrix}0 & -1\\ a_3^- & b_3^-\end{pmatrix}\begin{pmatrix}\dfrac{a_2^-}{a_2^+} & \dfrac{\Delta b_2}{a_4^+}\\[2mm] 0 & 1\end{pmatrix} A \begin{pmatrix}\dfrac{c_1}{a_1^+}\\[1mm] 0\end{pmatrix} p_2+ \atop \begin{pmatrix}0 & -1\\ a_3^- & b_3^-\end{pmatrix} B \begin{pmatrix}\dfrac{c_2}{a_2^+}\\[1mm] 0\end{pmatrix}(^2q_1 + p_1 p_2)\right)\overline{\delta s_1}\delta s_1$$

$$+(1,0)\cdot\left(B\begin{pmatrix}0 & -1\\ a_2^+ & b_2^+\end{pmatrix}\begin{pmatrix}a_2^- & \Delta b_2\\ a_2^+ & a_4^+\\ 0 & 1\end{pmatrix}A\begin{pmatrix}\frac{c_1}{a_1^+}\\ 0\end{pmatrix}-B\begin{pmatrix}0 & -1\\ a_2^+ & b_2^+\end{pmatrix}\begin{pmatrix}\frac{c_2}{a_2^+}\\ 0\end{pmatrix}2p_1\right)\overline{\delta s_1}\delta s_1$$

$$-(1,0)\frac{c_1}{a_1^+}\frac{c_2}{a_2^+}A_2\begin{pmatrix}-2h_1\dfrac{B_4}{B_2}+l_1\\[2mm] 2h_2\dfrac{B_4}{B_2}+l_2\end{pmatrix}\overline{\delta s_1}\delta s_1$$

$$=\left(-c_3\frac{B_1}{B_2}p_2q_1-c_2p_1\frac{\det B}{B_2}-2d\bar a\frac{c_2}{a_2^+}q_1B_1\right)\overline{\delta s_1}\delta s_1$$

$$+\frac{c_1}{a_1^+}\frac{c_2}{a_2^+}A_2\left((2h_1\frac{B_1}{B_2}-2h_2)\frac{B_4}{B_2}+l_2-l_1\right)\overline{\delta s_1}\delta s_1.$$

Summing them up, we find

$$\partial^2 J.u_2.u_3$$

$$=\frac{c_3}{B_2}p_2q_1+c_2p_1\frac{B_4}{B_2}+2d\bar a\frac{c_2}{a_2^+}q_1-c_3\frac{B_1}{B_2}p_2q_1-c_2p_1\frac{\det B}{B_2}-2d\bar a\frac{c_2}{a_2^+}q_1B_1$$

$$+\frac{c_1}{a_1^+}\frac{c_2}{a_2^+}A_2\left((2h_1-2h_2)\frac{B_4}{B_2}+l_2-l_1\right)$$

$$=q_1\underbrace{\left(\frac{c_3}{B_2}p_2+2d\bar a\frac{c_2}{a_2^+}-2d\bar a\frac{c_2}{a_2^+}B_1-c_3\frac{B_1}{B_2}p_2\right)}_{(I)}$$

$$+\underbrace{c_2p_1\frac{B_4}{B_2}-c_2p_1\frac{\det B}{B_2}c_2p_1\frac{\det B}{B_2}}_{(II)}+\underbrace{\frac{c_1}{a_1^+}\frac{c_2}{a_2^+}A_2((2h_1\frac{B_1}{B_2}-2h_2)\frac{B_4}{B_2}+l_2-l_1)}_{(III)}.$$

2.11.10 *Details involved in 2.8*

2.11.11 *Proof of Lemma 52*

It is clear that the x,y coordinates of x_6 read $\begin{pmatrix}0\\ 0\end{pmatrix}$, and the z coordinate is δs_3. And in the neighborhood of point x_6, the vector v reads $\dfrac{\partial}{\partial z}-y\dfrac{\partial}{\partial x}+$

$$(a_3^+ x + b_3^+ y + Q_2^3(x,y,z))\frac{\partial}{\partial y}, \text{ so at } x_6, \ v \text{ is parallel with } \begin{pmatrix} 0 \\ Q_2^3(0,0,\delta s_3) \\ 1 \end{pmatrix},$$

so x_6' is $\begin{pmatrix} 0 \\ 0 \\ \delta s_3 \end{pmatrix} + \overline{\delta s_3}\begin{pmatrix} 0 \\ Q_2^3(0,0,\delta s_3) \\ 1 \end{pmatrix} = \begin{pmatrix} 0 \\ 0 \\ \delta s_3 + \overline{\delta s_3} \end{pmatrix} + U\delta s_1^2 + V\overline{\delta s_1}^2 +$

$O(\delta s_1^3 + \overline{\delta s_1}^3).$

2.11.12 *Proof of Lemma 53*

Following $u_2(u_2)$, we write only the first order terms in a first step,

x_1' is $\begin{pmatrix} 0 \\ 0 \\ \delta s_1 + \overline{\delta s_1} \end{pmatrix}$; the transport matrix along the first ξ-piece is

$\begin{pmatrix} a_1^- & b_1^- - b_1^+ \\ a_1^+ & a_1^+ \\ 0 & 1 \end{pmatrix}$ and there is a small increment $\begin{pmatrix} c_1 \\ a_1^+ \\ 0 \end{pmatrix}(\delta s_1 + \overline{\delta s_1})$ due to

the effect of the $z-$coordinate, therefore x_2' reads $\begin{pmatrix} c_1 \\ a_1^+ \\ 0 \\ 1 \end{pmatrix}(\delta s_1 + \overline{\delta s_1})$; now

we go to point x_3', it reads $\begin{pmatrix} A\begin{pmatrix} c_1 \\ a_1^+ \\ 0 \end{pmatrix}(\delta s_1 + \overline{\delta s_1}) \\ \overline{\delta s_2} + \delta s_2 \end{pmatrix}$; x_4' is , using the same

argument

$$\begin{pmatrix} \left(\begin{pmatrix} a_2^- & b_2^- - b_2^+ \\ a_2^+ & a_2^+ \\ 0 & 1 \end{pmatrix} A\begin{pmatrix} c_1 \\ a_1^+ \\ 0 \end{pmatrix}(\delta s_1 + \overline{\delta s_1}) + \begin{pmatrix} c_2 \\ a_2^+ \\ 0 \end{pmatrix}(\delta s_2 + \overline{\delta s_2}) \right) \\ \overline{\delta s_2} + \delta s_2 \end{pmatrix}.$$

Therefore x_5' reads

$$\begin{pmatrix} \left(B\begin{pmatrix} a_2^- & b_2^- - b_2^+ \\ a_2^+ & a_2^+ \\ 0 & 1 \end{pmatrix} A\begin{pmatrix} c_1 \\ a_1^+ \\ 0 \end{pmatrix}(\delta s_1 + \overline{\delta s_1}) + B\begin{pmatrix} c_2 \\ a_2^+ \\ 0 \end{pmatrix}(\delta s_2 + \overline{\delta s_2}) \right) \\ \overline{\delta s_3} + \delta s_3 \end{pmatrix}.$$

The transport matrix along the third ξ-piece is $\begin{pmatrix} a_3^- & \dfrac{\Delta b_3 - 2d\overline{a}_c}{a_3^+} \\ a_3^+ & \\ 0 & 1 \end{pmatrix}$ and there is a small increment

$\begin{pmatrix} \overline{c_3} \\ a_3^+ \\ 0 \end{pmatrix} (\delta s_3 + \overline{\delta s_3})$, so the x, y- coordinates of x_6 read

$$\begin{pmatrix} a_3^- & \dfrac{\Delta b_3 - 2d\overline{a}_c}{a_3^+} \\ a_3^+ & \\ 0 & 1 \end{pmatrix} B \begin{pmatrix} a_2^- & \dfrac{b_2^- - b_2^+}{a_2^+} \\ a_2^+ & \\ 0 & 1 \end{pmatrix} A \begin{pmatrix} \dfrac{c_1}{a_1^+} \\ 0 \end{pmatrix} (\delta s_1 + \overline{\delta s_1})$$

$$+ \begin{pmatrix} a_3^- & \dfrac{\Delta b_3 - 2d\overline{a}_c}{a_3^+} \\ a_3^+ & \\ 0 & 1 \end{pmatrix} B \begin{pmatrix} \dfrac{c_2}{a_2^+} \\ 0 \end{pmatrix} (\delta s_2 + \overline{\delta s_2}) + \begin{pmatrix} \dfrac{c_3}{a_3^+} \\ 0 \end{pmatrix} (\delta s_2 + \overline{\delta s_3}),$$

and the z–coordinate remains the same $\overline{\delta s_3} + \delta s_3$.

Using Lemma 52, we have

$$\begin{pmatrix} a_3^- & \dfrac{\Delta b_3 - 2d\overline{a}_c}{a_3^+} \\ a_3^+ & \\ 0 & 1 \end{pmatrix} B \begin{pmatrix} a_2^- & \dfrac{b_2^- - b_2^+}{a_2^+} \\ a_2^+ & \\ 0 & 1 \end{pmatrix} A \begin{pmatrix} \dfrac{c_1}{a_1^+} \\ 0 \end{pmatrix} (\delta s_1 + \overline{\delta s_1})$$

$$+ \begin{pmatrix} a_3^- & \dfrac{\Delta b_3 - 2d\overline{a}_c}{a_3^+} \\ a_3^+ & \\ 0 & 1 \end{pmatrix} B \begin{pmatrix} \dfrac{c_2}{a_2^+} \\ 0 \end{pmatrix} (\delta s_2 + \overline{\delta s_2}) + \begin{pmatrix} \dfrac{c_3}{a_3^+} \\ 0 \end{pmatrix} (\delta s_3 + \overline{\delta s_3}) = \begin{pmatrix} 0 \\ 0 \end{pmatrix}.$$

Observe that these are exactly the equations satisfied by $\delta s_1, \delta s_2$ relative to δs_3. At first order, there is no difference. Solving it for $\overline{\delta s_2}$ and $\overline{\delta s_3}$, we derive the lemma.

2.11.13　*Proof of Claim 3*

Since x_1 is $\begin{pmatrix} 0 \\ 0 \\ \delta s_1 \end{pmatrix}$, and the vector v reads, up to a multiplicative factor,

$$\frac{\partial}{\partial z} - y\frac{\partial}{\partial x} + (a_1^+ x + b_1^+ y + c_1 z + P_2^1(x, y, z))\frac{\partial}{\partial y} \quad \text{around } x_1, \text{ along the } v$$

trajectory, it satisfies

$$\dot{x} = -y,$$
$$\dot{y} = a_1^+ x + b_1^+ y + c_1 z + P_2^1(x, y, z),$$
$$\dot{z} = 1.$$

Integrating it, we get

$$x = x_0 - y_0 t - \left(a_1^+ x_0 + b_1^+ y_0 + c_1 z_0\right) \frac{t^2}{2} + O_2(x, y, z)t + O(t^3),$$
$$y = y_0 + \left(a_1^+ x_0 + b_1^+ y_0 + c_1 z_0\right) t + \left(-a_1^+ y_0 + b_1^+ \left(a_1^+ x_0 + b_1^+ y_0 + c_1 z_0\right)\right.$$
$$\left. + c_1\right) \frac{t^2}{2} + O_2(x, y, z)t + O(t^3),$$
$$z = z_0 + t.$$

Here $t = \delta s_1 + \overline{\delta s_1}$, therefore

$$x = O(\overline{\delta s_1}^3 + \delta s_1^3),$$
$$y = c_1 \delta s_1 \overline{\delta s_1} + O(\overline{\delta s_1}^3 + \delta s_1^3) + c_1 \frac{\overline{\delta s_1}^2 + \delta s_1^2}{2},$$
$$z = \overline{\delta s_1} + \delta s_1.$$

2.11.14 *Proof of Claim 4*

We write only the first order and $\delta s_1 \overline{\delta s_1}$ terms. The $x, y-$coordinates of x_3' read:

$$A \begin{pmatrix} \frac{c_1}{a_1^+} \\ 0 \end{pmatrix} (\delta s_1 + \overline{\delta s_1}) + A \begin{pmatrix} a_1^- & b_1^- - b_1^+ \\ a_1^+ & a_1^+ \\ 0 & 1 \end{pmatrix} 0 c_1 \delta s_1 \overline{\delta s_1}$$

$$+ 2p \begin{pmatrix} 0 & -1 \\ a_2^- & b_2^- \end{pmatrix} A \begin{pmatrix} \frac{c_1}{a_1^+} \\ 0 \end{pmatrix} \delta s_1 \overline{\delta s_1}$$

$$- 2A \begin{pmatrix} 0 & -1 \\ a_1^+ & b_1^+ \end{pmatrix} \begin{pmatrix} \frac{c_1}{a_1^+} \\ 0 \end{pmatrix} \delta s_1 \overline{\delta s_1} + \begin{pmatrix} 0 \\ c_2 \end{pmatrix} p^2 \overline{\delta s_1} \delta s_1 + A \begin{pmatrix} N_1 \\ 0 \end{pmatrix},$$

and the $z-$coordinate remains $\overline{\delta s_2} + \delta s_2$.

The x, y−coordinates of x_4' read

$$\begin{pmatrix} a_2^- & \Delta b_2 \\ \overline{a_2^+} & \overline{a_2^+} \\ 0 & 1 \end{pmatrix} A \begin{pmatrix} \overline{c_1} \\ a_1^+ \\ 0 \end{pmatrix} (\delta s_1 + \overline{\delta s_1}) + \begin{pmatrix} \overline{c_2} \\ a_2^+ \\ 0 \end{pmatrix} (\delta s_2 + \overline{\delta s_2})$$

$$+ \begin{pmatrix} a_2^- & \Delta b_2 \\ \overline{a_2^+} & \overline{a_2^+} \\ 0 & 1 \end{pmatrix} A \begin{pmatrix} N_1 \\ 0 \end{pmatrix} + \begin{pmatrix} N_2 \\ 0 \end{pmatrix} + \begin{pmatrix} a_2^- & \Delta b_2 \\ \overline{a_2^+} & \overline{a_2^+} \\ 0 & 1 \end{pmatrix} A \begin{pmatrix} a_1^- & \Delta b_1 \\ \overline{a_1^+} & \overline{a_1^+} \\ 0 & 1 \end{pmatrix} (0 c_1) \, \delta s_1 \overline{\delta s_1}$$

$$+ \begin{pmatrix} a_2^- & \Delta b_2 \\ \overline{a_2^+} & \overline{a_2^+} \\ 0 & 1 \end{pmatrix} 2p \begin{pmatrix} 0 & -1 \\ a_2^- & b_2^- \end{pmatrix} A \begin{pmatrix} \overline{c_1} \\ a_1^+ \\ 0 \end{pmatrix} \delta s_1 \overline{\delta s_1}$$

$$- 2 \begin{pmatrix} a_2^- & \Delta b_2 \\ \overline{a_2^+} & \overline{a_2^+} \\ 0 & 1 \end{pmatrix} A \begin{pmatrix} 0 & -1 \\ a_1^+ & b_1^+ \end{pmatrix} \begin{pmatrix} \overline{c_1} \\ a_1^+ \\ 0 \end{pmatrix} \delta s_1 \overline{\delta s_1}$$

$$+ \begin{pmatrix} a_2^- & \Delta b_2 \\ \overline{a_2^+} & \overline{a_2^+} \\ 0 & 1 \end{pmatrix} \begin{pmatrix} 0 \\ c_2 \end{pmatrix} p^2 \overline{\delta s_1} \delta s_1 + \begin{pmatrix} a_2^- & \Delta b_2 \\ \overline{a_2^+} & \overline{a_2^+} \\ 0 & 1 \end{pmatrix} A \begin{pmatrix} N_1 \\ 0 \end{pmatrix}$$

and the z−coordinate remains the same $\overline{\delta s_2} + \delta s_2$. Here, $\Delta b_i = b_i^- - b_i^+$ and N_2 is the second order term of $a_c(x_3') - a_c(\tilde{x}_3)$.

The modification of the v transport contributes,

$$2 \left(\frac{c_1}{a_1^+} A_2 \right)^2 \frac{B_4}{B_2} \begin{pmatrix} -l_1 + \dfrac{B_4}{B_2} h_1 \\[2mm] -l_2 + \dfrac{B_4}{B_2} h_2 \end{pmatrix} \overline{\delta s_1} \delta s_1$$

if we only count $\overline{\delta s_1} \delta s_1$ terms.

Therefore, writing only the first order and $\delta s_1 \overline{\delta s_1}$ terms, the x, y−coordinates of x_5' read

$$B \begin{pmatrix} a_2^- & \Delta b_2 \\ \overline{a_2^+} & \overline{a_2^+} \\ 0 & 1 \end{pmatrix} A \begin{pmatrix} \overline{c_1} \\ a_1^+ \\ 0 \end{pmatrix} (\delta s_1 + \overline{\delta s_1}) + B \begin{pmatrix} \overline{c_2} \\ a_2^+ \\ 0 \end{pmatrix} (\delta s_2 + \overline{\delta s_2})$$

$$+ B \begin{pmatrix} a_2^- & \Delta b_2 \\ \overline{a_2^+} & \overline{a_2^+} \\ 0 & 1 \end{pmatrix} A \begin{pmatrix} N_1 \\ 0 \end{pmatrix} + B \begin{pmatrix} N_2 \\ 0 \end{pmatrix} + B \begin{pmatrix} a_2^- & \Delta b_2 \\ \overline{a_2^+} & \overline{a_2^+} \\ 0 & 1 \end{pmatrix} A \begin{pmatrix} a_1^- & \Delta b_1 \\ \overline{a_1^+} & \overline{a_1^+} \\ 0 & 1 \end{pmatrix} \begin{pmatrix} 0 \\ c_1 \end{pmatrix} \delta s_1 \overline{\delta s_1}$$

$$+B\begin{pmatrix} \frac{a_2^-}{a_2^+} & \frac{\Delta b_2}{a_2^+} \\ 0 & 1 \end{pmatrix} 2p \begin{pmatrix} 0 & -1 \\ a_2^- & b_2^- \end{pmatrix} A \begin{pmatrix} \frac{c_1}{a_1^+} \\ 0 \end{pmatrix} \delta s_1 \overline{\delta s_1}$$

$$-2B\begin{pmatrix} \frac{a_2^-}{a_2^+} & \frac{\Delta b_2}{a_2^+} \\ 0 & 1 \end{pmatrix} A \begin{pmatrix} 0 & -1 \\ a_1^+ & b_1^+ \end{pmatrix} \begin{pmatrix} \frac{c_1}{a_1^+} \\ 0 \end{pmatrix} \delta s_1 \overline{\delta s_1} B \begin{pmatrix} \frac{a_2^-}{a_2^+} & \frac{\Delta b_2}{a_2^+} \\ 0 & 1 \end{pmatrix} \begin{pmatrix} 0 \\ c_2 \end{pmatrix} p^2 \overline{\delta s_1}\delta s_1$$

$$+ \begin{pmatrix} 0 & -1 \\ a_2^- & b_2^- \end{pmatrix} B \left(2q \begin{pmatrix} \frac{a_2^-}{a_2^+} & \frac{\Delta b_2}{a_2^+} \\ 0 & 1 \end{pmatrix} A \begin{pmatrix} \frac{c_1}{a_1^+} \\ 0 \end{pmatrix} + 2pq \begin{pmatrix} \frac{c_2}{a_2^+} \\ 0 \end{pmatrix} \right) \delta s_1 \overline{\delta s_1}$$

$$-B \begin{pmatrix} 0 & -1 \\ a_1^+ & b_1^+ \end{pmatrix} \left(2p \begin{pmatrix} \frac{a_2^-}{a_2^+} & \frac{\Delta b_2}{a_2^+} \\ 0 & 1 \end{pmatrix} A \begin{pmatrix} \frac{c_1}{a_1^+} \\ 0 \end{pmatrix} + 2p^2 \begin{pmatrix} \frac{c_2}{a_2^+} \\ 0 \end{pmatrix} \right) \delta s_1 \overline{\delta s_1} + \begin{pmatrix} 0 \\ c_3 \end{pmatrix} q^2 \delta s_1 \overline{\delta s_1}$$

$$+2 \left(\frac{c_1}{a_1^+} A_2 \right)^2 \frac{B_4}{B_2} \begin{pmatrix} -l_1 + \frac{B_4}{B_2} h_1 \\ -l_2 + \frac{B_4}{B_2} h_2 \end{pmatrix} \delta s_1 \overline{\delta s_1},$$

and the $z-$coordinate is $\overline{\delta s_3} + \delta s_3$.

The x and y coordinates of x_5' at the first order are,

$$B \begin{pmatrix} \frac{a_2^-}{a_2^+} & \frac{\Delta b_2}{a_2^+} \\ 0 & 1 \end{pmatrix} A \begin{pmatrix} \frac{c_1}{a_1^+} \\ 0 \end{pmatrix} (\delta s_1 + \overline{\delta s_1}) + B \begin{pmatrix} \frac{c_2}{a_2^+} \\ 0 \end{pmatrix} (\delta s_2 + \overline{\delta s_2})$$

$$= \begin{pmatrix} A_1 \frac{c_1}{a_1^+} (\delta s_1 + \overline{\delta s_1}) \\ 0 \end{pmatrix}.$$

Using the Lemma 47, along the third ξ-piece,

$$\delta x = N_3,$$
$$\delta y = 0,$$
$$\delta z = d(\delta s_3 + \overline{\delta s_3})^2 \overline{a_c}.$$

Writing only $\delta s_1 \overline{\delta s_1}$ terms,

$$\delta x = N_3,$$
$$\delta y = 0,$$
$$\delta z = 2d\overline{a_c}q^2 \delta s_1 \overline{\delta s_1},$$

here, N_3 is the second order term of $a_c(x_5') - a_c(\tilde{x}_5)$.

$$x_6 \text{ is } \begin{pmatrix} a_3^- & \dfrac{\Delta b_3 - 2d\overline{a_c}}{a_3^+} \\ a_3^+ & \\ 0 & 1 \end{pmatrix} \cdot x_5' + \begin{pmatrix} c_3 \\ a_3^+ \\ 0 \end{pmatrix} (\overline{\delta s_3} + \delta s_3) + \begin{pmatrix} \delta x \\ \delta y \end{pmatrix}$$

so the $x, y-$ coordinates of x_6' are

$$\begin{pmatrix} a_3^- & \dfrac{\Delta b_3 - 2d\overline{a_c}}{a_3^+} \\ a_3^+ & \\ 0 & 1 \end{pmatrix} B \begin{pmatrix} a_2^- & \dfrac{\Delta b_2}{a_2^+} \\ a_2^+ & \\ 0 & 1 \end{pmatrix} A \begin{pmatrix} c_1 \\ a_1^+ \\ 0 \end{pmatrix} (\delta s_1 + \overline{\delta s_1})$$

$$+ \begin{pmatrix} a_3^- & \dfrac{\Delta b_3 - 2d\overline{a_c}}{a_3^+} \\ a_3^+ & \\ 0 & 1 \end{pmatrix} B \begin{pmatrix} c_2 \\ a_2^+ \\ 0 \end{pmatrix} (\delta s_2 + \overline{\delta s_2}) + \begin{pmatrix} c_3 \\ a_3^+ \\ 0 \end{pmatrix} (\overline{\delta s_3} + \delta s_3)$$

$$+ \begin{pmatrix} a_3^- & \dfrac{\Delta b_3 - 2d\overline{a_c}}{a_3^+} \\ a_3^+ & \\ 0 & 1 \end{pmatrix} B \begin{pmatrix} a_2^- & \dfrac{\Delta b_2}{a_2^+} \\ a_2^+ & \\ 0 & 1 \end{pmatrix} A \begin{pmatrix} N_1 \\ 0 \end{pmatrix}$$

$$+ \begin{pmatrix} a_3^- & \dfrac{\Delta b_3 - 2d\overline{a_c}}{a_3^+} \\ a_3^+ & \\ 0 & 1 \end{pmatrix} B \begin{pmatrix} N_2 \\ 0 \end{pmatrix} + \begin{pmatrix} N_3 \\ -6aq^2\delta s_1\overline{\delta s_1} \end{pmatrix}$$

$$+ \begin{pmatrix} a_3^- & \dfrac{\Delta b_3 - 2d\overline{a_c}}{a_3^+} \\ a_3^+ & \\ 0 & 1 \end{pmatrix} \begin{pmatrix} 0 \\ c_3 \end{pmatrix} q^2\delta s_1\overline{\delta s_1}$$

$$+ \begin{pmatrix} a_3^- & \dfrac{\Delta}{b}\,3 \; - 2d\overline{a_c}a_3^+ \\ a_3^+ & \\ 0 & 1 \end{pmatrix} B \begin{pmatrix} a_2^- & \dfrac{\Delta b_2}{a_2^+} \\ a_2^+ & \\ 0 & 1 \end{pmatrix} A \begin{pmatrix} a_1^- & \dfrac{\Delta b_1}{a_1^+}\,0\;1 \\ a_1^+ & \end{pmatrix} \begin{pmatrix} 0 \\ c_1 \end{pmatrix} \delta s_1\overline{\delta s_1}$$

$$+ \begin{pmatrix} a_3^- & \dfrac{\Delta b_3 - 2d\overline{a_c}}{a_3^+} \\ a_3^+ & \\ 0 & 1 \end{pmatrix} B \begin{pmatrix} a_2^- & \dfrac{\Delta b_2}{a_2^+} \\ a_2^+ & \\ 0 & 1 \end{pmatrix} \begin{pmatrix} 2p \begin{pmatrix} 0 & -1 \\ a_2^- & b_2^- \end{pmatrix} A \begin{pmatrix} c_1 \\ a_1^+ \\ 0 \end{pmatrix} \\ -2A \begin{pmatrix} 0 & -1 \\ a_1^+ & b_1^+ \end{pmatrix} \begin{pmatrix} c_1 \\ a_1^+ \\ 0 \end{pmatrix} \end{pmatrix} \delta s_1\overline{\delta s_1}$$

$$+\begin{pmatrix} \dfrac{a_3^-}{a_3^+} & \dfrac{\Delta b_3 - 2d\overline{a_c}}{a_3^+} \\ 0 & 1 \end{pmatrix} B \begin{pmatrix} \dfrac{a_2^-}{a_2^+} & \dfrac{\Delta b_2}{a_2^+} \\ 0 & 1 \end{pmatrix} \begin{pmatrix} 0 \\ c_2 \end{pmatrix} p^2 \overline{\delta s_1} \delta s_1$$

$$+\begin{pmatrix} \dfrac{a_3^-}{a_3^+} & \dfrac{\Delta b_3 - 2d\overline{a_c}}{a_3^+} \\ 0 & 1 \end{pmatrix} \begin{pmatrix} 0 & -1 \\ a_3^- & b_3^- \end{pmatrix} B \left(2q \begin{pmatrix} \dfrac{a_2^-}{a_2^+} & \dfrac{\Delta b_2}{a_2^+} \\ 0 & 1 \end{pmatrix} A \begin{pmatrix} \dfrac{c_1}{a_1^+} \\ 0 \end{pmatrix} \right.$$

$$\left. +2pq \begin{pmatrix} \dfrac{c_2}{a_2^+} \\ 0 \end{pmatrix} \right) \delta s_1 \overline{\delta s_1}$$

$$-\begin{pmatrix} \dfrac{a_3^-}{a_3^+} & \dfrac{\Delta b_3 - 2d\overline{a_c}}{a_3^+} \\ 0 & 1 \end{pmatrix} B \begin{pmatrix} 0 & -1 \\ a_2^+ & b_2^+ \end{pmatrix} \left(2p \begin{pmatrix} \dfrac{a_2^-}{a_2^+} & \dfrac{\Delta b_2}{a_2^+} \\ 0 & 1 \end{pmatrix} A \begin{pmatrix} \dfrac{c_1}{a_1^+} \\ 0 \end{pmatrix} \right.$$

$$\left. +2p^2 \begin{pmatrix} \dfrac{c_2}{a_2^+} \\ 0 \end{pmatrix} \right) \delta s_1 \overline{\delta s_1}$$

$$+2 \left(\dfrac{c_1}{a_1^+} A_2 \right)^2 \dfrac{B_4}{B_2} \begin{pmatrix} \dfrac{a_3^-}{a_3^+} & \dfrac{\Delta b_3 - 2d\overline{a_c}}{a_3^+} \\ 0 & 1 \end{pmatrix} \begin{pmatrix} -l_1 + \dfrac{B_4}{B_2} h_1 \\ -l_2 + \dfrac{B_4}{B_2} h_2 \end{pmatrix}$$

Comparing this with Lemma 52, we derive the equation

$$R \begin{pmatrix} \dfrac{a_3^-}{a_3^+} & \dfrac{\Delta b_3 - 2d\overline{a_c}}{a_3^+} \\ 0 & 1 \end{pmatrix} B \begin{pmatrix} \dfrac{c_2}{a_2^+} \\ 0 \end{pmatrix} \delta s_1 \overline{\delta s_1} + T \begin{pmatrix} \dfrac{c_3}{a_3^+} \\ 0 \end{pmatrix} \delta s_1 \overline{\delta s_1}$$

$$+\begin{pmatrix} \dfrac{a_3^-}{a_3^+} & \dfrac{\Delta b_3 - 2d\overline{a_c}}{a_3^+} \\ 0 & 1 \end{pmatrix} B \begin{pmatrix} \dfrac{a_2^-}{a_2^+} & \dfrac{\Delta b_2}{a_2^+} \\ 0 & 1 \end{pmatrix} A \begin{pmatrix} N_1 \\ 0 \end{pmatrix} + \begin{pmatrix} N_3 \\ -6aq^2 \delta s_1 \overline{\delta s_1} \end{pmatrix}$$

$$+\begin{pmatrix} \dfrac{a_3^-}{a_3^+} & \dfrac{\Delta b_3 - 2d\overline{a_c}}{a_3^+} \\ 0 & 1 \end{pmatrix} B \begin{pmatrix} N_2 \\ 0 \end{pmatrix} + \begin{pmatrix} \dfrac{a_3^-}{a_3^+} & \dfrac{\Delta b_3 - 2d\overline{a_c}}{a_3^+} \\ 0 & 1 \end{pmatrix} \begin{pmatrix} 0 \\ c_3 \end{pmatrix} q^2 \delta s_1 \overline{\delta s_1}$$

$$+\begin{pmatrix} \dfrac{a_3^-}{a_3^+} & \dfrac{\Delta b_3 - 2d\overline{a_c}}{a_3^+} \\[4pt] 0 & 1 \end{pmatrix} B \begin{pmatrix} \dfrac{a_2^-}{a_2^+} & \dfrac{\Delta b_2}{a_2^+} \\[4pt] 0 & 1 \end{pmatrix} A \begin{pmatrix} \dfrac{a_1^-}{a_1^+} & \dfrac{\Delta b_1}{a_1^+} \\[4pt] 0 & 1 \end{pmatrix} \begin{pmatrix} 0 \\ c_1 \end{pmatrix} \delta s_1 \overline{\delta s_1}$$

$$+\begin{pmatrix} \dfrac{a_3^-}{a_3^+} & \dfrac{\Delta b_3 - 2d\overline{a_c}}{a_3^+} \\[4pt] 0 & 1 \end{pmatrix} B \begin{pmatrix} \dfrac{a_2^-}{a_2^+} & \dfrac{\Delta b_2}{a_2^+} \\[4pt] 0 & 1 \end{pmatrix} \begin{pmatrix} 2p \begin{pmatrix} 0 & -1 \\ a_2^- & b_2^- \end{pmatrix} A \begin{pmatrix} \dfrac{c_1}{a_1^+} \\ 0 \end{pmatrix} \\[12pt] -2A \begin{pmatrix} 0 & -1 \\ a_1^+ & b_1^+ \end{pmatrix} \begin{pmatrix} \dfrac{c_1}{a_1^+} \\ 0 \end{pmatrix} \end{pmatrix} \delta s_1 \overline{\delta s_1}$$

$$+\begin{pmatrix} \dfrac{a_3^-}{a_3^+} & \dfrac{\Delta b_3 - 2d\overline{a_c}}{a_3^+} \\[4pt] 0 & 1 \end{pmatrix} B \begin{pmatrix} \dfrac{a_2^-}{a_2^+} & \dfrac{\Delta b_2}{a_2^+} \\[4pt] 0 & 1 \end{pmatrix} \begin{pmatrix} 0 \\ c_2 \end{pmatrix} p^2 \overline{\delta s_1} \delta s_1$$

$$+\begin{pmatrix} \dfrac{a_3^-}{a_3^+} & \dfrac{\Delta b_3 - 2d\overline{a_c}}{a_3^+} \\[4pt] 0 & 1 \end{pmatrix} \begin{pmatrix} 0 & -1 \\ a_3^- & b_3^- \end{pmatrix} B \left(2q \begin{pmatrix} \dfrac{a_2^-}{a_2^+} & \dfrac{\Delta b_2}{a_2^+} \\[4pt] 0 & 1 \end{pmatrix} A \begin{pmatrix} \dfrac{c_1}{a_1^+} \\ 0 \end{pmatrix} + 2pq \begin{pmatrix} \dfrac{c_2}{a_2^+} \\ 0 \end{pmatrix} \right)$$

$$\times \delta s_1 \overline{\delta s_1}$$

$$-\begin{pmatrix} \dfrac{a_3^-}{a_3^+} & \dfrac{\Delta b_3 - 2d\overline{a_c}}{a_3^+} \\[4pt] 0 & 1 \end{pmatrix} B \begin{pmatrix} 0 & -1 \\ a_2^+ & b_2^+ \end{pmatrix} \left(2p \begin{pmatrix} \dfrac{a_2^-}{a_2^+} & \dfrac{\Delta b_2}{a_2^+} \\[4pt] 0 & 1 \end{pmatrix} A \begin{pmatrix} \dfrac{c_1}{a_1^+} \\ 0 \end{pmatrix} + 2p^2 \begin{pmatrix} \dfrac{c_2}{a_2^+} \\ 0 \end{pmatrix} \right)$$

$$\times \delta s_1 \overline{\delta s_1}$$

$$+2\left(\frac{c_1}{a_1^+} A_2 \right)^2 \frac{B_4}{B_2} \begin{pmatrix} \dfrac{a_3^-}{a_3^+} & \dfrac{\Delta b_3 - 2d\overline{a_c}}{a_3^+} \\[4pt] 0 & 1 \end{pmatrix} \begin{pmatrix} -l_1 + \dfrac{B_4}{B_2} h_1 \\[8pt] -l_2 + \dfrac{B_4}{B_2} h_2 \end{pmatrix}$$

$$= \begin{pmatrix} 0 \\ 0 \end{pmatrix}. \tag{$**$}$$

Comparing the y-coordinates, we get

$$c_1 \frac{\Delta b_1}{a_1^+} B_2 \frac{a_2^- A_1 + \Delta b_2 A_2}{a_2^+} - c_1 B_2 \frac{a_2^- A_3 + \Delta b_2 A_4}{a_2^+} + c_1 B_4 \left(\frac{\Delta b_1}{a_1^+} A_2 - A_4 \right)$$

$$+ 2p A_2 B_2 \frac{c_1}{a_1^+} \left(1 - \frac{a_2^-}{a_2^+} \right) + 2p \frac{c_1}{a_1^+} B_2 \frac{\Delta b_2}{a_2^+} \left(a_2^- A_1 + b_2^- A_2 \right)$$

$$+ c_2 p^2 \left(\frac{\Delta b_2}{a_2^+} B_2 - B_4 \right) - c_3 q^2 + \frac{c_2}{a_2^+} B_2 R + B_2 N_2$$

$$+ \left(B_4 A_2 + B_2 \frac{a_2^- A_1 + \Delta b_2 A_2}{a_2^+} \right) N_1$$

$$+ 2 \left(\frac{c_1}{a_1^+} A_2 \right)^2 \frac{B_4}{B_2} \left(-l_2 + \frac{B_4}{B_2} h_2 \right) = 0.$$

2.11.15 *Details of the Calculation of $\partial^2 J.u_2.u_2$*

In order to calculate the $\overline{\delta s_1} \delta s_1$ terms in $J(\overline{x} + \overline{\delta s_1} u_2(u_2) + \delta s_1 u_2) - J(\overline{x} + \delta s_1 u_2)$, we need to count the $\overline{\delta s_1} \delta s_1$ terms contributed by all the ξ-pieces.

Computing the first ξ-piece contributes

$$N_1 + (1,0) \cdot \begin{pmatrix} \dfrac{a_1^- - a_1^+}{a_1^+} & \dfrac{b_1^- - b_1^+}{a_1^+} \\ 0 & 1 \end{pmatrix} \begin{pmatrix} 0 \\ c_1 \end{pmatrix} \delta s_1 \overline{\delta s_1} = N_1 + \frac{\Delta b_1}{a_1^+} c_1 \delta s_1 \overline{\delta s_1}.$$

For the second ξ-piece, as before, the $\overline{\delta s_1} \delta s_1$ terms come from three sources. Since $\overline{\delta s_2} = p_1 \overline{\delta s_1} + R \overline{\delta s_1} \delta s_1$, so R is one of the three sources. Also the second order terms of $da_c(x_3')$, N_2 contribute to the $\overline{\delta s_1} \delta s_1$ terms too. All the other $\overline{\delta s_1} \delta s_1$ terms come from the v-transport. Computing, we find:

$$R(1,0) \begin{pmatrix} \dfrac{c_2}{a_2^+} \\ 0 \end{pmatrix} \delta s_1 \overline{\delta s_1} + (1,0) \begin{pmatrix} \dfrac{a_2^- - a_2^+}{a_2^+} & \dfrac{\Delta b_2}{a_2^+} \\ 0 & 1 \end{pmatrix} A \begin{pmatrix} N_1 \\ 0 \end{pmatrix} + N_2$$

$$+ (1,0) \cdot \begin{pmatrix} \dfrac{a_2^- - a_2^+}{a_2^+} & \dfrac{\Delta b_2}{a_2^+} \\ 0 & 1 \end{pmatrix} A \begin{pmatrix} \dfrac{a_1^-}{a_1^+} & \dfrac{\Delta b_1}{a_1^+} \\ 0 & 1 \end{pmatrix} \begin{pmatrix} 0 \\ c_1 \end{pmatrix} \delta s_1 \overline{\delta s_1}$$

$$+(1,0) \cdot \begin{pmatrix} \dfrac{a_2^- - a_2^+}{a_2^+} & \dfrac{\Delta b_2}{a_2^+} \\ 0 & 1 \end{pmatrix} 2p \begin{pmatrix} 0 & -1 \\ a_2^- & b_2^- \end{pmatrix} A \begin{pmatrix} \dfrac{c_1}{a_1^+} \\ 0 \end{pmatrix} \delta s_1 \overline{\delta s_1}$$

$$-2(1,0) \cdot \begin{pmatrix} \dfrac{a_2^- - a_2^+}{a_2^+} & \dfrac{\Delta b_2}{a_2^+} \\ 0 & 1 \end{pmatrix} A \begin{pmatrix} 0 & -1 \\ a_1^+ & b_1^+ \end{pmatrix} \begin{pmatrix} \dfrac{c_1}{a_1^+} \\ 0 \end{pmatrix} \delta s_1 \overline{\delta s_1}$$

$$+(1,0) \cdot \begin{pmatrix} \dfrac{a_2^- - a_2^+}{a_2^+} & \dfrac{\Delta b_2}{a_2^+} \\ 0 & 1 \end{pmatrix} \begin{pmatrix} 0 \\ c_2 \end{pmatrix} p^2 \overline{\delta s_1} \delta s_1$$

$$= \begin{pmatrix} c_1 \left(A_3 - \dfrac{\Delta b_1}{a_1^+} A_1 \right) - c_1 \dfrac{B_4}{B_2} \left(\dfrac{\Delta b_1}{a_1^+} A_2 - A_4 \right) + c_2 p^2 \dfrac{B_4}{B_2} - \left(A_1 + \dfrac{B_4 A_2}{B_2} \right) N_1 \\ +c_3 \dfrac{1}{B_2} q^2 - 2 \left(\dfrac{c_1}{a_1^+} A_2 \right)^2 \dfrac{B_4}{B_2^2} \left(-l_2 + \dfrac{B_4}{B_2} h_2 \right) \end{pmatrix}$$

$$\times \delta s_1 \overline{\delta s_1}.$$

For the third $\xi-$piece, the $\overline{\delta s_1} \delta s_1$ terms come from three sources, the R term of $\overline{\delta s_3}$, the second order terms of $da_c(x_5')$ $i.e. N_3$, and those from the v-transport. Computing, we find

$$R(1,0) \cdot \begin{pmatrix} \dfrac{a_3^-}{a_3^+} & \dfrac{\Delta b_3 - 2d\overline{a}_c}{a_3^+} \\ 0 & 1 \end{pmatrix} B \begin{pmatrix} \dfrac{c_2}{a_2^+} \\ 0 \end{pmatrix} \delta s_1 \overline{\delta s_1} + T(1,0) \cdot \begin{pmatrix} \dfrac{c_3}{a_3^+} \\ 0 \end{pmatrix} \delta s_1 \overline{\delta s_1}$$

$$+(1,0) \cdot \begin{pmatrix} \dfrac{a_3^-}{a_3^+} & \dfrac{\Delta b_3 - 2d\overline{a}_c}{a_3^+} \\ 0 & 1 \end{pmatrix} B \begin{pmatrix} \dfrac{a_2^-}{a_2^+} & \dfrac{\Delta b_2}{a_2^+} \\ 0 & 1 \end{pmatrix} A \begin{pmatrix} N_1 \\ 0 \end{pmatrix} + N_3$$

$$+(1,0) \cdot \begin{pmatrix} \dfrac{a_3^-}{a_3^+} & \dfrac{\Delta b_3 - 2d\overline{a}_c}{a_3^+} \\ 0 & 1 \end{pmatrix} B \begin{pmatrix} N_2 \\ 0 \end{pmatrix}$$

$$+(1,0) \cdot \begin{pmatrix} \dfrac{a_3^-}{a_3^+} & \dfrac{\Delta b_3 - 2d\overline{a}_c}{a_3^+} \\ 0 & 1 \end{pmatrix} \begin{pmatrix} 0 \\ c_3 \end{pmatrix} q^2 \delta s_1 \overline{\delta s_1}$$

$$+(1,0)\cdot\begin{pmatrix}\dfrac{a_3^-}{a_3^+} & \dfrac{\Delta b_3 - 2d\overline{a_c}}{a_3^+}\\[2mm] 0 & 1\end{pmatrix} B \begin{pmatrix}\dfrac{a_2^-}{a_2^+} & \dfrac{\Delta b_2}{a_2^+}\\[2mm] 0 & 1\end{pmatrix}\begin{pmatrix}\dfrac{a_1^-}{a_1^+} & \dfrac{\Delta b_1}{a_1^+}\\[2mm] 0 & 1\end{pmatrix}\begin{pmatrix}0\\ c_1\end{pmatrix}\delta s_1\overline{\delta s_1}$$

$$+(1,0)\cdot\begin{pmatrix}\dfrac{a_3^-}{a_3^+} & \dfrac{\Delta b_3 - 2d\overline{a_c}}{a_3^+}\\[2mm] 0 & 1\end{pmatrix} B \begin{pmatrix}\dfrac{a_2^-}{a_2^+} & \dfrac{\Delta b_2}{a_2^+}\\[2mm] 0 & 1\end{pmatrix}\begin{pmatrix}2p\begin{pmatrix}0 & -1\\ a_2^- & b_2^-\end{pmatrix} A\begin{pmatrix}\dfrac{c_1}{a_1^+}\\ 0\end{pmatrix}\\[4mm] +A\begin{pmatrix}0 & -1\\ a_1^+ & b_1^+\end{pmatrix}\begin{pmatrix}\dfrac{c_1}{a_1^+}\\ 0\end{pmatrix}\end{pmatrix}\delta s_1\overline{\delta s_1}$$

$$+(1,0)\cdot\begin{pmatrix}\dfrac{a_3^-}{a_3^+} & \dfrac{\Delta b_3 - 2d\overline{a_c}}{a_3^+}\\[2mm] 0 & 1\end{pmatrix} B \begin{pmatrix}\dfrac{a_2^-}{a_2^+} & \dfrac{\Delta b_2}{a_2^+}\\[2mm] 0 & 1\end{pmatrix}\begin{pmatrix}0\\ c_2\end{pmatrix}p^2\overline{\delta s_1}\delta s_1$$

$$+(1,0)\cdot\begin{pmatrix}\dfrac{a_3^-}{a_3^+} & \dfrac{\Delta b_3 - 2d\overline{a_c}}{a_3^+}\\[2mm] 0 & 1\end{pmatrix}\begin{pmatrix}0 & -1\\ a_3^- & b_3^-\end{pmatrix} B\begin{pmatrix}2q\begin{pmatrix}\dfrac{a_2^-}{a_2^+} & \dfrac{\Delta b_2}{a_2^+}\\[2mm] 0 & 1\end{pmatrix} A\begin{pmatrix}\dfrac{c_1}{a_1^+}\\ 0\end{pmatrix}\end{pmatrix}$$

$$+2pq\begin{pmatrix}\dfrac{c_2}{a_2^+}\\ 0\end{pmatrix}\Bigg)\delta s_1\overline{\delta s_1}$$

$$-(1,0)\cdot\begin{pmatrix}\dfrac{a_3^-}{a_3^+} & \dfrac{\Delta b_3 - 2d\overline{a_c}}{a_3^+}\\[2mm] 0 & 1\end{pmatrix} B\begin{pmatrix}0 & 1\\ a_2^+ & b_2^+\end{pmatrix}\begin{pmatrix}2p\begin{pmatrix}\dfrac{a_2^-}{a_2^+} & \dfrac{\Delta b_2}{a_2^+}\\[2mm] 0 & 1\end{pmatrix} A\begin{pmatrix}\dfrac{c_1}{a_1^+}\\ 0\end{pmatrix}\end{pmatrix}$$

$$+2p^2\begin{pmatrix}\dfrac{c_2}{a_2^+}\\ 0\end{pmatrix}\Bigg)\delta s_1\overline{\delta s_1}$$

$$+(1,0)\cdot 2\left(\dfrac{c_1}{a_1^+}A_2\right)^2\dfrac{B_2}{B_4}\begin{pmatrix}\dfrac{a_3^-}{a_3^+} & \dfrac{\Delta b_3 - 2d\overline{a_c}}{a_3^+}\\[2mm] 0 & 1\end{pmatrix}\begin{pmatrix}l_1 - \dfrac{B_2}{B_4}h_1\\[2mm] l_2 - \dfrac{B_2}{B_4}h_2\end{pmatrix}\delta s_1\overline{\delta s_1}.$$

Comparing with $(**)$, we find

$$-(1,0)\cdot\left(RB\begin{pmatrix}\frac{c_2}{a_2^+}\\0\end{pmatrix}s_1\overline{\delta s_1}+B\begin{pmatrix}\frac{a_2^-}{a_2^+}&\frac{\Delta b_2}{a_2^+}\\0&1\end{pmatrix}A\begin{pmatrix}N_1\\0\end{pmatrix}+B\begin{pmatrix}N_2\\0\end{pmatrix}\right)$$

$$-(1,0)\cdot B\begin{pmatrix}\frac{a_2^-}{a_2^+}&\frac{\Delta b_2}{a_2^+}\\0&1\end{pmatrix}\left(\begin{matrix}A\begin{pmatrix}\frac{a_1^-}{a_1^+}&\frac{\Delta b_1}{a_1^+}\\0&1\end{pmatrix}\begin{pmatrix}0\\c_1\end{pmatrix}\delta s_1\overline{\delta s_1}\\+2p\begin{pmatrix}0&-1\\a_2^-&b_2^-\end{pmatrix}A\begin{pmatrix}\frac{c_1}{a_1^+}\\0\end{pmatrix}-2A\begin{pmatrix}0&-1\\a_1^+&b_1^+\end{pmatrix}\begin{pmatrix}\frac{c_1}{a_1^+}\\0\end{pmatrix}\end{matrix}\right)$$

$$\times\delta s_1\overline{\delta s_1}$$

$$-(1,0)\cdot B\begin{pmatrix}\frac{a_2^-}{a_2^+}&\frac{\Delta b_2}{a_2^+}\\0&1\end{pmatrix}\begin{pmatrix}0\\c_2\end{pmatrix}p^2\overline{\delta s_1}\delta s_1$$

$$-(1,0)\cdot\begin{pmatrix}0&-1\\a_3^-&b_3^-\end{pmatrix}B\left(2q\begin{pmatrix}\frac{a_2^-}{a_2^+}&\frac{\Delta b_2}{a_2^+}\\0&1\end{pmatrix}A\begin{pmatrix}\frac{c_1}{a_1^+}\\0\end{pmatrix}+2pq\begin{pmatrix}\frac{c_2}{a_2^+}\\0\end{pmatrix}\right)\delta s_1\overline{\delta s_1}$$

$$+(1,0)\cdot B\begin{pmatrix}0&-1\\a_2^+&b_2^+\end{pmatrix}\left(2p\begin{pmatrix}\frac{a_2^-}{a_2^+}&\frac{\Delta b_2}{a_2^+}\\0&1\end{pmatrix}A\begin{pmatrix}\frac{c_1}{a_1^+}\\0\end{pmatrix}+2p^2\begin{pmatrix}\frac{c_2}{a_2^+}\\0\end{pmatrix}\right)\delta s_1\overline{\delta s_1}$$

$$-(1,0)\cdot 2\left(\frac{c_1}{a_1^+}A_2\right)^2\frac{B_4}{B_2}\begin{pmatrix}l_1-\frac{B_4}{B_2}h_1\\l_2-\frac{B_4}{B_2}h_2\end{pmatrix}$$

$$=\left(\begin{matrix}c_1(\frac{\Delta b_1}{a_1^+}A_2-A_4)\left(\frac{B_4B_1}{B_2}-B_3\right)+c_2p^2\left(B_3-\frac{B_4B_1}{B_2}\right)\\+A_2N_1\left(\frac{B_4B_1}{B_2}-B_3\right)-c_3\frac{B_1}{B_2}q^2\end{matrix}\right)\delta s_1\overline{\delta s_1}$$

$$+2\left(\frac{c_1}{a_1^+}A_2\right)^2\frac{B_4}{B_2}\left(\frac{B_1}{B_2}\left(-l_2+\frac{B_4}{B_2}h_2\right)-\left(-l_1+\frac{B_4}{B_2}h_1\right)\right)\delta s_1\overline{\delta s_1}.$$

Summing them together, we have

$$\partial^2 J.u_1.u_1 = N_1 + \frac{\Delta b_1}{a_1^+}c_1 + c_1(A_3 - \frac{\Delta b_1}{a_1^+}A_1) - c_1\frac{B_4}{B_2}(\frac{\Delta b_1}{a_1^+}A_2 - A_4)$$

$$+ c_2 p^2 \frac{B_4}{B_2} - (A_1 + \frac{B_4 A_2}{B_2})N_1 + c_3\frac{1}{B_2}q^2 + c_1(\frac{\Delta b_1}{a_1^+}A_2 - A_4)\left(\frac{B_4 B_1}{B_2} - B_3\right)$$

$$+ c_2 p^2 \left(B_3 - \frac{B_4 B_1}{B_2}\right) + A_2 N_1\left(\frac{B_4 B_1}{B_2} - B_3\right)$$

$$- 2\left(\frac{c_1}{a_1^+}A_2\right)^2 \frac{B_4}{B_2^2}\left(-l_2 + \frac{B_4}{B_2}h_2\right) - c_3\frac{B_1}{B_2}q^2$$

$$+ 2\left(\frac{c_1}{a_1^+}A_2\right)^2 \frac{B_4}{B_2}\left(\frac{B_1}{B_2}\left(-l_2 + \frac{B_4}{B_2}h_2\right) + \left(l_1 - \frac{B_4}{B_2}h_1\right)\right)\delta s_1\overline{\delta s_1}$$

$$= \underbrace{(1 - A_1 + \frac{A_2}{B_2}(\det B - B_4))N_1 + c_1(\frac{\Delta b_1}{a_1^+}A_2 - A_4)\frac{1}{B_2}(\det B + B_4)}_{(I)}$$

$$\underbrace{-c_2 p^2 \frac{1}{B_2}(\det B - B_4) + c_1(A_3 - \frac{\Delta b_1}{a_1^+}A_1) + \frac{\Delta b_1}{a_1^+}c_1 + c_3\frac{1 - B_1}{B_2}q^2}_{(I)}$$

$$\underbrace{- 2\left(\frac{c_1}{a_1^+}A_2\right)^2\frac{B_4}{B_2^2}\left(-l_2 + \frac{B_4}{B_2}h_2\right) + 2\left(\frac{c_1}{a_1^+}A_2\right)^2\frac{B_4}{B_2}\left(\frac{B_1}{B_2}\left(l_2 - \frac{B_4}{B_2}h_2\right)\right.}_{}$$
$$\underbrace{\left.- \left(l_1 - \frac{B_4}{B_2}h_1\right)\right)}_{(II)} .$$

Bibliography

Bahri, A., Pseudo-Orbits of Contact Forms, Pitman Research Notes in Mathematics Series No. 173 Longman Scientific and Technical, Longman, London (1988)

Bahri, A., Classical and Quantic periodic motions of multiply polarized spin-manifolds, Pitman Research Notes in Mathematics Series No. 378 Longman and Addison - Wesley, London and Reading, MA (1998)

Bahri, A., Flow-lines and Algebraic invariants in Contact Form Geometry PNLDE, **53** (2003), Birkhauser, Boston

Bahri, A., Recent Progress in Conformal Geometry, Adv. Nonlinear Stud. **3** (2003), no. 1, 65–150.

Bahri, A., (to appear)